Multiphase Permanent Magnet Synchronous Machines and Drives

To Prof. Tim Miller

Thanks for your support!

From Qiang

2025.5.1

Other related titles:

You may also like

- Condition Monitoring Of Rotating Electrical Machines, 3rd Edition | Tavner | 2020 | PBPO145
- Signal Processing For Fault Detection And Diagnosis In Electric Machines And Systems | Benbouzid | 2020 | PBPO153
- The Control Techniques Drives And Controls Handbook, 2nd Edition | Drury | 2009 | PBPO057
- Diagnosis And Fault Tolerance Of Electrical Machines, Power Electronics And Drives | Marques-Cardoso | 2018 | PBPO126

We also publish a wide range of books on the following topics:
Computing and Networks
Control, Robotics and Sensors
Electrical Regulations
Electromagnetics and Radar
Energy Engineering
Healthcare Technologies
History and Management of Technology
IET Codes and Guidance
Materials, Circuits and Devices
Model Forms
Nanomaterials and Nanotechnologies
Optics, Photonics and Lasers
Production, Design and Manufacturing
Security
Telecommunications
Transportation

All books are available in print via https://shop.theiet.org or as eBooks via our Digital Library https://digital-library.theiet.org.

IET ENERGY ENGINEERING SERIES 243

Multiphase Permanent Magnet Synchronous Machines and Drives

Principles, design, and control

Zi Qiang Zhu, Shensheng Wang, Bo Shao,
Luocheng Yan, and Dawei Liang

The Institution of Engineering and Technology

About the IET

This book is published by the Institution of Engineering and Technology (The IET).

We inspire, inform and influence the global engineering community to engineer a better world. As a diverse home across engineering and technology, we share knowledge that helps make better sense of the world, to accelerate innovation and solve the global challenges that matter.

The IET is a not-for-profit organisation. The surplus we make from our books is used to support activities and products for the engineering community and promote the positive role of science, engineering and technology in the world. This includes education resources and outreach, scholarships and awards, events and courses, publications, professional development and mentoring, and advocacy to governments.

To discover more about the IET please visit https://www.theiet.org/

About IET books

The IET publishes books across many engineering and technology disciplines. Our authors and editors offer fresh perspectives from universities and industry. Within our subject areas, we have several book series steered by editorial boards made up of leading subject experts.

We peer review each book at the proposal stage to ensure the quality and relevance of our publications.

Get involved

If you are interested in becoming an author, editor, series advisor, or peer reviewer please visit https://www.theiet.org/publishing/publishing-with-iet-books/ or contact author_support@theiet.org.

Discovering our electronic content

All of our books are available online via the IET's Digital Library. Our Digital Library is the home of technical documents, eBooks, conference publications, real-life case studies and journal articles. To find out more, please visit https://digital-library.theiet.org.

In collaboration with the United Nations and the International Publishers Association, the IET is a Signatory member of the SDG Publishers Compact. The Compact aims to accelerate progress to achieve the Sustainable Development Goals (SDGs) by 2030. Signatories aspire to develop sustainable practices and act as champions of the SDGs during the Decade of Action (2020–2030), publishing books and journals that will help inform, develop, and inspire action in that direction.

In line with our sustainable goals, our UK printing partner has FSC accreditation, which is reducing our environmental impact to the planet. We use a print-on-demand model to further reduce our carbon footprint.

Published by The Institution of Engineering and Technology, London, United Kingdom

The Institution of Engineering and Technology (the "**Publisher**") is registered as a Charity in England & Wales (no. 211014) and Scotland (no. SC038698).

First published 2024

The Institution of Engineering and Technology
Futures Place,
Kings Way, Stevenage,
Herts, SG1 2UA, United Kingdom

www.theiet.org

British Library Cataloguing in Publication Data
A catalogue record for this product is available from the British Library

ISBN 978-1-83953-772-1 (hardback)
ISBN 978-1-83953-773-8 (PDF)

Typeset in India by MPS Limited
Printed in the UK by CPI Group (UK) Ltd, Eastbourne

Cover image: Bosca78/E+ via Getty Images

Contents

Preface

Among various electrical machine technologies, permanent magnet synchronous machines (PMSMs) are inherently more efficient and have high power/torque density due to permanent magnet excitation.

To date, single three-phase electrical machines and drives have prevailed. However, multiphase electrical machines and drives have many advantages, such as excellent fault-tolerant capability, good power sharing capability, high torque/power density, low torque ripple, additional degrees of freedom for control, reduced dc-link capacitor requirement, good modularity, and easy maintenance and repairing, etc. They are particularly attractive for applications requiring high reliability and high power, such as power generation, aerospace, electrified transportations including electric vehicles, aircrafts, trains, and ships, etc. Multi-three-phase electrical machines, particularly dual three-phase electrical machines, have become more popular among multiphase electrical machines since mature three-phase inverters and control techniques can be utilized.

This book aims to provide a comprehensive reference on the principles, machine topologies, winding configurations, modelling and design optimization techniques, various modulation strategies and control methods including field oriented control, direct torque control, model predictive control, sensorless control, and fault-tolerant control, as well as current harmonic and torque ripple minimization techniques for multiphase PMSMs, with particular reference to dual three-phase PMSMs. We have strived to highlight the global research achievements and many new techniques developed at the University of Sheffield.

I would like to acknowledge the contributions from my PhD students and post-doctoral researchers at Sheffield who worked on multiphase and multi-three-phase PMSMs, including Dr Arwyn Thomas, Dr Khoa Dang Hoang, Dr Kai Wang, Dr Yuan Ren, Dr Yashan Hu, Dr Beomseok Lee, Dr Ali Almarhoon, Dr Peilin Xu, Dr Yanxin Li, Dr Xue Liu, Dr Jin Xu, Dr Javier Riedemann, Dr Tianyi Liu, Dr Dawei Liang, Dr Bo Shao, Dr Luocheng Yan, Dr Shensheng Wang, Dr Yang Chen, Mr Jialu He, and Mr Furkan Tokgoz. I also thank Mr Ian Parmenter for English proofreading.

I dedicate my thanks to our sponsors, including the UK government, particularly the UK EPSRC Prosperity Partnership ("A New Partnership in Offshore Wind" under Grant No. EP/R004900/1), and our industrial partners, e.g. Siemens on wind power, Rolls-Royce on aerospace, CRRC on fast train propulsion, NEVC, Midea, Protean, Nissan, and Toyota on electric vehicles.

We sincerely hope this book will be useful to industrial engineers and researchers, university professionals, post-doctoral researchers, and students alike.

Professor Zi Qiang (Z.Q.) Zhu
FREng, FIEEE, FIET, FCSEE, FCES
August 2024, Sheffield, UK

List of abbreviations

2D	Two-dimension
3D	Three-dimension
ac	Alternating current
AE	Autoencoder
AI	Artificial intelligence
AIPM	Asymmetric interior permanent magnet
AlNiCo	Aluminum-nickel-cobalt
ANN	Artificial neural networks
Back-EMF	Back electromotive force
BLAC	Brushless alternating current
BLDC	Brushless direct current
CNN	Convolutional Neural Network
CPSR	Constant power speed range
D-DTC	Duty-ratio-based ST-DTC
DBN	Deep belief Network
dc	Direct current
DL	Deep learning
DL	Double layer
DP	Demagnetization percentage
DS	Double layer short-pitched
DTC	Direct torque control
DTP	Dual three-phase
DTP-DS	Dual three-phase with double layer short-pitched windings
DTP-SF	Dual three-phase with single layer full-pitched windings
EMF	Electromotive force
E-EMF	Extended EMF
EMI	Electromagnetic interference
EV	Electric vehicle
EKF	Extended Kalman filter
FFT	Fast Fourier transform
FE	Finite element

FEA	Finite element analysis
FIS	Fuzzy inference system
FLC	Fuzzy logic controller
FOC	Field oriented control
FP	Frozen permeability
FSCW	Fractional-slot concentrated winding
GA	Genetic algorithms
GAN	Generative adversarial network
GCD	Greatest common divisor
HEV	Hybrid electric vehicle
HF	High frequency
HHT	Hilbert–Huang transform
ID	Inner diameter
IM	Induction machines
IPM	Interior permanent magnet
IPMSM	Interior permanent magnet synchronous machine
ITSC	Inter-turn short circuit
LCM	Least common multiple
LPF	Low pass filter
LUT	Look-up tables
MCSA	Machine current signature analysis
ML	Machine learning
MLT	Mean length per turn
MMF	Magnetomotive force
MF	Membership function
MPC	Model-based predictive control
MP	Multiphase
MSRF	Multiple synchronous reference frame
MPC	Model predictive control
MPCC	Model predictive current control
MPTC	Model predictive torque control
MPSC	Model predictive speed control
MTP	Multi-three-phase
MTPA	Maximum torque per ampere
NdFeB	Neodymium-iron-boron
Non-OW/NOLP	Non-overlapping
NSC	Negative sequence component
OC	Open-circuit

OD	Outer diameter
OSPC	One step modulation predictive current control
OW/OLP	Overlapping
PI	Proportional integral
PIR	Proportional integral resonant
PLL	Phase-locked loop
PM	Permanent magnet
PMA-SynRM	Permanent magnet assisted-synchronous reluctance machine
PMSM	Permanent magnet synchronous machine
PMSG	Permanent magnet synchronous generator
PP	Peak-to-peak
PR	Proportional resonant
PSC	Positive sequence component
PTP	Pentuple three-phase
PWM	Pulse width modulation
QTP	Quadruple three-phase
QPR	Quasi-proportional resonant
RMS	Root-mean-square
RNN	Recurrent neural network
RT	Redundant tooth
RVV	Reference VV
SC	Short-circuit
SF	Single layer full-pitched
STFT	Short-time Fourier transform
STSM	Super-twisting sliding mode
SL	Single layer
SLG	Single line-to-ground
SMC	Sliding mode controller
SmCo	Samarium-cobalt
SPM	Surface permanent magnet
SPMSM	Surface permanent magnet synchronous machine
SRF-PI	Synchronous reference frame PI
SRFF	Synchronous reference frame filter
SRM	Switched reluctance machine
ST	Switching-table
ST-DTC	Switching-table based DTC
STP	Single three-phase
SV	Space vector

SVM	Space vector modulation
SVM-DTC	Space vector modulation DTC
SVPWM	Space vector PWM
SynRM	Synchronous reluctance machine
T–S	Takagi–Sugeno
THD	Total harmonic distortion
TTP	Triple three-phase
UKF	Unscented Kalman filters
UMF	Unbalanced magnetic force
UMP	Unbalanced magnetic pull
VSD	Vector space decomposition
VSI	Voltage source inverter
VV	Voltage vector
VVV	Virtual voltage vector
VVVS	Virtual voltage vector synthesis
WD	Wigner–Ville distribution

List of symbols

Symbol	Description	Unit
A, A_{coil}	Cross-section area of conductor	mm^2
A1, B1, C1	Three phases of winding set 1 for DTP PMSMs	
A2, B2, C2	Three phases of winding set 2 for DTP PMSMs	
α	Angular position in stator reference frame	°
α_{shift}	Phase shift between axes of two three-phase winding sets	°
B_{r}	Remanence	T
B_m	Friction coefficient	N.m.s
β	Angle displacement of DTP PMSM	°
β_a	Current advancing angle	°
$\boldsymbol{C}$	Park transformation	
C	Value of capacitance	F
d_n	Zero diagnostic variable	
$\Delta\theta$	Estimation position error	°
Δg	Control constraint in cost function of MPC	
ΔT	Half period of injected signal	s
ΔV_c	Change in capacitor voltage	V
ΔQ_c	Change in capacitor charge	V
δ_0	Initial current angle	°
δ_f	Faulty angle	°
δ_h	Health angle	°
e_a, e_b, e_c	Three-phase back-EMFs	V
e_α, e_β	α-axis and β-axis back-EMFs	V
e_d^e, e_q^e	Estimated d-axis and q-axis back-EMFs	V
e_n	Reference current error	A
e_{sT}, e_{sT}	T-axis back-EMF	V
E_{ex}	Extended back-EMF	V
$E_{ex,d}$, $E_{ex,q}$	d-axis and q-axis extended back-EMFs	V
$E_{ex,\alpha}$, $E_{ex,\beta}$	α-axis and β-axis extended back-EMFs	V
f	State function of nonlinear model	
f_{base}	Frequency at based speed	Hz
f_{cog}	Frequency of cogging torque	Hz
f_k	State equation of discrete model	
f_s	Sampling frequency	Hz
F_{sk}	Stator MMF harmonic (kth)	AT
F_{rk}	Rotor MMF harmonic (kth)	AT
F_{sum}	Stator magnetomotive force	AT
F_r	Rotor PM MMF	AT
F_v	vth MMF component of windings	AT
F_{rv}	vth harmonic rotor PM MMF	AT

(Continues)

(*Continued*)

Symbol	Description	Unit
γ_s	Angle difference of $\alpha\beta$-axis stator fluxes	°
γ	Split ratio	
g	Cost function in MPC	
G_{R}	Transfer function of ideal resonant controller	
G_{PR}	Transfer function of proportional resonant controller	
G_{QR}	Transfer function of quasi-proportional resonant controller	
h	Output function of nonlinear model	
h_k	Output equation of discrete model	
$H_{\mathrm{1st\text{-}LPF}}$	First-order low-pass filter	
$H_{\mathrm{2nd\text{-}LPF}}$	Second-order low-pass filter	
H_c	Coercivity	A/m
H_{PR}	Discrete transfer function of proportional resonant controller	
g_{MPCC}, g_{MPTC}, g_{MPSC}	Cost functions in model predictive current control (MPCC), model predictive torque control (MPTC), model predictive speed control (MPSC).	
i_d^e, i_q^e	d-axis and q-axis currents in estimated rotating reference frame	A
I_5, I_7	5th and 7th current harmonics in z_1z_2 subspace	A
I_c	Capacitor current	A
I_S	Phase current	A
$I_{p,max}$	Maximum phase current	A
I_P	Peak current	A
I_A, I_B, I_C	Three-phase currents	A
I_m	Amplitude of phase current	A
I_{P-N}	Phase current of N-phase winding	A
$i_{\alpha h}$, $i_{\beta h}$	α-axis and β-axis HF currents	A
i_d, i_q	d-axis and q-axis currents	A
i_{d1}, i_{q1}	d-axis and q-axis currents of winding set 1 for DTP-PMSM	A
i_{d2}, i_{q2}	d-axis and q-axis currents of winding set 2 for DTP-PMSM	A
i_{dh}, i_{qh}	d-axis and q-axis HF currents	A
$\Delta i_{dh}^{e'}$, $\Delta i_{qh}^{e'}$	Current responses after polarity normalization	A
i_{dh}^e, i_{qh}^e	Estimated d-axis and q-axis HF currents	A
i_{o1}, i_{o2}	Zero-sequence current path in o_1o_2 subspace	A
I_p, I_n	Amplitudes of HF current responses	A
$[I_{\mathrm{s/c}}]_{dq}$	Short-circuit current vector in dq reference frame	A
I_{sz1z2}	Currents in z_1z_2 subspace	A
J	System inertia	kg m^2
k	Number of spokes per phase in star of slots	
K_a	Attenuation factor	
K_{av}	Attenuation factor of vth harmonic	
K_d	Distribution factor	
k_{ec}	Excess loss coefficient	W/m^2
k_{ex}	Eddy current loss coefficient	W/m^2
k_{hy}	Hysteresis loss coefficient	W/m^2
K_p	Pitch factor	
K_{pv}	Pitch factor of vth harmonic	
K_w	Winding factor	

(Continues)

(*Continued*)

Symbol	Description	Unit
$K_{wv\text{-}initial}$	Initial winding factor of vth harmonic in one winding set	
K_{pv}	Winding factor of the vth harmonic	
K_p	Proportional gain coefficient	
k_p, k_i	Parameters of SMC controller	
K_r	Resonant coefficient	
l	Length of conductor	mm
l_a	Stator axial length	mm
l_{cd}	Distance between two coil sides	mm
l_{ew}	Length of end-winding	mm
l_{ex}	Extra part of end-winding length	mm
L_{aa}, L_{bb}, L_{cc}	Three-phase self-inductances	H
L_{ad}, L_{aq}	d-axis and q-axis armature reaction inductances	H
L_{ab}, L_{ba}, L_{ca}, L_{ac}, L_{bc}, L_{cb}	Three-phase mutual-inductances	H
L_{ave}	Average of dq-axis incremental inductances	H
L_{dif}	Difference of dq-axis incremental inductances	H
L_d, L_q	d-axis and q-axis inductances	H
L_{d1}, L_{q1}	d-axis and q-axis inductances of winding set 1 for DTP-PMSM	H
L_{d2}, L_{q2}	d-axis and q-axis inductances of winding set 2 for DTP-PMSM	H
L_{dh}, L_{qh}	d-axis and q-axis HF incremental self-inductances	H
L_{dqh}, L_{qdh}	d-axis and q-axis HF incremental mutual inductances	H
L_D, L_Q	Equivalent d-axis and q-axis inductances of DTP PMSM	H
L_{end}	End-winding length	mm
L_{motor}	Machine axial length	mm
L_s	Stator inductance	H
L_σ	Stator leakage inductance	H
L_v	Virtual stator impedance	Ω
λ	Coupling factor	
λ_1, λ_2, λ_3	Weighting coefficients of control objectives	
m	Modulation index	
m	Number of phases in multiphase machine	
M_d, M_q	d-axis and q-axis mutual inductances	H
M_{d12}, M_{d21}	d-axis mutual inductances between two sets of DTP machine	H
M_{q12}, M_{q21}	q-axis mutual inductances between two sets of DTP machine	H
N_{A1}, N_{B1}, N_{C1}	Turn functions of PMSM	
N_s	Number of slots	
N_{turn}	Number of series turns in one phase	
n_s	Number of spokes in star of slots	
n_{s/c_A}, n_{s/c_B}, n_{s/c_C}	Three-phase short-circuited turns ratios	
ω	Rotor electrical angular speed	Rad/s
ω_0	Resonant frequency	Rad/s
ω_m	Mechanical rotor speed	Rad/s
ω_{high}	High-speed border	Rad/s
ω_{low}	Low-speed border	Rad/s
ω_{prev}	Rotor speed of previous step	Rad/s

(*Continues*)

(*Continued*)

Symbol	Description	Unit
ω_{HF}	Frequency of HF signal	Rad/s
ω_r^e	Estimated rotor electrical angular speed	Rad/s
ω_c	Cut-off frequency	Rad/s
ω_n	Undamped natural frequency of second-order LPF	Rad/s
p	Pole pair number	
P_{Cu}	Copper loss	W
p_{slot}	Radius of coil pitch	mm
P_{Eddy}	Eddy current loss	W
P_{Hyst}	Hysteresis loss	W
P_{Fe}	Iron loss	W
P_{Me}	Mechanical loss	W
P_{in}	Input power	W
P_{out}	Output power	W
Q	Covariance matrix of state noise vector	
$Q(\theta_{s/c})$	Fault localization matrix	
R	Covariance matrix of measurement noise vector	
ψ_f	PM flux linkage	Wb
ψ_{fa}, ψ_{fb}, ψ_{fc},	Three-phase PM flux linkages	Wb
ψ_{fd}, ψ_{fq}	d-axis and q-axis PM flux linkages	Wb
ψ_{fd1}, ψ_{fq1}	d-axis and q-axis PM flux linkages of winding set 1	Wb
ψ_{fd2}, ψ_{fq2}	d-axis and q-axis PM flux linkages of winding set 2	Wb
ψ_α^r, ψ_β^r	Observed rotor flux vector	Wb
ψ_{AF}	Active flux	Wb
$\psi_{d,AF}$, $\psi_{q,AF}$	d-axis and q-axis active fluxes	Wb
ψ_d^e, ψ_q^e	Estimated d-axis and q-axis flux linkages	Wb
ψ_{sav}	Average values of stator flux	Wb
$\psi_{s\alpha\beta}$	Stator flux in $\alpha\beta$ reference frame	Wb
ψ_{s_ripple}	Stator flux ripple	Wb
ψ_{si}	Instant value of stator flux	Wb
ψ_{sM}, ψ_{sT},	M-axis and T-axis stator flux linkages	Wb
ψ_{sz1z2}	Stator flux in z_1z_2 subspace	Wb
ψ_s	Stator flux vector	Wb
r_e	Equivalent radius of end-winding	mm
r_r	Rotor outer radius	mm
r_s	Stator outer radius	mm
r_{slot}	Radius of middle of slot	mm
R_{ph}	Phase resistance	Ω
R_s	Stator resistance	Ω
R_{phase}	Phase resistance	Ω
$\rho_{cu,0}$, $\rho_{cu,s}$	Copper resistivity at reference and actual temperatures	Ω m
S	Sliding mode function	
S_A, S_B, S_C	Switching states in 2-level 3-phase inverter	
S_A, S_B, S_C,S_X, S_Y, S_Z	Switching states in 2-level dual 3-phase inverter	
σ_u, σ_y	Variances of noises on input and output signals	
t	Electrical periodicity number	
t_{coil}	Throw of coil	
T_0, T_1, T_2	Time durations of zero voltage vector	s

(Continues)

(*Continued*)

Symbol	Description	Unit
$\boldsymbol{T}_{\alpha\beta}$	Clarke transformation operator	
$\boldsymbol{T}_{5\text{-}dqz}$	Synchronously rotating transformation matrix of 5th harmonic	
$\boldsymbol{T}_{7\text{-}dqz}$	Synchronously rotating transformation matrix of 7th harmonic	
T_{direct}	Torque obtained from finite element analysis directly	Nm
T_e	Electromagnetic torque	Nm
T_L	Load torque	Nm
T_{ave}	Average value of electromagnetic torque	Nm
T_{e_ripple}	Torque ripple	Nm
$T_{5\text{-}dqz}$, $T_{7\text{-}dqz}$	Rotating transformation matrices of 5th and 7th harmonics	
T_{pm}	PM torque component	Nm
T_{rel}	Reluctance torque component	Nm
T_{mr}	Torque component caused by magnets and stator reluctance	Nm
T_s	Sampling period	s
T_{on}	Switch on time of 3-phase inverter	s
$\widehat{\theta}_r$, θ_r^e	Estimated rotor position	°
$\widehat{\theta}_{r,\text{HF}}$	Rotor position estimated by HF signal injection method	°
$\theta_{r,\text{FO}}$	Rotor position estimated by flux-linkage based method	°
θ_m	Position error caused by cross-coupling saturation effect	°
ϑ_0, ϑ_s	Reference and actual temperatures of copper	°C
u_a, u_b, u_c	Three-phase voltages	V
u_α, u_β	α-axis and β-axis voltages	V
u_α, u_β, u_z	Voltages in $\alpha\beta z$ plane	V
$u_{\alpha h}$, $u_{\beta h}$	α-axis and β-axis HF voltages	V
u_d, u_q	d-axis and q-axis voltages	V
u_{d1}, u_{q1}	d-axis and q-axis voltages of winding set 1	V
u_{d2}, u_{q2}	d-axis and q-axis voltages of winding set 2	V
u_{dh}, u_{qh}	d-axis and q-axis HF voltages	V
u_d^e, u_q^e	Estimated d-axis and q-axis HF voltages	V
u_A, u_B, u_C	Three-phase voltages	V
u_{HF}	High frequency injected voltage	V
u_d^e, u_q^e	d-axis and q-axis voltages in estimated rotating reference frame	V
u_{dh}^e, u_{qh}^e	Injected d-axis and q-axis square wave voltages	V
U_{dc}	Direct current voltage	V
$U_{\text{ref, phase}}$	Fundamental magnitude of output phase voltage	V
$U_{\text{ref, line}}$	Fundamental magnitude of output line voltage	V
u_{z1}, u_{z2}	Voltages in z_1z_2 subspace	V
v	Harmonic order	
vv_i	Selected VVV	
V	Measurement noise vector	
$V_{0\text{-}7}$	Switch patterns in 2-level 3-phase inverter	
V_A, V_B, V_C	Three-phase voltage vectors	V
V_{AB}, V_{BC}, V_{CA}	Three-phase line voltage vectors	V
V_{dc}	dc-link voltage	V
V_{ref}	Reference voltage vector	V
V_{HF}	Amplitude of high-frequency voltage	V

(Continues)

(*Continued*)

Symbol	Description	Unit
$V_{s\alpha\beta}$	Voltage vector in $\alpha\beta$ reference frame	V
V_{sM}^{*}, V_{sT}^{*}	M-axis and T-axis reference voltages	V
V_{sz1z2}	Voltage vectors in z_1z_2 subspace	V
$V_{\%ripple}$	Voltage ripple	V
W	State noise vector	
X_e	Extended state vector	
y	Coil pitch number	
α	Spatial shifting angle in space	°
β	Phase shifting angle in time	°
β	Current advancing angle	°
θ_{sk}	Skew angle	°
θ_r	Electrical rotor angle	°
μ_r	Magnetic permeability	H/m
ρ	Electrical resistivity of conductor	Ω/m
X_e	Extended state vector	
y	Coil pitch number	
ζ	Damping ratio of second-order LPF	
$\left[Z_{s/c}\right]_{dq}^{-1}$	Fault impedance	Ω
$Z_{s/cA}$, $Z_{s/cB}$, $Z_{s/cC}$	Three-phase short-circuit impedances	Ω

About the authors

Zi Qiang Zhu is a professor at the School of Electrical and Electronic Engineering of the University of Sheffield, UK. He is the founding academic director of the Sheffield Siemens Wind Power Research Centre, a fellow of the UK Royal Academy of Engineering, of the IEEE and of the IET. His research interests include the design and control of permanent magnet machines and drives for applications spanning electrified transportation (including electric vehicles, fast trains and aircraft), domestic appliances, and wind power generation. His awards include the 2019 IEEE Industry Application Society Outstanding Achievement Award, the 2021 IEEE Nikola Tesla Award, the 2024 Global Energy Prize, and 39 Best Paper Awards, and he is the author or co-author of >200 patents and >1,500 papers.

Shensheng Wang is a PhD student at the School of Electrical and Electronic Engineering of the University of Sheffield, UK. His research interests include the design and analysis of permanent magnet synchronous machines.

Bo Shao is a postdoctoral research associate at the School of Electrical and Electronic Engineering of the University of Sheffield, UK. His research interests include the control of permanent magnet machine drives.

Luocheng Yan is a PhD student at the School of Electrical and Electronic Engineering of the University of Sheffield, UK. His research interests include the control of permanent magnet synchronous machine drives.

Dawei Liang is a postdoctoral research associate at the School of Electrical and Electronic Engineering of the University of Sheffield, UK. His research interests include thermal modelling and the management of permanent magnet synchronous machines.

Chapter 1

General introduction

1.1 Introduction

Electrical machines convert electrical energy into mechanical energy or vice versa. Among various electrical machine technologies, permanent magnet (PM) synchronous machines (PMSMs) are inherently more efficient and have high power/torque density due to PM excitation [1]. The development of new PMSMs is always affiliated with the development of new PM materials [2,3]. In 1931, the first aluminum-nickel-cobalt (AlNiCo) magnet was invented, and in the early 1950s, the first ferrite magnet was developed. Since then, AlNiCo magnets and ferrite magnets have been extensively researched and widely employed in PMSMs. For example, some PMSMs utilizing AlNiCo and ferrite magnets were reported in [4–7]. A milestone in the performance of PMs occurred in the late 1960s when the samarium-cobalt (SmCo) magnet was discovered. Furthermore, in the early 1980s, the neodymium-iron-boron (NdFeB) magnet was invented. NdFeB magnets have inherent high coercivity, high remanence, and high energy products. PMSMs equipped with NdFeB magnets can achieve high efficiency, high power/torque density, high reliability, and robustness, and are eminently suitable and have been widely used for power generation, aerospace, automotive, automation, robots, industrial drives, household appliances, electrified transportation including electric vehicles, aircraft, trains, and ships, etc. [8,9].

Up to date, single three-phase (STP) electrical machines and drives have prevailed in a wide variety of applications. However, multiphase (MP) electrical machines and drives have the advantages of excellent fault-tolerant characteristics, good power/torque sharing capability, high torque/power density, smooth torque, reduced dc-link capacitor requirement, and easy maintenance and repair, etc. [1,10,11]. They are particularly attractive for high fault-tolerant, high power, and/or high current applications since high power and high switching frequency semiconductor devices may be very expensive or not available.

The early application of multiphase machines was in the field of ship propulsion. Five-phase, six-phase, and other multiphase machines have all been successfully applied [12–14]. The increased emphasis on environmental protection has accelerated the development of electrified transportation and has led to the rise of renewable energy industries. The advantages of multiphase machines have attracted researchers to utilize them in electric and hybrid electric vehicles (EVs) [1,15,16], e.g., five-phase [17], six-phase [18], and nine-phase [19] machines.

Besides, they are also considered preferable solutions for wind power generation systems due to their power-sharing capability, modularity, additional degrees of freedom, and fault-tolerant capability provided by their multiphase topologies [20–24]. Furthermore, electric aircraft have received much attention in the modern commercial and military aerospace industries [25–29], where multiphase machines have exhibited irreplaceable fault tolerance capabilities compared with conventional three-phase machines.

Generally, multiphase (MP) machines can be classified into two categories according to the number of phases, i.e., multi-three-phase (MTP) (6-phase [30], 9-phase [31], 12-phase [32], etc.) machines and other multiphase (4-phase [25], 5-phase [12], 7-phase [33], 11-phase [34], etc.) machines. Since commercial three-phase inverters and many advanced control techniques can be utilized in MTP machines, MTP machines have become more popular. Meanwhile, dual three-phase (DTP) machines, which are composed of two three-phase winding sets with a specific displaced angle (most commonly 30 electrical degrees) between two sets, are especially attractive since the main advantages of MTP machines are maintained and the phase number does not increase too much compared to conventional STP machines. More importantly, most machine and control technologies of the DTP machines are generic and also applicable to other MTP machines [35–39]. Besides, due to developments in PM materials as mentioned above, DTP PMSMs are widely studied and applied. Therefore, this book will focus on DTP PMSMs and the corresponding control strategies although other MTP and MP machines are also covered.

This book aims to provide a comprehensive reference on the principles, machine topologies, winding configurations, modeling and design optimization techniques, various modulation strategies and control methods, including field-oriented control (FOC), direct torque control (DTC), model predictive control (MPC), sensorless control, and fault-tolerant control, as well as current harmonic and torque ripple minimization techniques of multiphase PMSMs, with particular reference to DTP PMSMs.

In this chapter, STP, MP, MTP, and DTP drives will be described in Section 1.2. The stator winding and rotor configurations in conventional STP PMSMs will be introduced in Sections 1.3 and 1.4. In Section 1.5, DTP PMSMs are compared with other multiphase PMSMs. Finally, the scope of this book will be outlined in Section 1.6. It should be mentioned that although numerous PMSM machine topologies are available, only the most popular type, i.e., radial field PMSMs, will be described.

1.2 Drives for single-, dual-, multi-three-phase and multiphase PMSMs

The schematics of STP and DTP PMSMs and voltage source inverters (VSIs) are shown in Figure 1.1(a) and (b), respectively. An STP VSI drive, as shown in Figure 1.1(a), consists of six semiconductor power switches, usually IGBTs or MOSFETs paralleled with power diodes. These six switches form a three-phase bridge, with each leg of the bridge consisting of two switches. The middle point of each leg is connected to the one-phase terminal of the PM machine stator winding.

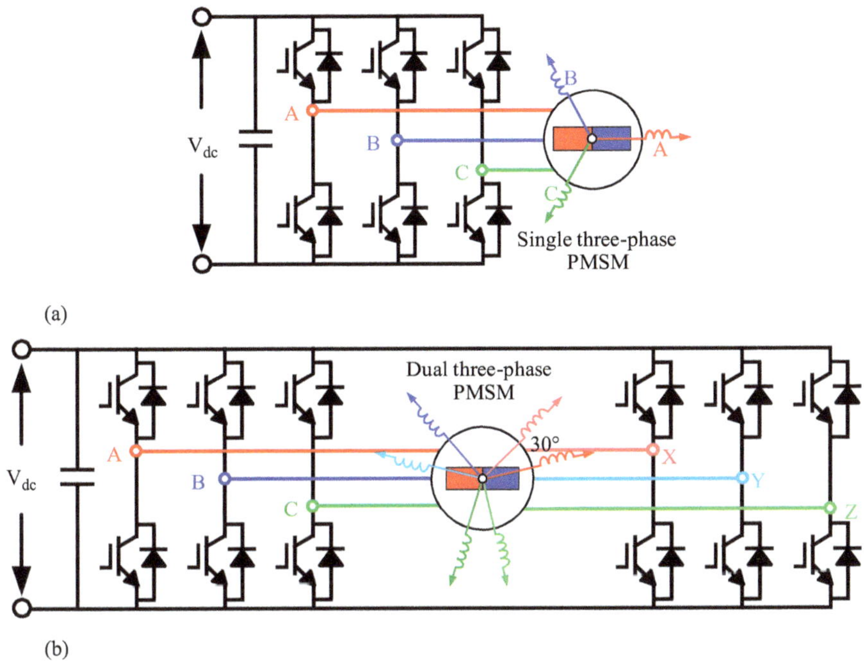

Figure 1.1 STP and DTP machines and drives. (a) STP and (b) DTP.

The three legs are connected to a DC voltage source with a DC-link capacitor which is used to smooth the DC-link voltage. The drive of DTP PMSMs can be simply derived from the drive of single three-phase machines, as shown in Figure 1.1(b). Two three-phase inverters are used to drive the two winding sets of the machine, respectively, and the DC links of the two inverters can be connected to the same or different DC power supplies. The DTP PMSM drives not only exhibit the merits of multiphase machines and retain the advantages of PMSMs, e.g., high torque density, high power density, and high efficiency, but also possess much simpler drive topology compared to other multiphase or MTP machines, such as five/seven-phase machines or triple/quadra three-phase machines.

The conventional STP PMSMs have one set of three-phase windings, and the spatial phase shifting angle between two adjacent phases is 120 electrical degrees, as shown in Figure 1.2(a). The DTP PMSMs have two sets of three-phase windings with isolated neutral points, and the spatial phase shifting angle between the two sets can be 0 electrical degrees in Figure 1.2(b), 30 electrical degrees in Figure 1.2(c), and 60 electrical degrees in Figure 1.2(d). As will be discussed in Chapters 2, 3, and 4, it is also possible to have other spatial phase shifting angles. Meanwhile, it is possible for two sets of three-phase windings with isolated or non-isolated neutral points, as shown in Figure 1.3(a) and (b), although it is more common to have isolated neutral points to minimize the interaction between the two sets. Compared to other kinds of DTP PMSMs, the machine with 30 electrical degrees phase shifting shows

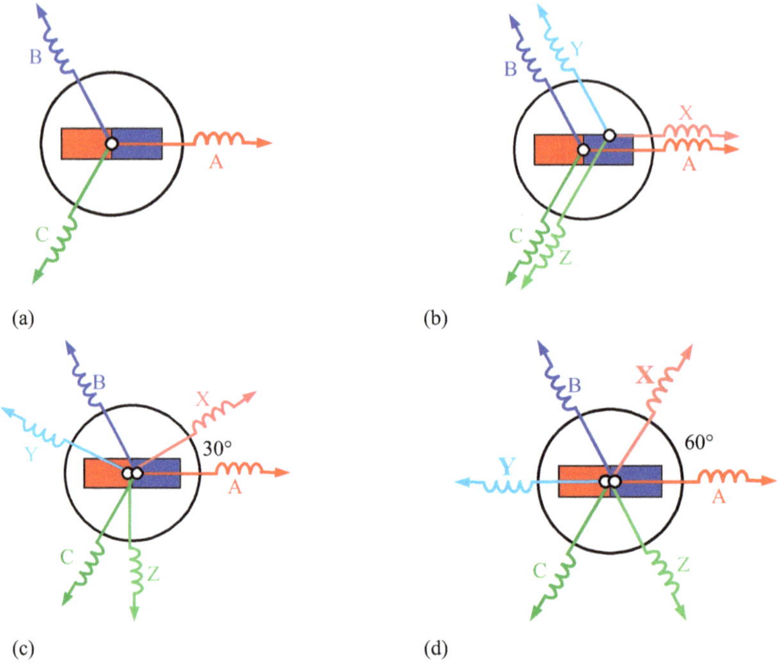

Figure 1.2 Spatial phase shifting angle between two sets of DTP windings. (a) STP, (b) DTP with 0-degree phase shifting, (c) DTP with 30-degree phase shifting, and (d) DTP with 60-degree phase shifting.

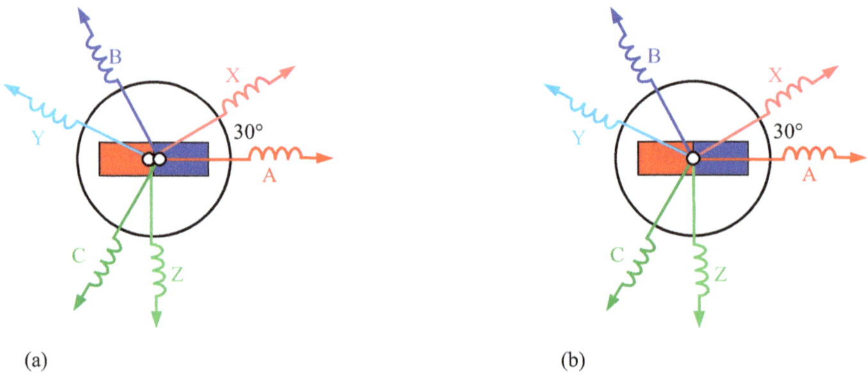

Figure 1.3 Isolated and non-isolated neutral points between DTP winding sets. (a) Isolated and (b) non-isolated.

advantageous performance, such as greatly reduced spatial harmonics and elimination of the 6th harmonic in the torque ripple.

Figure 1.4 illustrates the spatial phase shifting angle between the winding axes of multiphase machines for 5-phase and MTP machines, respectively.

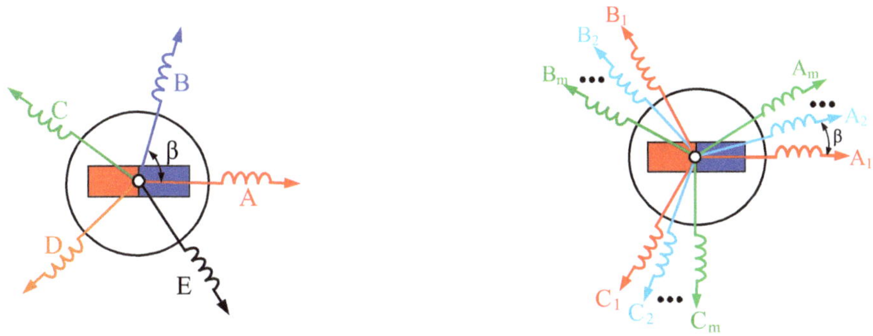

(a) (b)

Figure 1.4 Spatial phase shifting angle between winding axes of multiphase machines. (a) Five phase and (b) MTP.

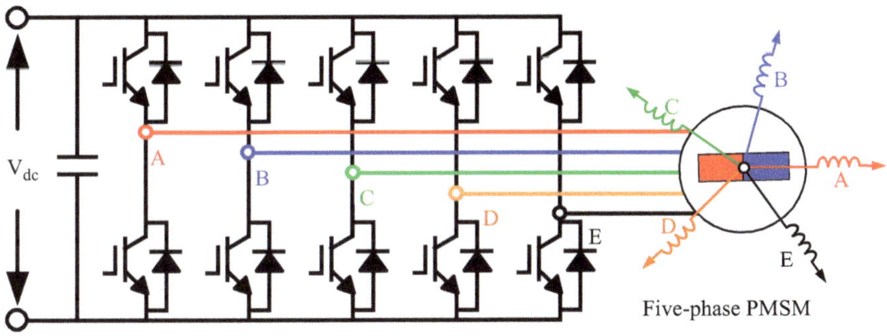

Figure 1.5 Inverter for 5-phase machines.

Figure 1.5 shows the VSI for five-phase machines, which is essentially the same as that of an STP VSI, Figure 1.1(a), albeit with a total of five legs. For MTP drives, each of the three-phase winding sets can be fed by multiple generic and modular three-phase inverters, as presented in Figure 1.6. The DC links of these inverters can be connected to the same power source, Figure 1.6(a), or isolated power sources, Figure 1.6(b). This means that the drive topologies do not need to be redesigned and can be flexibly connected to fit different applications. For the common power source, it is usually employed in the applications that only have a single power source, such as electric vehicles. This drive topology makes it possible to achieve interleave techniques to reduce the DC link ripples and minimize the DC link capacitor. For the isolated power source, it is feasible for applications with multiple power sources, such as the propulsion system of a ship or wind power generation, and the power can be freely exchanged among several isolated systems through multiphase windings.

The most popular control strategies for STP, DTP, and MTP PMSMs, including field-oriented control (FOC or vector control), direct torque control (DTC), and model predictive control (MPC), will be described in Chapter 8. Chapter 7 focuses on

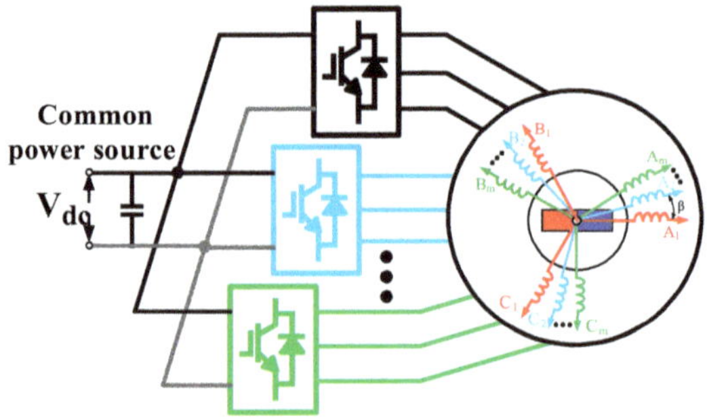

(a)

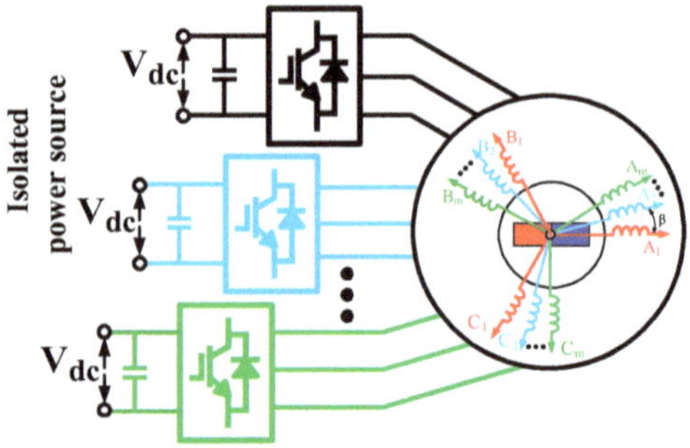

(b)

Figure 1.6 Drive topologies of MTP machines. (a) Common power source and (b) isolated power sources.

modulation strategies, such as the selection of optimal voltage vectors for FOC and DTC, as well as space vector modulation (SVM) based, switching table-based, and duty-ratio based DTCs, with focus on minimizing current harmonics and torque ripples. Sensorless control methods and fault tolerant control methods will be presented in Chapters 9 and 10, respectively.

1.3 Rotor PM configurations

It should be mentioned that the rotor configurations are usually the same for STP, DTP, MTP, and MP PMSMs. According to the rotor PM configurations, surface-mounted PM (SPM) and interior PM (IPM) PMSMs (i.e., SPMSM and IPMSM) are shown in Figures 1.7 and 1.8, respectively.

Among all PM rotor topologies in PMSMs, surface PM (SPM) is the most widely used one due to its simple structure. In SPM rotors, magnets are bonded to the outer surface of rotors using retaining sleeves or adhesives.

According to the shape of the rotor lamination, SPM rotors can be further categorized into surface-mounted and surface-inset types, as shown in Figure 1.7(a) and (b), respectively. Since the relative permeability of magnets is very similar to that of air, the *d*- and *q*-axis inductances in SPM rotors are almost the same, but the

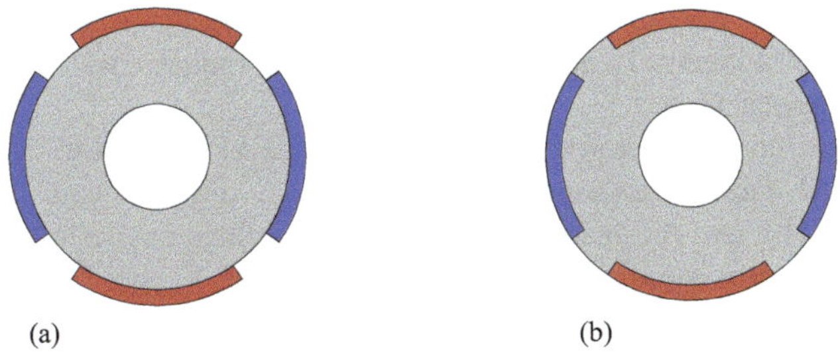

Figure 1.7 *Rotor configurations of SPM rotors. (a) Surface-mounted PM and (b) surface inset PM.*

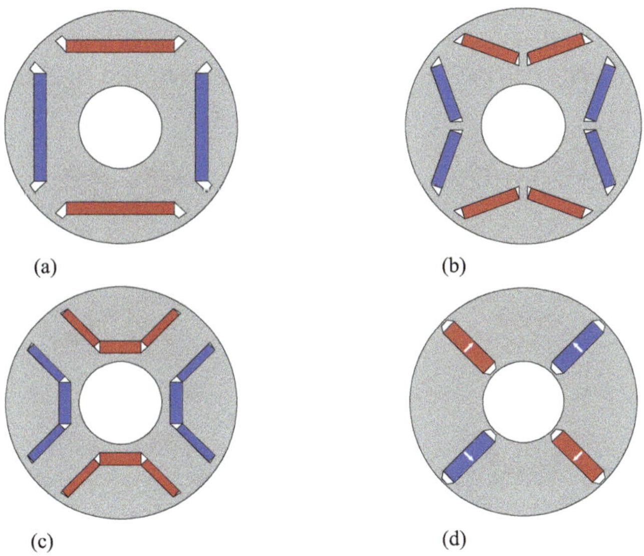

Figure 1.8 *Rotor configurations of single-layer IPM rotors. (a) I-type, (b) V-type, (c) U-type, and (d) spoke-type.*

d- and *q*-axis inductances in surface-inset PM rotors are different. Hence, SPM rotor PMSMs exhibit very limited magnetic saliency and reluctance torque and rely almost completely on the PM torque component to produce torque, but surface inset PM rotors can benefit from a reluctance torque component.

In IPM rotors, PMs are embedded inside rotors, and hence, an IPM rotor has better mechanical PM retainment than its SPM counterpart. Another prominent feature of the IPM rotor is its high rotor saliency ratio. In the IPM rotor, the *q*-axis inductance is larger than the *d*-axis inductance. Hence, IPM machines can utilize both PM and reluctance torque components.

After the invention and commercialization of NdFeB magnets in the 1980s, numerous developments and analyses have been made on PMSMs [5,40–47].

In this section, various IPM rotor topologies are briefly reviewed according to the number of PM layers. To ease manufacturing and reduce costs, it is desirable to employ rectangular PMs.

The single-layer IPM rotor is the simplest type of IPM rotor. The most conventional IPM rotor arrangements include I-, V-, and U-shape, as shown in Figure 1.8(a), (b), and (c), respectively. Among the three single-layer IPM rotors, the I-type IPM rotor has a simple construction, but the V- and U-type IPM rotors can enhance the flux-focusing effect and improve torque capability. The spoke-type IPM rotor as shown in Figure 1.8(d), in which the magnets are circumferentially magnetized, can also achieve flux focusing effect when the number of poles is relatively high but suffers from its relatively higher flux leakages towards the shaft.

When the PM layer number is 2, the double-layer magnets can be combined with two single-layer magnets. Based on the single layer I-, V-, and U-type magnet layouts, double I-, double V-, double U-, I+V-, I+U-, and V+U-type IPM rotors were proposed and analyzed in papers [48–55], as presented in Figure 1.9. With

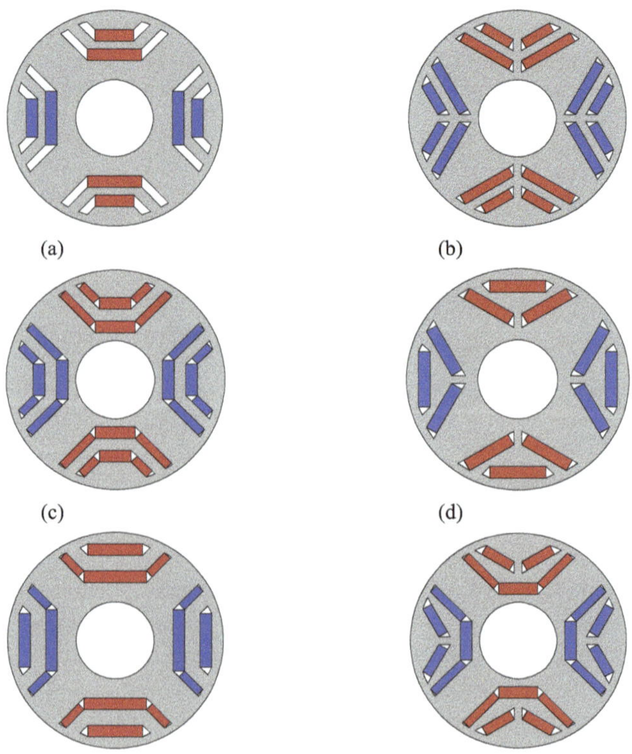

Figure 1.9 Rotor configurations of double-layer IPM rotors. (a) Double I-type, (b) double V-type, (c) double U-type, (d) I+V(delta)-type, (e) I+U-type, and (f) V+U-type.

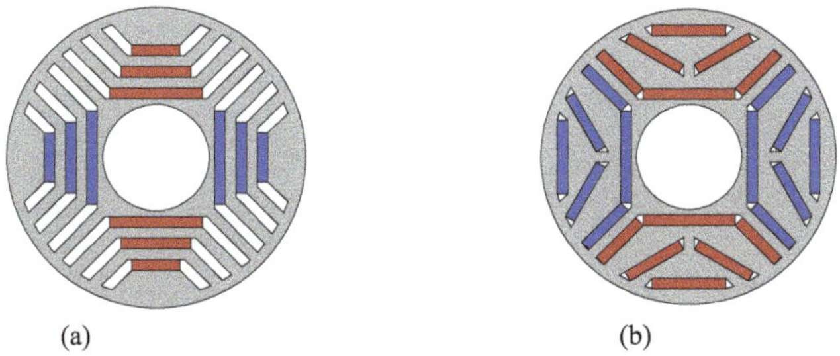

Figure 1.10 Rotor configurations of triple layer IPM rotors. (a) Triple I-type and (b) I+V+U-type.

the double-layer PMs, the rotor saliency ratio can be enhanced. Besides the double-layer IPM rotor, the PM layer number can further be increased to 3 or more [8,55–61], as shown in Figure 1.10(a).

Although the reluctance torque increases with the number of PM layers, the benefits diminish when the PM layer number is higher than 3, albeit with a large number of PM pieces which increases the cost of materials and manufacturing. Therefore, the PM layer number is usually limited to lower than 3 in practice, while 1 or 2 layers are the most common.

1.4 Stator winding configurations

The main differences between STP, DTP, MTP, and MP machines are in the stator winding configurations. It should be noted that various stator winding configurations for DTP, MTP, and MP PMSMs will only be briefly introduced with examples in this section since more detailed descriptions will be given in Chapters 3, 4, and 5.

In the slotted stator of STP PMSM, the coils of three-phase armature windings are accommodated in slots to produce a rotating magnetic field. For any STP PMSM, the detailed winding connection is determined by its slot/pole number combination and can be obtained by using the star of slots [62]. In general, stator windings can be categorized into two groups, i.e., overlapping winding (OW) and non-overlapping winding (non-OW).

In OWs, the coil pitch almost matches the pole pitch, and hence, OWs can be widely utilized in STP PMSMs with the slot number per pole per phase ≥ 1. According to the winding layout, OWs can be further classified into distributed winding and concentrated winding, as shown in Figure 1.11(a) and (b), respectively. It can be seen that in distributed windings, the coils of the same phase and/or different phases are overlapped with each other. In general, the coil pitch of OWs is ≥ 2 slot pitches. Hence, OWs always result in relatively long end-windings which cause high copper loss and reduce torque density, but OWs can have a high winding factor and can utilize the reluctance torque component.

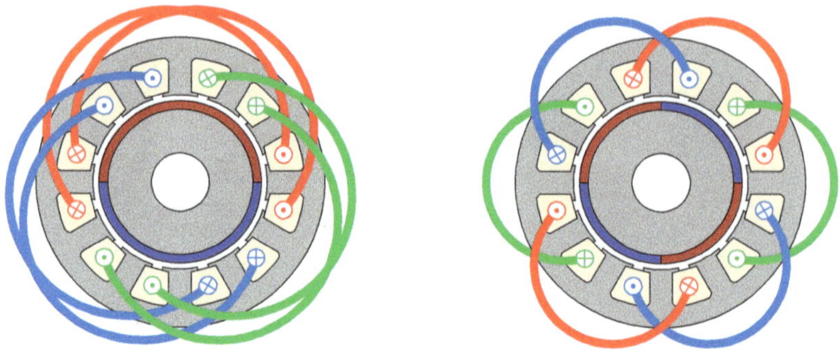

(a) (b)

Figure 1.11 Configurations of overlapping windings for STP PMSMs. (a) Distributed windings (12-slot/2-pole) and (b) concentrated windings (12-slot/4-pole).

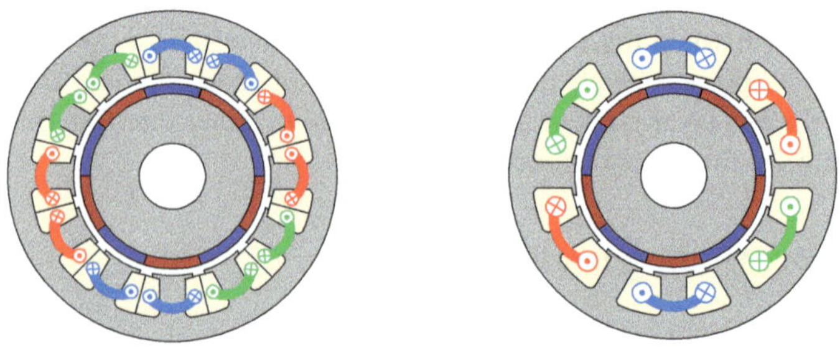

(a) (b)

Figure 1.12 Configurations of non-overlapping (concentrated) windings for STP PMSMs. (a) All teeth wound (double layer) and (b) alternate teeth wound (single layer).

Non-OWs are suitable for STP PMSMs with the slot number per pole per phase <1. In non-OWs, all the coils are wound around stator teeth without any overlapping, as shown in Figure 1.12. Since the slot number per pole per phase is fractional and the coils are all concentrated, this kind of winding is also known as fractional-slot concentrated winding (FSCW). Compared with OWs, an FSCW configuration can shorten end-windings, reduce copper loss, improve slot fill factor, and ease manufacturing, hence, a FSCW configuration is very popular and has been widely researched all over the world since the early 2000s [63–68]. It should also be noted that a FSCW can be further classified into all teeth wound (double layer, DL) or alternate teeth wound (single layer, SL) types, as shown in Figure 1.12(a) and (b), respectively [65,69].

An example of four-phase PMSMs can be found in [25]. A four-phase 8-slot/6-pole SPM machine was designed for an engine fuel pump in an aircraft, as shown in

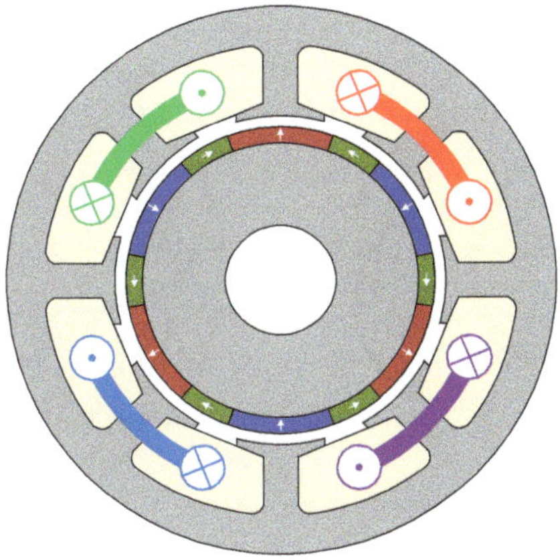

Figure 1.13 Cross-section of a four-phase 8-slot/6-pole SPM (Halbach array) machine.

Figure 1.13. A series of tests were carried out on this machine in a fully functional pump, and the fault-tolerant capability of this four-phase SPM machine was validated.

For five-phase PMSMs, Toliyat and Parsa made some important contributions. In [12], they designed three five-phase PMSMs, i.e., a 20-slot stator with single-layer windings, a 40-slot stator with single-layer windings, and a 40-slot stator with double-layer windings, all combined with 4-pole rotors, as shown in Figure 1.14 (a)–(c). The mathematical model and the vector control method of five-phase PMSMs were also presented in [12]. Then, a fault-tolerant control strategy for five-phase PMSMs was introduced in [70], based on a 15-slot/4-pole IPM machine, as shown in Figure 1.14(d). With the proposed fault-tolerant scheme, the five-phase PMSMs can continue to work safely even under the loss of up to two phases without any additional hardware requirements.

Besides distributed windings in [12,70], five-phase PMSMs can also be equipped with concentrated windings. In [71], two five-phase 20-slot/18-pole SPM machines with single-layer and double-layer windings respectively, were introduced, see Figure 1.15. Proper current control strategies for five-phase PMSMs were derived analytically and adopted. The performances of the two machines were compared under various post-fault conditions. It was found that with the proposed current control strategies, both single-layer and double-layer windings can achieve satisfactory post-fault performances.

In [72], to illustrate the effects of the phase number on the fault-tolerant capability in PMSMs, three PMSMs with three, four, and five phases were designed and compared, as shown in Figure 1.16. Various faults that could occur were discussed and some possible control strategies to deal with these faults were presented. The three PMSMs were compared under normal and open-circuit conditions. The results demonstrate that the fault-tolerant capability of PMSMs can be significantly improved by increasing the phase number but considering the high cost and the

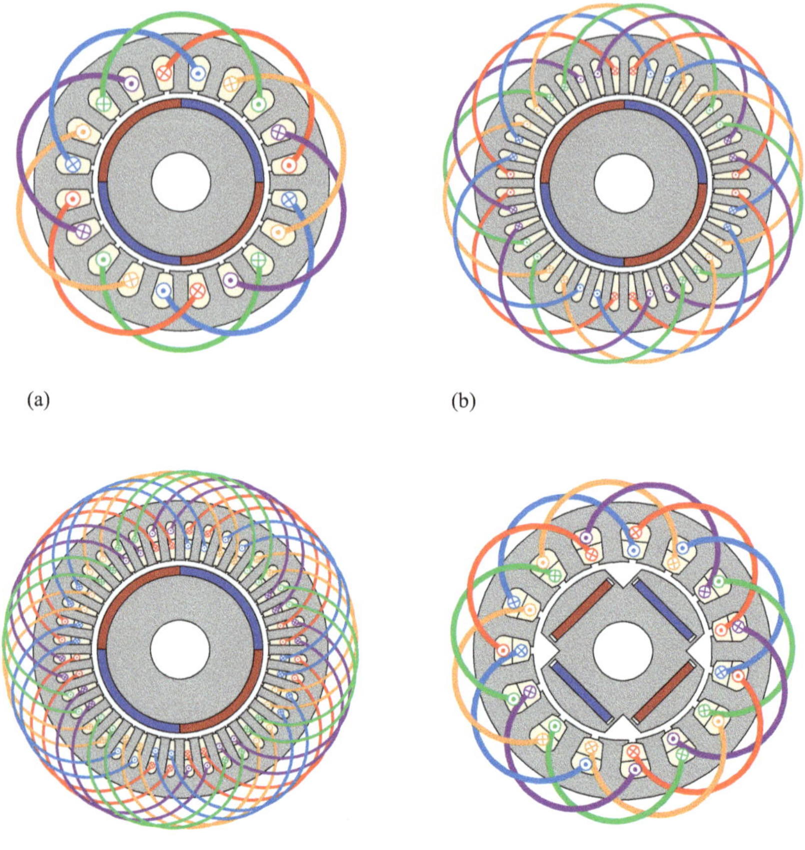

(a) (b) (c) (d)

Figure 1.14 Cross-sections of five-phase PMSMs with OWs. (a) 20-slot/4-pole SPM machine with single-layer windings. (b) 40-slot/4-pole SPM machine with single-layer windings. (c) 40-slot/4-pole SPM machine with double-layer windings. (d) 15-slot/4-pole IPM machine with double-layer windings.

additional complexity when using a high phase number, the optimal phase number in fault-tolerant PMSMs is suggested to be five.

Another comparison of three, four, and five-phase PMSMs was presented in [73], by considering the effects of modular stator design. Three-phase 12-slot/10-pole, four-phase 16-slot/12-pole, and five-phase 20-slot/18-pole SPM machines with modular and non-modular stators were designed and compared, as shown in Figure 1.17. It is found that the main sub-harmonics in multiphase PMSMs can be reduced by employing modular stators. It also demonstrates that in multiphase PMSMs with modular stators, the resultant electromagnetic performances are

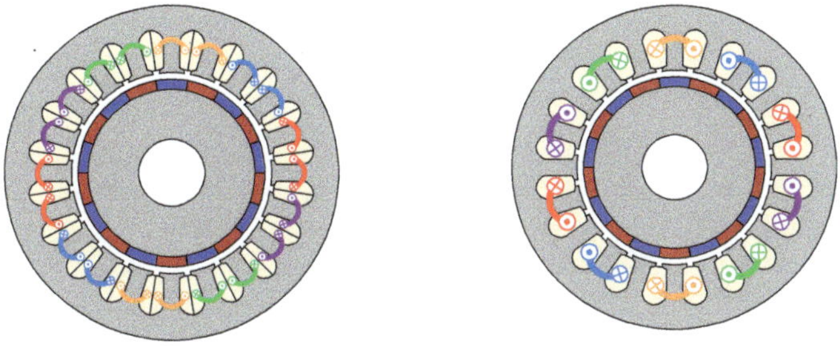

(a) (b)

Figure 1.15 *Cross-sections of five-phase PMSMs with non-OWs. (a) 20-slot/18-pole SPM machine with double-layer windings. (b) 20-slot/18-pole SPM machine with single-layer windings.*

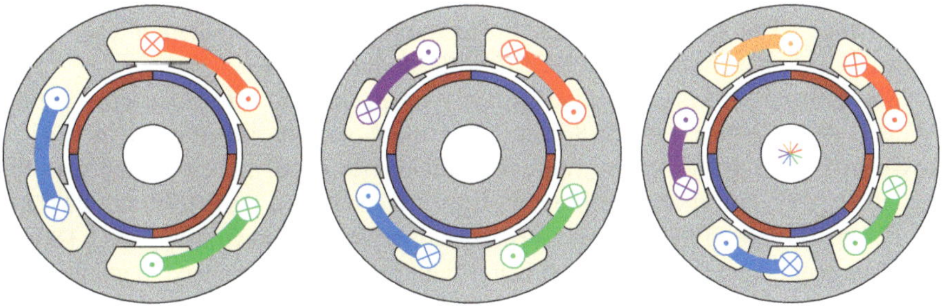

(a) (b) (c)

Figure 1.16 *Cross-sections of single-layer winding multiphase SPMSMs. (a) 3-phase 6-slot/4-pole, (b) 4-phase 8-slot/4-pole, and (c) 5-phase 10-slot/8-pole.*

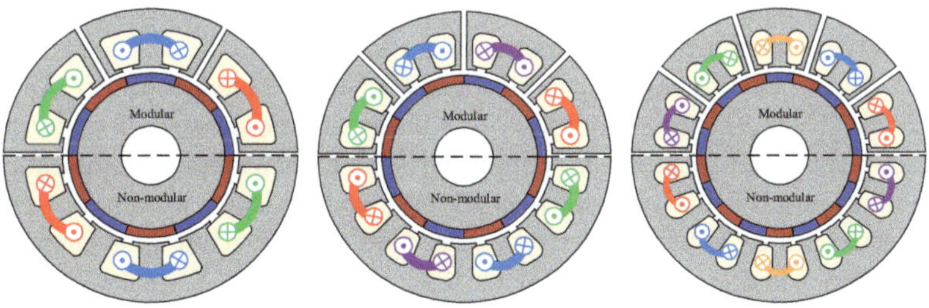

(a) (b) (c)

Figure 1.17 *Cross-sections of modular multiphase PMSMs. (a) 3-phase 12-slot/10-pole, (b) 4-phase 16-slot/12-pole, and (c) 5-phase 20-slot/18-pole.*

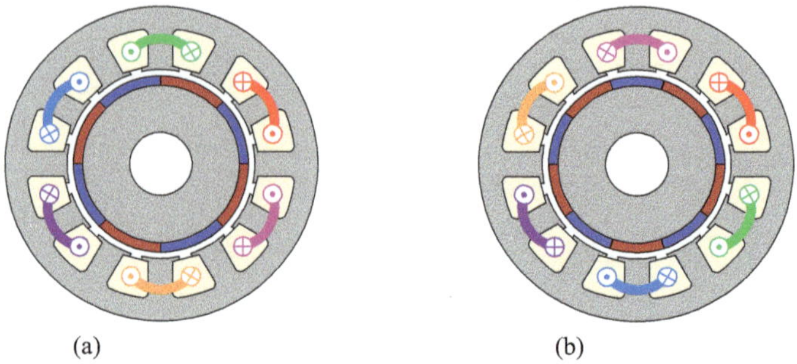

Figure 1.18 Cross-section of six-phase SPM machines with single-layer windings. (a) 12-slot/8-pole and (b) 12-slot/10-pole.

affected significantly by the flux gap between stator modules. When slot number > pole number, the flux gap has negative effects on the electromagnetic performance, but when slot number < pole number, the flux gap can improve the electromagnetic performance. The general conclusions summarized in this chapter are widely applicable to other multiphase PMSMs with modular stators.

Six-phase PMSMs were also initially proposed for aerospace applications [74–76]. A six-phase 12-slot/8-pole SPM machine was introduced in [74], as shown in Figure 1.18(a). The fault-tolerant operation of this machine was discussed in [75]. Another feasible topology of six-phase fault-tolerant PMSMs, i.e., 12-slot/10-pole SPM machine, is presented in [76], as shown in Figure 1.18(b). It should be mentioned that to achieve better fault-tolerant performance, both six-phase PMSMs shown in Figure 1.18 utilize single-layer windings and each phase was supplied from a separate H bridge, to realize more effective electrical, magnetic, thermal, and physical isolation between phases.

Although multiphase PMSMs have superior fault-tolerant capabilities than their conventional three-phase counterparts, the requirements of customized multiphase drives increased the cost and limited the development of multiphase PMSMs. In recent decades, as DTP machine topologies can make a balance between fault tolerance and the adoption of standard three-phase components, DTP PMSMs have attracted much more attention.

In [30], the electromagnetic performance of the 12-slot/10-pole PMSM with different DTP winding configurations was analyzed and compared under healthy and three-phase open-circuit (OC) conditions. The DTP winding configurations are shown in Figure 1.19. Both single-layer and double-layer windings were considered, and both SPM and IPM rotors were also considered in this chapter. It can be found that whether with an SPM or IPM rotor, the DL-3 (phase shift = 30°) configuration shows the best performance under both healthy and faulty conditions.

More feasible slot/pole number combinations of DTP PMSMs were analyzed based on the star of slots in [77], but only concentrated windings (coil pitch = 1) and phase shift = 30° were considered. It was found that among all slot/pole number combinations, 24-slot/20-pole and 24-slot/22-pole are two more promising

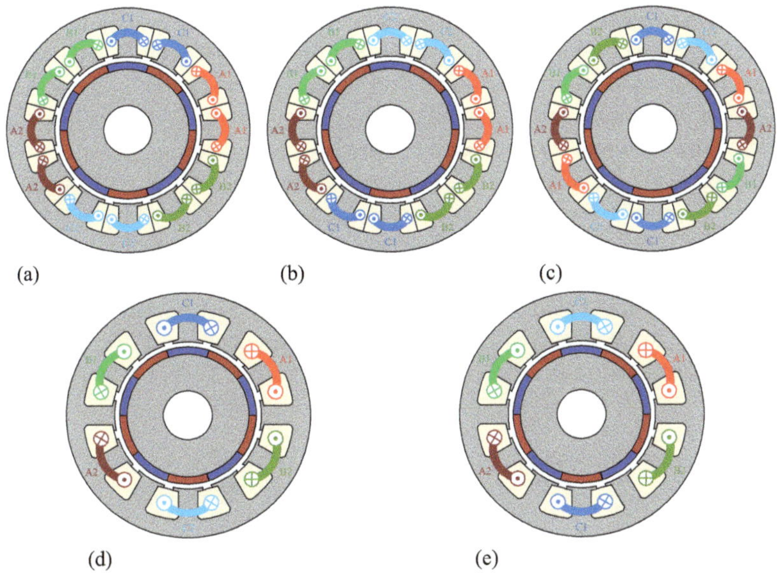

Figure 1.19 Cross-sections of DTP 12-slot/10-pole SPM machine with different DTP winding configurations. (a) DL-1, (b) DL-2, (c) DL-3, (d) SL-1, and (e) SL-2.

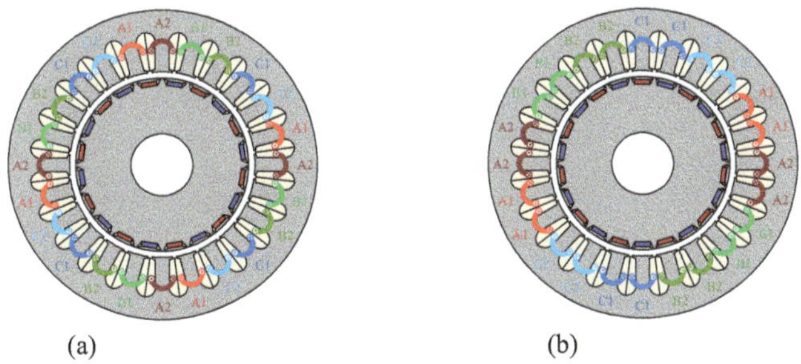

Figure 1.20 Cross-sections of DTP IPM machines with double-layer windings. (a) 24-slot/20-pole and (b) 24-slot/22-pole.

candidates, as shown in Figure 1.20. The electromagnetic performances of DTP 24-slot/20-pole and 24-slot/22-pole IPM machines are further compared. The 24-slot/22-pole machine can be equipped with single-layer and double-layer windings, but the 24-slot/20-pole machine can only use double-layer windings. The comparison results show that the 24-slot/22-pole machine with double-layer windings exhibits higher average torque, lower torque ripple, and lower mutual coupling. The 24-slot/20-pole machine exhibits lower rotor loss in healthy conditions and lower temperature rise in faulty conditions.

The DTP 24-slot/22-pole combination was also analyzed in [78], in an external SPM rotor machine, both single-layer and double-layer windings were considered, as shown in Figure 1.21. It was found that, compared with single-layer windings, double-layer windings can suppress magnetomotive force (MMF) harmonics, reduce iron and PM loss, and lower thermal rise in magnets. To achieve a feasible post-fault control, the machine was designed to be supplied with six separate H-bridges. Then, the fault-tolerant control strategies for the DTP 24-slot/22-pole SPM machine were discussed. With the proposed control strategies, the machine can be kept operating with up to two phases open-circuited.

Besides the DTP PMSMs shown in Figure 1.21, a six-phase 24-slot/14-pole machine with an external SPM rotor is also proposed in [79], as shown in Figure 1.22. Similar to the machines shown in Figure 1.21, the proposed machine

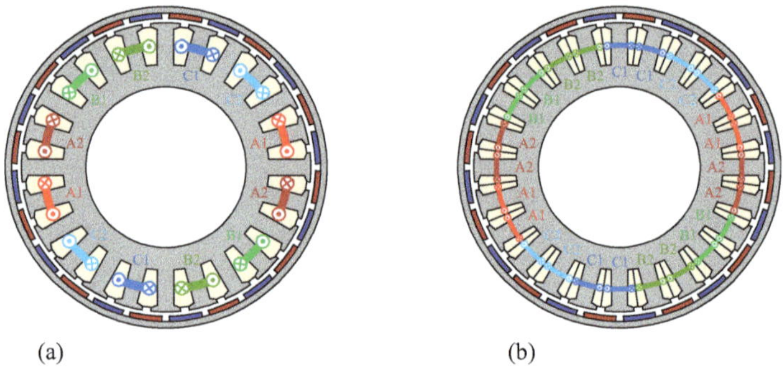

Figure 1.21 Cross-sections of DTP 24-slot/22-pole machines with external SPM rotors. (a) Single-layer windings and (b) double-layer windings.

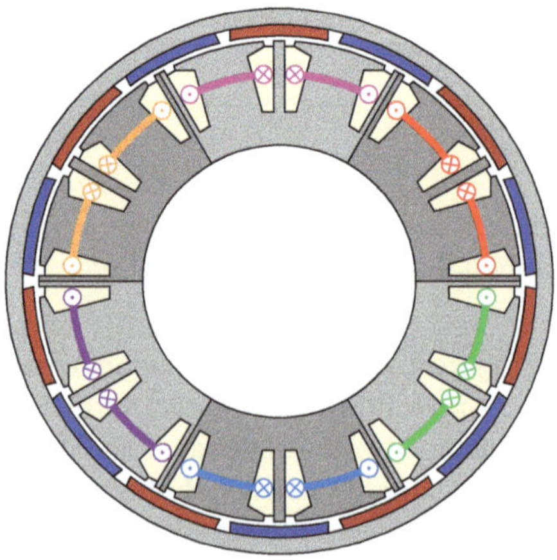

Figure 1.22 Cross-section of a six-phase 24-slot/14-pole machine with single-layer windings, modular stator, unequal stator tooth widths, and external SPM rotor.

has single-layer windings, and two adjacent coils belonging to the same phase, thus, making the automatic coil winding and modular stator feasible. Additionally, in the proposed machine, different phases are electrically, magnetically, thermally, and physically isolated, which is beneficial to their fault-tolerant capability. Furthermore, unequal stator tooth widths [80] were utilized in the machine, which can improve the winding factor of the working harmonic significantly. Thus, the proposed six-phase 24-slot/14-pole external rotor machine can be a suitable candidate for electric vehicle applications.

Another example of DTP PMSMs was given in [79]. It was found that for STP PMSMs with slot number = pole number $\pm$ 1, although the MMF harmonic contents are relatively abundant, all the odd or even harmonics can be eliminated by doubling the slots and three-phase windings. Thus, a DTP 18-slot/8-pole machine and a DTP 36-slot/8-pole machine can be proposed based on the STP 9-slot/8-pole machine, as shown in Figure 1.23. For the STP 9-slot/8-pole PMSM, the working harmonic is the 4th harmonic; with the proposed method, all the odd MMF harmonics in the 9-slot/8-pole PMSM can be eliminated in the DTP 18-slot/8-pole PMSM. Furthermore, all the 2nd, 6th, 10th, 14th, 18th, 22nd, etc. MMF harmonics can be eliminated by changing the DTP 18-slot/8-pole PMSM into a DTP 36-slot/8-pole PMSM. Hence, the undesirable MMF harmonics can be effectively reduced by the proposed DTP winding configuration.

It should also be mentioned that although the DTP winding connections are obtained using different methods in different papers, the winding arrangements can still be generally summarized based on the star of slots. The more detailed guidelines about arranging DTP windings will be provided in Chapters 2, 3, and 4.

As mentioned in [28,72], a system of more than five phases is generally considered undesirable, because it becomes overly complex and expensive. Multiphase PMSMs with phase number >5 can only be practical when using multiple three-phase winding configurations, such as six- (dual three) phase, nine- (triple three) phase, and twelve- (quadruple three) phase winding configurations. For practical nine-phase PMSMs, the windings are supplied from triple three-phase (TTP)

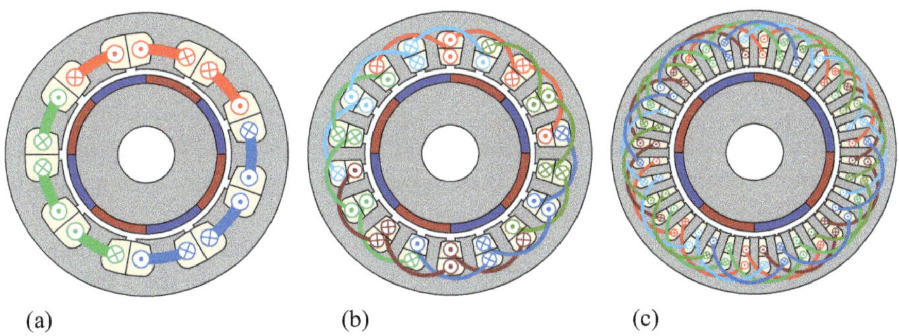

Figure 1.23 Cross-sections of DTP SPM machines. (a) 3-phase 9-slot/8-pole, (b) DTP 18-slot/8-pole, and (c) DTP 36-slot/8-pole.

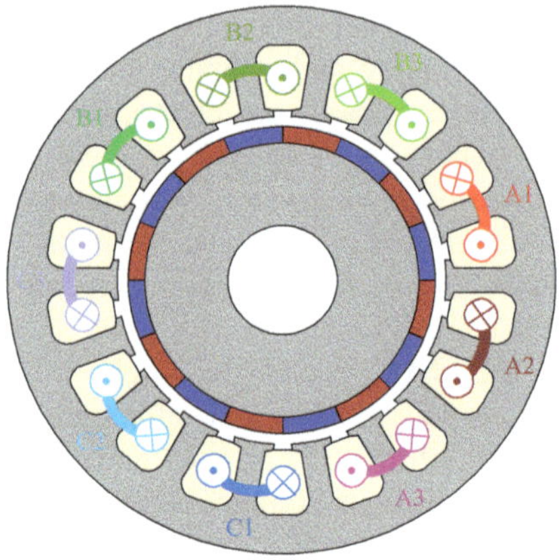

Figure 1.24 Cross-section of a TTP 18-slot/16-pole SPM machine with single-layer windings.

Figure 1.25 Cross-section of TTP 18-slot/14-pole IPM machine with double-layer windings (phase shift = 20°).

drives. A typical TTP machine topology was given in [25], as shown in Figure 1.24. The fault-tolerant control strategies for the TTP 18-slot/16-pole SPM machine were derived and validated. The machine is equipped with single layer windings and the phase shift between the 3 three-phase winding sets is 40°.

In [31], a TTP 18-slot/14-pole IPM machine with double-layer windings was designed, as shown in Figure 1.25. The phase shift angle between the 3 three-phase winding sets is selected as 20° to reduce MMF harmonics. With the TTP winding configuration, all the subharmonics and some higher-order harmonics in the TTP 18-slot/14-pole IPM machine are eliminated. The proposed machine exhibits a high efficiency over a wide torque and speed range.

1.5 Summary of multiphase PMSMs in literature

Some key papers in the literature on multiphase PMSMs are summarized in Table 1.1, according to the number of phases. As can be seen, the most popular number of phases for investigation are five and six phases in terms of the number of published papers.

Table 1.1 Summary of multiphase PMSMs in literature

Phase number	Reference	Slot/pole numbers	Rotor type	Winding layout	Remark
4	[1]	8/6	SPM	SL/Concentrated	–
	[73]	16/12	SPM	SL/Concentrated	Modular
5	[12]	20/4 and 40/4	SPM	SL and DL/Distributed	–
	[70]	15/4	IPM	DL	Coil pitch = 3 slot pitches
	[71]	20/18	SPM	SL and DL/Concentrated	–
	[83]	180/58	SPM	SL/Distributed	–
	[84]	40/42	SPM	SL/Concentrated	Phase separation physically
	[85]	20/18	IPM	SL/Concentrated	Outer spoke-type IPM rotor
	[86]	20/62	SPM	SL/Concentrated	Vernier machine
	[87]	10/8	SPM	DL/Concentrated	Rotor PM shaping
	[88]	General	SPM/IPM	DL/Coil pitch = 1 or 2 slot pitches	–
	[89]	20/18	SPM	DL	Dual five-phase
	[90]	40/18	IPM	DL/Coil pitch = 2 slot pitches	–
	[91]	20/18	IPM + EE	DL/Concentrated	Hybrid-excitation
	[92]	20/14	IPM	DL/Concentrated	Saliency ratio improvement
	[93]	60/22	SPM	Mixed SL and DL/Coil pitch = 2 or 3 slot pitches	–
6	[30]	12/10	SPM/IPM	SL and DL/Concentrated DTP (0/30°)	–
	[94]	24/10	IPM	DL/Coil pitch = 2 slot pitches STP or DTP (0°)	Unequal tooth width
	[78]	24/22	SPM	SL and DL/Concentrated DTP (30°)	
	[79]	24/14	SPM	SL/Concentrated DTP (30°)	Unequal tooth width
	[95]	18/8	IPM	DL/Coil pitch = 2 slot pitches DTP, (0°/20°/40°)	–
	[89]	12/10	SPM	DL/Star-delta DTP (30°)	–
	[96]	24/10	SPM	DL/Coil pitch = 2 slot pitches DTP (30°)	–

(Continues)

Table 1.1 (*Continued*)

Phase number	Reference	Slot/pole numbers	Rotor type	Winding layout	Remark
	[97]	12/10	IPM	Four-layer/Concentrated DTP (30°)	–
	[98]	78/12	SPM	SL/Distributed DTP (27.7°)	–
	[99]	78/24	SPM	SL/Coil pitch = 2 slot pitches DTP (0°)	–
	[100]	24/10 and 24/14	SPM	DL/Coil pitch = 2 slot pitches DTP (0°/15°/30°)	–
	[101]	24/22	SPM	DL/Concentrated DTP (0°/15°/30°)	–
	[102]	24/14	IPM	DL/Concentrated DTP (30°)	–
	[103]	42/32	SPM	DL/Concentrated DTP (30°)	Modular
	[104]	42/32 and 192/32	SPM	DL/Concentrated and SL/ Distributed DTP (30°)	Modular
	[105]	General	SPM	DL/Concentrated DTP (30°)	Modular
	[106]	24/22	SPM	DL/Concentrated DTP (0°/ 15°/30°/60°)	Consequent pole rotor
	[107]	30/14	SPM	DL/Coil pitch = 2 slot pitches DTP (0°)	Unequal turns per coil
7	[33]	14/4	SPM	DL/Concentrated	–
	[108]	14/4	SPM	DL/Concentrated	Axial-flux machine
9	[82]	18/16	SPM	DL/Concentrated TTP (40°)	
	[109]	18/14	SPM	DL/Concentrated TTP (40°)	
	[110]	36/6	IPM	SL/Full-pitched TTP (0°)	9-phase 6-terminal connection
	[111]	117/36	SPM	DL/Coil pitch = 3 slot pitches TTP (27.6°)	–
	[112]	54/6	IPM	DL/Distributed TTP (20°)	–
11	[113]	11/10	SPM	DL/Concentrated	–
12	[114]	48/44	SPM	DL/Concentrated	Four 3-phase

The stator shifting technique was introduced in [94] to reduce the undesirable MMF harmonics in FSCW PMSMs. The technique is achieved by doubling the number of slots and doubling the windings, then shifting the two winding sets in a specific shifting angle, which can naturally form DTP windings. There are many papers on stator shifting in literature, e.g., [96,102,109,115].

In DTP PMSMs, in general, the feasible phase shift angles between two sets of three-phase windings are mainly determined by the slot/pole number combinations. According to the literature, the phase shift angles for PMSMs with different slot numbers are summarized in Table 1.2. The detailed techniques about how to achieve these phase shift angles will be given in Chapters 2, 3, and 4.

From Table 1.2, it can be seen that the 30° phase shift angle is the most popular in DTP PMSMs. Besides the advantages of electromagnetic performance, another

Table 1.2 Summary of feasible phase shift angles between two sets of three-phase windings for DTP PMSMs in literature

Slot number	Pole number	Phase shift angles	Reference
12-slot	10-pole	0°/30°	[30]
	10-pole	30°	[89]
18-slot	8-pole	0°/20°/40°	[95]
	10-pole	20°	[116]
	14-pole	0°/20°/40°	[109]
24-slot	10-pole	30°	[96]
	10-pole	30°	[102]
	14-pole	0°/15°/30°	[100]
	10-pole	30°	[77]
	22-pole	30°	[77,78]
	22-pole	0°/15°/30°	[101]
36-slot	34-pole	10°/20°/30°/40°/50°/60°	[117]
48-slot	8-pole	30°	[16,117]
	22-pole	0°/7.5°/15°/30°/45°	[118,119]

advantage of a DTP winding configuration with a 30° phase shift angle is that the research on the control of the 30° phase shift angle is much more mature. Numerous control strategies were proposed based on DTP PMSMs with the phase shift angle = 30° [1,87,120–127].

Star-delta winding connection has been analyzed in induction machines for a long time [128–130]. In recent years, the star-delta winding connection was introduced in PMSMs to reduce space MMF harmonics [89,131,132]. A generic analysis of the effects of the star-delta winding connection on stator MMF harmonics was reported in [133], considering different phase numbers. In [134], the star-delta winding connection was combined with a conventional dual inverter DTP winding configuration. The proposed 24-slot/10-pole PMSM has two winding sets with a 15° phase shift angle, which can eliminate the 7th MMF harmonic, and each winding set has a star-delta connection, which can eliminate the 1st MMF harmonic. Thus, the two largest MMF harmonics that cause iron loss can be totally canceled by the combination of DTP and star-delta winding connections. The proposed winding configuration shows comparable average torque and efficiency to the 12-slot/10-pole DTP counterpart, but lower torque ripple and lower PM eddy current losses.

The machine topologies and features of DTP PMSMs are summarized in Table 1.3 according to slot number and pole number. It should be noted that Table 1.3 only summarizes some typical DTP machine designs in the literature. A more detailed general analysis of all feasible slot/pole number combinations, winding configurations, and feasible phase shift angles will be presented in Chapters 2, 3, and 4.

Considering the high cost of customized multiphase drives and the complexities in control strategies, MTP, especially DTP winding configuration, is

Table 1.3 Various DTP PMSM topologies in literature

Slot number	Reference	Topology	Features
12-slot	[30]	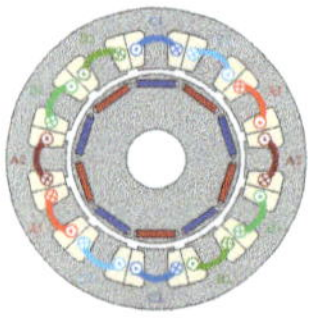	• 12-slot/10-pole • Phase shift = 30° • Coil pitch = 1 slot pitch • IPM and SPM rotor
12-slot	[89]	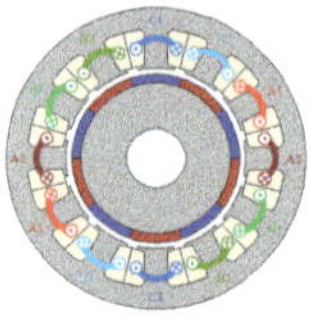	• 12-slot/10-pole • Phase shift = 30° • Coil pitch = 1 slot pitch • SPM rotor • Star-delta connection
18-slot	[95]		• 18-slot/8-pole • Phase shift = 0°/20°/40° • Coil pitch = 2 slot pitches • IPM rotor
18-slot	[116]	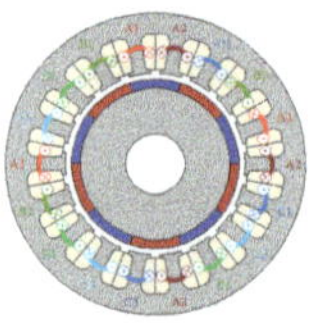	• 18-slot/10-pole • Phase shift = 20° • Coil pitch = 1 slot pitch • SPM rotor
18-slot	[109]	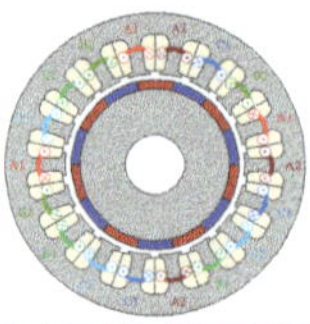	• 18-slot/14-pole • Phase shift = 0°/20°/40° • Coil pitch = 1 slot pitch • SPM rotor
24-slot	[96]	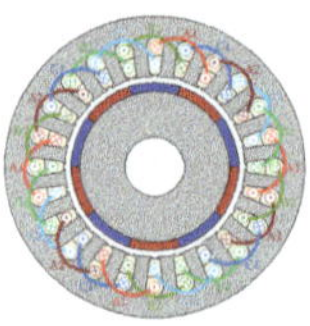	• 24-slot/10-pole • Phase shift = 30° • Stator shifting = 75° • Coil pitch = 2 slot pitches • SPM rotor
24-slot	[102]	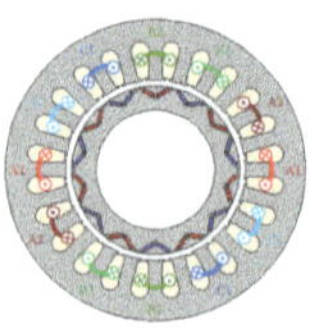	• 24-slot/14-pole • Phase shift = 30° • Stator shifting = 105° • Coil pitch = 1 slot pitch • IPM rotor

(Continues)

Table 1.3 (Continued)

Slot number	Reference	Topology	Features
24-slot	[100]	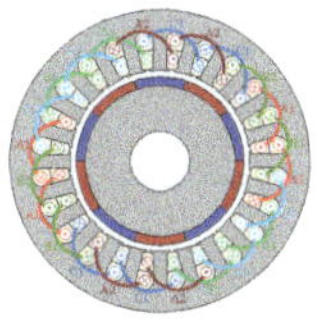	• 24-slot/10-pole • Phase shift = 0°/15°/30° • Coil pitch = 2 slot pitches • SPM rotor
24-slot	[77]	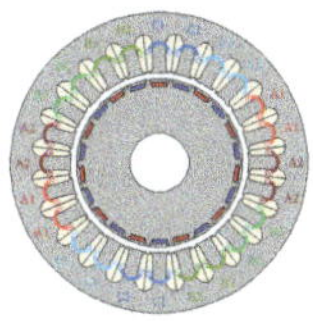	• 24-slot/20- and 22-pole • Phase shift = 30° • Coil pitch = 1 slot pitch • IPM rotor
24-slot	[78]	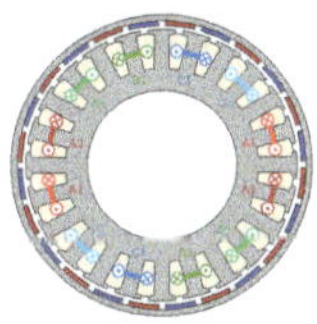	• 24-slot/22-pole • Phase shift = 30° • Coil pitch = 1 slot pitch • Outer SPM rotor • Single and double layer
24-slot	[101]	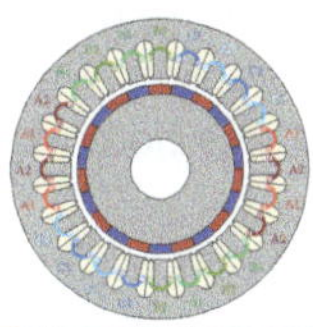	• 24-slot/22-pole • Phase shift = 0°/15°/30° • Coil pitch = 1 slot pitch • SPM rotor
48-slot	[16]	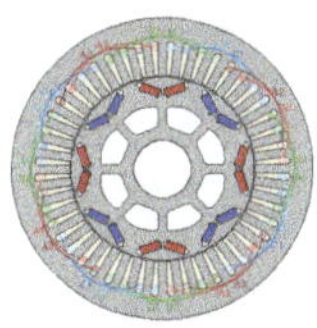	• 48-slot/8-pole • Phase shift = 30° • Coil pitch = 6 (SL) or 5 (DL) slot pitches • IPM rotor
48-slot	[118,119]	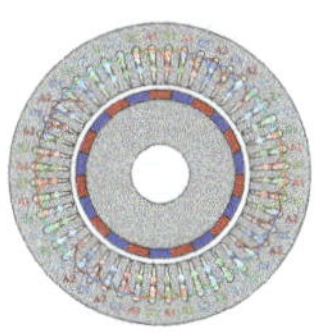	• 48-slot/22-pole • Phase shift = 7.5°/15°/30°/45° • Coil pitch = 2 slot pitches • SPM rotor

advantageous compared with other multiphase counterparts. The advantages of DTP winding configuration in PMSMs can be generally summarized as:

- Good balance between machine fault-tolerant capability and standard three-phase components.
- Improved torque performance in both average torque and torque ripple compared with conventional three-phase counterparts.

- Reduced space MMF harmonic contents, which can help to reduce eddy current loss, lower iron loss, and improve efficiency.
- When a fault occurs in one three-phase winding set, the machine can be run with the remaining three-phase winding set, with a much less complicated control scheme.
- It is possible to realize a power-sharing between DTP winding sets.

1.6 Scope of this book

A comprehensive state-of-the-art resource on multiphase PMSMs will be presented in this book. It will systematically describe the principles in Chapter 1; various rotor PM topologies in Chapter 1 and different stator winding configurations in Chapters 1, 2, 3, and 4; modeling techniques based on a multi-individual three-phase *dq* model and vector space decomposition (VSD) model, together with separation of torque components in Chapter 5; examples of design optimization by finite element method for electric vehicle application for 3-, 5-, 6-, 9-phase PMSMs in Chapter 6; modulation strategies covering linear modulation and over-modulation, voltage vector selection, and carrier phase shifting in Chapter 7; control methods including FOC, DTC, MPC in Chapter 8; sensorless control in Chapter 9; and fault-tolerant control in Chapter 10; as well as current harmonic and torque ripple minimization techniques in Chapters 7 and 8, with particular reference to DTP PMSMs. The overall scope of this book is summarized in Table 1.4.

Table 1.4 Scope of this book

Chapters	Subjects	Features
1	General introduction	• Drives for STP, DTP, and MTP and MP PMSMs • Rotor PM topologies • Stator winding configurations • Multiphase PMSMs in literature
2	General winding configurations of multiphase PMSMs	• General winding configurations • Winding factors • Attenuation factors
3	DTP PMSMs with integer slot overlapping windings	• Integer slot overlapping windings • Winding configurations • Examples of electromagnetic performance
4	DTP PMSMs with fractional slot non-overlapping windings	• Fractional slot DTP PMSMs with coil pitch = 1 or 2 slot pitches • Examples of electromagnetic performance
5	Modeling of multiphase PMSMs	• Multi-individual three-phase *dq* model • Vector space decomposition model • Separation of torque components

(Continues)

Table 1.4 (Continued)

Chapters	Subjects	Features
6	Design optimization of multiphase PMSMs	• 3-, 5-, 6-, and 9-phase PMSMs • Design optimization by finite element method • Comparison of electromagnetic performance, power switches, DC-link capacitors, and costs for electric vehicles
7	Modulation strategies for multiphase PMSMs	• Modulation strategies of FOC • Modulation strategies of SVM DTC • Modulation strategies of ST DTC • Over modulation strategies
8	Control strategies for multiphase PMSMs	• Field-oriented control • Direct torque control • Model predictive control
9	Sensorless control of multiphase PMSMs	• STP and DTP PMSMs • Fundamental model-based methods • Saliency tracking-based methods • Transition between low and high speeds
10	Fault-tolerant control for multiphase PMSMs	• Fault classification • Fault diagnosis • Fault-tolerant control

References

[1] Z. Q. Zhu, S. S. Wang, B. Shao, L. C. Yan, P. L. Xu, and Y. Ren, "Advances in dual-three-phase permanent magnet synchronous machines and control techniques," *Energies*, vol. 14, no. 22, pp. 7508, 2021.

[2] M. McCaig and A.G. Clegg, *Permanent Magnets in Theory and Practice*, 2nd edn, Pentech Press, 1987.

[3] R. J. Parker, *Advances in Permanent Magnet Magnetism*, Wiley, New York, 1990.

[4] F. W. Merrill, "AlNiCo materials for small motor and generator fields," *Electr. Manuf.*, vol. 39, pp. 78–83, 180–190, 1947.

[5] R.M. Saunders and R.H. Weakley, "Design of permanent-magnet alternators," *AIEE Trans.*, vol. 70, pt. II, pp. 1578–1581, 1951.

[6] M. W. Brainard, "Synchronous machines with rotating permanent-magnet fields, part I, characteristics and mechanical construction," *AIEE Trans.*, vol. 71, pt. III, pp. 670–676, 1952.

[7] J. F. H. Douglas, "Induction motors with permanent magnet excitation," *Electr. Eng.*, vol. 78, no. 12, pp. 1195–1195, 1959.

[8] J.R. Hendershot and T.J.E. Miller, *Design of Brushless Permanent-Magnet Motors*, Motor Design Books LLC, Venice, FL, 2010.
[9] J.F. Gieras and M. Wing, *Permanent Magnet Motor Technology: Design and Applications*, CRC Press, Boca Raton, FL, 2009.
[10] R. Bojoi, F. Farina, F. Profumo, and A. Tenconi, "Dual-three phase induction machine drives control – a survey," *IEEE Trans. Ind. Appl.*, vol. 126, no. 4, pp. 420–429, 2006.
[11] E. Levi, "Multiphase electric machines for variable-speed applications," *IEEE Trans. Ind. Electron.*, vol. 55, no. 5, pp. 1893–1909, 2008.
[12] L. Parsa and H. A. Toliyat, "Five-phase permanent-magnet motor drives," *IEEE Trans. Ind. Appl.*, vol. 41, no. 1, pp. 30–37, 2005.
[13] M. Qiao, C. Jiang, Y. Zhu, and G. Li, "Research on design method and electromagnetic vibration of six-phase fractional-slot concentrated winding PM motor suitable for ship propulsion," *IEEE Access*, vol. 4, pp. 8535–8543, 2016.
[14] K. Nounou, J. F. Charpentier, K. Marouani, M. Benbouzid, and A. Kheloui, "Emulation of an electric naval propulsion system based on a multiphase machine under healthy and faulty operating conditions," *IEEE Trans. Veh. Tech.*, vol. 67, no. 8, pp. 6895–6905, 2018.
[15] R. Yang, N. Schofield, N. Zhao, and A. Emadi, "Dual three-phase permanent magnet synchronous machine investigation for battery electric vehicle power-trains," *J. Eng.*, vol. 2019, no. 17, pp. 3981–3985, 2019.
[16] S. S. Wang, "Investigation of torque and torque ripple in dual three-phase permanent magnet synchronous machines for electric vehicles," *Ph.D. dissertation*, Univ. Sheffield, Sheffield, UK, 2023.
[17] I. Subotic, N. Bodo, E. Levi, M. Jones, and V. Levi, "Isolated chargers for EVs incorporating six-phase machines," *IEEE Trans. Ind. Electron.*, vol. 63, no. 1, pp. 653–664, 2016.
[18] M. S. Diab, A. A. Elserougi, A. S. Abdel-Khalik, A. M. Massoud, and S. Ahmed, "A nine-switch-converter-based integrated motor drive and battery charger system for EVs using symmetrical six-phase machines," *IEEE Trans. Ind. Electron.*, vol. 63, no. 9, pp. 5326–5335, 2016.
[19] N. Bodo, E. Levi, I. Subotic, J. Espina, L. Empringham, and C. M. Johnson, "Efficiency evaluation of fully integrated on-board EV battery chargers with nine-phase machines," *IEEE Trans. Energy Convers.*, vol. 32, no. 1, pp. 257–266, 2017.
[20] Y. X. Li, "Novel modular dual 3-phase permanent magnet machines for wind power application," *Ph.D. dissertation*, Univ. Sheffield, Sheffield, UK, 2023.
[21] Y. S. Hu, "Control of asymmetric permanent magnet synchronous generator systems," *Ph.D. dissertation*, Univ. Sheffield, Sheffield, UK, 2015.
[22] T. Y. Liu, "Sensorless control of PM wind power generators with parameter variation and asymmetries," *Ph.D. dissertation*, Univ. Sheffield, Sheffield, UK, 2022.
[23] H. S. Che, E. Levi, M. Jones, W. Hew, and N. A. Rahim, "Current control methods for an asymmetrical six-phase induction motor drive," *IEEE Trans. Power Electron.*, vol. 29, no. 1, pp. 407–417, 2014.

[24] I. González-Prieto, M. J. Duran, and F. J. Barrero, "Fault-tolerant control of six-phase induction motor drives with variable current injection," *IEEE Trans. Power Electron.*, vol. 32, no. 10, pp. 7894–7903, 2017.

[25] B. C. Mecrow, A. G. Jack, D. J. Atkinson, *et al.*, "Design and testing of a four-phase fault-tolerant permanent-magnet machine for an engine fuel pump," *IEEE Trans. Energy Convers.*, vol. 19, no. 4, pp. 671–678, 2004.

[26] J. W. Bennett, B. C. Mecrow, D. J. Atkinson, and G. J. Atkinson, "Safety-critical design of electromechanical actuation systems in commercial aircraft," *IET Electr. Power Appl.*, vol. 5, no. 1, pp. 37–47, 2011.

[27] A. S. Thomas, Z. Q. Zhu, R. L. Owen, G. W. Jewell, and D. Howe, "Multiphase flux-switching permanent-magnet brushless machine for aerospace application," *IEEE Trans. Ind. Appl.*, vol. 45, no. 6, pp. 1971–1981, 2009.

[28] W. P. Cao, B. C. Mecrow, G. J. Atkinson, J. W. Bennett, and D. J. Atkinson, "Overview of electric motor technologies used for more electric aircraft (MEA)," *IEEE Trans. Ind. Electron.*, vol. 59, no. 9, pp. 3523–3531, 2012.

[29] B. Sarlioglu and C.T. Morris, "More electric aircraft: review, challenges, and opportunities for commercial transport aircraft," *IEEE Trans. Transp. Electrif.*, vol. 1, no. 1, pp. 54–64, 2015.

[30] M. Barcaro, N. Bianchi, and F. Magnussen, "Analysis and tests of a dual three-phase 12-slot 10-pole permanent-magnet motor," *IEEE Trans. Ind. Appl.*, vol. 46, no. 6, pp. 2355–2362, 2010.

[31] X. Chen, J. Wang, V. I. Patel, and P. Lazari, "A nine-phase 18-slot 14-pole interior permanent-magnet machine with low space harmonics for electric vehicle applications," *IEEE Trans. Energy Convers.*, vol. 31, no. 3, pp. 860–871, 2016.

[32] L. Shao, W. Hua, N. Dai, M. Tong, and M. Cheng, "Mathematical modeling of a 12-phase flux-switching permanent-magnet machine for wind power generation," *IEEE Trans. Ind. Electron.*, vol. 63, no. 1, pp. 504–516, 2016.

[33] F. Scuiller, F. Becker, H. Zahr, and E. Semail, "Design of a bi-harmonic 7-phase PM machine with tooth-concentrated winding," *IEEE Trans. Energy Convers.*, vol. 35, no. 3, pp. 1567–1576, 2020.

[34] A. Abdelkhalik, M. Masoud, and W. Barry, "Eleven-phase induction machine: Steady-state analysis and performance evaluation with harmonic injection," *IET Electr. Power Appl.*, vol. 4, no. 8, pp. 670–685, 2010.

[35] Y. Ren, "Direct torque control of dual three-phase permanent magnet synchronous machine drives," *Ph.D. dissertation*, Univ. Sheffield, Sheffield, UK, 2015.

[36] A. Almarhoon, "Sensorless control of dual three-phase permanent magnet synchronous machine drives," *Ph.D. dissertation*, Univ. Sheffield, Sheffield, UK, 2016.

[37] B. Shao, "Direct torque control of dual three-phase permanent magnet synchronous machine drives with particular reference to reduction of current harmonics and torque ripples," *Ph.D. dissertation*, Univ. Sheffield, Sheffield, UK, 2022.

[38] L. C. Yan, "Control of current harmonics in dual three-phase PMSM systems," *Ph.D. dissertation*, Univ. Sheffield, Sheffield, UK, 2022.

[39] Y. Chen, "Sensorless control of single and dual three-phase interior permanent magnet synchronous machines with inductance asymmetries," *Ph.D. dissertation*, Univ. Sheffield, Sheffield, UK, 2024.

[40] D. J. Hanrahan and D. S. Toffolo, "Permanent magnet generators, part I-theory," *AIEE Trans.*, vol. 76, pt. III, pp. 1098–1103, 1957.

[41] V. B. Honsinger, "Performance of polyphase permanent magnet machines," *IEEE Trans. Power App. Syst.*, vol. PAS-99, no. 4, pp. 1510–1518, 1980.

[42] V. B. Honsinger, "The fields and parameters of interior type AC permanent magnet machines," *IEEE Trans. Power App. Syst.*, vol. PAS-101, no. 4, pp. 867–876, 1982.

[43] M. Rahman, T. Little, and G. Slemon, "Analytical models for interior-type permanent magnet synchronous motors," *IEEE Trans. Magn.*, vol. 21, no. 5, pp. 1741–1743, 1985.

[44] T. M. Jahns, G. B. Kliman, and T. W. Neumann, "Interior permanent-magnet synchronous motors for adjustable-speed drives," *IEEE Trans. Ind. Appl.*, vol. IA-22, no. 4, pp. 738–747, 1986.

[45] T. M. Jahns, "Flux-weakening regime operation of an interior permanent-magnet synchronous motor drive," *IEEE Trans. Ind. Appl.*, vol. IA-23, no. 4, pp. 681–689, 1987.

[46] R. F. Schiferl and T. A. Lipo, "Power capability of salient pole permanent magnet synchronous motor in variable speed drive applications," *IEEE Trans. Ind. Appl.*, vol. 26, no. 1, pp. 115–123, 1990.

[47] W. L. Soong and N. Ertugrul, "Field-weakening performance of interior permanent-magnet motors," *IEEE Trans. Ind. Appl.*, vol. 38, no. 5, pp. 1251–1258, 2002.

[48] E. C. Lovelace, T. M. Jahns, and J. H. Lang, "A saturating lumped-parameter model for an interior PM synchronous machine," *IEEE Trans. Ind. Appl.*, vol. 38, no. 3, pp. 645–650, 2002.

[49] H. S. Che, E. Levi, M. Jones, M. J. Duran, W. P. Hew, and N. A. Rahim, "Operation of a six-phase induction machine using series-connected machine-side converters," *IEEE Trans. Ind. Electron.*, vol. 61, no. 1, pp. 164–176, 2014.

[50] X. Chen, J. Wang, B. Sen, P. Lazari, and T. Sun, "A high-fidelity and computationally efficient model for interior permanent-magnet machines considering the magnetic saturation, spatial harmonics, and iron loss effect," *IEEE Trans. Ind. Electron.*, vol. 62, no. 7, pp. 4044–4055, 2015.

[51] K. C. Kim and S. H. Jeon, "Analysis on correlation between cogging torque and torque ripple by considering magnetic saturation," *IEEE Trans. Magn.*, vol. 49, no. 5, pp. 2417–2420, 2013.

[52] Y. Honda, T. Higaki, S. Morimoto, and Y. Takeda, "Rotor design optimisation of a multi-layer interior PM synchronous motor," *IEE Proc. Electr. Power Appl.*, vol.145, no. 2, pp.119–124, 1998.

[53] Y. Hu, Z.Q. Zhu, and M. Odavic, "Comparison of two-individual current control and vector space decomposition control for dual three-phase PMSM," *IEEE Trans. Ind. Appl.*, vol.53, no. 5, pp. 4483–4492, 2017.

[54] S. H. Han, T. M. Jahns, W. L. Soong, M. K. Güven, and M. S. Illindala, "Torque ripple reduction in interior permanent magnet synchronous machines using stators with odd number of slots per pole pair," *IEEE Trans. Energy Convers.*, vol. 25, no. 1, pp. 118–127, 2010.

[55] S. Zhu, Y. Hu, C. Liu, K. Wang and W. Chen, "Reduction of stator core loss in interior PM machines for electric vehicle applications," *IEEE Trans. Magn.*, vol. 54, no. 11, Art no. 8700405, 2018.

[56] I. Boldea, L. Tutelea and C. I. Pitic, "PM-assisted reluctance synchronous motor/generator (PM-RSM) for mild hybrid vehicles: electromagnetic design," *IEEE Trans. Ind. Appl.*, vol. 40, no. 2, pp. 492–498, 2004.

[57] N. Bianchi, M. D. Pre, and S. Bolognani, "Design of a fault-tolerant IPM motor for electric power steering," *IEEE Trans. Veh. Technol.*, vol. 55, no. 4, pp. 1102–1111, 2006.

[58] E. Armando, P. Guglielmi, G. Pellegrino, M. Pastorelli, and A. Vagati, "Accurate modeling and performance analysis of IPM-PMASR motors," *IEEE Trans. Ind. Appl.*, vol. 45, no. 1, pp. 123–130, 2009.

[59] G. Pellegrino, P. Guglielmi, A. Vagati, and F. Villata, "Core losses and torque ripple in IPM machines: dedicated modeling and design tradeoff," *IEEE Trans. Ind. Appl.*, vol. 46, no. 6, pp. 2381–2391, 2010.

[60] M. Barcaro, N. Bianchi, and F. Magnussen, "Permanent-magnet optimization in permanent-magnet-assisted synchronous reluctance motor for a wide constant-power speed range," *IEEE Trans. Ind. Electron.*, vol. 59, no. 6, pp. 2495–2502, 2012.

[61] Y. Hu, S. Zhu, C. Liu, and K. Wang, "Electromagnetic performance analysis of interior PM machines for electric vehicle applications," *IEEE Trans. Energy Convers.*, vol. 33, no. 1, pp. 199–208, 2018.

[62] N. Bianchi and M. Dai Pre, "Use of the star of slots in designing fractional-slot single-layer synchronous motors," *IEE Electr. Power Appl.*, vol. 153, no. 3, pp. 459–466, 2006.

[63] J. Pyrhonen, T. Jokinen, and V. Hrabovcova, *Design of Rotating Electrical Machines*, Wiley, New York, 2008.

[64] J. Cros and P. Viarouge, "Synthesis of high-performance PM motors with concentrated windings," *IEEE Trans. Energy Convers.*, vol. 17, no. 2, pp. 248–253, 2002.

[65] D. Ishak, Z. Q. Zhu, and D. Howe, "Comparison of PM brushless motors, having either all teeth or alternate teeth wound," *IEEE Trans. Energy Convers.*, vol. 21, no. 1, pp. 95–103, 2006.

[66] N. Bianchi, S. Bolognani, and G. Grezzani, "Design considerations for fractional-slot winding configurations of synchronous machines," *IEEE Trans. Ind. Appl.*, vol. 42, no. 4, pp. 997–1006, 2006.

[67] F. Magnussen and H. Lendenmann, "Parasitic effects in PM machines with concentrated windings," *IEEE Trans. Ind. Appl.*, vol. 43, no. 5, pp. 1223–1232, 2007.

[68] A. M. EL-Refaie, "Fractional-slot concentrated-windings synchronous permanent magnet machines: opportunities and challenges," *IEEE Trans. Ind. Electron.*, vol. 57, no. 1, pp. 107–121, 2010.

[69] A. M. El-Refaie and T. M. Jahns, "Impact of winding layer number and magnet type on synchronous surface PM machines designed for wide constant-power speed range operation," *IEEE Trans. Energy Convers.*, vol. 23, no. 1, pp. 53–60, 2008.

[70] L. Parsa and H. A. Toliyat, "Fault-tolerant interior-permanent-magnet machines for hybrid electric vehicle applications," *IEEE Trans. Veh. Technol.*, vol. 56, no. 4, pp. 1546–1552, 2007.

[71] N. Bianchi, S. Bolognani, and M. D. Dai Pre, "Impact of stator winding of a five-phase permanent-magnet motor on postfault operations," *IEEE Trans. Ind. Electron.*, vol. 55, no. 5, pp. 1978–1987, 2008.

[72] T. Gopalarathnam, H. A. Toliyat, and J. C. Moreira, "Multi-phase fault-tolerant brushless DC motor drives," *Conf. Rec. IAS Annu. Meeting*, Rome, Italy, Oct. 2000, pp. 1683–1688.

[73] G. J. Li, B. Ren, and Z. Q. Zhu, "Design guidelines for fractional slot multi-phase modular permanent magnet machines," *IET Electr. Power Appl.*, vol. 11, no. 6, pp. 1023–1031, 2017.

[74] A. G. Jack, B. C. Mecrow, and J. A. Haylock, "A comparative study of permanent magnet and switched reluctance motors for high-performance fault-tolerant applications," *IEEE Trans. Ind. Appl.*, vol. 32, no. 4, pp. 889–895, 1996.

[75] B. C. Mecrow, A. G. Jack, J. A. Haylock, and J. Coles, "Fault-tolerant permanent magnet machine drives," *IEE Proc. Electr. Power Appl.*, vol. 143, no. 6, pp. 437–442, 1996.

[76] G. M. Raimondi, R. D. McFarlane, C. M. Bingham, *et al.*, "Large electro-mechanical actuation systems for flight control surfaces," *Proc. IEE Colloq. All Electr. Aircr.*, London, UK, 1998, Art no. 260.

[77] M. Barcaro, N. Bianchi, and F. Magnussen, "Six-phase supply feasibility using a PM fractional-slot dual winding machine," *IEEE Trans. Ind. Appl.*, vol. 47, no. 5, pp. 2042–2050, 2011.

[78] P. Zheng, F. Wu, Y. Sui, P. Wang, Y. Lei, and H. Wang, "Harmonic analysis and fault-tolerant capability of a semi-12-phase permanent-magnet synchronous machine used for EVs," *Energies*, vol. 5, no. 9, pp. 3586–3607, 2012.

[79] P. Zheng, F. Wu, Y. Lei, Y. Sui, and B. Yu, "Investigation of a novel 24-slot/14-pole six-phase fault-tolerant modular permanent-magnet in-wheel motor for electric vehicles," *Energies*, vol. 6, no. 10, pp. 4980–5002, 2013.

[80] D. Ishak, Z. Q. Zhu, and D. Howe, "Permanent-magnet brushless machines with unequal tooth widths and similar slot and pole numbers," *IEEE Trans. Ind. Appl.*, vol. 41, no. 2, pp. 584–590, 2005.

[81] J. Wang, V. I. Patel, and W. Wang, "Fractional-slot permanent magnet brushless machines with low space harmonic contents," *IEEE Trans. Magn.*, vol. 50, no. 1, Art no. 8200209, 2014.
[82] M. Ruba and D. Fodorean, "Analysis of fault-tolerant multiphase power converter for a nine-phase permanent magnet synchronous machine," *IEEE Trans. Ind. Appl.*, vol. 48, no. 6, pp. 2092–2101, 2012.
[83] F. Scuiller, E. Semail, J. F. Charpentier, and P. Letellier, "Multi-criteria based design approach of multi-phase permanent magnet low-speed synchronous machines," *IET Electr. Power Appl.*, vol. 3, no. 2, pp. 102110, 2009.
[84] P. Zheng, Y. Sui, J. Zhao, C. Tong, T. A. Lipo, and A. Wang, "Investigation of a novel five-phase modular permanent-magnet in-wheel motor," *IEEE Trans. Magn.*, vol. 47, no. 10, pp. 4084–4087, 2011.
[85] Q. Chen, G. Liu, W. Gong, and W. Zhao, "A new fault-tolerant permanent-magnet machine for electric vehicle applications," *IEEE Trans. Magn.*, vol. 47, no. 10, pp. 4183–4186, 2011.
[86] G. Liu, J. Yang, W. Zhao, J. Ji, Q. Chen, and W. Gong, "Design and analysis of a new fault-tolerant permanent-magnet vernier machine for electric vehicles," *IEEE Trans. Magn.*, vol. 48, no. 11, pp. 4176–4179, 2012.
[87] K. Wang, Z. Q. Zhu, and G. Ombach, "Torque improvement of five-phase surface-mounted permanent magnet machine using third-order harmonic," *IEEE Trans. Energy Convers.*, vol. 29, no. 3, pp. 735–747, 2014.
[88] K. Wang, Z. Q. Zhu, and G. Ombach, "Synthesis of high-performance fractional-slot permanent-magnet machines with coil-pitch of two slot-pitches," *IEEE Trans. Energy Convers.*, vol. 29, no. 3, pp. 758–770, 2014.
[89] A. S. Abdel-Khalik, S. Ahmed, and A. M. Massoud, "Low space harmonics cancelation in double-layer fractional slot winding using dual multiphase winding," *IEEE Trans. Magn.*, vol. 51, no. 5, pp. 1–10, 2015.
[90] A. S. Abdel-Khalik, S. Ahmed, and A. Massoud, "Application of stator shifting to five-phase fractional-slot concentrated winding interior permanent magnet synchronous machine," *IET Electr. Power Appl.*, vol. 10, no. 7, pp. 681–690, 2016.
[91] L. Zhang, Y. Fan, R. D. Lorenz, R. Cui, C. Li, and M. Cheng, "Design and analysis of a new five-phase brushless hybrid-excitation fault-tolerant motor for electric vehicles," *IEEE Trans. Ind. Appl.*, vol. 53, no. 4, pp. 3428–3437, 2017.
[92] Q. Chen, Y. Yan, G. Liu, and G. Xu, "Design of a new fault-tolerant permanent magnet machine with optimized salient ratio and reluctance torque ratio," *IEEE Trans. Ind. Electron.*, vol. 67, no. 7, pp. 6043–6054, 2020.
[93] J. Huang, Y. Sui, Z. Yin, S. Yang, and P. Zheng, "Five-phase hybrid single/double layer fractional slot winding PMSM for torque improvement under third harmonic current injection condition," *IEEE Trans. Magn.*, vol. 58, no. 8, Art no. 8204206, 2022.
[94] G. Dajaku and D. Gerling, "A novel 24-slots/10-poles winding topology for electric machines," *Proc. Int. Electr. Mach. Drives Conf. (IEMDC)*, Niagara Falls, ON, Canada, pp. 65–70, 2011.

[95] V. I. Patel, J. Wang, W. Wang, and X Chen, "Six-phase fractional-slot-per-pole-per-phase permanent-magnet machines with low space harmonics for electric vehicle application," *IEEE Trans. Ind. Appl.*, vol. 50, no. 4, pp. 2554–2563, 2014.

[96] A. S. Abdel-Khalik, S. Ahmed, and A. M. Massoud, "A six-phase 24-slot/10-pole permanent-magnet machine with low space harmonics for electric vehicle applications," *IEEE Trans. Magn.*, vol. 52, no. 6, Art no. 8700110, 2016.

[97] A. S. Abdel-Khalik, S. Ahmed, and A. M. Massoud, "Effect of multilayer windings with different stator winding connections on interior PM machines for EV applications," *IEEE Trans. Magn.*, vol. 52, no. 2, Art no. 8100807, 2016.

[98] Y. Demir and M. Aydin, "A novel dual three-phase permanent magnet synchronous motor with asymmetric stator winding," *IEEE Trans. Magn.*, vol. 52, no. 7, Art no. 8105005, 2016.

[99] Y. Demir and M. Aydin, "A novel asymmetric and unconventional stator winding configuration and placement for a dual three-phase surface PM motor," *IEEE Trans. Magn.*, vol. 53, no. 11, Art no. 8111805, 2017.

[100] P. L. Xu, J. H. Feng, S. Y. Guo, *et al.*, "Analysis of dual three-phase permanent-magnet synchronous machines with different angle displacements," *IEEE Trans. Ind. Electron.*, vol. 65, no. 3, pp. 1941–1954, 2018.

[101] P. L. Xu, Z. Q. Zhu, B. Shao, *et al.*, "Analysis of dual 3-phase fractional-slot non-overlapping winding pm synchronous machines with different angle displacements," *2019 IEEE Energy Convers. Cong. Expo. (ECCE)*, Baltimore, MD, USA, Sept.–Oct. 2019, pp. 5616–5623.

[102] S. Zhu, T. Cox, Z. Xu, and C. Gerada, "Novel 24-slots14-poles fractional-slot concentrated winding topology with low-space harmonics for electrical machine," *J. Eng.*, vol. 2019, no. 17, pp. 3784–3788, 2019.

[103] Y. X. Li, Z. Q. Zhu, X. M. Wu, A. S. Thomas, and Z. Y. Wu, "Comparative study of modular dual 3-phase permanent magnet machines with overlapping/non-overlapping windings," *IEEE Trans. Ind. Appl.*, vol. 55, no. 4, pp. 3566–3576, 2019.

[104] Y. X. Li, Z. Q. Zhu, A. S. Thomas, Z. Y. Wu, and X. M. Wu, "Novel modular fractional slot permanent magnet machines with redundant teeth," *IEEE Trans. Magn.*, vol. 55, no. 9, Art no. 8204810, 2019.

[105] Y. X. Li, Z. Q. Zhu, and A. S. Thomas, "Generic slot and pole number combinations for novel modular permanent magnet dual 3-phase machines with redundant teeth," *IEEE Trans. Energy Convers.*, vol. 35, no. 3, pp. 1676–1687, 2020.

[106] F. Li, K. Wang, J. Zhang, and F. Zeng, "Effect of angle displacements on electromagnetic performance of dual three-phase consequent-pole permanent magnet machine," *IET Electr. Power Appl.*, vol. 14, no. 7, pp. 1177–1185, 2020.

[107] L. Guo, J. Xu, S. Wu, X. Xie, and H. Wang, "Analysis and design of dual three-phase fractional-slot permanent magnet motor with low space harmonic," *IEEE Trans. Magn.*, vol. 58, no. 1, Art no. 8100112, 2022.

[108] F. Locment, E. Semail, and F. Piriou, "Design and study of a multiphase axial-flux machine," *IEEE Trans. Magn.*, vol. 42, no. 4, pp. 1427–1430, 2006.
[109] H. S. Che, M. J. Duran, E. Levi, M. Jones, W. P. Hew, and N. A. Rahim, "Postfault operation of an asymmetrical six-phase induction machine with single and two isolated neutral points," *IEEE Trans. Power Electron.*, vol. 29, no. 10, pp. 5406–5416, 2014.
[110] B. Wang, J. Wang, A. Griffo and Y. Shi, "Investigation into fault-tolerant capability of a triple redundant PMA SynRM drive," *IEEE Trans. Power Electron.*, vol. 34, no. 2, pp. 1611–1621, 2019.
[111] Y. Demir, E. Yolacan, A. M. EL-Refaie and M. Aydin, "Investigation of different winding configurations and displacements of a nine-phase permanent-magnet-synchronous motor with unbalanced AC winding structure," *IEEE Trans. Ind. Appl.*, vol. 55, no. 4, pp. 3660–3670, 2019.
[112] L. Kang and Y. Chen, "Optimal design of 9-phase permanent magnet synchronous generator with low voltage change rate for diesel railway vehicles," *IEEE Trans. Veh. Technol.*, vol. 71, no. 3, pp. 2681–2690, 2022.
[113] G. Dajaku, "Advanced multi-phase fractional slot concentrated windings: characteristics and potentials," *Electr. Eng.*, vol. 103, no. 1, pp. 397–406, 2021.
[114] L. Shao, W. Hua, J. Soulard, Z. Q. Zhu, Z. Wu, and M. Cheng, "Electromagnetic performance comparison between 12-phase switched flux and surface-mounted PM machines for direct-drive wind power generation," *IEEE Trans. Ind. Appl.*, vol. 56, no. 2, pp. 1408–1422, 2020.
[115] P. B. Reddy, K. K. Huh, and A. M. EL-Refaie, "Generalized approach of stator shifting in interior permanent-magnet machines equipped with fractional-slot concentrated windings," *IEEE Trans. Ind. Electron.*, vol. 61, no. 9, pp. 5035–5046, 2014.
[116] G. Dajaku and D. Gerling, "Efficiency improvements of electric machines for automotive application," *World Electr. Veh. J.*, vol. 5, no. 2, pp. 520–526, 2012.
[117] H. Dhulipati, S. Mukundan, W. Li, J. Tjong, and N. C. Kar, "Investigation of phase angle displacements in six-phase PMSM with concentrated windings for reduced MMF harmonics," *2018 21st Int. Conf. Electr. Mach. Syst. (ICEMS)*, Jeju, South Korea, 2018, pp. 308–313.
[118] S. Zhu, W. Zhao, G. Liu, Y. Mao, and Y. Sun, "Effect of phase shift angle on radial force and vibration behavior in dual three-phase PMSM," *IEEE Trans. Ind. Electron.*, vol. 68, no. 4, pp. 2988–2998, 2021.
[119] Y. Sun, W. Zhao, J. Ji, J. Zheng, and X. Song, "Effect of phase shift on inductance and short-circuit current in dual three-phase 48-slot/22-pole permanent-magnet machines," *IEEE Trans. Ind. Electron.*, vol. 69, no. 2, pp. 1135–1145, 2022.
[120] Y. Zhao and T. A. Lipo, "Space vector PWM control of dual three-phase induction machine using vector space decomposition," *IEEE Trans. Ind. Appl.*, vol. 31, no. 5, pp. 1100–1109, 1995.
[121] S. Kallio, M. Andriollo, A. Tortella, and J. Karttunen, "Decoupled d-q model of double-star interior-permanent-magnet synchronous machines," *IEEE Trans. Ind. Electron.*, vol. 60, no. 6, pp. 2486–2494, 2013.

[122] Y. Ren and Z. Q. Zhu, "Enhancement of steady-state performance in direct-torque-controlled dual three-phase permanent-magnet synchronous machine drives with modified switching table," *IEEE Trans. Ind. Electron.*, vol. 62, no. 6, pp. 3338–3350, 2015.

[123] J. Karttunen, S. Kallio, J. Honkanen, P. Peltoniemi, and P. Silventoinen, "Partial current harmonic compensation in dual three-phase PMSMs considering the limited available voltage," *IEEE Trans. Ind. Electron.*, vol. 64, no. 2, pp. 1038–1048, 2017.

[124] Y. Hu, Z. Q. Zhu, and M. Odavic, "Torque capability enhancement of dual three-phase PMSM drive with fifth and seventh current harmonics injection," *IEEE Trans. Ind. Appl.*, vol. 53, no. 5, pp. 4526–4535, 2017.

[125] L. C. Yan, Z. Q. Zhu, J. Qi, *et al.*, "Suppression of major current harmonics for dual three-phase PMSMs by virtual multi three-phase systems," *IEEE Trans. Ind. Electron.*, vol. 69, no. 6, pp. 5478–5490, 2022.

[126] L. C. Yan, Z. Q. Zhu, J. Qi, *et al.*, "Enhancement of disturbance rejection capability in dual three-phase PMSM system by using virtual impedance," *IEEE Trans. Ind. Appl.*, vol.57, no. 5, pp. 4901–4912, 2021.

[127] B. Shao, Z. Q. Zhu, L. Yan, *et al.*, "Torque ripple reduction for direct torque control of dual three-phase PMSM based on multiple virtual voltage vectors," *IEEE Trans. Energy Convers.*, vol. 38, no. 1, pp. 296–309, 2023.

[128] A. Hughes, "New 3-phase winding of low m.m.f.-harmonic content," *Proc. IEE*, vol. 117, no. 8, pp. 1657–1666, 1970.

[129] J. Y. Chen and C. Z. Chen, "Investigation of a new AC electrical machine winding," *IEE Proc. Electr. Power Appl.*, vol. 145, no. 2, pp. 125–132, 1998.

[130] M. V. Cistelecan, F. J. T. E. Ferreira, and M. Popescu, "Adjustable flux three-phase AC machines with combined multiple-step star-delta winding connections," *IEEE Trans. Energy Convers.*, vol. 25, no. 2, pp. 348–355, 2010.

[131] G. Dajaku and D. Gerling, "A novel tooth concentrated winding with low space harmonic contents," *2013 Int. Electr. Mach. Drives Conf. (IEMDC)*, Chicago, IL, USA, 2013, pp. 755–760.

[132] M. S. Islam, M. A. Kabir, R. Mikail, and I. Husain, "Space-shifted wye–delta winding to minimize space harmonics of fractional-slot winding," *IEEE Trans. Ind. Appl.*, vol. 56, no. 3, pp. 2520–2530, 2020.

[133] W. Zhao, J. Zheng, J. Ji, S. Zhu, and M. Kang, "Star and delta hybrid connection of a FSCW PM machine for low space harmonics," *IEEE Trans. Ind. Electron.*, vol. 65, no. 12, pp. 9266–9279, 2018.

[134] I. A. Rudden, G. J. Li, D. K. Kana Padinharu, *et al.*, "Space harmonic cancellation in a dual three-phase SPM machine with star-delta windings," *IEEE Trans. Energy Convers.*, vol. 39, no. 1, pp. 457–468, 2024,

[135] F. Barrero, M. R. Arahal, R. S. Toral, and M. J. Duran, "A proof of concept study of predictive current control for VSI-driven asymmetrical dual three-phase AC machines," *IEEE Trans. Ind. Electron.*, vol. 56, no. 6, pp. 1937–1954, 2009.

Chapter 2
General winding configurations of multiphase PMSMs

2.1 Introduction

Dual three-phase (DTP) permanent magnet synchronous machines (PMSMs) offer numerous advantages over conventional single three-phase (STP) PMSMs, including enhanced fault-tolerant capacity, improved torque characteristics, reduced torque pulsation, and decreased power rating per inverter [1–5]. However, the extent to which these advantages could be realized depends on a specific winding configuration of the DTP PMSMs. This chapter identifies feasible winding configurations for DTP PMSMs and illustrates many examples by considering different slot/pole number combinations, coil pitches, and phase-shift angles (β) between the two winding sets. The corresponding stars of slots and the resultant winding layouts are presented. The winding factors for different DTP winding configurations are calculated, with the distribution factors accounting for different phase-shift angles and the pitch factors considering different coil pitch numbers. The interaction of two winding sets in the resultant stator MMF harmonics is analyzed by using attenuation factors, in which both spatial and temporal shifts between two winding sets are considered.

The arrangement of this chapter is shown in Figure 2.1 and is organized as follows. Section 2.2 shows the winding assignments of DTP windings in DTP

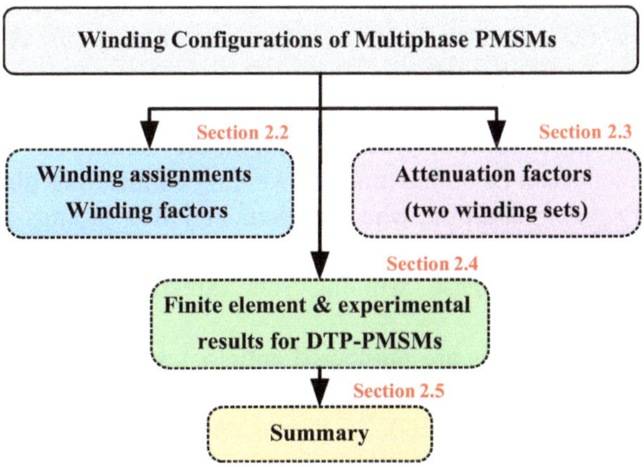

Figure 2.1 Structure of Chapter 2

PMSMs. The calculations of winding factors are also given in Section 2.2. Section 2.3 focuses on the attenuation factor for two winding sets in DTP PMSMs. An example of DTP PMSMs with different phase-shift angles is given in Section 2.4, the calculated winding factors and attenuation factors are validated by finite element (FE) and experimental results as examples.

2.2 Winding assignments and winding factors of DTP PMSMs

2.2.1 Star of slots

In a PMSM, the electromotive force (EMF) phasors induced in conductors in different slots can be geometrically represented as a star of slots. Since each coil has two coil sides, when all coils share the same coil pitch (also known as coil span or coil throw), the star of slots can further represent the EMF phasors induced in different coils. Hence, the star of slots has been widely utilized in determining the winding assignments of different phases in PMSMs with different slot/pole number combinations [6,7]. Based on the resultant winding assignments, distribution factors can also be calculated from the star of slots. In this chapter, the winding assignments and winding factors of DTP windings in DTP PMSMs are obtained based on the star of slots. Different phase-shift angles (the angle displacement in electrical degree between the two winding sets) are considered. It also should be mentioned that to simplify the descriptions, the following analyses are conducted in PMSMs with double-layer (DL) windings. For the PMSMs with single-layer (SL) windings, the number of spokes in the star of slots will be halved [8], while the described approaches are still applicable.

For a PMSM with slot number (N_s) and pole number ($2p$), the electrical periodicity number (t) can be computed as the greatest common divisor (GCD) between N_s and p,

$$t = \mathrm{GCD}(N_s, p) \tag{2.1}$$

For a N_s-slot/$2p$-pole DL-PMSM, the number of coils is N_s. Due to the electrical periodicity, the number of spokes in the star of slots (n_s) is N_s/t, and each spoke stands for t EMF phasors.

In the following analyses, the winding assignments in DTP PMSMs will mainly focus on the conditions with $t = 1$. For the conditions with $t \geq 2$, the winding assignments can be duplicating the winding connections obtained from the conditions with $t = 1$. In addition, when t is even, and the windings in one electrical periodicity can be assigned into three phases, the whole armature winding can also be split into two three-phase winding sets directly according to electrical periodicity.

In this chapter, the slots are numbered anticlockwise, and the coils are numbered according to the number of slots where their upper coil sides are located. As adjacent slots are displaced by $360°/N_s$ in mechanical and $360°/n_s$ in electric, the EMF phasors of adjacent coils are displaced by $360°\cdot p/n_s$ in electric.

The star of slots for any PMSM can be illustrated based on the slot number (N_s) and the pole number ($2p$). Two examples are shown in Figure 2.2 ($N_s = 12$, $2p = 10$) and Figure 2.3 ($N_s = 9$, $2p = 10$), where both cross-sections and stars of slots are provided. In the 12-slot/10-pole and 9-slot/10-pole PMSMs, t=1, and thus, the number of spokes in the star of slots (n_s) equals 12 and 9 respectively, as shown in Figures 2.2(b) and 2.3(b). Thus, the EMF phasors belonging to the same phase can be selected according to the number of phases indicated by the star of slots. To ensure the balanced resultant EMF per phase, the number of EMF phasors in different phases should be identical, and hence, to achieve DTP winding configuration, the number of spokes (n_s) should be a multiple of 6. Since the slot number (N_s) is the product of the electrical periodicity (t) and the number of spokes (n_s), N_s in DTP PMSMs should also be a multiple of 6.

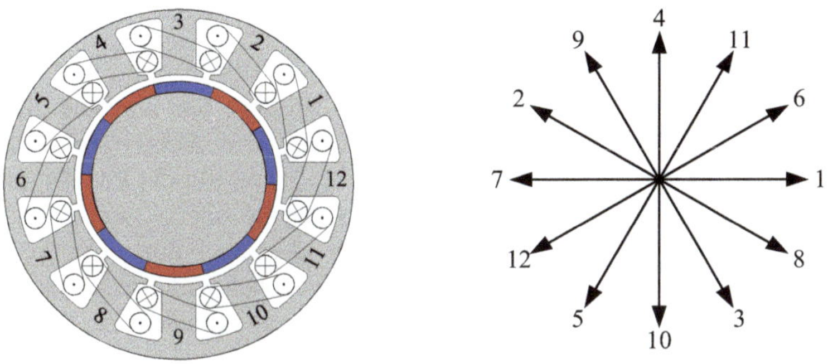

(a) (b)

Figure 2.2 Cross-section and star of slots of 12-slot/10-pole DTP PMSM with DL windings. (a) Cross-section and (b) star of slots.

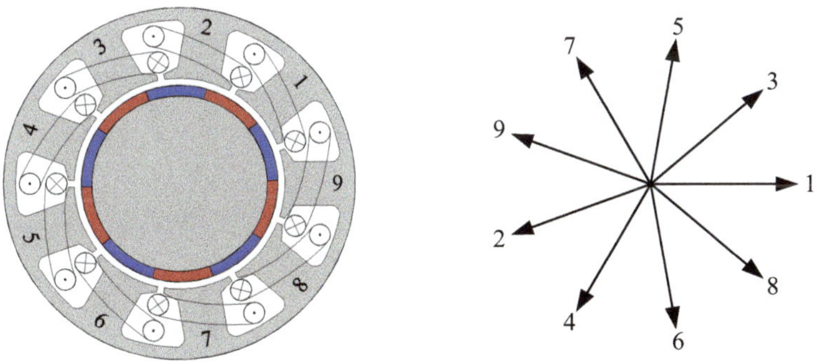

(a) (b)

Figure 2.3 Cross-section and star of slots of 9-slot/10-pole DTP PMSM with DL windings. (a) Cross-section and (b) star of slots.

2.2.2 Distribution factor and phase-shift angle

Before assigning the coils in the phase winding in DTP PMSMs, the two winding sets in the DTP PMSMs are named Winding 1 and Winding 2 for clarity. The first winding set, Winding 1, consists of Phases A1, B1, and C1, which are marked with light yellow, light green, and light red, respectively. Similarly, the second winding set, Winding 2, consists of Phases A2, B2, and C2, which are marked with dark yellow, dark green, and dark red, respectively.

As mentioned above, in DTP PMSM, the number of spokes in the star of slots (n_s) is a multiple of 6. Assuming $n_s = 6k$ (k is an integer), k denotes the number of spokes per phase. To select the k spokes of the same phase in the star of slots, various phase-shift angles should be considered. The resultant distribution factor also varies with different phase-shift angles. In this subsection, the analyses will focus on the PMSMs with electrical periodicity number, $t = 1$, wherein $n_s = N_s/t = N_s$, $k = n_s/6 = N_s/6$.

2.2.2.1 Phase-shift angle = 0°

In DTP PMSMs, selecting the coils of the two winding sets with the phase-shift angle = 0°, the 60° phase belt can be utilized in the star of slots, which is the same as that found in STP PMSMs. For example, in a 12-slot/10-pole PMSM, as shown in Figure 2.4(a), the spokes in the star of slots can be selected based on the 60° phase belt, as shown in Figure 2.4(b). It should be mentioned that when the coils are wound in the reverse direction, the corresponding spokes will also be shown in the opposite direction. Hence, Spoke 1 in Figure 2.4(b) can represent the EMF phasors of Coil 1 and Coil −7 simultaneously.

Based on the star of slots given in Figure 2.4(b), various approaches can achieve DTP windings with a phase-shift angle = 0°. For instance, Phase A1 can be composed of Coil 1 and Coil 6, Coil 1 and Coil −12, Coil −7 and Coil 6, or Coil −7 and Coil −12. Similarly, there are various options to choose corresponding coils for other phases. Considering that each three-phase winding set contains three

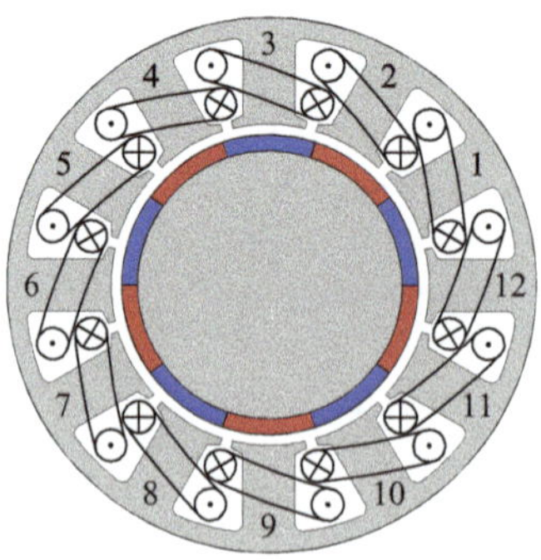

(a)

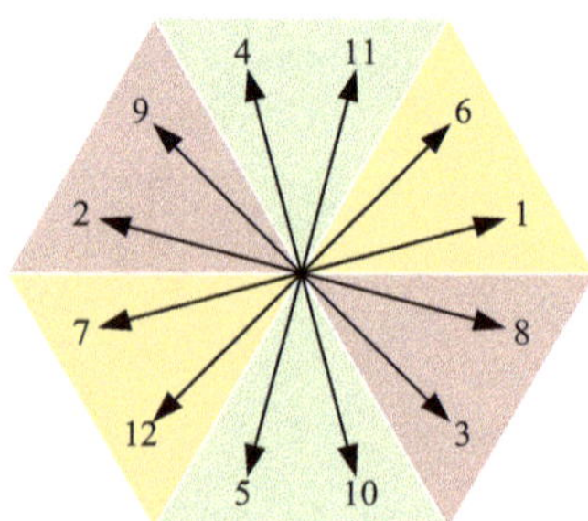

(b)

Figure 2.4 Cross-section and star of slots of 12-slot/10-pole DTP PMSM with 60° phase belt. (a) Cross-section and (b) star of slots.

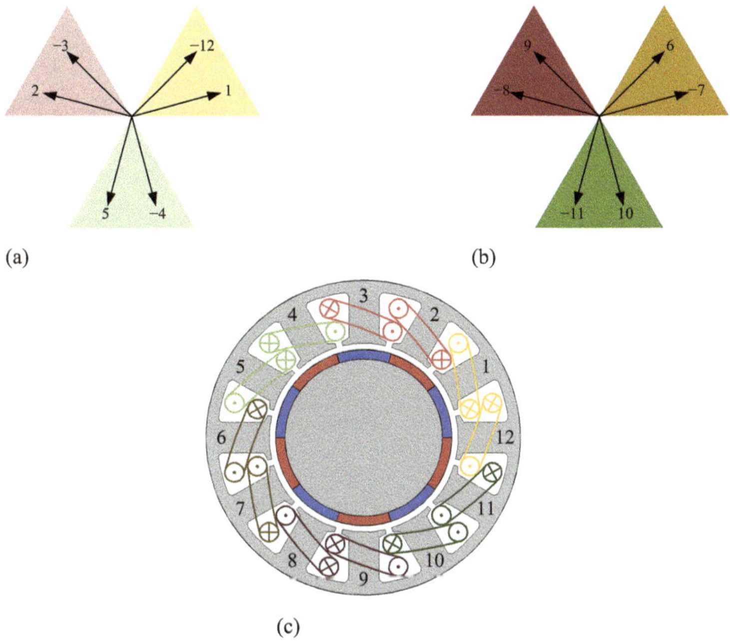

Figure 2.5 Star of slots and winding layouts of DTP 12-slot/10-pole DTP PMSM with phase-shift angle = 0° (DL-1). (a) Star of slots (Winding 1), (b) star of slots (Winding 2), and (c) winding layouts.

phases, the resultant choice of coils of the two three-phase winding sets becomes even more complex.

In [9], two feasible DTP winding configurations with the phase-shift angle = 0° are discussed and compared for a 12-slot/10-pole PMSM. Two DTP winding configurations, equipped with DL windings, are named DL-1 and DL-2, respectively. For DL-1, the stars of slots for the two winding sets are shown in Figure 2.5(a) and (b), and the resultant winding layouts are depicted in Figure 2.5(c). Similarly, the detailed information for DL-2 can be found in Figure 2.6.

Since the phase-shift angle = 0°, the currents in Phases A1, B1, and C1 are equal to those in Phases A2, B2, and C2 under healthy conditions. Hence, it can be expected that the electromagnetic performances of DTP PMSMs with DL-1 and DL-2 winding configurations are identical under healthy conditions. However, when one three-phase winding set is open-circuited and the DTP PMSMs are operated by the remaining winding set, the electromagnetic performances of DL-1 and DL-2 winding configurations will be different. Assuming Winding 2 is open-circuited, the remaining windings in DL-1 and DL-2 machines are shown in Figure 2.7(a) and (b), respectively. In this case, the DL-2 winding configuration can produce less unbalanced radial force than the DL-1 counterpart, as reported in [9].

In the above 12-slot/10-pole PMSM, the number of spokes per phase $k = 2$, i.e., an even value. As another example of 18-slot/14-pole DTP PMSM, its $k = 3$, i.e., an

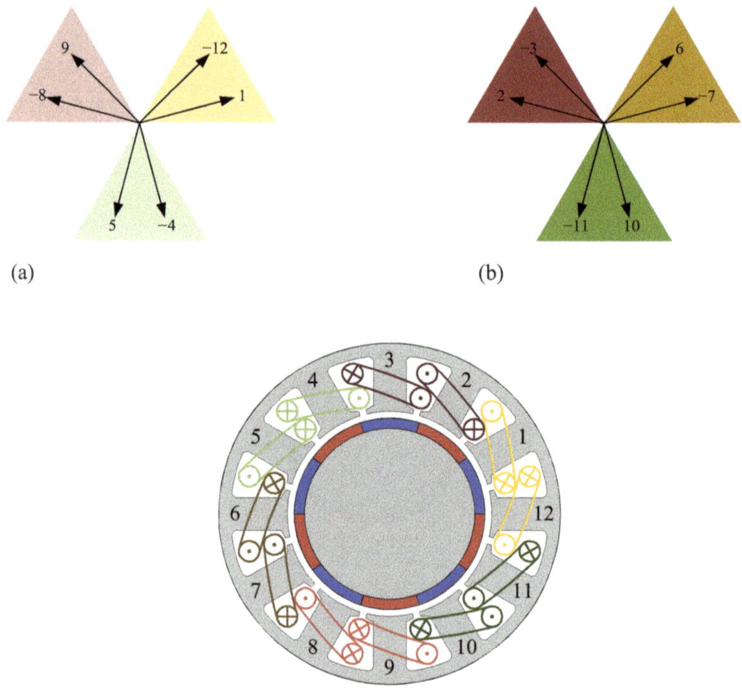

Figure 2.6 Star of slots and winding layouts of DTP 12-slot/10-pole DTP PMSM with phase-shift angle = 0° (DL-2). (a) Star of slots (Winding 1), (b) star of slots (Winding 2), and (c) winding layouts.

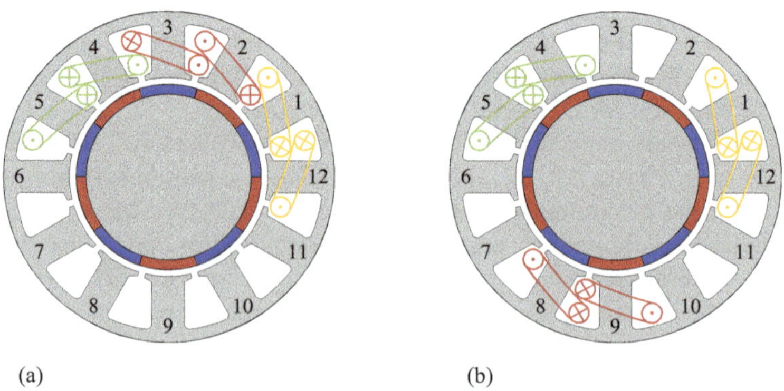

Figure 2.7 Winding layouts of a 12-slot 10-pole DTP PMSM with phase-shift angle = 0° under a single three-phase open-circuit condition. (a) DL-1 and (b) DL-2.

odd value. Its winding layouts with DL concentrated windings are shown in Figure 2.8(a). To achieve phase-shift angle = 0°, the spokes in the star of slots are selected using the 60° phase belt, as shown in Figure 2.8(b).

Similarly, various feasible methods exist to distinguish the coils for the two winding sets based on the star of slots, as shown in Figure 2.8(b). Considering the symmetricity of the two three-phase winding sets, one feasible winding configuration is selected as shown in Figure 2.9, where adjacent coils always belong to

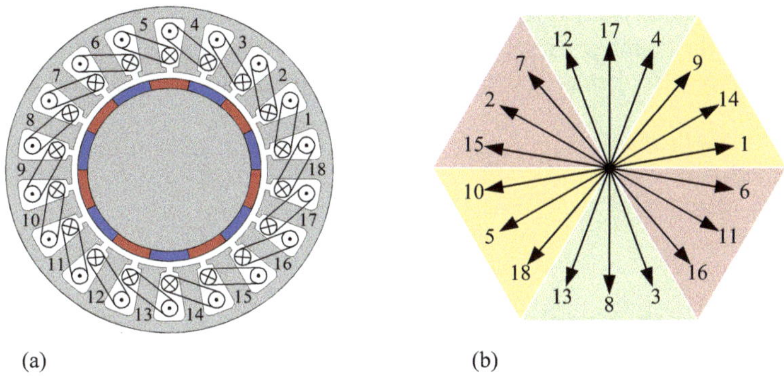

Figure 2.8 Cross-section and star of slots of 18-slot/14-pole DTP PMSM with 60° phase belt. (a) Cross-section and (b) star of slots.

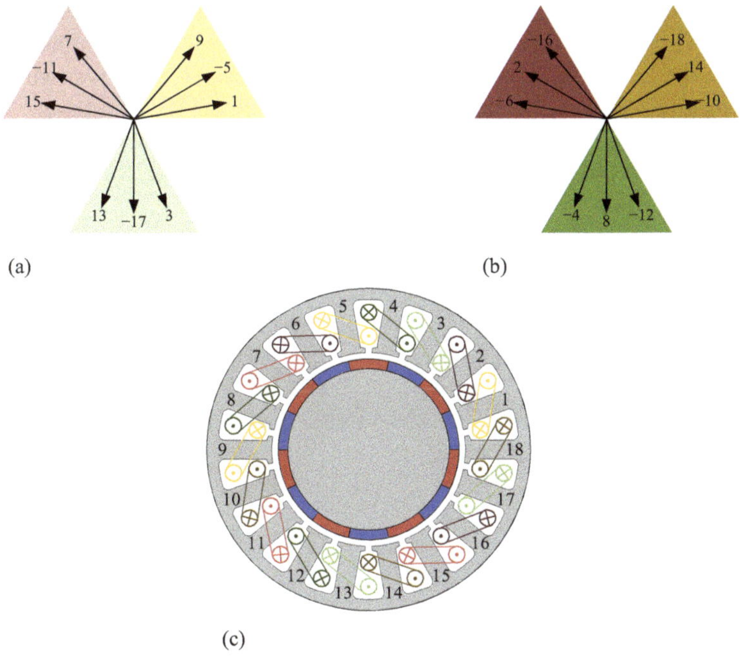

Figure 2.9 Star of slots and winding layouts of 18-slot/14-pole DTP PMSM with phase-shift angle = 0°. (a) Star of slots (Winding 1), (b) star of slots (Winding 2), and (c) winding layouts.

different winding sets. Compared with other winding configurations, where adjacent coils may belong to the same winding set, the winding configuration, illustrated in Figure 2.9, can exhibit better torque capability under a single three-phase open-circuit condition.

For each winding set, the distribution factors can be computed according to the selections of spokes in the star of slots. From Figure 2.4 to Figure 2.9, it is evident that in DTP PMSMs, when the phase-shift angle = 0°, the spokes are selected using the 60° phase belt, which aligns with the method used in STP PMSMs. Hence, in this case, the calculations of distribution factors in DTP and STP PMSMs are the same. When the number of spokes per phase is k, and the angular displacement between adjacent spokes is $360°/N_s$, the distribution factor can be obtained as

$$K_d = \frac{\sin\left(k\frac{360^\circ}{2N_s}\right)}{k\sin\left(\frac{360^\circ}{2N_s}\right)} = \frac{\sin(30^\circ)}{k\sin\left(\frac{30^\circ}{k}\right)} \tag{2.2}$$

2.2.2.2 Phase-shift angle = 30°

Another popular phase-shift angle in DTP PMSMs is 30°. To achieve the 30° phase-shift angle, the star of slots should be analyzed with the 30° phase belt, and hence, the star of slots is divided into 12 sectors. To ensure balanced resultant EMF phasors, N_s should be a multiple of 12. Since $N_s = 6k$, the 30° phase-shift angle is only feasible when k is even.

For instance, the DL concentrated windings in a 12-slot/10-pole PMSM are shown in Figure 2.10(a), and the star of slots can be analyzed with the 30° phase belt, as shown in Figure 2.10(b). The resultant star of slots of the two winding sets and the resultant winding layouts are shown in Figure 2.11.

Taking a 24-slot/22-pole DTP PMSM as another example, since $N_s = 24$, $2p = 22$, it can be obtained that $k = 24/6 = 4$. Since k is even, the 30° phase-shift

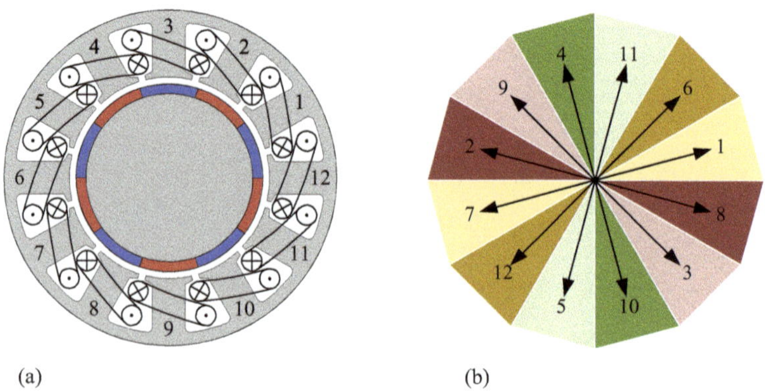

Figure 2.10 Cross-section and star of slots of 12-slot/10-pole DTP PMSM with 30° phase belt. (a) Cross-section and (b) star of slots.

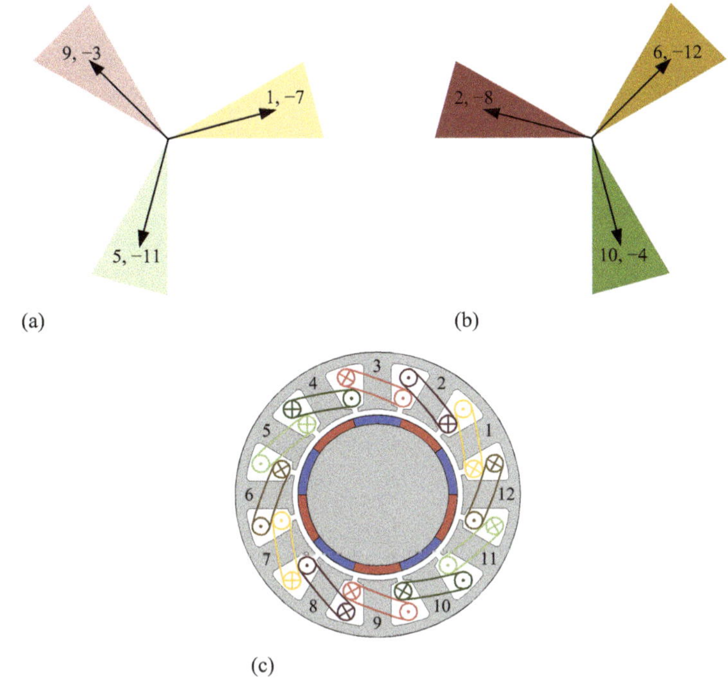

Figure 2.11 Star of slots and winding layouts of 12-slot/10-pole DTP PMSM with phase-shift angle = 30°. (a) Star of slots (Winding 1), (b) star of slots (Winding 2), and (c) winding layouts.

angle is feasible. The cross-section of the 24-slot/22-pole DTP PMSM with DL concentrated windings is shown in Figure 2.12(a), and the star of slots with the 30° phase belt is shown in Figure 2.12(b). Accordingly, the stars of slots of the two winding sets and the resultant winding layouts of the DTP 24-slot/22-pole PMSM with phase-shift angle = 30° can be obtained, as shown in Figure 2.13.

Overall, it can be concluded that in DTP PMSMs, the 30° phase-shift angle is only feasible when N_s is a multiple of 12. As can be observed from Figures 2.10(b) and 2.12(b), when the phase-shift angle = 30°, the opposite spokes belong to the same phase. The distribution factor can be computed by halving the number of spokes per phase, which is $k/2$, and the angular displacement between adjacent spokes is still $360°/N_s$. Thus, the resultant distribution factor can be calculated as

$$K_d = \frac{\sin\left(\frac{k}{2}\cdot\frac{360^\circ}{2N_s}\right)}{\frac{k}{2}\sin\left(\frac{360^\circ}{2N_s}\right)} = \frac{2\sin(15^\circ)}{k\sin\left(\frac{30^\circ}{k}\right)} \tag{2.3}$$

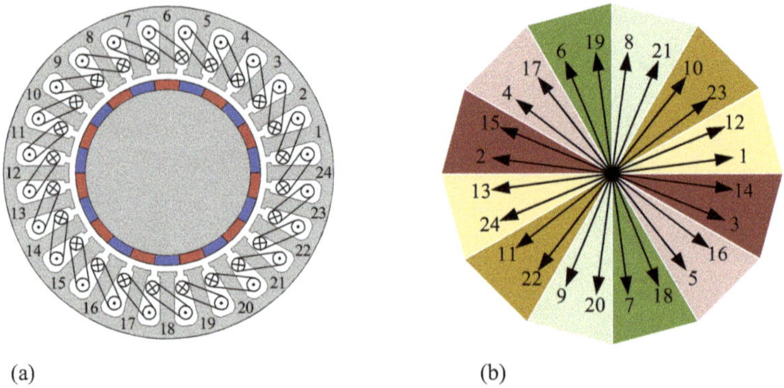

Figure 2.12 *Cross-section and star of slots of 24-slot/22-pole DTP PMSM with 30° phase belt. (a) Cross-section and (b) star of slots.*

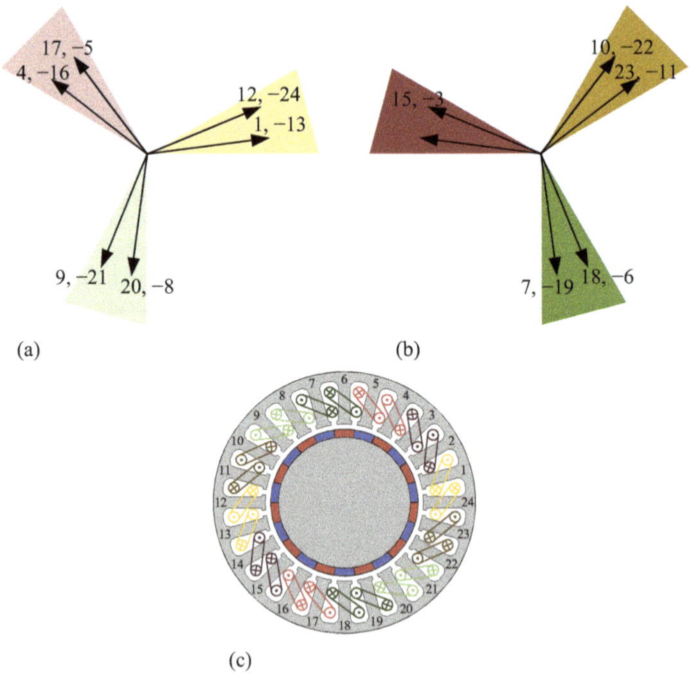

Figure 2.13 *Star of slots and winding layouts of 24-slot/22-pole DTP PMSM with phase-shift angle = 30°. (a) Star of slots (Winding 1), (b) star of slots (Winding 2), and (c) winding layouts.*

2.2.2.3 Phase-shift angle = slot-pitch angle

To achieve the conventional 0° and 30° phase-shift angles, the spokes in the star of slots are typically selected using the 60° and 30° phase belts, respectively. Besides, the spokes of the two winding sets can be selected alternately from the star of slots. This results in the two winding sets being shifted by 360°/N_s, which represents the angular displacement between the adjacent spokes in the star of slots. Referred to as the "slot-pitch angle" in this chapter, the phase-shift angle under this condition equals 360°/N_s.

In 12-slot/10-pole DTP PMSMs, N_s is 12, and hence, the slot-pitch angle is calculated as 360°/12 = 30°. Thus, the DTP winding configuration features a phase-shift angle = 30°, as depicted in Figures 2.10 and 2.11, in which the phase-shift angle is equal to the slot-pitch angle.

Another example is a 24-slot/22-pole DTP PMSM, of which N_s is 24 and the slot-pitch angle is 15°. When the machine is configured with DL concentrated windings, as shown in Figure 2.14(a), the spokes of the two winding sets can be selected from the star of slots using a 15° phase belt, as shown in Figure 2.14(b). The star of slots of the two winding sets and the resultant winding layouts of the DTP 24-slot/22-pole PMSM with a phase-shift angle = 15° can be obtained, as shown in Figure 2.15.

From the comparison between Figure 2.13 and, it can be observed that in the DTP 24-slot/22-pole PMSM, when the phase-shift angle = 30°, adjacent coils may belong to the same winding set, but when the phase-shift angle = 15°, adjacent coils always belong to different winding sets. Hence, it can be anticipated that when the DTP 24-slot/22-pole PMSM is operated under a single three-phase open-circuit condition, the winding configuration with a 15° phase-shift angle can exhibit better torque performance than the counterpart with a 30° phase-shift angle, especially when the currents are high [10,11].

Besides, an 18-slot/14-pole PMSM is further taken as an example, of which k = 3 (odd), N_s is 18, and the slot-pitch angle is 20°, as shown in Figure 2.16(a), where DL concentrated winding is employed. In this case, the spokes of the two

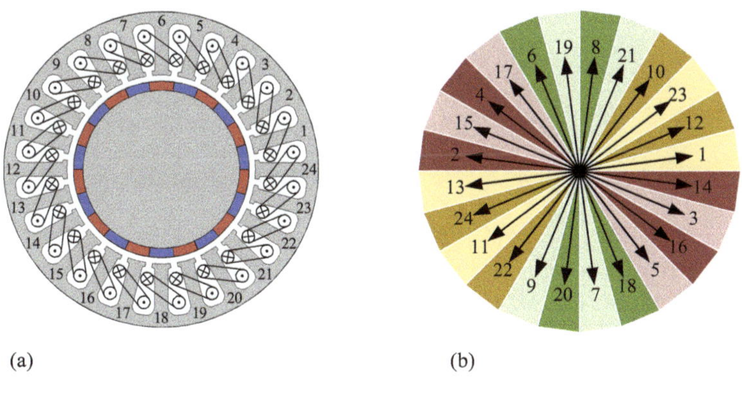

Figure 2.14 Cross-section and star of slots of 24-slot/22-pole PMSM with 15° phase belt. (a) Cross-section and (b) star of slots.

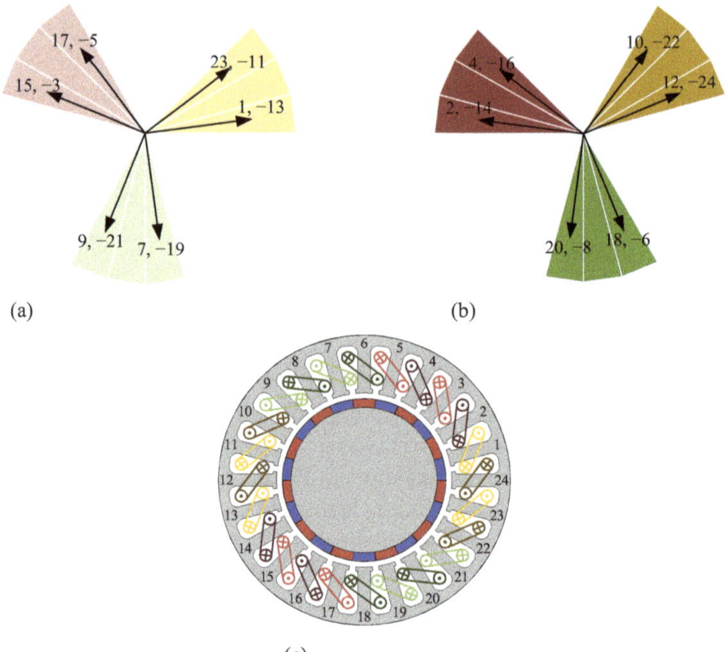

Figure 2.15 Star of slots and winding layouts of a DTP 24-slot/22-pole PMSM with phase-shift angle = 15°. (a) Star of slots (Winding 1), (b) star of slots (Winding 2), and (c) winding layouts.

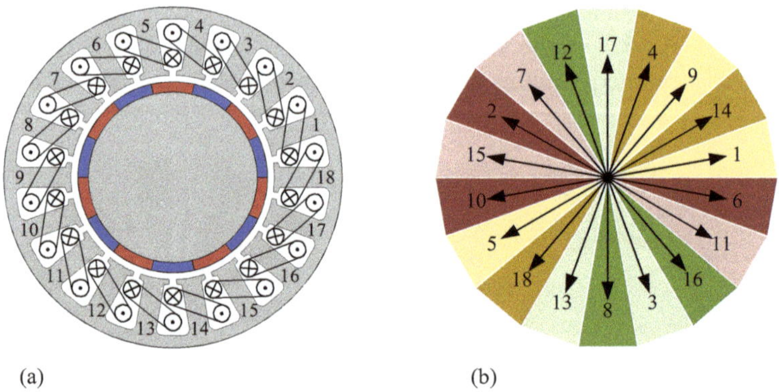

Figure 2.16 Cross-section and star of slots of 18-slot/14-pole PMSM with 20° phase belt. (a) Cross-section and (b) star of slots.

winding sets can be selected from the star of slots using a 20° phase belt, as shown in Figure 2.16(b). The DTP winding configuration with a phase-shift angle is 20°, the star of slots of two winding sets, and the resultant winding layouts are shown in Figure 2.17.

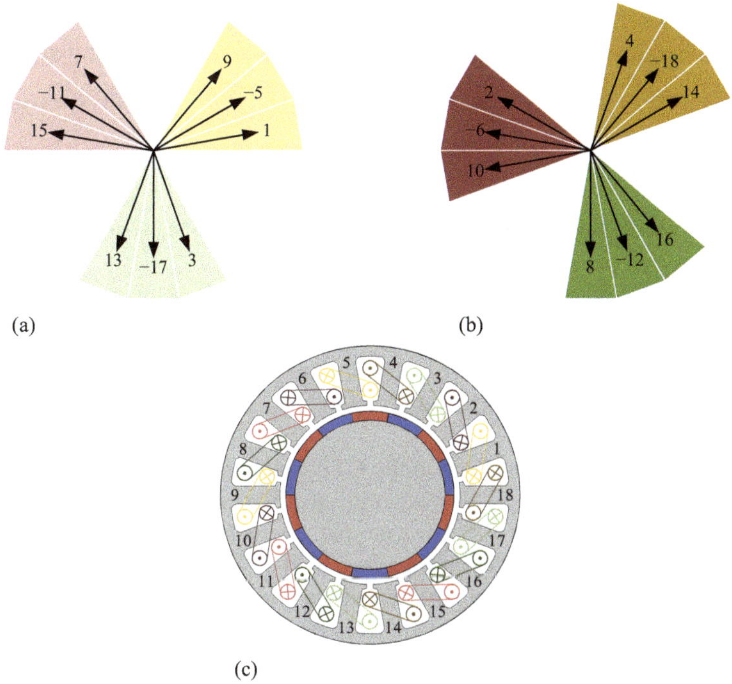

Figure 2.17 Star of slots and winding layouts of a DTP 18-slot/14-pole PMSM with phase-shift angle = 20°. (a) Star of slots (Winding 1), (b) star of slots (Winding 2), and (c) winding layouts.

From Figures 2.15(b) and 2.17(b), it can be observed that for the conditions with the phase-shift angle equal to the slot-pitch angle, the calculations of distribution factors are different when k is even or odd. On the one hand, when k is even, opposite spokes belong to the same phase, and thus, the distribution factor can be computed based on $k/2$ spokes and the angular displacement between the adjacent spokes is $2 \times 360°/N_s$. On the other hand, when k is odd, opposite spokes belong to different phases, and thus, the distribution factor still needs to be computed based on k spokes, but the angular displacement between the adjacent spokes is $360°/N_s$. Hence, in DTP PMSMs where the phase-shift angle = slot-pitch angle, the calculation of distribution factors depends on the number of spokes per phase, k, as

$$K_d = \begin{cases} \dfrac{\sin\left(\dfrac{k}{2}\dfrac{360°}{2N_s}\right)}{\dfrac{k}{2}\sin\left(\dfrac{360°}{2N_s}\right)} = \dfrac{2\sin(15°)}{k\sin\left(\dfrac{30°}{k}\right)}, & \text{when } k \text{ is even} \\[2ex] \dfrac{\sin\left(k\dfrac{360°}{2N_s}\right)}{k\sin\left(\dfrac{360°}{2N_s}\right)} = \dfrac{\sin(30°)}{k\sin\left(\dfrac{30°}{k}\right)}, & \text{when } k \text{ is odd} \end{cases} \tag{2.4}$$

2.2.2.4 Discussions about other phase-shift angles

Based on the analyses in the previous sub-sections, it can be found that:

1. The 30° phase-shift angle is feasible for all DTP PMSMs with N_s as a multiple of 12.
2. The 20° phase-shift angle is feasible for all DTP PMSMs with N_s as a multiple of 18.
3. The 15° phase-shift angle is feasible for all DTP PMSMs with N_s as a multiple of 24.
4. The 10° phase-shift angle is feasible for all DTP PMSMs with N_s as a multiple of 36.

Overall, it can be further derived that when $N_s = N_{s0}$, the slot-pitch angle is $360°/N_{s0}$, and then the phase-shift angle = $360°/N_{s0}$, is feasible for all DTP PMSMs with N_s as a multiple of N_{s0}.

It should also be mentioned that in the DTP windings with phase-shift angle = 0°, when all the coils in Winding 2 are wound in the reverse direction, the two winding sets can be seen as shifted by 60°. For example, the winding configuration of 12-slot/10-pole DTP PMSM with phase-shift angle = 0° (DL-2) shown in Figure 2.6, can be changed into a DTP winding configuration with a phase-shift angle = 60°, as shown in Figure 2.18. Hence, it can also be concluded in the DTP winding configuration, 0° and 60° phase-shift angles are equivalent. To achieve the 0° and 60° phase-shift angles, N_s, should be a multiple of 6.

In general, the feasibility of the DTP winding configurations in PMSMs with different slot/pole number combinations mainly depends on the number of spokes in the star of slots (n_s). When the electrical periodicity number, $t = 1$, the number of spokes in the star of slots (n_s) is equal to the number of slots (N_s), and the number of spokes per phase ($k = N_s/6$). When k is an integer, the DTP winding configuration is feasible, and the phase-shift angle could be 0°/60° and 60°/k (slot-pitch angle). Besides, when k is even (a multiple of 2), the phase-shift angle can also be 30° (60°/2), and when k is a multiple of 3, the phase-shift angle can also be 20° (60°/3), etc.

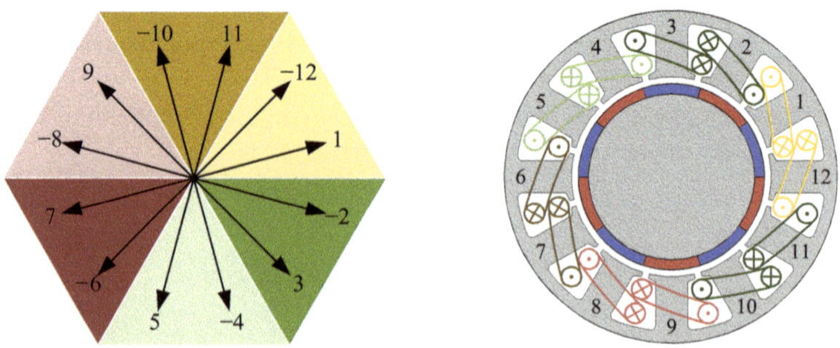

(a) (b)

Figure 2.18 Star of slots and winding layouts of a DTP 12-slot/10-pole PMSM with phase-shift angle = 60°. (a) Star of slots and (b) winding layouts.

For different phase-shift angles, the requirements on k and N_s in DTP PMSMs are summarized in Table 2.1. It should be mentioned that although the requirements on the pole pair number, p, are not listed, there is an implicit restriction: to achieve the phase-shift angles shown in the table, t should be 1, in other words, the selection of p should make GCD(N_s, p) equal to 1.

For DTP PMSMs with $N_s = 36$ and $t = 1$ (e.g., 36-slot with 2-pole, 10-pole, 14-pole, 22-pole, 26-pole, 30-pole, 34pole, etc.), the feasible phase-shift angles include 0°/60°, 30°, 20°, and 10°.

For DTP PMSMs with $N_s = 36$ and $t \neq 1$ (e.g., 36-slot with 4-pole, 6-pole, 8-pole, 12-pole, 16-pole, etc.), the feasible phase-shift angles should be found from their one electrical periodicity counterparts. Similarly, when $t = 1$, for different N_s, the feasible phase-shift angles can be summarized, as shown in Table 2.2.

In addition, the coil EMF phasors of the two winding sets can also be selected using the unbalanced phase belt method. With this method, the coils of Winding 1 are selected based on the 60° phase belt without considering reverse spokes. Meanwhile, the coils of Winding 2 are then selected according to the phase-shift angle and utilizing reverse spokes. Two examples illustrate this approach based on the 24-slot/22-pole PMSM, as shown in Figure 2.19 (phase-shift angle = 15°) and Figure 2.20 (phase-shift angle = 30°), respectively. The unbalanced phase belt method allows for the phase-shift angle to be any multiple of the slot-pitch angle.

Table 2.1 Requirements on k and N_s in DTP PMSMs for different phase-shift angles

Phase-shift angle	***k***	**N_s ($t = 1$)**
0°/60°	Integer (Multiple of 1)	6, 12, 18, 24, 30, 36, 42, 48, 54, 60, 66, 72, ...
30°	Even (Multiple of 2)	12, 24, 36, 48, 60, 72, ...
20°	Multiple of 3	18, 36, 54, 72, 90, 108, ...
15°	Multiple of 4	24, 48, 72, 96, 120, 144, ...
12°	Multiple of 5	30, 60, 90, 120, 150, 180, ...
10°	Multiple of 6	36, 72, 108, 144, 180, 216, ...
...	...	...

Table 2.2 Feasible phase-shift angles for different N_s ($t = 1$) in DTP PMSMs

N_s ($t = 1$)	***k***	**Feasible phase-shift angles (°)**
6	1	0/60
12	2	0/60, 30
18	3	0/60, 20
24	4	0/60, 15, 30
30	5	0/60, 12
36	6	0/60, 10, 20, 30
...	...	...

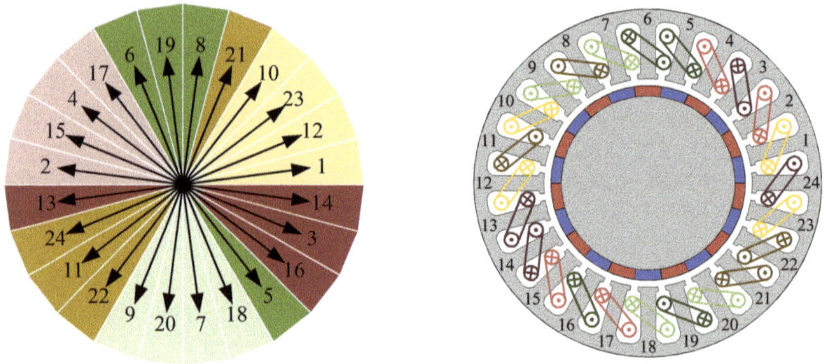

(a) (b)

Figure 2.19 Star of slots and winding layouts of a DTP 24-slot/22-pole PMSM with phase-shift angle = 15° using an unbalanced phase belt. (a) Star of slots and (b) winding layouts.

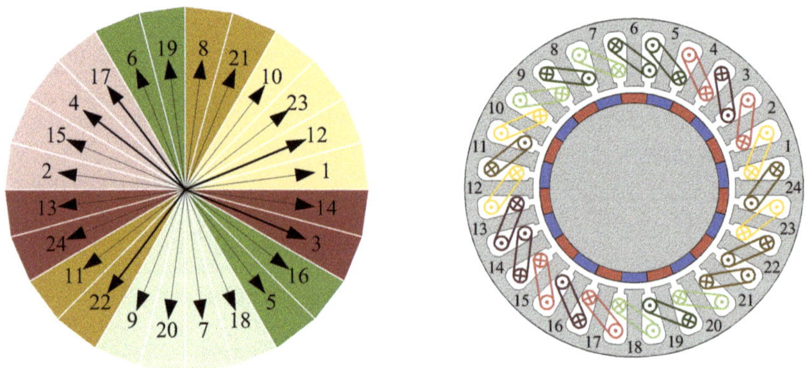

(a) (b)

Figure 2.20 Star of slots and winding layouts of a DTP 24-slot/22-pole PMSM with phase-shift angle = 30° using unbalanced phase belt. (a) Star of slots and (b) winding layouts.

Table 2.2 summarizes the feasible phase-shift angles for different N_s with the conventional balanced phase belt, while the unbalanced phase belt method enables even more phase-shift angles. However, since the spokes of the same phase are still distributed in one 60° sector with the unbalanced phase belt method, the distribution factors are still identical to those with a phase-shift angle = 0°/60°. Hence, while the unbalanced phase belt method provides some more freedom in control strategies, it cannot improve the distribution factor. Therefore, in the following analyses, the windings are assigned using the conventional balanced phase belt method.

2.2.2.5 Summary

This subsection describes the determination of the feasibility of all DTP winding configurations with different slot/pole number combinations, which depends on the electrical periodicity number, t, and the number of spokes in the star of slots, n_s. In addition, the feasible phase-shift angles are obtained from the number of spokes per phase in the star of slots, k.

In summary, for the PMSMs with slot number ≤ 36, pole number ≤ 30, the electrical periodicity number (t) and the number of spokes in the star of slots (n_s) are listed in Table 2.3. When n_s is a multiple of 6, the numbers of spokes per phase (k) are also provided. In the table, the cases when n_s is a multiple of 6 are marked with light blue, and the cases when t is even and N_s is an odd multiple of 3 are marked with light yellow. In addition, the DTP winding configurations are feasible in PMSMs with the highlighted slot/pole number combinations.

Table 2.3 Summary of electrical periodicity number (t), number of spokes in the star of slots (n_s), number of spokes per phase (k) for PMSMs with different slot/pole number combinations

N_s / $2p$		6	12	18	24	30	36
2	t	1	1	1	1	1	1
	n_s	6	12	18	24	30	36
	k	1	2	3	4	5	6
4	t	2	2	2	2	2	2
	n_s	3	6	9	12	15	18
	k	-	1	-	2	-	3
6	t	3	3	3	3	3	3
	n_s	2	4	6	8	10	12
	k	-	-	1	-	-	2
8	t	2	4	2	4	2	4
	n_s	3	3	9	6	15	9
	k	-	-	-	1	-	-
10	t	1	1	1	1	5	1
	n_s	6	12	18	24	6	36
	k	1	2	3	4	1	6
12	t	6	6	6	6	6	6
	n_s	1	2	3	4	5	6
	k	-	-	-	-	-	1
14	t	1	1	1	1	1	1
	n_s	6	12	18	24	30	36
	k	1	2	3	4	5	6
16	t	2	4	2	8	2	4
	n_s	3	3	9	3	15	9
	k	-	-	-	-	-	-

(Continues)

Table 2.3 (Continued)

N_s **2p**		**6**	**12**	**18**	**24**	**30**	**36**
18	t	3	3	9	3	3	9
	n_s	2	4	2	8	10	4
	k	-	-	-	-	-	-
20	t	2	2	2	2	10	2
	n_s	3	6	9	12	3	18
	k	-	1	-	2	-	3
22	t	1	1	1	1	1	1
	n_s	6	12	18	24	30	36
	k	1	2	3	4	5	6
24	t	6	12	6	12	6	12
	n_s	1	1	3	2	5	3
	k	-	-	-	-	-	-
26	t	1	1	1	1	1	1
	n_s	6	12	18	24	30	36
	k	1	2	3	4	5	6
28	t	2	2	2	2	2	2
	n_s	3	6	9	12	15	18
	k	-	1	-	2	-	3
30	t	3	3	3	3	15	3
	n_s	2	4	6	8	2	12
	k	-	-	1	-	-	2

Notes:

n_s is a multiple of 6; t is even and n_s is an odd multiple of 3

The feasible phase-shift angles for DTP PMSMs with different slot/pole number combinations are summarized in Table 2.4. It should be mentioned that since a 0° phase-shift angle is equivalent to a 60° phase-shift angle, the 60° phase-shift angles will not be presented.

Accordingly, the distribution factors can also be computed based on the slot/pole number combination and the specific phase-shift angle, which will be described as follows.

When t is even and n_s is an odd multiple of 3, the windings are selected with the same method as that in conventional STP PMSMs and the phase-shift angle is 0°. Considering that n_s is odd in these cases, the distribution factor is calculated as [8]

$$K_d = \frac{\sin\left(\frac{n_s}{3}\frac{180^\circ}{2n_s}\right)}{\frac{n_s}{3}\sin\left(\frac{180^\circ}{2n_s}\right)} = \frac{3\sin(30^\circ)}{n_s\sin\left(\frac{90^\circ}{n_s}\right)} \tag{2.5}$$

Table 2.4 Feasible phase-shift angles for DTP PMSMs with different slot/pole number combinations

N_s / $2p$	6	12	18	24	30	36
2	0	0/30	0/20	0/15/30	0/12	0/10/20/30
4	0	0	0	0/30	0	0/20
6	**	**	0	**	**	0/30
8	0	0	0	0	0	0
10	0	0/30	0/20	0/15/30	0	0/10/20/30
12	**	**	0	**	**	0
14	0	0/30	0/20	0/15/30	0/12	0/10/20/30
16	0	0	0	0	0	0
18	**	**	**	**	**	**
20	0	0	0	0/30	0	0/20
22	0	0/30	0/20	0/15/30	0/12	0/10/20/30
24	**	**	0	**	**	0
26	0	0/30	0/20	0/15/30	0/12	0/10/20/30
28	0	0	0	0/30	0	0/20
30	**	**	0	**	**	0/30

Notes:
** Unfeasible slot/pole combinations for DTP PMSMs.

t is even and n_s is an odd multiple of 3

$k = 1$; $k = 2$; $k = 3$; $k = 4$; $k = 5$; $k = 6$

When n_s is a multiple of 6, i.e., an even multiple of 3, the 0° phase-shift angle is always feasible and the distribution factor is calculated as:

$$K_d = \frac{\sin\left(\frac{n_s}{6}\frac{360^\circ}{2n_s}\right)}{\frac{n_s}{6}\sin\left(\frac{360^\circ}{2n_s}\right)} = \frac{6\sin(30^\circ)}{n_s\sin\left(\frac{180^\circ}{n_s}\right)} = \frac{\sin(30^\circ)}{k\sin\left(\frac{30^\circ}{k}\right)} \tag{2.6}$$

When n_s is a multiple of 6, the phase-shift angle = slot-pitch angle ($360^\circ/n_s$, $60^\circ/k$) is also feasible. However, the calculation of the distribution factor needs to consider whether k is odd or even.

- When k is odd, the distribution factor is identical to that in the 0° phase-shift angle condition.
- When k is even, the distribution factor is different from that in the 0° counterpart.

Hence, when the slot-pitch angle is utilized in the DTP winding configuration, the distribution factor is calculated as

$$K_d = \begin{cases} \dfrac{\sin\left(\dfrac{n_s}{12}\dfrac{360^\circ}{n_s}\right)}{\dfrac{n_s}{12}\sin\left(\dfrac{360^\circ}{n_s}\right)} = \dfrac{12\sin(30^\circ)}{n_s\sin\left(\dfrac{360^\circ}{n_s}\right)} = \dfrac{2\sin(30^\circ)}{k\sin\left(\dfrac{60^\circ}{k}\right)}, & \text{when } k \text{ is even} \\[2ex] \dfrac{\sin\left(\dfrac{n_s}{6}\dfrac{360^\circ}{2n_s}\right)}{\dfrac{n_s}{6}\sin\left(\dfrac{360^\circ}{2n_s}\right)} = \dfrac{6\sin(30^\circ)}{n_s\sin\left(\dfrac{180^\circ}{n_s}\right)} = \dfrac{\sin(30^\circ)}{k\sin\left(\dfrac{30^\circ}{k}\right)}, & \text{when } k \text{ is odd} \end{cases} \tag{2.7}$$

In addition, when n_s is a multiple of 12, the 30° phase-shift angle is also feasible, and the distribution factor under this condition is:

$$K_d = \frac{\sin\left(\frac{n_s}{12}\frac{360^\circ}{2n_s}\right)}{\frac{n_s}{12}\sin\left(\frac{360^\circ}{2n_s}\right)} = \frac{12\sin(15^\circ)}{n_s\sin\left(\frac{180^\circ}{n_s}\right)} = \frac{2\sin(15^\circ)}{k\sin\left(\frac{30^\circ}{k}\right)} \tag{2.8}$$

It can also be summarized that when $n_s = 6k$ and k is odd, the feasible phase-shift angles include 0° and slot-pitch angle ($60^\circ/k$). In comparison, when $n_s = 6k$ but k is even, the feasible phase-shift angles include 0°, 30°, and slot-pitch angle ($60^\circ/k$). The corresponding distribution factors are calculated by (2.9) and (2.10), respectively.

When $n_s = 6k$ and k is odd:

$$K_d = \frac{\sin\left(\frac{n_s}{6}\frac{360^\circ}{2n_s}\right)}{\frac{n_s}{6}\sin\left(\frac{360^\circ}{2n_s}\right)} = \frac{6\sin(30^\circ)}{n_s\sin\left(\frac{180^\circ}{n_s}\right)} = \frac{\sin(30^\circ)}{k\sin\left(\frac{30^\circ}{k}\right)} \tag{2.9}$$

When $n_s = 6k$ and k is even:

$$K_d = \begin{cases} \dfrac{6\sin(30^\circ)}{n_s\sin\left(\dfrac{180^\circ}{n_s}\right)} = \dfrac{\sin(30^\circ)}{k\sin\left(\dfrac{30^\circ}{k}\right)}, & \text{when } \beta = 0^\circ \\[2ex] \dfrac{12\sin(30^\circ)}{n_s\sin\left(\dfrac{360^\circ}{n_s}\right)} = \dfrac{2\sin(30^\circ)}{k\sin\left(\dfrac{60^\circ}{k}\right)}, & \text{when } \beta = \dfrac{60^\circ}{k} \\[2ex] \dfrac{12\sin(15^\circ)}{n_s\sin\left(\dfrac{180^\circ}{n_s}\right)} = \dfrac{2\sin(15^\circ)}{k\sin\left(\dfrac{30^\circ}{k}\right)}, & \text{when } \beta = 30^\circ \end{cases} \tag{2.10}$$

From (2.10), it can be further deduced that when n_s is a multiple of 12,

$$\frac{K_d(\beta = 30^\circ)}{K_d(\beta = 0^\circ)} = \frac{2\sin(15^\circ)}{\sin(30^\circ)} = \frac{1}{\cos(15^\circ)} \approx 1.0353 \tag{2.11}$$

Table 2.5 Feasible phase-shift angles and corresponding distribution factors for DTP PMSMs

n_s	Phase-shift angle (°)	Distribution factor, K_d
n_s is an odd multiple of 3		
3	0	1.000
9	0	0.960
15	0	0.957
n_s is an odd/even multiple of 6		
6	0	1.000
12	0	0.966
	30	1.000
18	0	0.960
	20	0.960
24	0	0.958
	15	0.968
	30	0.991
30	0	0.957
	12	0.957
36	0	0.956
	10	0.960
	20	0.956
	30	0.990

Hence, it can be concluded that in DTP PMSMs, when n_s is a multiple of 12, both 0° and 30° phase-shift angles are feasible. However, the DTP winding configuration with a 30° phase-shift angle can produce higher fundamental back EMF amplitude and hence higher average torque than the counterpart with a 0° phase-shift angle by about 3.53% due to the improvement in winding factor.

Overall, when t is even and n_s is an odd multiple of 3, the 0° phase-shift angle is always feasible. The corresponding distribution factors can be computed based on (2.5). In comparison, when n_s is a multiple of 6, the distribution factors can be calculated by (2.9) and (2.10). They are summarized in Table 2.5.

2.2.3 Coil pitch and pitch factor

The calculation of pitch factor in DTP PMSMs is the same as that in conventional STP PMSMs. The pitch factor is independent of winding assignments and can be calculated solely based on the slot/pole numbers and coil pitch number (y). The diagram of coil pitch with coil pitch number is shown in Figure 2.21. It can be seen that the coil span angle is $360° \cdot y/N_s$ in mechanical, and hence, for the νth harmonic, the coil span angle is $360° \cdot \nu y/N_s$. As the pitch factor is calculated based on the coil span angle, the pitch factor for the νth harmonic can be obtained as

$$K_{p\nu} = \sin\left(\frac{360° \nu y}{2N_s}\right) = \sin\left(\frac{180° \nu y}{N_s}\right) \tag{2.12}$$

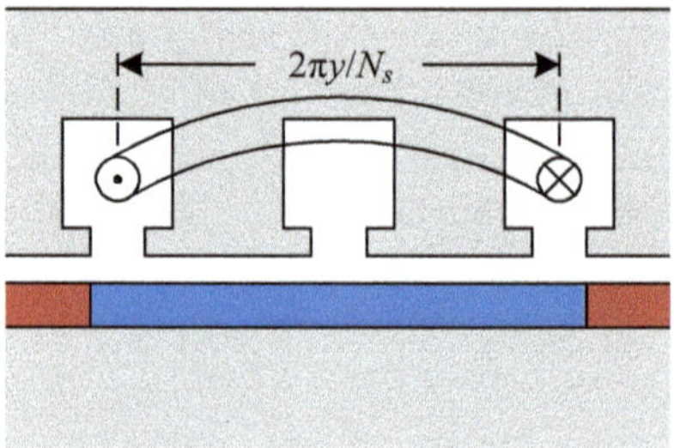

Figure 2.21 Diagram of pole-pitch and coil-pitch in PMSMs

Since the working harmonic of the N_s-slot/$2p$-pole PMSM is the pth harmonic, the pitch factor of the working harmonic is

$$K_{pv} = \sin\left(\frac{180^\circ py}{N_s}\right) \tag{2.13}$$

According to (2.13), the pitch factors of PMSMs with different slot/pole number combinations considering the coil pitch number varying from 1 to 4 are listed in Table 2.6. In the table, the pitch factors higher than 0.866 (cos(30°)) are marked with a pink background. It should be mentioned that as the feasibility of the DTP winding configuration is not considered in the calculation of pitch factor, some cells in Table 2.6 are zero, indicating that the specific combinations of slot, pole, and coil-pitch are unfeasible.

Based on (2.13) and Table 2.6, the optimized coil pitch number for PMSMs with different slot/pole number combinations can be obtained from the view of the pitch factor, as shown in Table 2.7. It can be seen that the optimized coil pitch number is around ($N_s/2p$) in terms of slot pitch. It also should be mentioned that considering the relatively shorter end-winding lengths in concentrated windings, for the conditions with $2p > N_s$, the optimized coil pitch numbers are all selected as 1 in the table.

With the optimized coil pitch numbers tabulated in Table 2.7, the pitch factors for different slot/pole number combinations are shown in Table 2.8. The pitch factors higher than 0.966 (cos (15°)) are marked with a pink background. However, it should be noted that the optimized coil pitch numbers and pitch factors shown in these two tables are selected from the view of enhancing pitch factors only.

In practice, it is usually preferred to use a lower coil pitch number, because the short-pitched windings can shorten the end-winding length and reduce high-order MMF harmonics. For example, $y = 1$ is selected for the 24-slot/14-pole PMSM in [12] (fractional-slot concentrated windings), and $y = 5$ is selected for the 48-slot/8-pole Prius 2010 machine (integer-slot distributed windings). Hence, it is necessary to consider the coil pitch number thoughtfully in the design of DTP PMSMs.

Table 2.6 Pitch factors (K_p) of PMSMs with different slot/pole number combinations and different coil pitch numbers

N_s / $2p$	6	12	18	24	30	36
			$y = 1$			
2	0.500	0.259	0.174	0.131	0.105	0.087
4	0.866	0.500	0.342	0.259	0.208	0.174
6	1.000	0.707	0.500	0.383	0.309	0.259
8	0.866	0.866	0.643	0.500	0.407	0.342
10	0.500	0.966	0.766	0.609	0.500	0.423
12	0.000	1.000	0.866	0.707	0.588	0.500
14	0.500	0.966	0.940	0.793	0.669	0.574
16	0.866	0.866	0.985	0.866	0.743	0.643
18	1.000	0.707	1.000	0.924	0.809	0.707
20	0.866	0.500	0.985	0.966	0.866	0.766
22	0.500	0.259	0.940	0.991	0.914	0.819
24	0.000	0.000	0.866	1.000	0.951	0.866
26	0.500	0.259	0.766	0.991	0.978	0.906
28	0.866	0.500	0.643	0.966	0.995	0.940
30	1.000	0.707	0.500	0.924	1.000	0.966
			$y = 2$			
2	0.866	0.500	0.342	0.259	0.208	0.174
4	0.866	0.866	0.643	0.500	0.407	0.342
6	0.000	1.000	0.866	0.707	0.588	0.500
8	0.866	0.866	0.985	0.866	0.743	0.643
10	0.866	0.500	0.985	0.966	0.866	0.766
12	0.000	0.000	0.866	1.000	0.951	0.866
14	0.866	0.500	0.643	0.966	0.995	0.940
16	0.866	0.866	0.342	0.866	0.995	0.985
18	0.000	1.000	0.000	0.707	0.951	1.000
20	0.866	0.866	0.342	0.500	0.866	0.985
22	0.866	0.500	0.643	0.259	0.743	0.940
24	0.000	0.000	0.866	0.000	0.588	0.866
26	0.866	0.500	0.985	0.259	0.407	0.766
28	0.866	0.866	0.985	0.500	0.208	0.643
30	0.000	1.000	0.866	0.707	0.000	0.500
			$y = 3$			
2	1.000	0.707	0.500	0.383	0.309	0.259
4	0.000	1.000	0.866	0.707	0.588	0.500
6	1.000	0.707	1.000	0.924	0.809	0.707
8	0.000	0.000	0.866	1.000	0.951	0.866
10	1.000	0.707	0.500	0.924	1.000	0.966
12	0.000	1.000	0.000	0.707	0.951	1.000
14	1.000	0.707	0.500	0.383	0.809	0.966
16	0.000	0.000	0.866	0.000	0.588	0.866
18	1.000	0.707	1.000	0.383	0.309	0.707
20	0.000	1.000	0.866	0.707	0.000	0.500
22	1.000	0.707	0.500	0.924	0.309	0.259
24	0.000	0.000	0.000	1.000	0.588	0.000

(Continues)

Table 2.6 (Continued)

N_s / $2p$	6	12	18	24	30	36
26	1.000	0.707	0.500	0.924	0.809	0.259
28	0.000	1.000	0.866	0.707	0.951	0.500
30	1.000	0.707	1.000	0.383	1.000	0.707
			$y = 4$			
2	0.866	0.866	0.643	0.500	0.407	0.342
4	0.866	0.866	0.985	0.866	0.743	0.643
6	0.000	0.000	0.866	1.000	0.951	0.866
8	0.866	0.866	0.342	0.866	0.995	0.985
10	0.866	0.866	0.342	0.500	0.866	0.985
12	0.000	0.000	0.866	0.000	0.588	0.866
14	0.866	0.866	0.985	0.500	0.208	0.643
16	0.866	0.866	0.643	0.866	0.208	0.342
18	0.000	0.000	0.000	1.000	0.588	0.000
20	0.866	0.866	0.643	0.866	0.866	0.342
22	0.866	0.866	0.985	0.500	0.995	0.643
24	0.000	0.000	0.866	0.000	0.951	0.866
26	0.866	0.866	0.342	0.500	0.743	0.985
28	0.866	0.866	0.342	0.866	0.407	0.985
30	0.000	0.000	0.866	1.000	0.000	0.866

Table 2.7 Optimised coil pitch numbers for PMSMs with different slot/pole number combinations

N_s / $2p$	6	12	18	24	30	36
2	3	6	9	12	15	18
4	1	3	4	6	7	9
6	1	2	3	4	5	6
8	1	1	2	3	4	4
10	1	1	2	2	3	4
12	1	1	1	2	2	3
14	1	1	1	2	2	3
16	1	1	1	1	2	2
18	1	1	1	1	2	2
20	1	1	1	1	1	2
22	1	1	1	1	1	2
24	1	1	1	1	1	1
26	1	1	1	1	1	1
28	1	1	1	1	1	1
30	1	1	1	1	1	1

Notes:

$y = 1$; $y = 2$; $y = 3$; $y = 4$

Table 2.8 *Pitch factors (K_p) of PMSMs with different slot/pole number combinations and different coil pitch numbers*

N_s / $2p$	6	12	18	24	30	36
2	1.000	1.000	1.000	1.000	1.000	1.000
4	0.866	1.000	0.985	1.000	0.995	1.000
6	1.000	1.000	1.000	1.000	1.000	1.000
8	0.866	0.866	0.985	1.000	0.995	0.985
10	0.500	0.966	0.985	0.966	1.000	0.985
12	0.000	1.000	0.866	1.000	0.951	1.000
14	0.500	0.966	0.940	0.966	0.995	0.966
16	0.866	0.866	0.985	0.866	0.995	0.985
18	1.000	0.707	1.000	0.924	0.951	1.000
20	0.866	0.500	0.985	0.966	0.866	0.985
22	0.500	0.259	0.940	0.991	0.914	0.940
24	0.000	0.000	0.866	1.000	0.951	0.866
26	0.500	0.259	0.766	0.991	0.978	0.906
28	0.866	0.500	0.643	0.966	0.995	0.940
30	1.000	0.707	0.500	0.924	1.000	0.966

2.2.4 Resultant winding factor

In summary, this section describes the winding assignments and winding factors of DTP winding configurations in PMSMs systemically based on the star of slots, where different slot/pole number combinations, phase-shift angles, and coil pitch numbers are all considered. For a DTP PMSM with any slot/pole number combination, the feasible phase-shift angles and corresponding winding layouts and distribution factors can be obtained from Section 2.2.2. The optimized coil pitch number and corresponding pitch factor are shown in Section 2.2.3. Section 2.2.4 provides a comprehensive table, including the feasible phase-shift angles and corresponding winding factors for different slot/pole number combinations. Thus, the potentials of different DTP winding configurations can be quickly compared from the viewpoint of winding factors, as listed in Table 2.9.

2.3 Attenuation factor for two three-phase winding sets

2.3.1 Attenuation factor caused by spatial shifting

In Section 2.2, the winding layouts of DTP winding configurations in PMSMs are obtained from the star of slots to enhance the winding factor of the working harmonic. In fact, the two winding sets in DTP PMSMs can also be distributed according to the concept of "stator shifting" [13], which involves the following steps.

- Divide the original windings into two separate winding sets.
- Both windings are identical and supplied by identical currents.

Table 2.9 Feasible phase-shift angles and corresponding winding factors for DTP PMSMs with different slot/pole number combinations and optimized coil pitch numbers

N_s	6		12		18		24		30		36	
$2p$	β(°)	K_w	β(°)	K_w	β(°)	K_w	β(°)	K_w	β(°)	K_w	β(°)	K_w
2	0	1.000	0 30	0.966 1.000	0/20	0.960	0 15 30	0.958 0.966 0.991	0/12	0.957	0/20 10 30	0.956 0.960 0.990
4	0	0.866	0	1.000	0	0.945	0 30	0.966 1.000	0	0.951	0/20	0.960
6	**		**		0	1.000	**		**		0 30	0.966 1.000
8	0	0.866	0	0.866	0	0.945	0	1.000	0	0.951	0	0.945
10	0	0.500	0 30	0.933 0.966	0/20	0.945	0 15 30	0.925 0.933 0.958	0	1.000	0/20 10 30	0.942 0.945 0.975
12	**		**		0	0.866	**		**		0	1.000
14	0	0.500	0 30	0.966 0.966	0/20	0.902	0 15 30	0.925 0.933 0.958	0/12	0.951	0/20 10 30	0.924 0.927 0.956
16	0	0.866	0	0.866	0	0.945	0	0.866	0	0.951	0	0.945
18	**		**		**		**		**		**	
20	0	0.866	0	0.500	0	0.945	0 30	0.933 0.966	0	0.866	0/20	0.945
22	0	0.500	0 30	0.250 0.259	0/20	0.902	0 15 30	0.949 0.958 0.983	0/12	0.874	0/20 10 30	
24	**		**		0	0.866	**		**		0	0.866
26	0	0.500	0 30	0.250 0.259	0/20	0.735	0 15 30	0.949 0.958 0.983	0/12	0.936	0/20 10 30	0.867 0.870 0.897
28	0	0.866	0	0.500	0	0.617	0 30	0.933 0.966	0	0.951	0/20	0.902
30	**		**		0	0.500	**		**		0 30	0.933 0.966

Notes:
** Unfeasible slot/pole combinations for DTP PMSMs.

t is even and N_s is an odd multiple of 3

$k = 1$; $k = 2$; $k = 3$; $k = 4$; $k = 5$; $k = 6$

- The second winding set is shifted mechanically for a specific angle relative to the first winding set.

Hence, when utilizing stator shifting in PMSMs, the two winding sets are shifted spatially but fed with identical currents without temporal shifting. If the two winding sets are connected in series and supplied by one inverter, the PMSM is still an STP PMSM. However, if the two winding sets are supplied by two separate inverters, the whole windings can be seen as DTP windings with a phase-shift angle = 0°. The effects of stator shifting on the resultant stator MMF harmonics are investigated in [14], and the effects of the spatial shifting on different MMF harmonics are explained by the "attenuation factor (K_a)".

For the νth harmonic, the final winding factor considering spatial shifting is $K_{w\nu\text{-}final}$,

$$K_{w\nu-final} = K_{w\nu-initial} \cdot K_{a\nu} \tag{2.14}$$

where $K_{w\nu\text{-}initial}$ is the initial winding factor of the νth harmonic in one winding set, and $K_{a\nu}$ is the attenuation factor of the νth harmonic.

Assuming the magnitudes of the νth harmonics of Windings 1 and 2 are $F_{\nu 1}$ and $F_{\nu 2}$, respectively, since the two winding sets are identical in their winding configurations, it can be obtained by

$$\left|\dot{F}_{\nu 1}\right| = \left|\dot{F}_{\nu 2}\right| \tag{2.15}$$

The resultant νth MMF component of the whole windings (F_ν) is the vectorial sum of $F_{\nu 1}$ and $F_{\nu 2}$:

$$\dot{F}_\nu = \dot{F}_{\nu 1} + \dot{F}_{\nu 2} \tag{2.16}$$

Assuming that the spatial shifting angle between the two winding sets is α, in electrical degree. For the νth MMF components, the spatial shifting angle is $\nu\alpha$. Hence, the relationship between F_ν, $F_{\nu 1}$, and $F_{\nu 2}$ can be reflected geometrically in Figure 2.22.

From Figure 2.22, the attenuation factor for the νth MMF components ($K_{a\nu}$) can be obtained as the ratio of the vectorial sum of $F_{\nu 1}$ and $F_{\nu 2}$ to their arithmetic sum, as

$$K_{a\nu} = \frac{\left|\dot{F}_{\nu 1} + \dot{F}_{\nu 1}\right|}{\left|\dot{F}_{\nu 1}\right| + \left|\dot{F}_{\nu 2}\right|} = \frac{2\left|\cos\left(\frac{\nu\alpha}{2}\right)\right|\left|\dot{F}_{\nu 1}\right|}{2\left|\dot{F}_{\nu 1}\right|} = \left|\cos\left(\frac{\nu\alpha}{2}\right)\right| \tag{2.17}$$

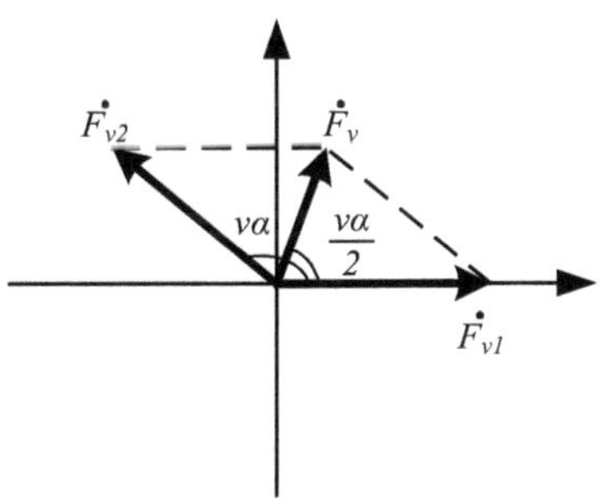

Figure 2.22 Geometrical relationship between the v-th MMF components of different winding sets in PMSM with spatial shifting angle = α.

Thus, in PMSMs with spatial shifted windings, the attenuation factors for different MMF harmonics can be computed based on the spatial shifting angle using (2.17). The variations of the attenuation factors of the 1st, 5th, and 7th MMF harmonics with spatial shifting angle are provided in Figure 2.23 as examples.

It can be observed from Figure 2.23 that when a proper spatial shifting angle is selected, it is feasible to minimize the non-working harmonics without much sacrifice of the working harmonic. For example, when the 5th harmonic is the working harmonic, the spatial shifting angle (α) can be set as 77.14°, and K_{a5} = 0.975 and K_{a7} = 0. Consequently, the 7th harmonics of the two winding sets can be canceled completely, while the 5th harmonics are reduced by only 2.5%. When the 7th harmonic is the working harmonic, the spatial shifting angle can be set as 108°, and K_{a5} = 0 and K_{a7} = 0.951. Therefore, the 5th harmonics are canceled completely, while the 7th harmonics are reduced by only 4.9%.

However, it should be mentioned that the actual spatial shifting angle between the two winding sets is limited by the number of slots and the stator configuration. For example, in the case of the 24-slot/10-pole PMSM in [13], a 77.14° spatial shifting is selected to reduce the 7th harmonics, while achieving this specific angle may require unequal stator teeth widths, which complicates the manufacturing process. Another example is the 12-slot/10-pole PMSM with DL windings, of which the whole armature windings can be seen as composed of two single-layer winding sets with a specific spatial shifting angle. The first winding set can be designed based on the star of slots, as shown in Figure 2.24. Subsequently, the second single-layer three-phase winding set can be arranged using different spatial shifting angles, as presented in Figure 2.25. To ensure the stator teeth widths are identical, the spatial shifting angle (α) can only be chosen from 30°, 90°, and 150°, and the corresponding attenuation factors of the 1st, 5th, and 7th harmonics can be

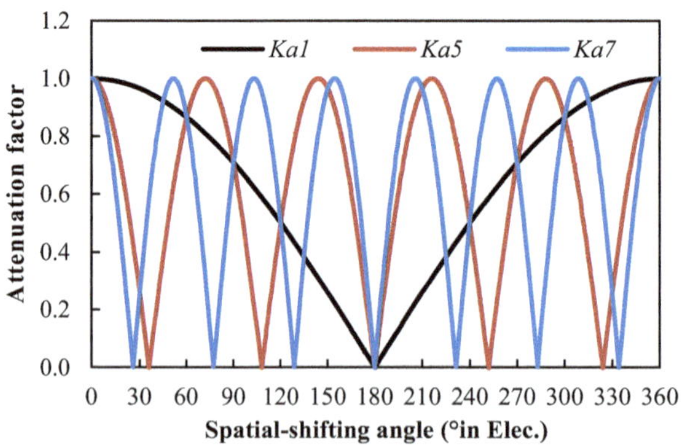

Figure 2.23 Variations of attenuation factors of different MMF harmonic orders with spatial shifting angle.

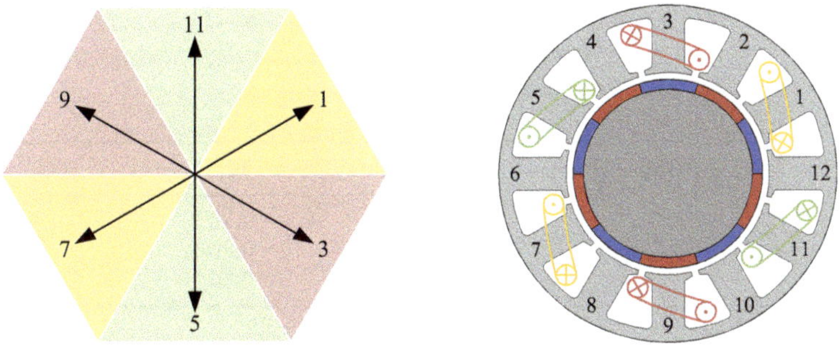

(a) (b)

Figure 2.24 Star of slots and winding layouts of 12-slot/10-pole PMSM with single-layer three-phase windings. (a) Star of slots and (b) winding layouts.

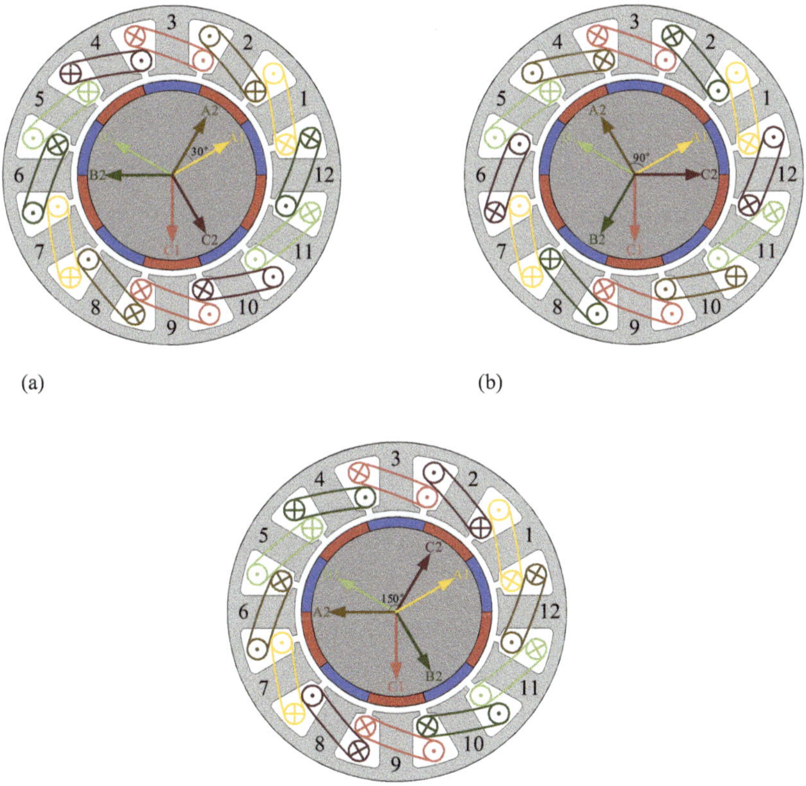

(a) (b)

(c)

Figure 2.25 Winding layouts of 12-slot/10-pole DTP PMSM with two sets of single-layer three-phase windings using different spatial shifting angles. (a) α = 30°, (b) α = 90°, and (c) α = 150°.

Table 2.10 Attenuation factors of 1st, 5th, and 7th harmonics in 12-slot/10-pole DTP PMSM with different spatial shifting angles

Spatial shifting angle α	**30°**	**90°**	**150°**
K_{a1}	0.966	0.707	0.259
K_{a5}	0.259	0.707	0.966
K_{a7}	0.259	0.707	0.966

obtained from Figure 2.23 and listed in Table 2.10. It can be seen that to enhance the working harmonic (the 5th harmonic), the 150° spatial shifting should be selected among 30°, 90°, and 150° candidates.

2.3.2 Attenuation factor caused by spatial and time shifting

In the 12-slot/10-pole DTP PMSM, selecting a spatial shifting angle of 150° results in winding layouts as shown in Figure 2.25(c), which looks similar to that depicted in Figure 2.11 (DTP windings with a phase-shift angle of 30°). Essentially, the DTP winding configurations with a phase-shift angle of 30° in Figure 2.11 are composed of two three-phase windings, and the two winding sets are shifted by 150° in spatial and 30° in time. Hence, it is still necessary to further investigate the effects of spatial and time-shifting angles on the resultant MMF production. In [15], a generic approach to calculate MMF in PMSMs with multiple three-phase winding sets is introduced. Based on [15], the concept "attenuation factor" is extended by considering spatial and time-shifting angles simultaneously in this sub-section.

The MMF acting across the air gap produced by Winding 1 can be obtained as

$$F_1(\theta,t) = N_{A1}(\theta)i_{A1}(t) + N_{B1}(\theta)i_{B1}(t) + N_{C1}(\theta)i_{C1}(t) \tag{2.18}$$

where $N_{A1}(\theta)$, $N_{B1}(\theta)$, and $N_{C1}(\theta)$ are the turn functions of the PMSM, $i_{A1}(t)$, $i_{B1}(t)$, and $i_{C1}(t)$ are the currents fed in the three-phase winding set. With the utilization of the winding factor, the windings are seen as full-pitch windings, and thus, the winding turn function and phase current of Phase A1 are shown in Figure 2.26.

From Figure 2.26, the Fourier series of the winding turn functions of Phase A1, B1, and C1 can be expressed as

$$\begin{aligned}
N_{A1}(\theta) &= \sum\nolimits_{v=1,3,5,\ldots}^{\infty} \frac{4N}{v\pi} \sin(v\theta) \\
N_{B1}(\theta) &= \sum\nolimits_{v=1,3,5,\ldots}^{\infty} \frac{4N}{v\pi} \sin\left[v\left(\theta + \frac{2\pi}{3}\right)\right] \\
N_{C1}(\theta) &= \sum\nolimits_{v=1,3,5,\ldots}^{\infty} \frac{4N}{v\pi} \sin\left[v\left(\theta - \frac{2\pi}{3}\right)\right]
\end{aligned} \tag{2.19}$$

where N is the number of turns per coil, and θ is the rotor position.

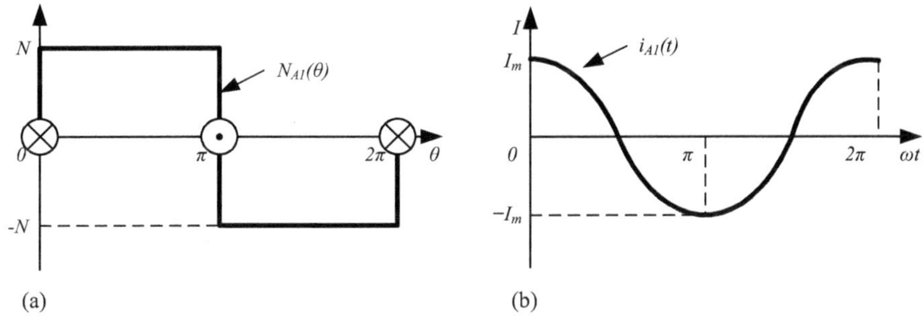

Figure 2.26 Winding turn function and current of Phase A1. (a) Turn function and (b) phase current.

Winding 1 is fed with balanced three-phase sinusoidal currents, i.e.,

$$\begin{cases} i_{A1}(t) = I_m\cos(\omega t) \\ i_{B1}(t) = I_m\cos(\omega t - 2\pi/3) \\ i_{C1}(t) = I_m\cos(\omega t + 2\pi/3) \end{cases} \tag{2.20}$$

where I_m is the peak value of phase currents, ω and t are the fundamental radian frequency and the time, respectively.

In a balanced three-phase winding set, the MMF components with different harmonic orders rotate either clockwise or anti-clockwise. Specifically, for the νth harmonics with $\nu = 1, 7, 13, 19$, ($6k+1$, k is a natural number) ..., the MMF rotates clockwise, and for the νth harmonics with $\nu = 5, 11, 17$, ($6k-1$, k is a positive integer) ..., the MMF rotates anti-clockwise. By using a negative number to denote negative rotation direction, the resultant MMF produced by Winding 1 can be expressed as

$$F_1(\theta, t) = \frac{6NI_m}{\pi}\sum_{\nu=1,-5,7,\ldots}^{\infty}\frac{k_{w\nu}}{\nu}\sin(\nu\theta - \omega t) \tag{2.21}$$

where $K_{w\nu}$ is the initial winding factor for the νth harmonic in Winding 1.

For the second three-phase winding set in the DTP PMSM, the spatial and time-shifting angles are illustrated in Figure 2.30. It can be seen that Winding 2 leads Winding 1 by α in space, and the currents in Winding 2 are advanced by β in time compared to those in Winding 1.

According to (2.21) and the spatial and time displacements shown in Figure 2.27, the resultant MMF produced by Winding 2 can be expressed as

$$F_2(\theta, t) = \frac{6NI_m}{\pi}\sum_{\nu=1,-5,7,\ldots}^{\infty}\frac{k_{w\nu}}{\nu}\sin[\nu(\theta + \alpha) - (\omega t - \beta)] \tag{2.22}$$

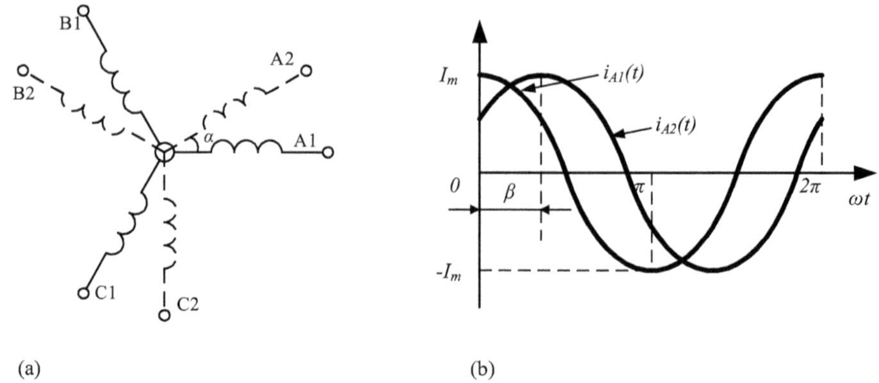

Figure 2.27 Scheme of space and time displacements between two 3-phase winding sets. (a) Spatial shifting angle in space (α) and (b) time shifting in currents (β).

Therefore, the overall MMF along the air gap is the sum of those produced by Winding 1 and Winding 2, as

$$\begin{aligned} F(\theta,t) &= F_1(\theta,t) + F_2(\theta,t) \\ &= \frac{6NI_m}{\pi}\sum_{v=1,-5,7,\ldots}^{\infty}\frac{k_{wv}}{v}\{\sin(v\theta-\omega t)+\sin[v(\theta+\alpha)-(\omega t+\beta)]\} \\ &= \frac{6NI_m}{\pi}\sum_{v=1,-5,7,\ldots}^{\infty}\frac{k_{wv}}{v}[\sin(v\theta-\omega t)+\sin(v\theta-\omega t+v\alpha+\beta)] \end{aligned} \tag{2.23}$$

In (2.23), it can be obtained that the resultant vth MMF component in the DTP PMSM can be calculated as

$$\begin{aligned} F_v(\theta,t) &= F_{v1}(\theta,t) + F_{v2}(\theta,t) \\ &= \frac{6NI_m}{\pi}\frac{k_{wv}}{v}[\sin(v\theta-\omega t)+\sin(v\theta-\omega t+v\alpha+\beta)] \\ &= \frac{6NI_m}{\pi}\frac{k_{wv}}{v}[\sin(v\theta-\omega t)+\sin(v\theta-\omega t)\cos(v\alpha+\beta) \\ &\quad +\cos(v\theta-\omega t)\sin(v\alpha+\beta)] \\ &= \frac{6NI_m}{\pi}\frac{k_{wv}}{v}\left\{\sin(v\theta-\omega t)\left[\sin^2\left(\frac{v\alpha+\beta}{2}\right)+\cos^2\left(\frac{v\alpha+\beta}{2}\right)\right.\right. \\ &\quad \left.+\cos^2\left(\frac{v\alpha+\beta}{2}\right)-\sin^2\left(\frac{v\alpha+\beta}{2}\right)\right]+\cos(v\theta-\omega t) \\ &\quad \left.\cdot 2\sin\left(\frac{v\alpha+\beta}{2}\right)\cos\left(\frac{v\alpha+\beta}{2}\right)\right\} \end{aligned}$$

$$= \frac{6NI_m}{\pi}\frac{k_{wv}}{v}\left[\sin(v\theta - \omega t)\cdot 2\cos^2\left(\frac{v\alpha+\beta}{2}\right) + \cos(v\theta - \omega t)\right.$$

$$\left.\cdot 2\sin\left(\frac{v\alpha+\beta}{2}\right)\cos\left(\frac{v\alpha+\beta}{2}\right)\right]$$

$$= \frac{6NI_m}{\pi}\frac{k_{wv}}{v}2\cos\left(\frac{v\alpha+\beta}{2}\right)$$

$$\cdot\left[\sin(v\theta - \omega t)\cos\left(\frac{v\alpha+\beta}{2}\right) + \cos(v\theta - \omega t)\sin\left(\frac{v\alpha+\beta}{2}\right)\right]$$

$$= \frac{6NI_m}{\pi}\frac{k_{wv}}{v}2\cos\left(\frac{v\alpha+\beta}{2}\right)\sin\left(v\theta - \omega t + \frac{v\alpha+\beta}{2}\right) \tag{2.24}$$

The resultant vth MMF component of the two winding sets obtained in (2.24) can also be expressed as

$$F_v(\theta,t) = \frac{6(2N)I_m}{\pi}\frac{k_{wv}\cos\left(\frac{v\alpha+\beta}{2}\right)}{v}\sin\left[v\left(\theta+\frac{\alpha}{2}\right) - \left(\omega t - \frac{\beta}{2}\right)\right] \tag{2.25}$$

Meanwhile, according to (2.21), the vth MMF component produced by Winding 1 is

$$F_{v1}(\theta,t) = \frac{6NI_m}{\pi}\frac{k_{wv}}{v}\sin(v\theta - \omega t) \tag{2.26}$$

Hence, with spatial and time-shifting angles simultaneously, the attenuation factor of the vth MMF harmonic can be calculated as

$$K_{av} = \frac{\left|\dot{F}_{v1} + \dot{F}_{v1}\right|}{\left|\dot{F}_{v1}\right| + \left|\dot{F}_{v2}\right|} = \frac{\left|\dot{F}_v\right|}{2\left|\dot{F}_{v1}\right|} = \frac{\left|2k_{wv}\cos\left(\frac{v\alpha+\beta}{2}\right)\right|}{2\left|k_{wv}\right|} = \left|\cos\left(\frac{v\alpha+\beta}{2}\right)\right| \tag{2.27}$$

In (2.27), both the spatial-shifting angle (α) and the time-shifting angle (β) are considered in the attenuation factor. In practice, the time-shifting angle between the two winding sets is usually obtained according to the spatial shifting between them. Specifically, to enhance the winding factor of the working harmonic, the time shifting angle is set as the advancing angle of Winding 1 to Winding 2.

For example, for the 12-slot/10-pole PMSM with a spatial shifting angle = 30°, 90°, and 150°, as presented in Figure 2.25, the stars of slots for the 5th harmonic (working harmonic) are shown in Figure 2.28. When the spatial-shifting angle (α) is 30°, the time-shifting angle (β) should be the advancing angle of Spoke 1 (Phase A1) to Spoke 2 (Phase A2), which is 150°. Similarly, for α = 90°, β = 90°; and for α = 150°, β = 30°, respectively.

Thus, in the 12-slot/10-pole PMSM with two winding sets, the attenuation factors resulting from various spatial and time-shifting angles can be calculated, as

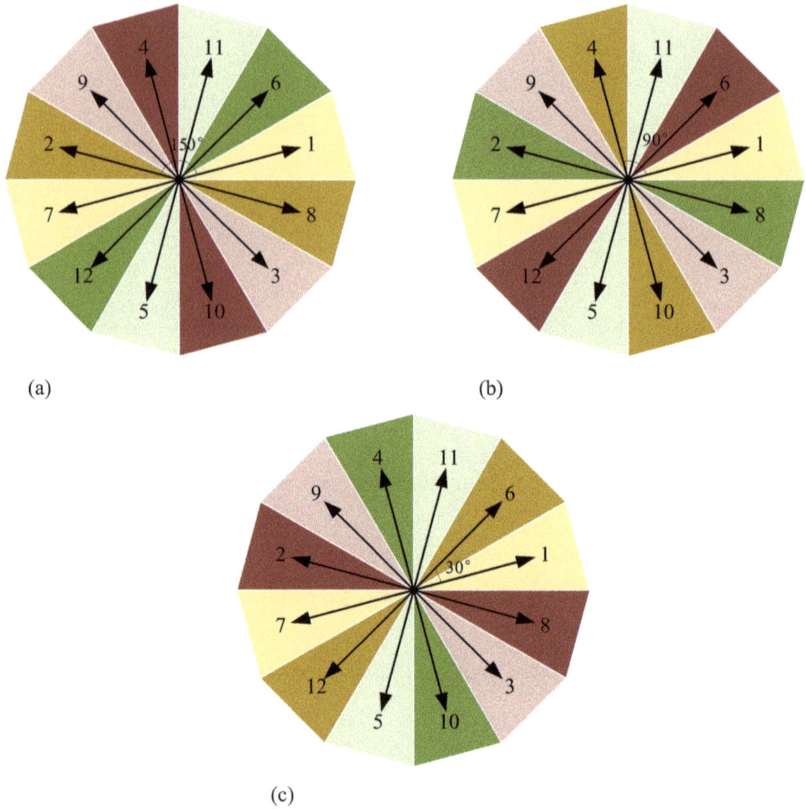

Figure 2.28 Winding layouts of 12-slot/10-pole PMSM with two sets of single-layer three-phase windings using different spatial shifting angles. (a) $\alpha = 30°$, $\beta = 150°$; (b) $\alpha = 90°$, $\beta = 30°$; and (c) $\alpha = 150°$, $\beta = 30°$.

Table 2.11 Attenuation factors of different harmonics in 12-slot/10-pole DTP PMSM with different spatial and time-shifting angles

Spatial-shifting angle α	**30**		**90**		**150**	
Time-shifting angle, β	**0**	**150**	**0**	**90**	**0**	**30**
Harmonic order, ν	**Attenuation factor, $K_{\alpha\nu}=\left\lvert\cos\left(\frac{\nu\alpha+\beta}{2}\right)\right\rvert$**					
1	0.966	0.000	0.707	0.000	0.259	0.000
−5	0.259	1.000	0.707	1.000	0.966	1.000
7	0.259	1.000	0.707	1.000	0.966	1.000
−11	0.966	0.000	0.707	0.000	0.259	0.000
13	0.966	0.000	0.707	0.000	0.259	0.000
−17	0.259	1.000	0.707	1.000	0.966	1.000
19	0.259	1.000	0.707	1.000	0.966	1.000

outlined in Table 2.11. For any harmonic order, when $K_{av} = 0$, the corresponding harmonic order will be eliminated. It can be seen that the DTP winding configurations with both spatial and time-shifting angles can eliminate some MMF harmonic contents effectively without sacrificing the working harmonic order.

In summary, this section introduced the concept of "attenuation factor" [14], which describes the effects of the two winding sets with specific spatial-shifting angles on the resultant MMF. This concept is extended in this section by considering both spatial and time-shifting angles. It also should be mentioned that when the two winding sets are shifted by a specific spatial-shifting angle (see Figure 2.28), the time-shifting angle between the currents in the two winding sets is determined accordingly. When the currents fed in the two winding sets obey the time-shifting angle obtained before, the currents can be described "in phase currents" [9]. The effect of DTP windings with "in phase currents" has been described in Section 2.2, while the attenuation factor introduced in this section makes it possible to investigate the effects of DTP windings on the resultant MMF field with "out phase currents". Overall, in DTP PMSMs, with the extended attenuation factor, the influences of any spatial and time-shifting angles between the two winding sets on the resultant MMF production can be predicted simply according to the different harmonic order, without the need for complicated computation or simulations.

2.4 Finite element and experimental results

2.4.1 Machine and winding configuration

In this section, the analyses of winding assignments, winding factors, and attenuation factors in DTP PMSMs are validated based on a 24-slot/22-pole DTP PMSM, where different phase-shift angles are considered. First, the effects of the different DTP winding configurations on different harmonic orders are calculated using winding factors and attenuation factors. Afterward, the calculated results are validated by FE and experimental results.

As introduced in Section 2.2 and detailed in Table 2.4, for the 24-slot/22-pole DTP PMSMs, their feasible phase-shift angles include 0°, 15°, and 30°. Specific winding layouts for different phase-shift angles can be obtained from the star of slots, as shown in Figure 2.29.

2.4.2 Finite element results

For the 24-slot/22-pole DTP PMSM in Figure 2.11, the winding factors of the working harmonics with 0°, 15°, and 30° phase-shift angles are 0.949, 0.958, and 0.983, respectively. Compared with the counterpart with a phase-shift angle = 0°, the amplitude of the back EMF can be improved by 0.83% when the phase-shift angle = 15°, or by 3.53% when the phase-shift angle = 30°.

As can be observed from Figure 2.29(a), when the phase-shift angle is 0°, the spatial shifting angle is the angular displacement from Slot 1 to Slot 13, which is 180°. For the conditions with a 0° phase-shift angle, the time-shifting angle should

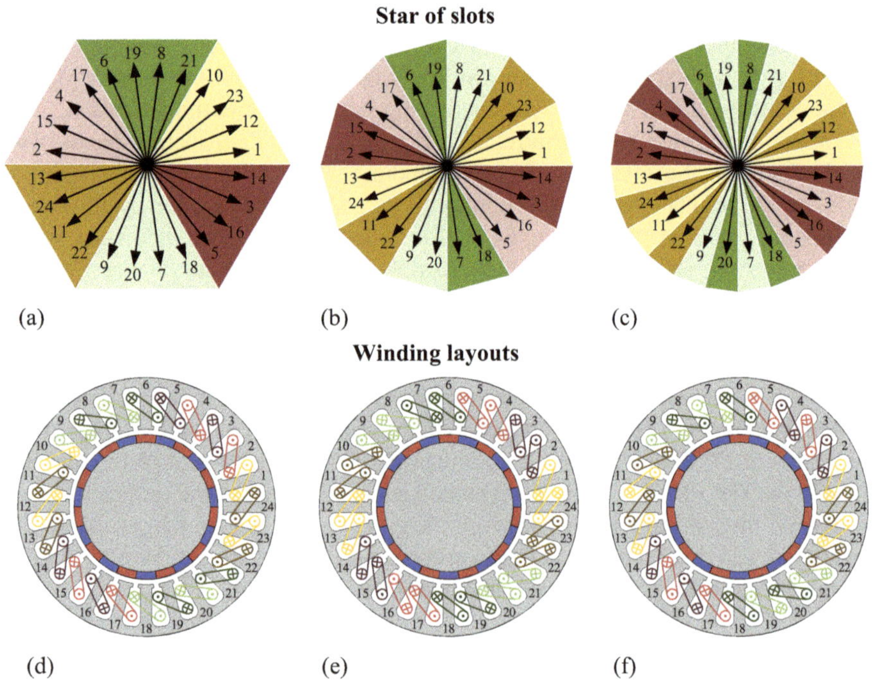

Figure 2.29 Star of slots ((a) phase-shift angle = 0°, (b) phase-shift angle = 15°, and (c) phase-shift angle = 30°) and winding layouts ((d) phase-shift angle = 0°, (e) phase-shift angle = 15°, and (f) phase-shift angle = 30°) of 24-slot/22-pole DTP PMSM with different phase-shift angles

be 0°. However, as the coils of Slots 13, 24, 11, and 22 are reverse-wound, the actual time-shifting angle is 180°. Similarly, the spatial and time-shifting angles with different phase-shift angles can be observed in Figure 2.29. When the phase-shift angle is 15°, the spatial-shifting angle is 165°, and the time-shifting angle is 15°. When the phase-shift angle is 30°, the spatial-shifting angle is 330°, and the time-shifting angle = 30°. Subsequently, the attenuation factors of different MMF harmonics in various DTP winding configurations are calculated, as shown in Table 2.12.

Table 2.12 indicates that the two winding sets in the DTP windings with a 0° phase shift cannot affect any of the resultant MMF harmonics. The DTP windings with a 15° phase shift can eliminate the 1st harmonic and reduce the 5th, 7th, 17th, and 19th harmonics. Similarly, the DTP windings with a 30° phase shift can eliminate the 5th, 7th, 17th, and 19th harmonics.

Considering different phase-shift angles, the air-gap MMF contents in the DTP PMSM are calculated by normalized currents and 2D FE simulations via JMAG FE software. The waveforms and spectra of the calculated and FE-simulated MMF

Table 2.12 Attenuation factors of different harmonics in 24-slot/22-pole DTP PMSM with different phase-shift angles

Phase-shift angle	**0°**	**15°**	**30°**
Spatial-shifting angle, α	**180°**	**165°**	**330°**
Time-shifting angle, β	**180°**	**15°**	**30°**
Harmonic order, ν	**Attenuation factor, $K_{\alpha\nu}=\left|\cos\left(\frac{\nu\alpha+\beta}{2}\right)\right|$**		
1	1	0	1
−5	1	0.707	0
7	1	0.707	0
−11	1	1	1
13	1	1	1
−17	1	0.707	0
19	1	0.707	0

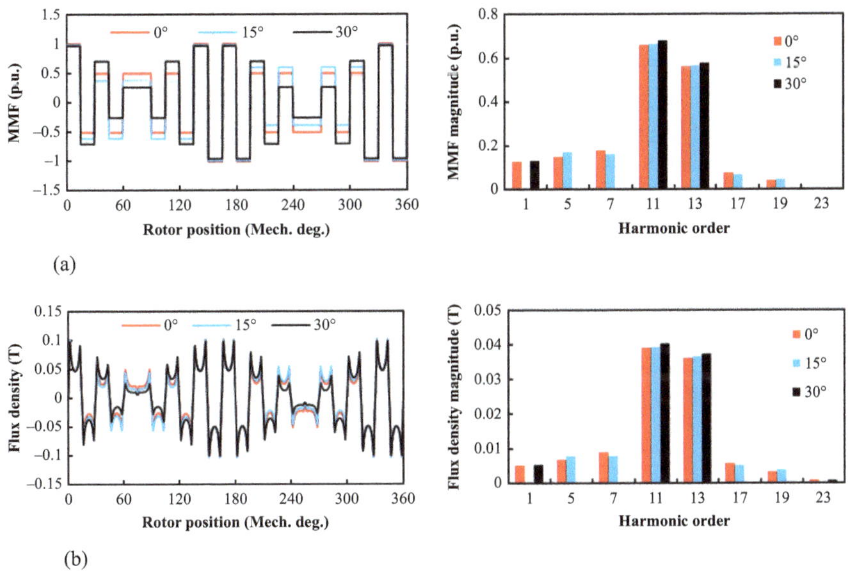

Figure 2.30 Comparison of calculated and FE simulated MMF contents in 24-slot/22-pole DTP PMSMs with different phase-shift angles. (a) Waveforms and spectra of calculated MMF. (b) Waveforms and spectra of FE simulated air-gap flux density.

distributions are shown in Figure 2.30. It is evident that the FE simulated results agree well with the calculated results.

Figure 2.30 illustrates that the 15° and 30° phase-shift angles can improve the amplitude of the 11th harmonic in the benchmark DTP PMSM with 0° phase shift. Additionally, the elimination effects of DTP windings on specific harmonics can also be observed and aligned with the prediction results in Table 2.12. Overall, it can be concluded that the effects of different DTP winding configurations on the stator MMF distributions and the resultant performance of PMSMs can be predicted by the calculations of winding factors and attenuation factors.

2.4.3 Experimental results

The benchmark 24-slot/22-pole DTP PMSM is fabricated and tested considering different phase-shift angles [10]. The main design specifications are given in Table 2.13. The stator and rotor of the prototype are shown in Figure 2.31(a) and (b), respectively.

As the experiments needed to be conducted with various DTP winding configurations (Phase-shift angle = 0°, 15°, and 30°), the terminals of all coils are connected to three connector blocks according to phases (Phases A1 and A2 in one connector block; Phases B1 and B2 in one connector block; and Phases C1 and C2 in one connector block) to switch the winding connections between the different DTP winding configurations.

Although the air-gap flux density distributions cannot be measured directly, the back EMF waveforms of Phases A1 and A2 with different phase-shift angles can be measured, as shown in Figure 2.32. The FE-simulated phase-back EMFs are also presented in Figure 2.32, where good agreement between FE simulation results and measured results can be found. Meanwhile, the influences of different phase-shift angles (0°, 15°, and 30°) on back EMFs can also be clearly observed.

Table 2.13 Main design specifications of benchmark 24-slot/22-pole DTP PMSM

Parameters	Value	Parameters	Value
Stator slot number	24	Rotor pole number	22
Stator OD, mm	90	Rotor OD, mm	60
Stator ID, mm	62.8	Rotor ID, mm	28
Tooth width, mm	3.4	PM thickness, mm	3
Yoke thickness, mm	3.3	PM remanence, T	1.2
Stack length, mm	25	PM relative permeability	1.05

(a) (b)

Figure 2.31 Photos of 24-slot/22-pole DTP PMSM prototype. (a) Stator and (b) rotor.

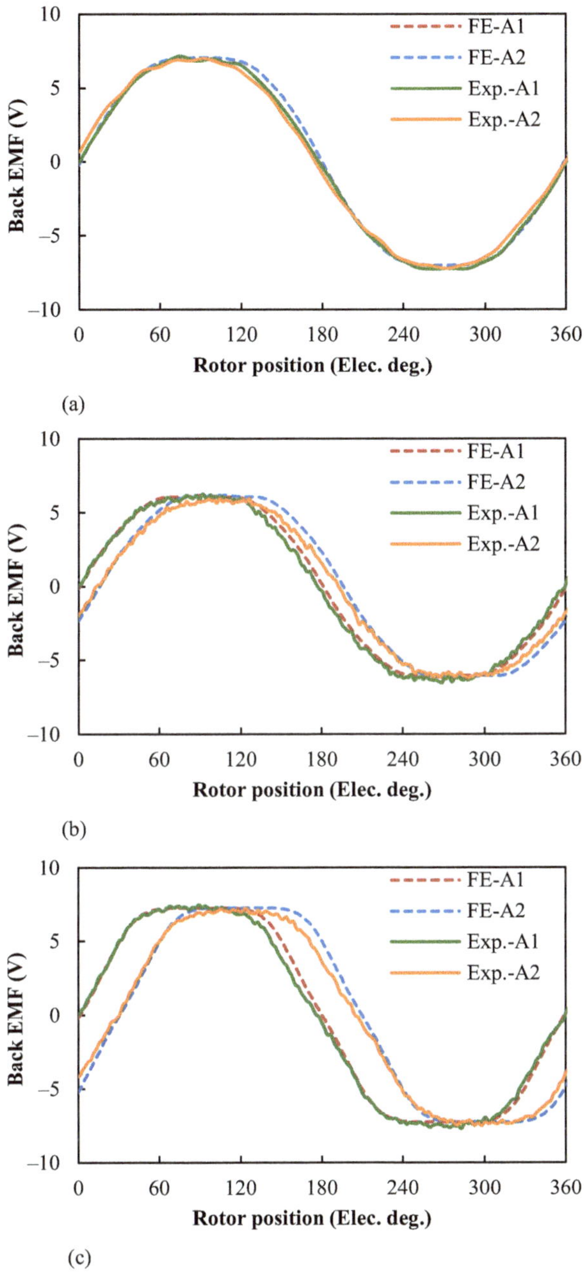

Figure 2.32 Comparison of FE simulated and measured phase back EMF waveforms of 24-slot/22-pole DTP PMSM with different phase-shift angles at 1,000 r/min. (a) Phase-shift angle = 0°, (b) phase-shift angle = 15°, and (c) phase-shift angle = 30°.

Thus, the correctness of the FE models utilized in this section is validated by the experimental results. Considering that the winding factor and attenuation factor calculations can be validated by the FE method, the calculation results can also be validated by the experimental results.

2.5 Summary

This chapter systematically analyses the feasibility of DTP winding configurations in PMSMs based on the star of slots, where different slot/pole number combinations, phase-shift angles, and coil pitch numbers are all considered. The feasible DTP winding configurations and corresponding winding factors are summarized and listed in Table 2.9.

In addition, in DTP PMSMs, the two winding sets are shifted by a spatial-shifting angle, and the currents in the two winding sets are shifted by a time-shifting angle. The effects of spatial and time-shifting angles on the resultant MMF harmonics are examined simultaneously by the concept of "attenuation factor". With the extended attenuation factor, the effects of any spatial and time-shifting angles on different MMF harmonic orders can be predicted directly without the need for complicated calculations and simulations.

Finally, the FE and experimental results are provided to illustrate the DTP winding assignments, winding factors, and attenuation factors based on a 24-slot/22-pole DTP PMSM.

References

[1] Z. Q. Zhu, S. S. Wang, B. Shao, L. C. Yan, P. L. Xu, and Y. Ren, "Advances in dual-three-phase permanent magnet synchronous machines and control techniques," *Energies*, vol. 14, no. 22, pp. 7508, 2021.
[2] E. Levi, "Multiphase electric machines for variable-speed applications," *IEEE Trans. Ind. Electron.*, vol. 55, no. 5, pp. 1893–1909, 2008.
[3] M. Barcaro, N. Bianchi, and F. Magnussen, "Six-phase supply feasibility using a PM fractional-slot dual winding machine," *IEEE Trans. Ind. Appl.*, vol. 47, no. 5, pp. 2042–2050, 2011.
[4] P. L. Xu, J. H. Feng, S. Y. Guo, *et al.*, "Analysis of dual three-phase permanent-magnet synchronous machines with different angle displacements," *IEEE Trans. Ind. Electron.*, vol. 65, no. 3, pp. 1941–1954, 2018.
[5] A. Salem and M. Narimani, "A review on multiphase drives for automotive traction applications," *IEEE Trans. Transp. Electrific.*, vol. 5, no. 4, pp. 1329–1348, 2019.
[6] J. R. Hendershot and T. J. E. Miller, *Design of Brushless Permanent Magnet Motors*, Magna Physics Publishing, Lebanon, OH, 1994.
[7] J. Pyrhonen, T. Jokinen, and V. Hrabovcova, *Design of Rotating Electrical Machines*,Wiley, New York, 2008.

[8] N. Bianchi and M. Dai Pre, "Use of the star of slots in designing fractional-slot single-layer synchronous motors," *IEE Electr. Power Appl.*, vol. 153, no. 3, pp. 459–466, 2006.

[9] M. Barcaro, N. Bianchi, and F. Magnussen, "Analysis and tests of a dual three-phase 12-slot 10-pole permanent-magnet motor," *IEEE Trans. Ind. Appl.*, vol. 46, no. 6, pp. 2355–2362, 2010.

[10] P. L. Xu, Z. Q. Zhu, B. Shao, *et al.*, "Analysis of dual 3-phase fractional-slot non-overlapping winding PM synchronous machines with different angle displacements," *2019 IEEE Energy Convers. Cong. Expo. (ECCE)*, Baltimore, MD, USA, Sept.–Oct. 2019, pp. 5616–5623.

[11] J. H. Feng, P. L. Xu, Z. Q. Zhu, *et al.*, "Comparative study of dual 3-phase permanent magnet machines with coil span of two slot-pitches," *IET Electr. Power Appl.*, vol. 16, no. 12, pp. 1426–1438, 2021.

[12] S. Zhu, T. Cox, Z. Xu, and C. Gerada, "Novel 24-slots14-poles fractional-slot concentrated winding topology with low-space harmonics for electrical machine," *J. Eng.*, vol. 2019, no. 17, pp. 3784–3788, 2019.

[13] G. Dajaku and D. Gerling, "A novel 24-slots/10-poles winding topology for electric machines," *Proc. Int. Electr. Mach. Drives Conf. (IEMDC)*, Niagara Falls, ON, Canada, pp. 65–70, 2011.

[14] P. B. Reddy, K. K. Huh, and A. M. EL-Refaie, "Generalized approach of stator shifting in interior permanent-magnet machines equipped with fractional-slot concentrated windings," *IEEE Trans. Ind. Electron.*, vol. 61, no. 9, pp. 5035–5046, 2014.

[15] X. Chen, J. Wang, and V. I. Patel, "A generic approach to reduction of magnetomotive force harmonics in permanent-magnet machines with concentrated multiple three-phase windings," *IEEE Trans. Magn.*, vol. 50, no. 11, pp. 1–4, 2014.

Chapter 3

Dual-three-phase PMSMs with integer slot overlapping windings

3.1 Introduction

In Chapter 2, various winding configurations of dual three-phase (DTP) permanent magnet synchronous machines (PMSMs) with different slot/pole number combinations were discussed. The electromagnetic performances of some specific DTP PMSMs with overlapping and non-overlapping windings will be described in Chapter 3 and Chapter 4, respectively, for illustrative purposes. In this chapter, DTP PMSMs with integer slot overlapping windings will be described and analyzed based on the topology of the 48-slot/8-pole Toyota Prius 2010 STP PMSM. Two DTP winding configurations, i.e., single-layer full-pitched (DTP-SF) and double-layer short-pitched (DTP-DS) windings [1,2], respectively, are compared with the original STP winding configuration.

The structure of this chapter is shown in Figure 3.1. The details of different winding configurations are given in Section 3.2. In Section 3.3, the electromagnetic performances of these winding configurations are compared, including output torque, torque ripple, torque-speed characteristic, efficiency, etc. The experimental verification is given in Section 3.4, followed by a summary in Section 3.5.

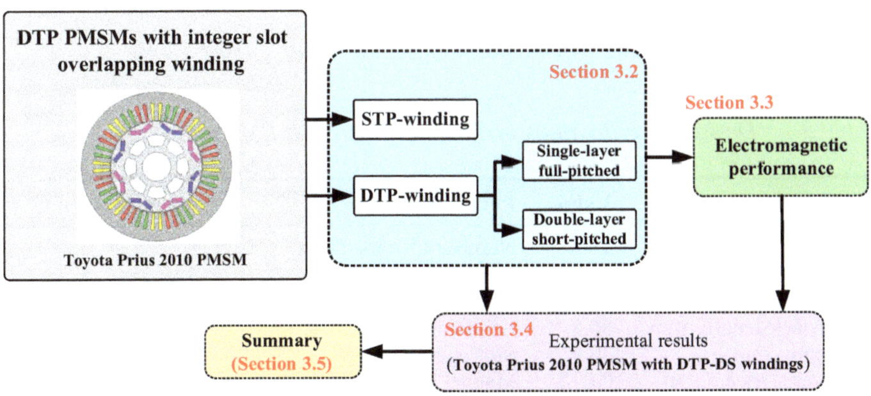

Figure 3.1 Structure of Chapter 3.

3.2 Winding configurations

The Toyota Prius 2010 PMSM has been widely investigated by researchers in recent years due to its excellent performance [3,4]. The cross-section of the machine, featuring a 48-slot stator and an 8-pole V-type IPM rotor, is presented in Figure 3.2. The detailed design specifications of the Prius 2010 PMSM can be found in [3], and the main parameters are shown in Table 3.1.

3.2.1 Winding arrangements

The stator windings in the Toyota Prius 2010 PMSM are 5-slot-pitched single-layer three-phase windings, as shown in Figure 3.3(a). In 48-slot/8-pole PMSMs, the number of slots per pole is 6. Hence, the original 5-slot-pitched windings in the Toyota Prius 2010 PMSM are short-pitched windings, which can reduce

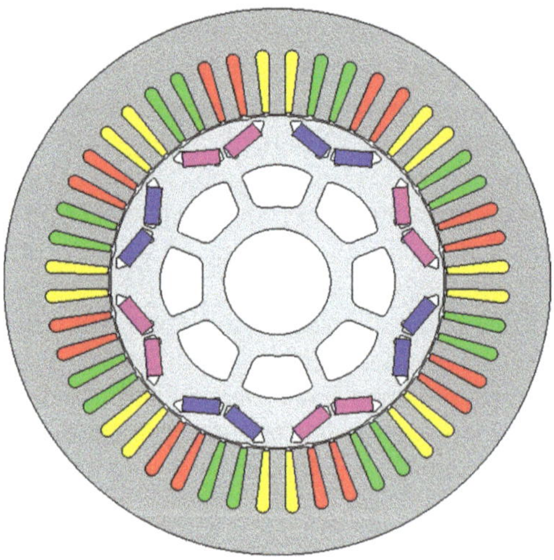

Figure 3.2 Cross-section of Toyota Prius 2010 PMSM.

Table 3.1 Design specifications of Toyota Prius 2010 PMSM

Parameter	**Value**	**Parameter**	**Value**
Number of stator slots	48	Number of rotor poles	8
Stator OD, mm	264	Rotor OD, mm	160.44
Stator ID, mm	161.9	Rotor ID, mm	51
Stator stack length, mm	50.8	Lamination thickness, mm	0.305
Slot depth, mm	30.9	Slot opening, mm	1.88
Phase resistance, Ohm	0.077	Magnet dimensions, mm	49.3 × 17.88 × 7.16
Magnet remanence, T	1.2	Magnet relative permeability	1.05

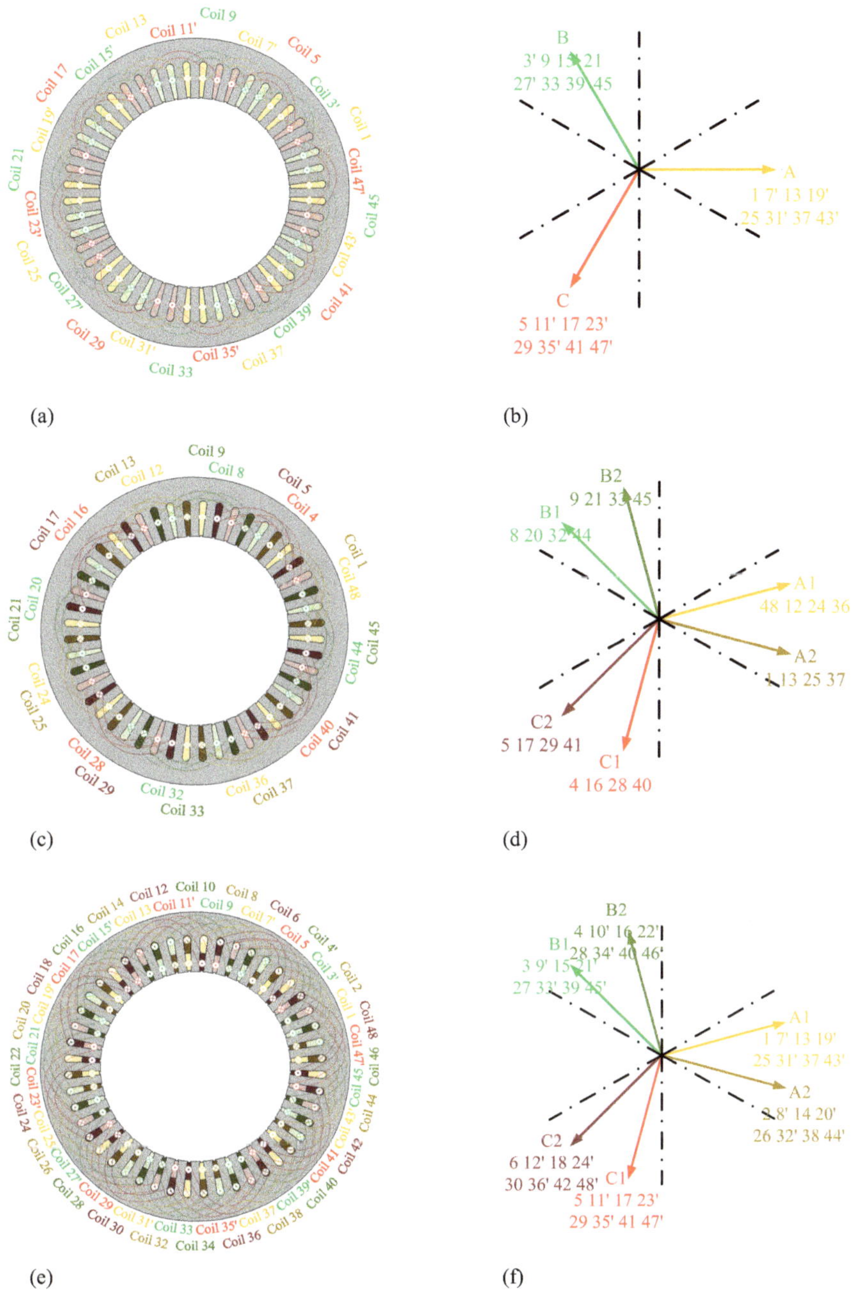

Figure 3.3 Winding layouts and coil EMF phasor diagrams of STP, DTP-SF, DTP-DS PMSMs. (a) STP-Winding layout, (b) STP-Coil EMF phasor diagram, (c) DTP-SF-Winding layout, (d) DTP-SF-Coil EMF phasor diagram, (e) DTP-DS-Winding layout, and (f) DTP-DS-Coil EMF phasor diagram.

electromotive force (EMF) and magnetomotive force (MMF) harmonics. The diagram of coil EMF phasors is shown in Figure 3.3(b). There is no phase shift between the coil EMF phasors of the same phase. In other words, the original three-phase windings cannot be directly transferred into two three-phase windings with a 30° phase shift.

As is well-known, the DTP winding configuration with a 30° phase shift can improve output torque and reduce torque harmonics at the same time in PMSMs [5,6]. To realize the advantages of 30° DTP configuration in 48-slot/8-pole PMSMs, two DTP winding configurations have been proposed in the literature. Different from the original single-layer short-pitched windings, the stator windings utilized in the two DTP winding configurations are SF windings and DS windings, respectively. In this section, the original STP and the two DTP winding configurations are designated as the STP, DTP-SF, and DTP-DS, respectively. The winding connections and coil EMF phasor diagrams of the STP, DTP-SF, and DTP-DS machines are shown in Figure 3.3.

First, in the STP machine, as shown in Figure 3.3(a), the coils of Phase A are Coils 1, 7', 13, 19', 25, 31', 37, and 43'. The interval between adjacent coils is six slot pitches in space, which is 180° in electrical degrees, and hence, compared with Coils 1, 13, 25, and 37, Coils 7, 19, 31, and 43 are opposite connected. Thus, all the coils of Phase A in the STP machine share the same back EMF phases, as shown in Figure 3.3(b).

Second, the detailed winding layouts of the DTP-SF machine are shown in Figure 3.3(c), it can be observed that the coil pitch number is 6 and the two three-phase winding sets are displaced by one slot in space, which is 30° in electric. Thus, the back EMFs induced in Coils 48, 12, 24, and 36 are advanced to those in Coils 1, 13, 25, and 37 by 30° in electric, as illustrated in Figure 3.3(d).

Third, in the DTP-DS machine, the double-layer windings can be seen as duplicated from the original windings in the STP machine, and the new winding set is shifted by one slot pitch, as shown in Figure 3.3(e) and (f). All the coils of Phase A in the STP machine, Coils 1, 7, 13, 19, 25, 31, 37, and 43, belong to Phase A1 in the DTP-DS machine, and the new Coils 2, 8, 14, 20, 26, 32, 38, and 44 belong to Phase A2. Thus, Phase A2 lags behind Phase A1 by one slot pitch, which is 30° in electric.

3.2.2 *Winding specifications*

The total output power of DTP PMSMs is split into two parts and provided by two separate inverters. When the inverters for STP and DTP PMSMs share the same DC-link voltage, the phase currents in DTP PMSMs can be halved from those in STP PMSMs for the equivalent output power.

In the Toyota Prius 2010 STP machine, the maximum phase current is set as 236 Apk, and the rated DC-link voltage is 650 V. Hence, for the DTP-SF and DTP-DS machines, the maximum phase current is 118 Apk and the dc-link voltage is 650 V. The lower current requirement suggests that cheaper power devices can be utilized in the inverters for DTP PMSMs. Thus, although the number of switches is doubled in the DTP drive system compared with the STP counterpart, the total cost is still lower than twice the original cost.

The detailed specifications of STP, DTP-SF, and DTP-DS winding configurations are summarized in Table 3.2. Since the current requirement is halved in the DTP-SF and DTP-DS windings compared with the STP counterpart, the number of wires per turn is changed from 12 (STP windings) to six (DTP-SF and DTP-DS windings). It should also be noticed that due to different winding layouts, as shown in Figure 3.3, the number of coils per phase is eight in STP and DTP-DS windings, but only four in DTP-SF windings. To keep the resultant ampere-turns the same as STP and DTP-DS counterparts, the number of turns per coil in DTP-SF windings is doubled from 11 to 22. Thus, the number of turns per phase keeps 88 in STP, DTP-SF, and DTP-DS windings. Another difference between the STP, DTP-SF, and DTP-DS winding configurations is the coil pitch number, which is five in STP and DTP-DS machines and six in DTP-SF machines.

Based on the specifications in Table 3.2, the axial length and the phase resistance of STP, DTP-SF, and DTP-DS winding configurations are summarized in Table 3.3.

As shown in Table 3.3, compared with DTP-DS windings, the phase resistance in DTP-SF machines is increased by 13.2%, and the total axial length is also increased by 11.1%. This suggests that the DTP-DS winding configuration is better than the DTP-SF counterpart in efficiency and machine size because the higher phase resistance implies higher copper loss under the same phase currents and the longer axial length indicates larger machine size. The total volumes of copper wires used in STP and DTP-DS machines are identical. However, the DTP-SF machine needs more copper wires due to the longer coil pitch, i.e., 13.2% higher than the counterparts. In other words, the cost of winding connections in the DTP-SF

Table 3.2 Specifications of STP, DTP-SF, and DTP-DS winding configurations

Parameter	STP	DTP-SF	DTP-DS
Number of phases	3	6	6
Wires in parallel per turn	12	6	6
Turns in series per coil	11	22	11
Coils in series per phase	8	4	8
Coil pitch number	5	6	5

Table 3.3 Estimated axial lengths and phase resistances of STP, DTP-SF, and DTP-DS winding configurations [7,8]

Parameter	STP	DTP-SF	DTP-DS
Total axial length (mm)	113.90	126.51	113.90
Phase resistance (Ohm)	0.077	0.174363	0.154

machine is higher than the other two counterparts. It can be concluded that compared with the DTP-SF counterpart, the DTP-DS winding configuration has inherent advantages in winding resistance and machine size due to shorter pitched windings.

3.3 Electromagnetic performance

To reveal the merits and demerits of different winding configurations, the electromagnetic performance of the Toyota Prius 2010 PMSMs with STP, DTP-SF, and DTP-DS windings are analyzed using the 2D finite element (FE) method via JMAG Designer [9]. Due to its 48-slot/8-pole number combination, the electrical periodicity within the machine is 4. Hence, the flux distributions are identical every quarter machine and the electromagnetic performance of different winding configurations can be obtained from 1/4 models, as shown in Figure 3.4.

3.3.1 Winding factor

For all winding configurations, the winding factor is the product of the distribution factor and pitch factor. Distribution factors can be obtained from the coil EMF phasor diagrams shown in Figure 3.3, and pitch factors can be determined by the slot/pole number combination and coil pitch number.

Specifically, for STP, DTP-SF, and DTP-DS windings, as illustrated in Figure 3.3(b), (d), and (f), respectively, the EMF phasors of the same phase overlap

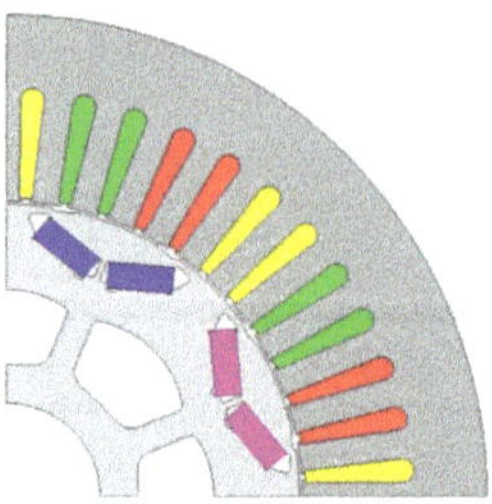

(a)

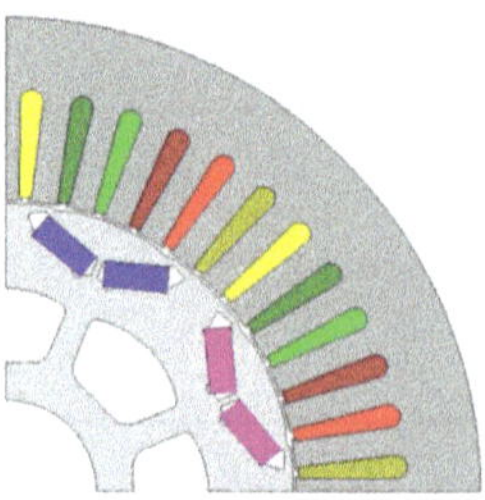

(b)

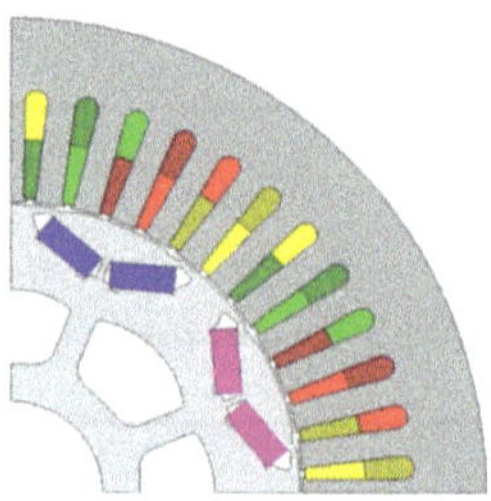

(c)

Figure 3.4 FE models of Prius 2010 PMSMs with different winding configurations. (a) STP, (b) DTP-SF, and (c) DTP-DS.

each other. Hence, the distribution factors of the three winding configurations are all 1. On the other hand, the pitch factor can be obtained from the ratio of coil-pitch to pole-pitch. Thus, for the full-pitched windings (DTP-SF), the pitch factor is 1, and for the short-pitched windings (STP and DTP-DS), the pitch factors are 0.966. The winding factors of STP, DTP-SF, and DTP-DS winding configurations are listed in Table 3.4.

It can be found that the resultant winding factors for STP, DTP-SF, and DTP-DS winding configurations are 0.966, 1, and 0.966, respectively. Compared with its STP counterpart, the DTP-SF winding configuration can improve the winding factor by about 3.53%, but the DTP-DS winding configuration remains the same.

However, it should be noted that Table 3.4 only shows the winding factors of the fundamental components in one winding set across different winding configurations. However, there are two winding sets in DTP-SF and DTP-DS machines, and the resultant stator MMF is produced by both sets. Therefore, it is still necessary to evaluate the winding factors of higher-order harmonics and mutual effects between the two winding sets. The detailed calculation methods for the winding factors of higher-order harmonics in one winding set and the resultant winding factors after two winding sets are fed with time-shifted phase currents have been discussed in Section 2.3 in Chapter 2.

The effects of the DTP winding configurations on the resultant stator MMF harmonics can also be obtained from the ideal stator MMF distributions. The waveforms and spectra of the ideal stator MMF distributions in STP, DTP-SF, and DTP-DS machines are shown in Figure 3.5. It is evident that the DTP-SF winding

Table 3.4 Winding factors of STP, DTP-SF, and DTP-DS winding configurations

	STP	DTP-SF	DTP-DS
Distribution factor, k_d	1	1	1
Pitch factor, k_p	0.966	1	0.966
Winding factor, k_w	0.966	1	0.966

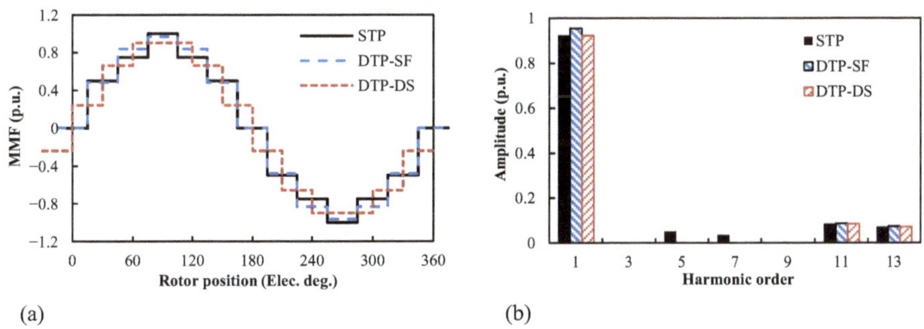

Figure 3.5 Ideal stator MMF distributions for STP, DTP-SF, and DTP-DS winding configurations. (a) Waveforms and (b) spectra.

configuration still has a higher fundamental component compared to the STP and DTP-DS counterparts. Furthermore, the 5th and 7th harmonics in both DTP-SF and DTP-DS winding configurations are eliminated by the 30°-shifted two winding sets. This implies that the 6th torque ripple can also be entirely eliminated in DTP-SF and DTP-DS machines [5,6,10].

3.3.2 Open-circuit performance

3.3.2.1 Flux distribution

Under open-circuit (OC) conditions, the flux in the PMSM is only produced by the rotor magnets, and thus, different winding configurations have no influence on the flux distribution performance under OC conditions. The OC flux density contour and equal potential distribution of the Prius 2010 PMSM are plotted in Figure 3.6 (a) and (b), respectively. The waveform and spectrum of the radial air-gap flux density are shown in Figure 3.7(a) and (b), respectively.

Although different winding configurations exhibit the same OC flux density distributions, their flux linkages are different due to variations in winding

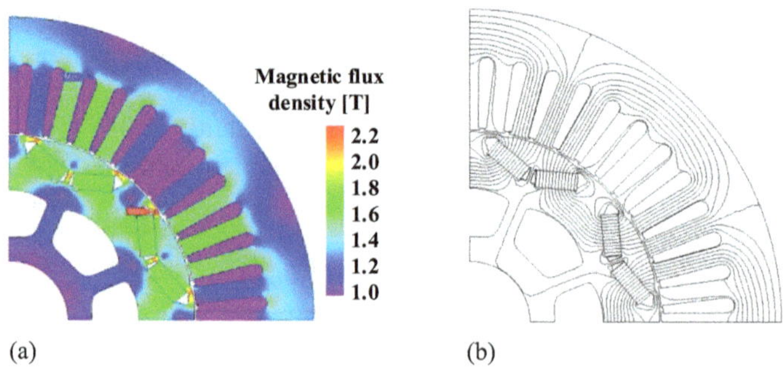

Figure 3.6 OC flux density and equal potential distributions of Prius 2010 PMSM. (a) Flux density distribution and (b) equal potential distribution.

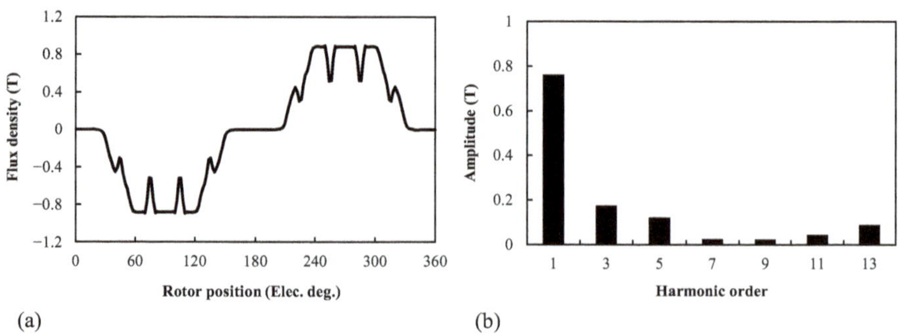

Figure 3.7 OC radial air-gap flux density distribution of Prius 2010 PMSM. (a) Waveform and (b) spectrum.

connections. The waveforms and spectra of OC phase flux linkages in STP, DTP-SF, and DTP-DS windings are shown in Figure 3.8. The phase shifting angle (30°) between the two winding sets can clearly be seen in Figure 3.8(a).

Notably, the flux linkage curve of Phase A in STP windings is completely overlapped with that of Phase A1 in DTP-DS windings because the first three-phase winding set in DTP-DS windings has identical layouts with that in the STP windings, as presented in Figure 3.3. Thus, STP and DTP-DS winding configurations have identical phase flux linkage harmonic contents. However, the DTP-SF winding configuration has slightly higher fundamental components than the STP and DTP-DS counterparts.

3.3.2.2 Back EMF

Under OC conditions, the phase-back EMFs in STP, DTP-SF, and DTP-DS machines at 3,000 r/min are compared in Figure 3.9. It can be seen that STP and DTP-DS winding configurations have identical harmonic contents in phase-back EMFs, and DTP-SF winding configuration has higher magnitudes in the 1st, 3rd, 5th, and 7th harmonics than STP and DTP-DS counterparts.

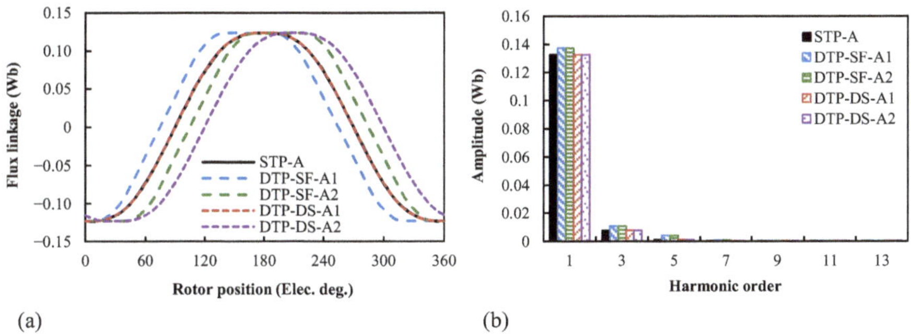

Figure 3.8 OC phase flux linkages of different winding configurations. (a) Waveforms and (b) spectra.

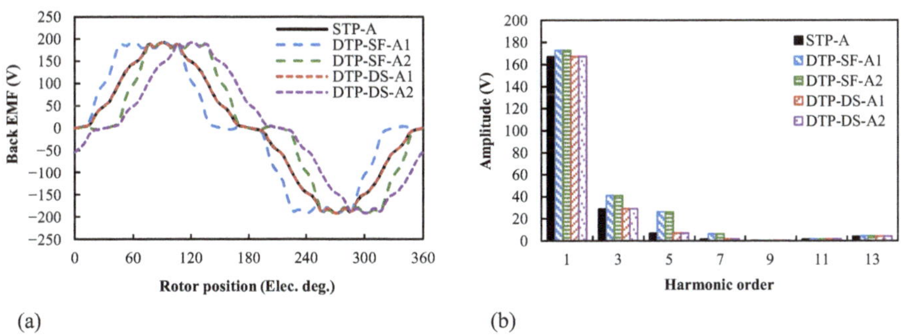

Figure 3.9 Phase-back EMFs of different winding configurations at 3,000 r/min. (a) Waveforms and (b) spectra.

Based on Figure 3.9, the line-to-line back EMFs of STP, DTP-SF, and DTP-DS winding configurations are compared in Figure 3.10. The harmonic contents in line-to-line back EMFs are almost the same as those in phase-back EMFs except that all the triple harmonics (the 3rd and 9th harmonics, etc.) are eliminated due to symmetrical three-phase winding layouts.

3.3.2.3 Cogging torque

As cogging torque is not influenced by armature windings, different winding configurations do not affect the cogging torque characteristic. The waveform and spectrum of the cogging torque in the Toyota Prius 2010 PMSM are shown in Figure 3.11. Due to the 48-slot/8-pole number combination, the fundamental order of cogging torque in the benchmark machine is 48 in mechanical angle (the smallest common multiple between slot number and pole number) and 12 in electrical angle [11]. It can be seen from Figure 3.11 that the dominant harmonic in cogging torque is the 12th harmonic.

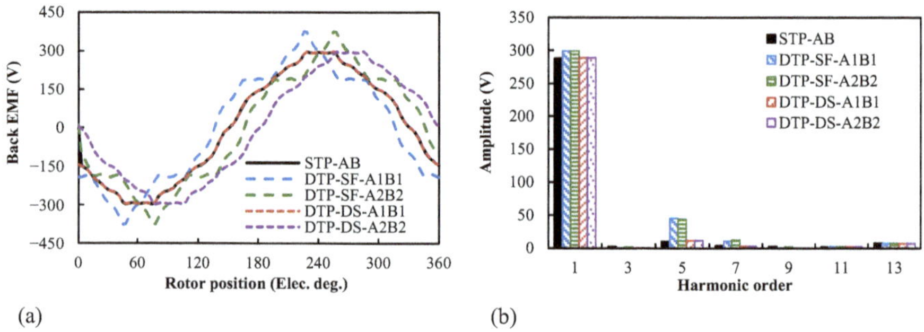

Figure 3.10 Line-to-line back EMFs of different winding configurations at 3,000 r/min. (a) Waveforms and (b) spectra.

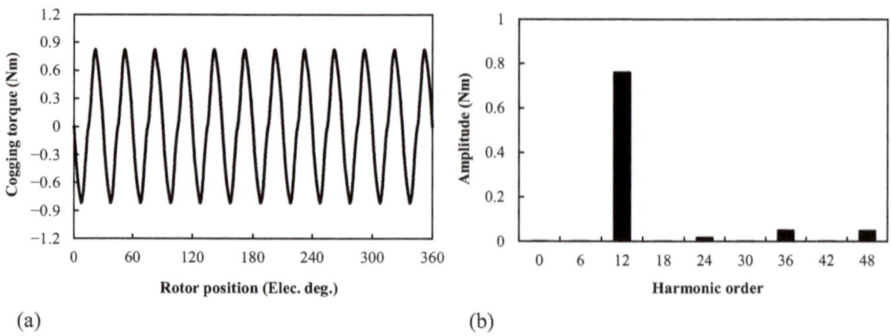

Figure 3.11 Cogging torque of Toyota Prius 2010 PMSM. (a) Waveform and (b) spectrum.

3.3.3 Torque performance

3.3.3.1 Average torque-current advancing angle characteristics

Since an IPM rotor is employed in the Toyota Prius 2010 PMSM, the output torque varies with the current advancing angle. Hence, the optimal current advancing angle for the maximum output torque should be chosen specially for STP, DTP-SF, and DTP-DS winding configurations to make full use of the PM and reluctance torque components in these PMSMs.

To produce the largest torque, phase currents should be set as the maximum values, i.e., 236 Apk in STP windings and 118 Apk in DTP-SF and DTP-DS windings. Since there are two winding sets in DTP-SF and DTP-DS winding configurations, and the two winding sets are shifted by 30° in electric, in the following sections, the current advancing angles for DTP winding configurations only denote the average values of the actual current advancing angles in the two winding sets. Thus, the average torque-current advancing angle characteristics of STP, DTP-SF, and DTP-DS winding configurations can be compared in Figure 3.12, and the detailed values are listed in Table 3.5. Compared with the STP counterpart, both DTP-SF and DTP-DS winding configurations can improve average torque when using the same current advancing angle. As underpinned by the calculated winding factors, the DTP-SF winding configuration can produce the highest average torque among the three winding configurations. However, it can also be found that in the STP, DTP-SF, and DTP-DS PMSMs, the optimal current angles for the maximum average torque are all about 50°. In other words, DTP winding configurations do not affect the optimal current advancing angle in STP PMSMs.

It should also be noted that although the DTP-DS winding configuration has identical winding factors to its STP counterpart, the average torque produced by the STP PMSM can still be improved slightly by using DTP-DS windings, especially when the current advancing angle is close to 90°. This phenomenon is caused by the different flux distributions and different winding layouts in STP and DTP

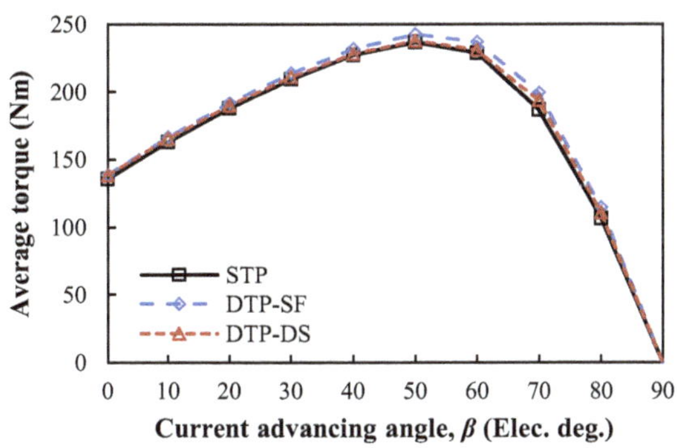

Figure 3.12 Average torque-current advancing angle characteristics of different winding configurations.

Table 3.5 Comparison of average torque-current advancing angle characteristics of different winding configurations (I_A = 236 Apk)

	STP	DTP-SF		DTP-DS	
Current advancing angle, β_a (°)	**Average torque (Nm)**	**Average torque (Nm)**	**Improvement from STP (%)**	**Average torque (Nm)**	**Improvement from STP (%)**
0	135.88	138.67	+2.06	138.48	+1.92
10	163.01	166.38	+2.06	165.43	+1.49
20	187.79	191.65	+2.06	189.87	+1.11
30	209.48	213.82	+2.07	211.06	+0.76
40	227.27	232.01	+2.08	228.14	+0.38
50	236.99	242.79	+2.45	237.77	+0.33
60	228.95	236.96	+3.50	231.18	+0.97
70	186.76	199.30	+6.72	193.35	+3.53
80	106.33	114.45	+7.64	110.31	+3.75
90	−0.01	−0.01	–	−0.05	–

PMSMs and can be explained as follows. The condition I_A = 236 A, β_a = 80° is taken as an example for illustration, where the average torque produced by the STP PMSM can be improved by 7.64% and 3.75%, respectively in DTP-SF and DTP-DS PMSMs. To further investigate how the STP and DTP winding configurations affect the resultant torque performance, the flux distributions of the STP, DTP-SF, and DTP-DS PMSMs under the condition I_A = 236 A, β_a = 80°, are shown in Figure 3.13. As can be seen, the flux density distributions of different winding configurations are similar to each other under this condition, and thus, the differences between different winding configurations cannot be distinguished.

Figure 3.14 shows the radial air-gap flux density waveforms and spectra of different winding configurations under the same condition.

Clearly, since STP, DTP-SF, and DTP-DS windings have different stator MMFs, the amplitudes of the fundamental components in DTP-SF and DTP-DS PMSMs are slightly higher than that in the STP counterpart. Meanwhile, the phases of the fundamental components in DTP PMSMs are shifted a little compared with that in the STP, which also contributes to the resultant average torque improvements.

3.3.3.2 Instantaneous torque characteristics under maximum torque condition

Based on Figure 3.12 and Table 3.5, the optimal current advancing angles are chosen as 50° for STP, DTP-SF, and DTP-DS PMSMs for the maximum torque conditions. The waveforms and spectra of the instantaneous torque in STP, DTP-SF, and DTP-DS PMSMs are shown in Figure 3.15. The average torques and torque ripples are listed in Table 3.6. Compared with the STP, not only can the average torque be improved by the DTP winding configurations, but also torque ripples can be reduced significantly.

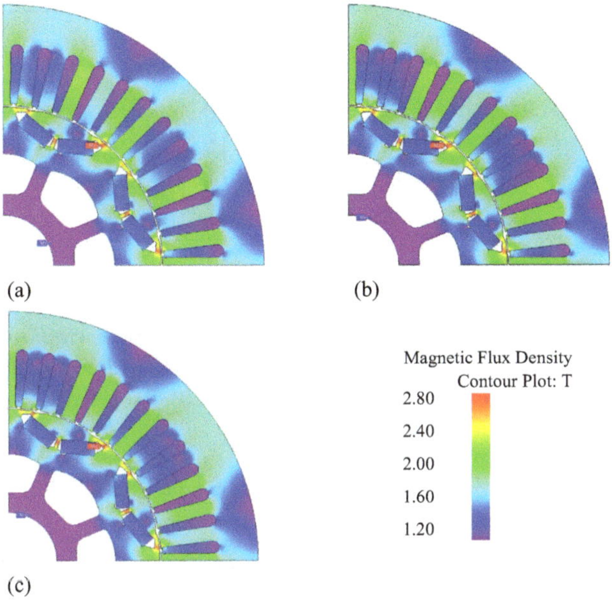

Figure 3.13 *Flux density distributions of different winding configurations at I_A = 236 A, β_a = 80°. (a) STP, (b) DTP-SF, and (c) DTP-DS.*

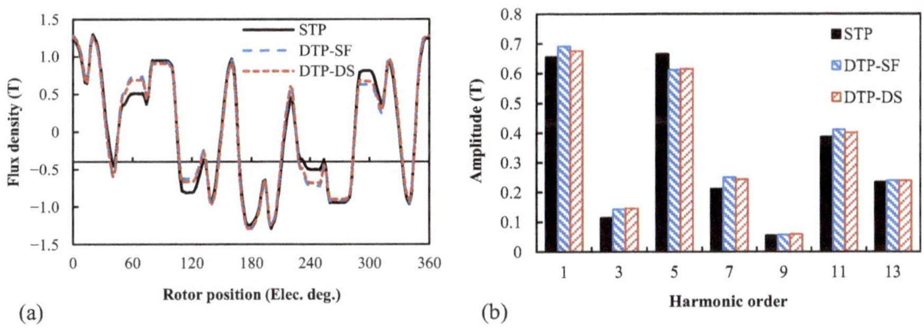

Figure 3.14 *Radial air-gap flux density distributions of different winding configurations at I_A = 236 A, β_a = 0°. (a) Waveforms and (b) spectra.*

From Figure 3.15(b), it can be seen that the dominant torque harmonics in STP PMSM are the 6th and 12th harmonics, while in DTP-SF and DTP-DS PMSMs, only the 12th harmonics are the dominant torque harmonics. In other words, with the utilization of DTP-SF and DTP-DS winding configurations, the 6th torque harmonics can be eliminated, and consequently, the on-load torque ripples are reduced significantly.

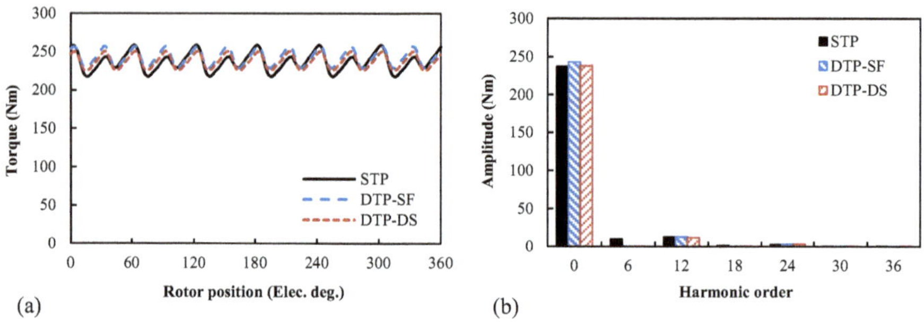

Figure 3.15 Instantaneous torque performance of different winding configurations under maximum torque condition. (a) Waveforms and (b) spectra.

Table 3.6 Comparison of instantaneous torque performance of different winding configurations under maximum torque condition

	STP	DTP-SF	DTP-DS
Average torque	236.98	242.80	237.75
Improvement with STP (%)	–	+2.45	+0.33
Torque ripple (Nm)	34.11	20.46	18.54
Torque ripple (%)	14.39	8.43	7.80

3.3.3.3 Average torque and torque ripple–current amplitude characteristics

Besides the maximum torque condition, the torque characteristics of STP, DTP-SF, and DTP-DS winding configurations under other excitation conditions are also investigated. It should be mentioned again that in the following sections, only the amplitudes and phases of the phase currents in STP PMSM are utilized to describe load conditions. The actual phase current amplitudes in the DTP-SF and DTP-DS PMSMs are only half of that in the STP counterpart, and the actual phase current phases in the two winding sets in the DTP-SF and DTP-DS PMSMs are shifted by 30° in electric. Assuming I_A is varied from 10, 50, 100, 150, 200 to 236 Apk, and β_a varied from 0°, 10°, 20°, 30°, 40°, 50°, 60°, 70°, 80°, and 90°, the torque characteristics of the STP, DTP-SF, and DTP-DS PMSMs can be obtained separately using the FE method. When the PMSMs are controlled using a maximum torque per ampere (MTPA) strategy, the variations of the average torque and torque ripple of different winding configurations with phase current amplitude are shown in Figure 3.16 and the detailed values are listed in Table 3.7.

According to Table 3.7, compared with the STP counterpart the variations of the average torque improvements in DTP-SF and DTP-DS PMSMs with phase current amplitude are shown in Figure 3.17. It can be concluded that both DTP-SF

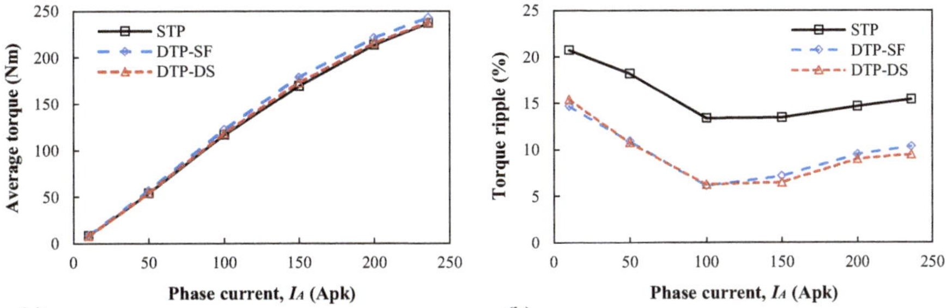

(a) (b)

Figure 3.16 Correlations of average torque/torque ripple with phase current amplitude in different winding configurations with MTPA control strategy. (a) Average torque and (b) torque ripple.

Table 3.7 Comparison of torque characteristics of different winding configurations with different phase current amplitudes

I_A (Apk)	Performance	STP	DTP-SF	DTP-DS
10	Average torque (Nm)	8.27	8.57	8.26
	Improvement with STP (%)	–	+3.73	−0.02
	Torque ripple (Nm)	1.71	1.26	1.27
	Torque ripple (%)	20.73	14.65	15.40
50	Average torque (Nm)	54.13	56.63	54.26
	Improvement with STP (%)	–	+4.63	+0.25
	Torque ripple (Nm)	9.85	6.16	5.82
	Torque ripple (%)	18.19	10.87	10.73
100	Average torque (Nm)	116.79	122.52	118.05
	Improvement with STP (%)	–	+4.91	+1.08
	Torque ripple (Nm)	15.62	7.50	7.37
	Torque ripple (%)	13.38	6.12	6.24
150	Average torque (Nm)	169.51	178.91	172.92
	Improvement with STP (%)	–	+5.55	+2.01
	Torque ripple (Nm)	22.80	12.77	11.19
	Torque ripple (%)	13.45	7.14	6.47
200	Average torque (Nm)	213.46	221.18	215.68
	Improvement with STP (%)	–	+3.61	+1.04
	Torque ripple (Nm)	31.24	20.88	19.33
	Torquc ripple (%)	14.64	9.44	8.96
236	Average torque (Nm)	236.99	242.79	237.77
	Improvement with STP (%)	–	+2.45	+0.33
	Torque ripple (Nm)	36.52	24.97	22.49
	Torque ripple (%)	15.41	10.28	9.46

and DTP-DS winding configurations can improve the average torque and/or reduce torque ripples compared to STP PMSM.

Specifically, both the DTP-SF and DTP-DS winding configurations can always reduce on-load torque ripples, but only the DTP-SF winding configuration can

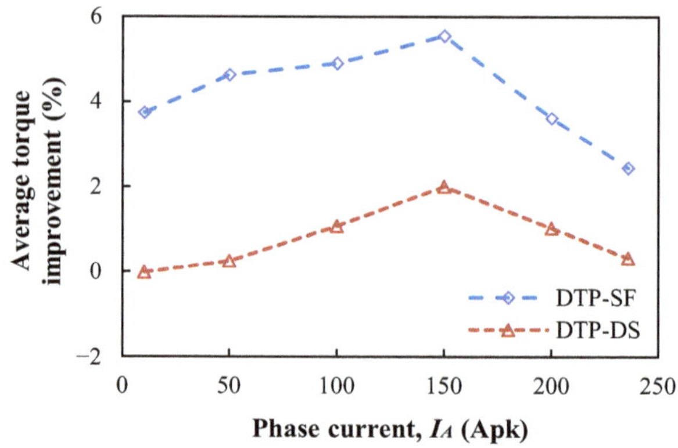

Figure 3.17 Average torque improvement-phase current amplitude characteristics of DTP-SF and DTP-DS winding configurations when using MTPA strategy.

always improve average torque, no matter what the excitation is. For the DTP-DS winding configuration, it can only improve average torque a little when using suitable phase currents. For instance, when I_A = 150 Apk, the torque improvements of DTP-SF and DTP-DS PMSMs are 5.59% and 2.02%, respectively, which are much higher than the theoretical increases in winding factors (3.53% in the DTP-SF winding configuration and 0% in the DTP-DS winding configuration). It indicates that the real torque improvements by the DTP winding configurations can be much higher than theoretical increases in winding factors under specific on-load conditions.

3.3.3.4 Torque and power speed characteristics

In the analyses above, the speeds of the STP, DTP-SF, and DTP-DS PMSMs are fixed at 3,000 r/min. In this subsection, the torque and power-speed envelopes of the STP, DTP-SF, and DTP-DS winding configurations are presented. Furthermore, since the PMSMs are constrained by DC voltage under high-speed conditions and back EMFs on load can be calculated by using *d*- and *q*-axis inductances in PMSMs, the *d*- and *q*-axis inductances of different winding configurations should be calculated first. With the method proposed in [12], the self- and mutual-saturation effects are considered to improve the accuracy of the calculations. Assuming that I_d varies from −240 A to 0 A and I_q varies from 0 A to 240 A, the variations of *d*- and *q*-axis inductances with different winding configurations and different *d*- and *q*-axis current combinations can be obtained, as shown in Figure 3.18. It can be observed that both *d*- and *q*-axis currents can affect *d*- and *q*-axis inductances.

In the calculation, the DC voltage is fixed at 650 V, and the phase currents are limited at 236 Apk in STP windings and 118 Apk in DTP windings. The torque and output power-speed envelopes of STP, DTP-SF, and DTP-DS winding configurations are shown in Figure 3.19. In terms of input powers of different PMSMs, the

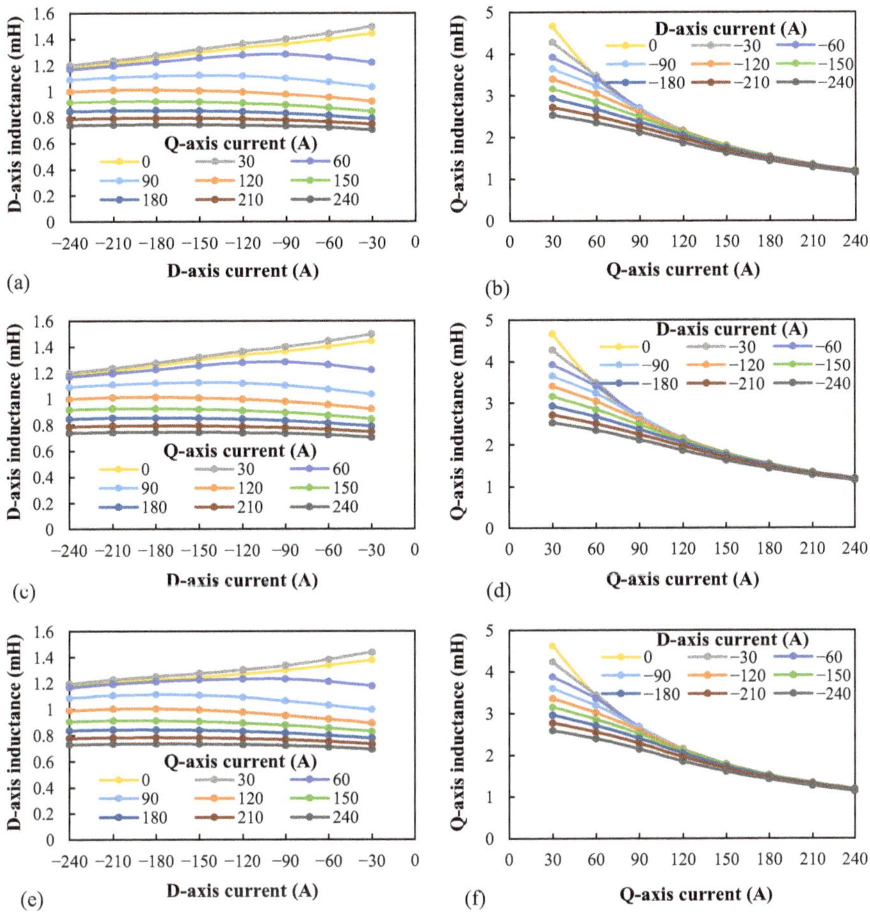

Figure 3.18 Variations of d- and q-axis inductances with different winding configurations and different d- and q-axis current combinations. (a) STP-D-axis, (b) STP-Q-axis, (c) DTP-SF-D-axis, (d) DTP-SF-Q-axis, (e) DTP-DS-D-axis, and (f) DTP-DS-Q-axis.

machine losses should be taken into consideration, as will be presented subsection 3.3.4. In Figure 3.19, the STP, DTP-SF, and DTP-DS PMSMs are operated at two ranges, i.e., constant torque range and constant power range. When the speed is lower than the base speed, the PMSM is operated to produce the highest average torque according to the limits of phase currents, but when the speed is higher than the base speed, the highest average torque is limited by DC voltage. The base speeds of STP, DTP-SF, and DTP-DS PMSMs are 3,500, 3,400, and 3,600 r/min, respectively. Thus, the DTP-DS winding configuration can only slightly increase the base speed.

From Figure 3.19, the DTP-SF winding configuration has the highest average torque in the constant torque range. However, the DTP-DS winding configuration

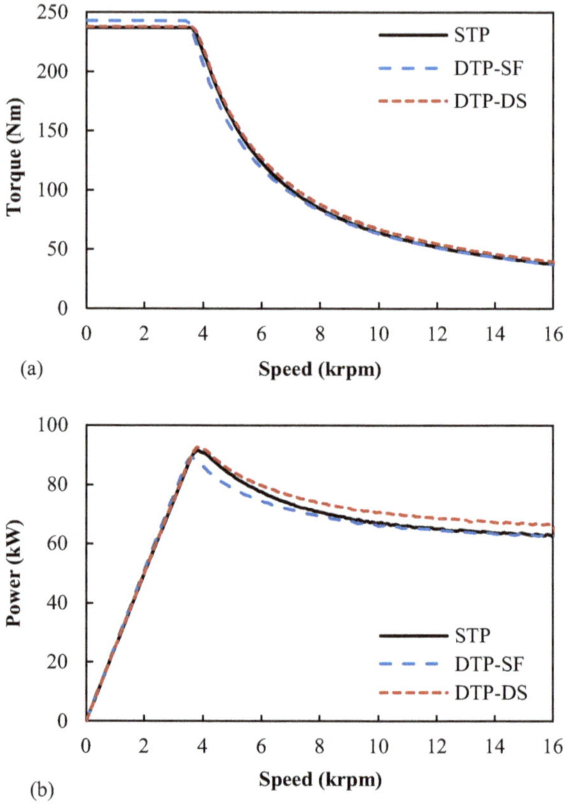

Figure 3.19 Torque- and output power-speed envelopes of different winding configurations. (a) Torque-speed envelope and (b) output power-speed envelope.

has the highest output power in the constant power range. Under the same speed, the higher output power of the DTP-DS PMSM suggests that the DTP-DS PMSM can produce the highest average torque among the three PMSMs.

Actually, to obtain the torque- and power-speed envelopes shown in Figure 3.19, the maximum d- and q-axis currents of STP, DTP-SF, and DTP-DS winding configurations at different speeds are calculated and shown in Figure 3.20.

According to the d- and q-axis currents shown in Figure 3.20, the amplitudes and current advancing angles of the phase currents in the STP, DTP-SF, and DTP-DS winding configurations over the entire speed range are shown in Figure 3.21(a) and (b), respectively. In Figure 3.21(a), the phase current amplitudes for the DTP-SF and DTP-DS windings are doubled from the actual values for the convenience of the comparison to the STP.

From Figure 3.21(a) and (b), it can be found that the amplitudes of phase currents in the DTP-DS windings can be higher than those in the DTP-SF and STP

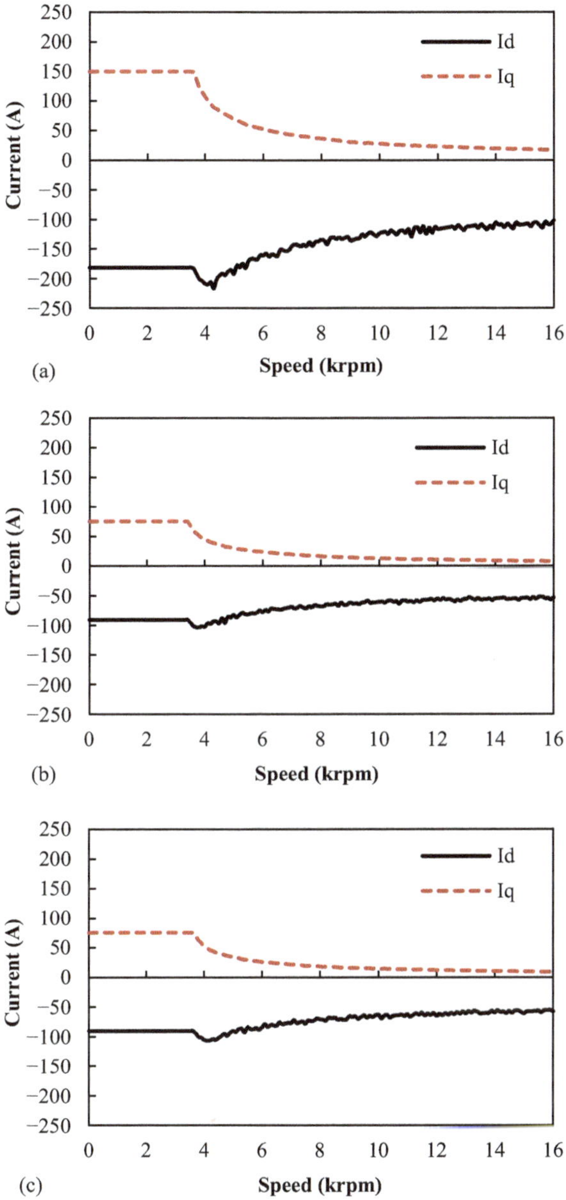

Figure 3.20 d- and q-Axis currents of different winding configurations over the entire speed range. (a) STP, (b) DTP-SF, and (c) DTP-DS.

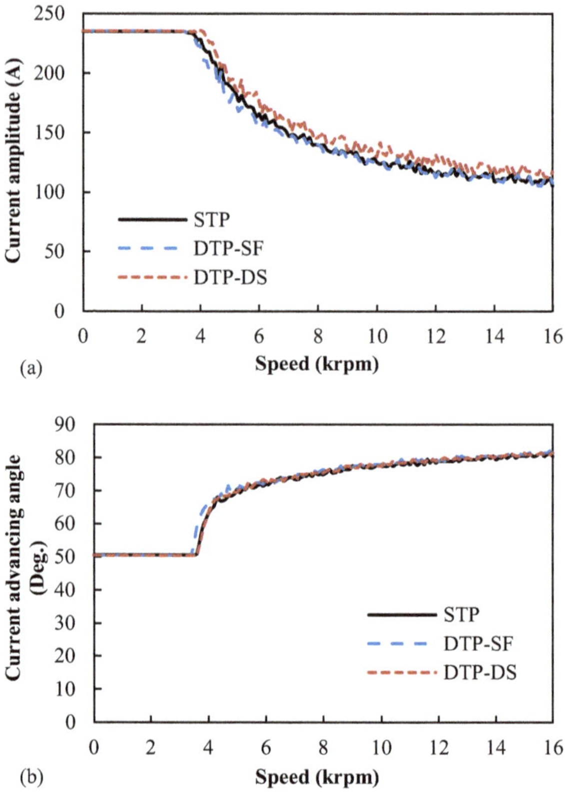

Figure 3.21 Amplitudes and advancing angles of phase currents in different winding configurations over the entire speed range. (a) Current amplitude and (b) current advancing angle.

windings under high-speed conditions, while the current angles of different winding configurations are very similar to each other. Considering that under high-speed conditions, the phase back EMFs are limited by dc voltage, the higher phase currents in the DTP-DS windings result in larger output torque in the DTP-DS PMSM in constant power range. Overall, it can be concluded that the DTP-SF winding configuration produces the highest average torque in the constant torque range and the DTP-DS winding configuration can produce the highest average torque in the constant power range, although both are not very significant.

3.3.4 Loss and efficiency

3.3.4.1 Loss and efficiency under maximum torque condition

In this sub-section, the losses of the STP, DTP-SF, and DTP-DS winding configurations under the maximum torque condition are calculated by using the method in [13], including copper loss, iron loss, PM eddy loss, and mechanical loss.

Copper loss is calculated by using Joule's law, as

$$P_{Cu} = 3R_0 {I_a}^2 \tag{3.1}$$

where R_0 is the phase resistance, and I_a is the phase current amplitude.

For a specific speed, iron loss and PM eddy loss are calculated by the FE software, JMAG, according to the properties of materials. In this example, the lamination material is "35H270", and the PM material is "NdFeB_Br = 1.2T". However, it should be mentioned that for other speeds, iron losses should be estimated based on the hysteresis and eddy current iron losses at the base speed, as

$$P_{Fe} = P_{Hyst-base}\frac{f}{f_{base}} + P_{Eddy-base}\left(\frac{f}{f_{base}}\right)^2 \tag{3.2}$$

where $P_{Hyst\text{-}base}$ and $P_{Eddy\text{-}base}$ are the hysteresis and eddy current iron losses at the base speed, respectively. f and f_{base} are the frequencies of the operating condition and the base speed, respectively.

Although the STP, DTP-SF, and DTP-DS PMSMs have different winding configurations, their mechanical structures are identical. Hence, all three PMSMs share the same mechanical loss characteristics, which can be estimated using the measured results in [14], as

$$P_{Me} = 0.26f + 0.00103f^2 \tag{3.3}$$

Overall, the loss and efficiency characteristics of STP, DTP-SF, and DTP-DS PMSMs under the maximum torque condition can be summarized and compared in Table 3.8. Copper loss is the most dominant loss among the different kinds of losses. Compared with the STP and DTP-DS windings, the larger phase resistance of the full-pitched windings in the DTP-SF winding results in larger copper loss and lower efficiency. From Table 3.8, it can also be seen

Table 3.8 Loss and efficiency of PMSMs with different winding configurations under maximum torque condition

	STP	**DTP-SF**	**DTP-DS**
Speed (r/min)	3000	3000	3000
Average torque (Nm)	236.99	242.77	237.78
Copper loss (W)	6,432.89	7,283.49	6,432.89
Iron loss (W)	373.12	373.12	370.24
PM eddy loss (W)	108.17	52.35	56.06
Mechanical loss (W)	93.20	93.20	93.20
Output power (W)	74,452.60	76,267.14	74,700.57
Total loss (W)	7,007.38	7,802.16	6,952.39
Input power (W)	81,459.98	84,069.30	81,652.96
Efficiency (%)	91.40	90.72	91.49

that compared with the STP, PM eddy losses can be reduced significantly by using DTP windings. This phenomenon can be explained by the reduced stator MMF harmonics produced by DTP windings. However, since PM eddy loss is much smaller than copper loss, the reduction in PM eddy loss in the DTP-DS winding configuration can only slightly increase its efficiency, but it may affect the rotor temperature significantly.

3.3.4.2 Loss and efficiency maps

Based on the above calculations, the losses and efficiencies of STP, DTP-SF, and DTP-DS winding configurations under the entire speed range can be obtained. The copper loss and iron loss maps of these winding configurations are calculated first, as shown in Figures 3.22 and 3.23, respectively.

As shown in Table 3.8, PM eddy current loss only accounts for a small part of total losses in all of the investigated PMSMs. Considering that it is difficult to obtain PM eddy current loss under different speeds, PM eddy current loss is neglected in the calculation of the efficiency maps. For mechanical loss, in the investigated PMSMs, the variation of mechanical loss with speed can be obtained from (3.3).

The efficiency maps of the STP, DTP-SF, and DTP-DS winding configurations are shown in Figure 3.24. The highest efficiency is in the STP and DTP-DS PMSMs and can reach 97%, while the highest efficiency in the DTP-SF counterpart

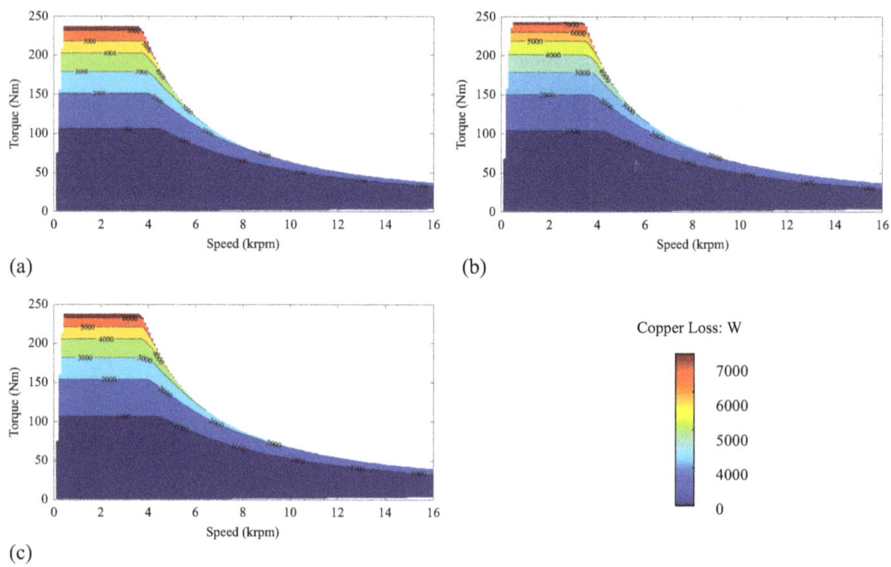

Figure 3.22 Copper loss maps of different winding configurations. (a) STP, (b) DTP-SF, and (c) DTP-DS.

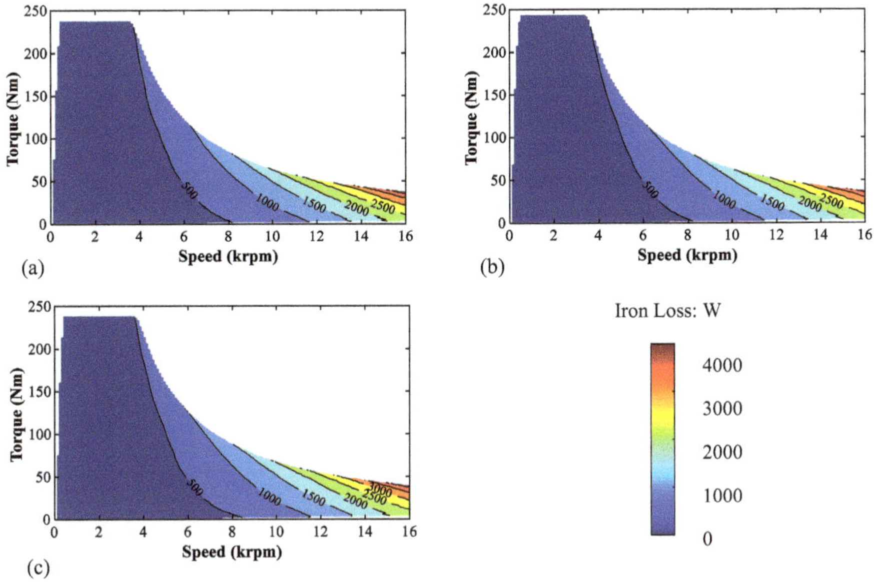

Figure 3.23 Iron loss maps of different winding configurations. (a) STP, (b) DTP-SF, and (c) DTP-DS.

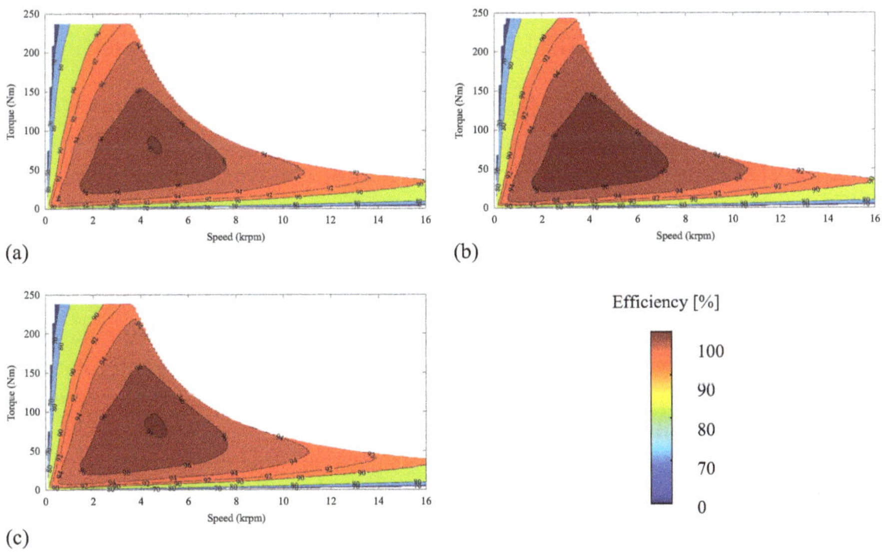

Figure 3.24 Efficiency maps of different winding configurations. (a) STP, (b) DTP-SF, and (c) DTP-DS.

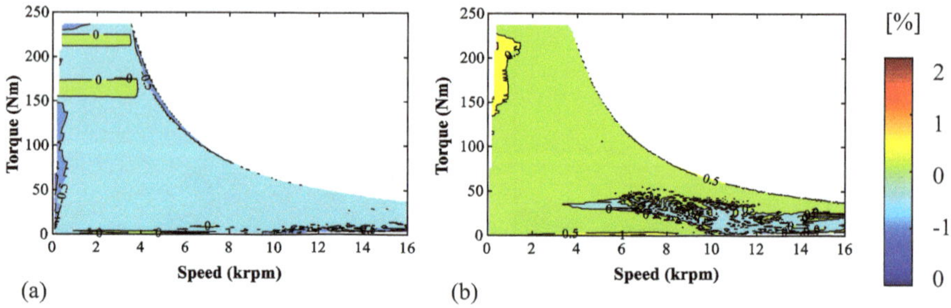

Figure 3.25 Differences between efficiency maps of DTP and STP winding configurations. (a) DTP-SF–STP and (b) DTP-DS–STP.

is 96%. Furthermore, the high-efficiency (97%) range in DTP-DS PMSM is slightly larger than that in the STP PMSM.

To better show the differences between the efficiency maps of DTP and STP PMSMs, the efficiency difference maps between the DTP and STP PMSMs are shown in Figure 3.25. For the DTP-SF PMSM, the efficiency difference is almost zero over the whole range. However, for the DTP-DS PMSM, the efficiency difference is positive in most of the whole range. It indicates that the overall efficiency of the DTP-SF PMSM is slightly lower than that of its STP counterpart, while a DTP-DS winding configuration can improve the efficiency of STP PMSM over the whole operating region. It should also be noticed that almost all of the efficiency differences shown in Figure 3.25 are lower than ±0.5%, which suggests that the influences of DTP-SF and DTP-DS winding configurations on the resultant efficiency are not obvious.

In summary, in this section, the electromagnetic performances of PMSMs with STP, DTP-SF, and DTP-DS winding configurations are compared comprehensively using the 2D FE method. The comparison includes flux density distribution, back EMF performance, torque and power characteristics, and loss and efficiency performance. It is found that both DTP-SF and DTP-DS winding configurations can improve average torque and reduce torque ripples in the researched STP PMSM. Specifically, the DTP-SF winding configuration is more suitable for constant torque range, as the torque improvement in the DTP-SF winding configuration is higher than that in the DTP-DS counterpart. The DTP-DS winding configuration is preferred for constant power range because it can produce higher output torque with lower torque pulsation under high-speed conditions.

3.4 Experimental results

To verify the above FE analyses, a Toyota Prius 2010 PMSM equipped with DTP-DS windings was manufactured and tested. The photos of the prototype and the test rig are shown in Figures 3.26 and 3.27, respectively.

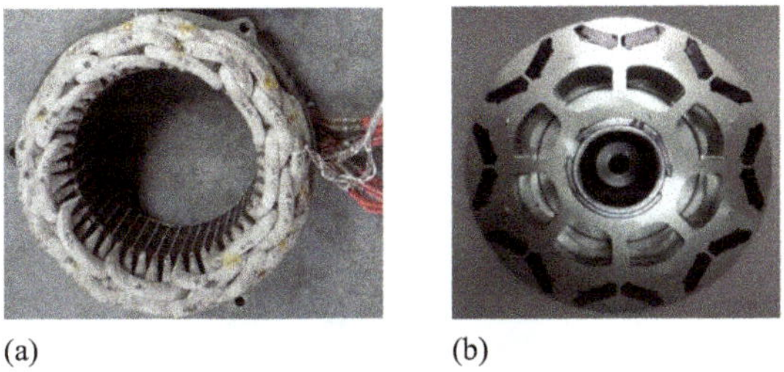

Figure 3.26 Photos of DTP-DS prototype. (a) Stator and (b) rotor.

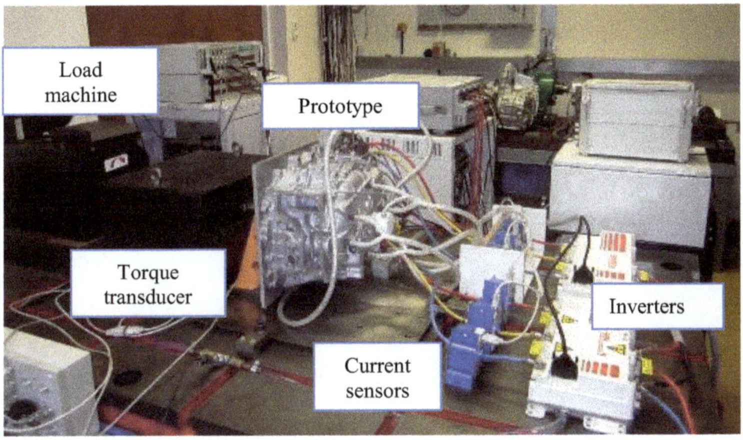

Figure 3.27 Photo of the test rig.

The 3D FE method was also utilized to improve the accuracy of the FE simulations. The 3D model of the researched DTP-DS machine is shown in Figure 3.28.

The back EMF performance of the prototype was tested at 1,000 r/min. The waveforms and spectra of FE-predicted and tested phase-back EMFs are shown in Figure 3.29. Although the measured back EMF is slightly lower than the FE-predicted values, it can still be found that the measured results agree well with the FE-predicted results, and the 3D FE results can agree with the experimental results even better. The agreements between simulated results and tested results shown in Figure 3.30 are good to confirm the correctness of the FE analyses.

The on-load experiment was carried out with 20 Apk phase currents first. The on-load torque is measured using a torque transducer with a load machine. Since the torque transducer is connected to the prototype machine and the load machine by shaft couplings, the measured torque is affected by shaft couplings and the load machine. Hence, the measurement of instantaneous torque is very difficult. What is more, the torque pulsations in the prototype machine can be reduced by the load

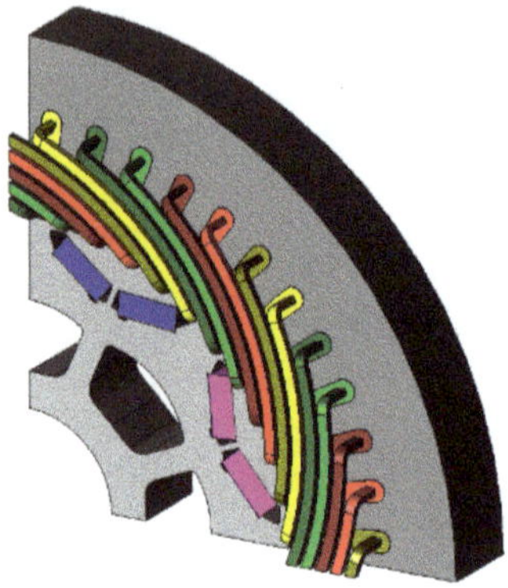

Figure 3.28 3D model of DTP-DS prototype.

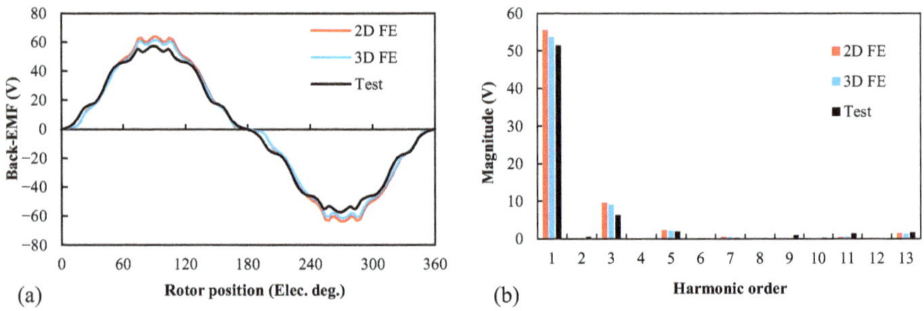

Figure 3.29 Comparison of FE predicted and measured phase-back EMFs at 1,000 r/min. (a) Waveforms and (b) spectra.

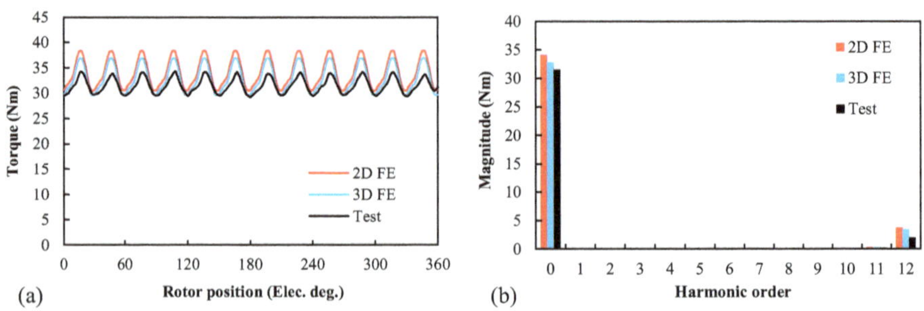

Figure 3.30 Comparison of FE predicted and measured instantaneous torque at 20 Apk phase currents. (a) Waveforms and (b) spectra (amplitude).

machine and the shaft couplings. The measured torque ripple is lower than the real torque ripple produced by the prototype machine. The waveforms and spectra of instantaneous torque under this condition are shown in Figure 3.30 and compared with the 2D and 3D FE analysis results, a good agreement can be found.

3.5 Summary

In this chapter, by way of example, DTP PMSMs with two different integer slot overlapping windings, i.e., DTP-SF and DTP-DS, based on the 48-slot/8-pole Toyota Prius 2010 STP PMSM have been described and compared with the original STP winding configuration. A prototype equipped with DTP-DS windings was fabricated and tested. The pros and cons of STP, DTP-SF, and DTP-DS PMSMs are summarized in Table 3.9.

Table 3.9 Pros and cons of different winding configurations under maximum torque conditions

	STP	DTP-SF	DTP-DS
Pros	• Simple winding connections • Simple drive system	• Highest torque under healthy condition • Lower torque pulsation under healthy conditions (compared with STP) • Ability to run under three-phase OC condition	• Slightly higher torque under healthy conditions (compared with STP) • Lower torque pulsation under healthy conditions (compared with STP) • Slightly higher efficiency (compared with STP) • Highest torque under high-speed healthy condition • Ability to run under three-phase OC condition • Higher average torque and lower torque pulsation under three-phase OC condition (compared with DTP-SF)
Cons	• Lower average torque and higher torque pulsation under healthy conditions • Cannot run under three-phase OC condition	• Complex winding connections • Complex drive system • Largest winding usage • Worst efficiency performance • Lowest torque under high-speed healthy condition • Worse torque performance under three-phase OC condition (Compared with DTP-DS)	• Complex winding connections • Complex drive system

Some findings are summarized as follows:

1) Only the DTP-SF winding configuration can improve the winding factor of the STP machine, while the DTP-DS windings have the same winding factors as the STP windings.
2) In the constant torque region, the DTP-SF windings can produce higher torque than the DTP-DS and STP windings. The output torque of the DTP-DS machine is only slightly higher than that of the STP machine in this region. Both DTP-SF and DTP-DS winding configurations can reduce torque ripples significantly, and the torque ripple in the DTP-DS PMSM is even lower.
3) In the constant power region, DTP-DS windings show the largest torque output capacity and the lowest torque pulsation. However, the output torque of the DTP-SF PMSM is even lower than that of the STP counterpart.
4) In DTP-SF windings, the full-pitched windings result in the highest copper loss among the three counterparts. Considering that the output torque of the DTP-DS machine is slightly higher than that of the STP machine, the DTP-DS PMSM has a better efficiency performance than its STP counterpart, and the DTP-SF PMSM shows the worst efficiency performance.
5) The stator and rotor cores and magnets of the STP, DTP-SF, and DTP-DS machines are identical. The differences in overall costs only lie in the drive system and winding topologies. Compared with the STP PMSM, the DTP-SF and DTP-DS PMSMs need another two half-rating inverters and the DTP-SF PMSM needs more copper wires due to the longer pitched windings.
6) Although the DTP-SF winding configuration can increase the winding factor and produce the highest torque in the constant torque region, considering the advantages of the DTP-DS winding configuration in average torque, torque ripple, and efficiency, especially under high-speed and one set of 3-phase OC conditions [7], the DTP-DS winding configuration is preferred for the DTP PMSMs utilized in EVs.

References

[1] V. I. Patel, J. Wang, W. Wang, and X. Chen, "Six-phase fractional-slot-per-pole-per-phase permanent-magnet machines with low space harmonics for electric vehicle application," *IEEE Trans. Ind. Appl.*, vol. 50, no. 4, pp. 2554–2563, 2014.

[2] R. Yang, N. Schofield, N. Zhao, and A. Emadi, "Dual three-phase permanent magnet synchronous machine investigation for battery electric vehicle power-trains," *J. Eng.*, vol. 2019, no. 17, pp. 3981–3985, 2019.

[3] T. A. Burress, S. L. Campbell, C. L. Coomer, *et al.*, "Evaluation of the 2010 Toyota Prius hybrid synergy drive system," *ORNL/TM-2010/253*, Oak Ridge National Laboratory, Oak Ridge, TN, USA, 2011.

[4] M. Taniguchi, T. Yashiro, K. Takizawa, *et al.*, "Development of new hybrid transaxle for compact-class vehicles," *SAE Int. Tech. Paper*, Art no. 2016-01-1163, 2016.

[5] Y. Zhao and T. A. Lipo, "Space vector PWM control of dual three-phase induction machine using vector space decomposition," *IEEE Trans. Ind. Appl.*, vol. 31, no. 5, pp. 1100–1109, 1995.

[6] M. Barcaro, N. Bianchi, and F. Magnussen, "Analysis and tests of a dual three-phase 12-slot 10-pole permanent-magnet motor," *IEEE Trans. Ind. Appl.*, vol. 46, no. 6, pp. 2355–2362, 2010.

[7] S.S. Wang, Z.Q. Zhu, A. Pride, J.T. Shi, R. Deodhar, and C. Umemura, "Comparison of different winding configurations for dual three-phase interior PM machines in electric vehicles," *World Electr. Veh. J.*, vol. 13, no. 1, pp. 1–51, 2022.

[8] S.S. Wang, "Investigation of torque and torque ripple in dual 3-phase permanent magnet machines for electric vehicles," *PhD thesis*, University of Sheffield, 2023.

[9] JMAG. (2019). *JSOL*. Accessed: June. 2019. [Online]. Available: https://www.jmag-international.com/.

[10] Z. Q. Zhu, S. S. Wang, B. Shao, L. C. Yan, P. L. Xu, and Y. Ren, "Advances in dual-three-phase permanent magnet synchronous machines and control techniques," *Energies*, vol. 14, no. 22, paper 7508, pp.1–46, 2021.

[11] Z. Q. Zhu and D. Howe, "Influence of design parameters on cogging torque in permanent magnet machines," *IEEE Trans. Energy Convers.*, vol. 15, no. 4, pp. 407–412, 2000.

[12] G. Qi, J. T. Chen, Z. Q. Zhu, D. Howe, L. B. Zhou, and C. L. Gu, "Influence of skew and cross-coupling on flux-weakening performance of permanent-magnet brushless AC machines," *IEEE Trans. Magn.*, vol. 45, no. 5, pp. 2110–2117, 2009.

[13] W. Q. Chu, Z. Q. Zhu, J. Zhang, X. Liu, D. A. Stone, and M. P. Foster, "Investigation on operational envelops and efficiency maps of electrically excited machines for electrical vehicle applications," *IEEE Trans. Magn.*, vol. 51, no. 4, pp. 1–10, 2015.

[14] J. S. Hsu, C. W. Ayers, C. L. Coomer, *et al.*, "Report on Toyota/Prius motor torque-capability, torque-property, no-load back EMF, and mechanical losses," Oak Ridge National Laboratory, U.S. Department of Energy, 2004.

Chapter 4

Dual three-phase PMSMs with fractional slot non-overlapping windings

4.1 Introduction

Fractional-slot permanent magnet synchronous machines (PMSMs) with non-overlapping concentrated windings have attracted significant research attention in recent decades [1–5]. In comparison to integral-slot PMSMs with overlapping distributed windings described in Chapters 2 and 3, fractional-slot PMSMs have advantages of high fault-tolerant capability, short end-windings, high slot packing-factor, and consequently reduced size and copper losses, resulting in higher torque density and efficiency.

Typically, for dual three-phase (DTP) machines, different angle displacements can result in significant differences in the resultant electromagnetic performances, where the most common angle displacements between two sets of three-phase windings are 0° and 30° [6–17]. Specifically, for fractional-slot 12-slot/10-pole DTP machines, it has been reported in [6,7] that the 30° configuration has larger torque density, reduced stator magnetomotive force (MMF) harmonics, and lower torque ripple compared to the 0° counterpart. Additionally, under fault conditions, the 30° configuration does not suffer from unbalanced magnetic force, unlike the 0° configuration. However, the 0° isolated configuration, having less mutual coupling between the two sets of three-phase windings, results in a smaller short-circuited current compared to the 30° counterpart [6,7].

In this book, various angle displacements between the two sets of three-phase windings depending on the slot/pole number combinations have been described in Chapter 2, and DTP PMSMs with integer-slot overlapping windings have been described with examples in Chapter 3. In this chapter, fractional-slot DTP PMSMs with non-overlapping windings will be described alongside examples, with references to STP PMSMs. The feasibilities and connections of fractional-slot DTP PMSMs can be obtained using the general winding configuration method from Chapter 2.

The structure of Chapter 4 is shown in Figure 4.1. In Section 4.2 and Section 4.3, the fractional-slot DTP PMSMs with non-overlapping windings and with one and two-slot pitches will be described, respectively, of which all feasible angle displacements between two sets of three-phase windings are discussed in general, accounting for various slot/pole number combinations. In these sections, the 24-slot/22-pole and 24-slot/10-pole DTP PMSMs with different angle

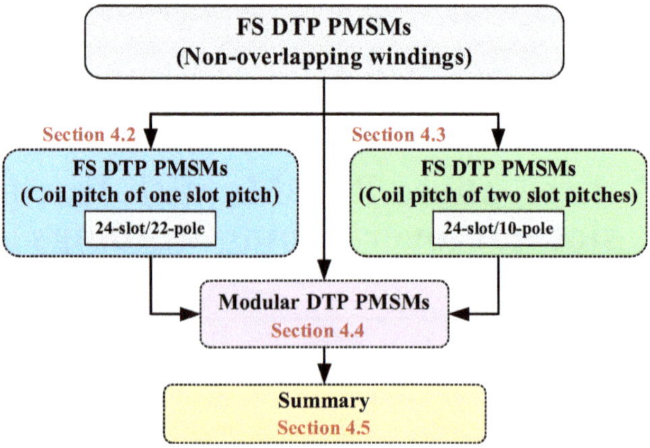

Figure 4.1 Structure of Chapter 4.

displacements, i.e., 0°, 15°, and 30°, are taken as examples for illustration. Section 4.4 will then discuss the modular PMSM topologies briefly. Section 4.5 summarizes this chapter.

4.2 Fractional-slot DTP PMSMs with coil pitch of one slot pitch

4.2.1 Feasible angle displacements

In this section, the slot/pole number combinations $N_s/2p$ for the fractional-slot DTP PMSMs with non-overlapping windings are summarized in Table 4.1, together with the fundamental winding factors K_{dp}, considering all feasible angle displacements. The possible angle displacements (β) between two sets of three-phase windings can be expressed as

$$\beta = \min\left(\frac{60}{n}, \frac{60(n-1)}{n}\right)\text{deg.} \tag{4.1}$$

where n is a positive integer number and can be determined from the star of slots.

As can be observed, with the increase of the slot number N_s (N_s=6k, k is an integer number), non-conventional angle displacements (i.e., not 0° and 30°) also exist. By way of example, for the 24-slot DTP PMSMs with 22 and 26 poles, the angle displacement can be 15°. Similarly, for 36-slot DTP PMSMs, the angle displacement can also be either 10° or 20°. Therefore, in this section, these non-conventional angle displacements will be examined to determine if they could potentially offer better overall electromagnetic performances than conventional angle displacements. Specifically, a 24-slot/22-pole DTP PMSM is used for analysis, as it can be designed for both conventional (0 and 30°) and non-conventional (15°) angle displacements.

Table 4.1 Typical slot/pole number combinations for fractional-slot DTP PMSMs with non-overlapping windings accounting for different angle displacements

Slot number N_s	**Pole number 2*p***	**Winding factor K_{dp}**	**Displacement angle β (deg.)**
6	4	0.866	0
	8	0.866	0
12	10	0.933	0/30
	14	0.933	0/30
18	14	0.902	0/20
	16	0.945	0/20
	20	0.945	0/20
	22	0.902	0/20
24	20	0.933	0/30
	22	0.949	**0/15/30**
	26	0.949	0/15/30
	28	0.933	0/30
30	28	0.951	0/12/24
	32	0.951	0/12/24
36	32	0.945	0/20
	34	0.953	0/10/20/30
	38	0.953	0/10/20/30
	40	0.945	0/20
48	44	0.949	0/15/30
	46	0.954	0/7.5/15/30
	50	0.954	0/7.5/15/30
	52	0.949	0/15/30

Note: The bold values indicate a 24-slot/22-pole DTP PMSM used for analysis.

4.2.2 *Winding configurations of 24-slot/22-pole DTP PMSMs*

To obtain the winding arrangements for the 24-slot/22-pole DTP PMSMs with different angle displacements, the single three-phase (STP) configuration and the coil EMF phasor diagram are illustrated in Figure 4.2. As can be seen, each phase consists of eight electromotive force (EMF) phasors, which can be categorized into three types of DTP winding configurations, as illustrated in Figure 4.3.

In addition to the conventional 0° and 30° angle displacements, the non-conventional 15° angle displacement is identified in Figure 4.3(c) and (d). Specifically, if the coils for phase A1 are connected as 1-3-13-15, and phase A2 are connected as 2-4-14-16, the angle between the two sets of three-phase windings will be exactly 15° (Figure 4.3(d)). The coil connections for the phases B1/B2, and C1/C2 can also be obtained for the three angle displacements from Figure 4.3(a)–(f).

4.2.3 *Electromagnetic performance of 24-slot/22-pole DTP PMSMs*

Based on the above three angle displacements for fractional-slot DTP PMSM, i.e., 0°, 15°, and 30°, the corresponding electromagnetic performance will be presented,

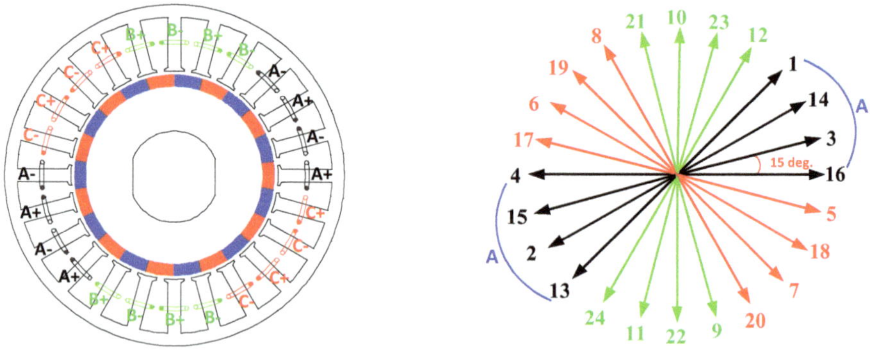

(a) (b)

Figure 4.2 Winding arrangement and EMF phasor diagram for 24-slot/22-pole STP PMSM. (a) Winding arrangement and (b) coil EMF phasor diagram.

such as open-circuit back-EMF, on-load torque, etc. The main machine design parameters and specifications for the 24-slot/22-pole surface-mounted PMSM (SPMSM) are given in Table 4.2. It is noted that the 24-slot/22-pole SPMSM has the same stator outer diameter as that of the Toyota Prius 2010 PMSM [18], and its current density is also similar to that of the Prius 2010 PMSM.

4.2.3.1 Winding factor and open-circuit back-EMF

With the winding arrangement and coil EMF phasors in Figure 4.3 for the three types of winding configurations, the winding factor for each set of three-phase winding can be obtained and expressed as follows.

For the angle displacement of 0°:

$$K_{\text{dpn}} = \sin^2\left(\frac{n\pi}{N_s}\right)\cos\left(\frac{2n\pi}{N_s}\right) \tag{4.2}$$

For the angle displacement of 15°:

$$K_{\text{dpn}} = \sin\left(\frac{n\pi}{N_s}\right)\cos\left(\frac{2n\pi}{N_s}\right)\sin\left(\frac{n\pi}{2}\right) \tag{4.3}$$

For the angle displacement of 30°:

$$K_{dpn} = \sin^2\left(\frac{n\pi}{N_s}\right)\cos\left(\frac{n\pi}{2}\right) \tag{4.4}$$

where K_{dpn} is the winding factor, n is the harmonic order in the mechanical period. Based on (4.2)–(4.4), for the 24-slot/22-pole machine, the main harmonic ($n = 11$) winding factors are 0.949, 0.958, and 0.980 for the 0°/15°/30° configurations, respectively. Thus, the 30° configuration has the highest fundamental back-EMF

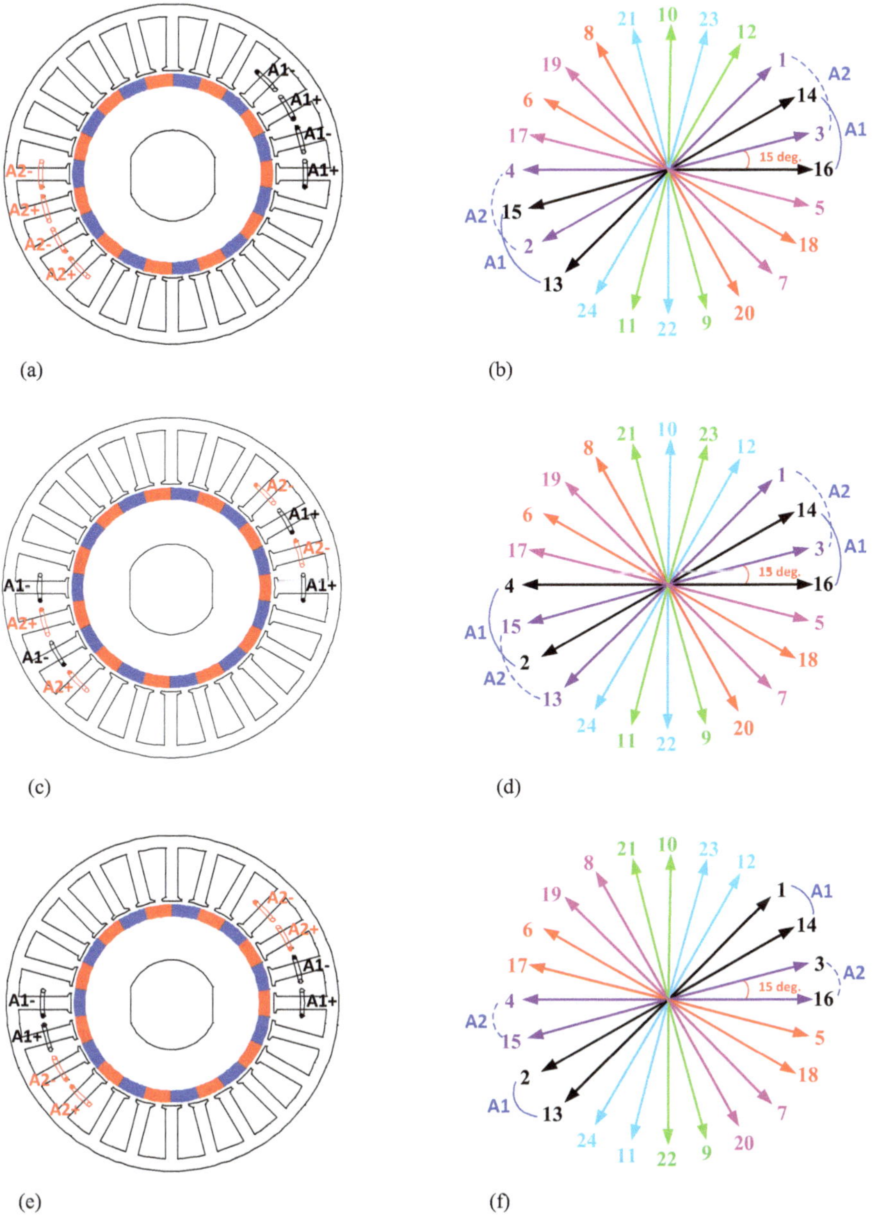

Figure 4.3 Winding arrangements and EMF phasor diagrams for 24-slot/22-pole DTP PMSMs with different angle displacements. (a) Winding arrangement, 0°. (b) Coil EMF phasor diagram, 0°. (c) Winding arrangement, 15°. (d) Coil EMF phasor diagram, 15°. (e) Winding arrangement, 30°. (f) Coil EMF phasor diagram, 30°.

Table 4.2 Main design parameters of 24-slot/22-pole PMSM

Parameter	Value	Parameter	Value
Number of poles	22	Stack length (mm)	50.8
Number of stator slots	24	Tooth width (mm)	12.4
Stator OD (mm)	264	Yoke thickness (mm)	6.8
Stator ID (mm)	204	PM remanence (T)	1.2
Rotor OD (mm)	202	SPM thickness (mm)	7.16
Air gap length (mm)	1	Turns per phase	44
Current (A)	264.2	Current density (J/mm^2)	21.97

amplitude and thus, the highest average torque output. In comparison, the 15° angle displacement is between the 0° and 30° configurations, as investigated below.

The waveforms and spectra of the finite element (FE) predicted phase- and line-back-EMFs for the three angle displacements (0°/15°/30°) at 400 r/min are shown in Figures 4.4 and 4.5, respectively. It can be found that the 30° angle displacement has the largest fundamental amplitude, followed by the 15° and 0° configurations.

The resultant stator MMF F_{sum} for the 24-slot/22-pole DTP PMSM can be generally expressed as

$$\begin{aligned} F_{\text{sum}} &= \sum_n K_{\text{dpn}} F_n \cos\left[n\alpha \pm (p\omega_r t + \delta_0)\right] \\ &\quad + \sum_n K_{\text{dpn}} F_n \cos\left[n(\alpha - \alpha_{\text{shift}}) \pm (p\omega_m t + \delta_0 + \delta_{\text{shift}}\right) \end{aligned} \tag{4.5}$$

where F_n is the nth harmonic stator MMF amplitude (for each group of three-phase winding). α is the angular position in the stator reference frame. ω_m is the mechanical rotor speed. α_{shift} is the phase shift between the axes of the two three-phase winding sets in the stator reference frame (e.g., $\alpha_{\text{shift}} = 195°$ (13 slot pitches) for the 15° configuration, Figures 3.2(a) and 4.3(c)). δ_0 is the initial current angle, δ_{shift} is the angle difference in the input current angle, i.e., 0°/15°/30° for the three configurations, respectively.

According to (4.2)–(4.5), the waveforms and spectra of stator MMFs for the STP and DTP machines are shown in Figure 4.6. Although the MMF distribution looks similar, the resultant MMF harmonic distributions are quite different. For the 30° angle displacement, the $(12k \pm 5)$th (k is an integer) harmonics are zero, Figure 4.6(b), which can also be easily derived from (4.5). In other words, the MMF harmonics contributing to the large 6th harmonic torque ripples are eliminated for the 30° angle displacement, leading to a much smaller torque ripple. In contrast, due to the similar distributions for the $(12k \pm 5)$th MMF harmonics, Figure 4.6(b), the 15° angle displacement has quite similar torque ripple as the 0° DTP and STP counterparts.

4.2.3.2 Electromagnetic torque performance

This subsection compares the electromagnetic torque performance for the 24-slot/22-pole DTP PMSMs with different angle displacements.

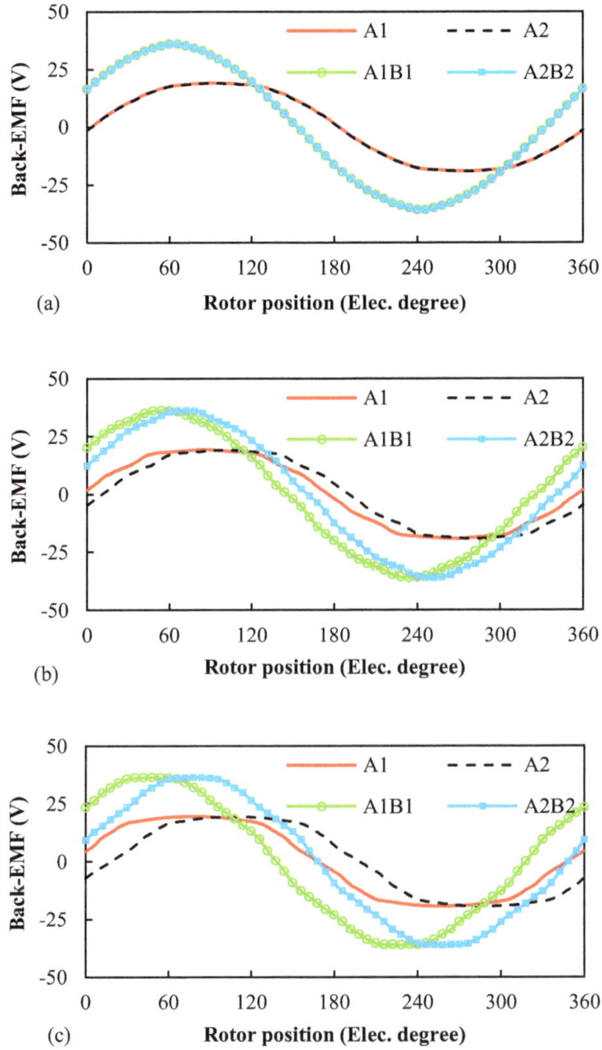

Figure 4.4 Waveforms of phase- and line-back-EMFs for 24-slot/22-pole DTP PMSM with different angle displacements at 400 r/min. (a) 0°, (b) 15°, and (c) 30°.

The average torque T_{ave} can be calculated by:

$$T_{\text{ave}} = \frac{e_{a1}i_{a1} + e_{b1}i_{b1} + e_{c1}i_{c1}}{\omega_r} + \frac{e_{a2}i_{a2} + e_{b2}i_{b2} + e_{c2}i_{c2}}{\omega_r} \tag{4.6}$$

where ω_r is the rotor's mechanical angular speed. e_k and i_k with $\{k = a/b/c1,\ a/b/c2\}$ are the phase back-EMFs and currents, respectively.

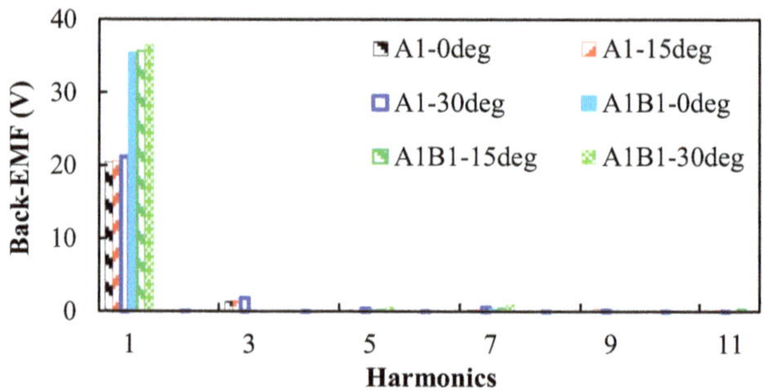

Figure 4.5 Spectra of phase- and line-back-EMFs for 24-slot/22-pole DTP PMSM with different angle displacements at 400 r/min.

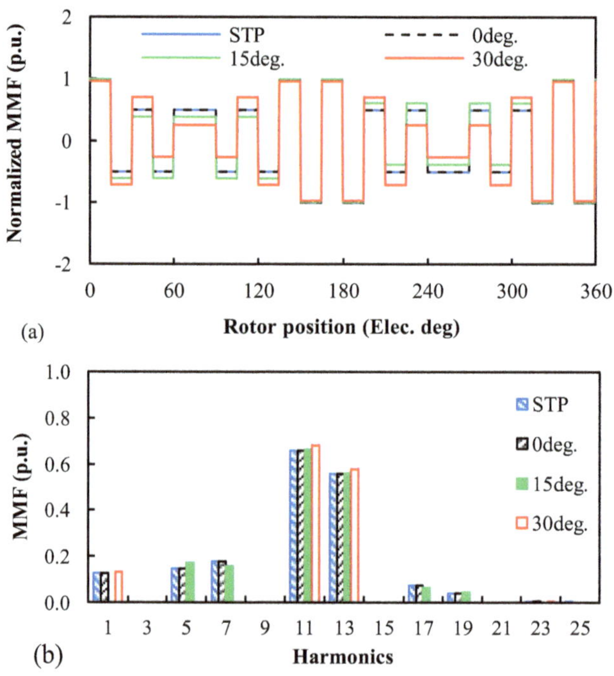

Figure 4.6 Waveforms and spectra of stator MMF distributions for 24-slot/22-pole STP and DTP with different angle displacements (0°, 15°, and 30°). (a) MMF distributions and (b) spectra.

Since the back-EMFs for phases A1 and A2 are displaced by different phase angles, as shown in Figure 4.5, the input current angles for each phase are then adjusted to maximize the torque output (i.e., 0°/15°/30° for the three angle displacements, respectively).

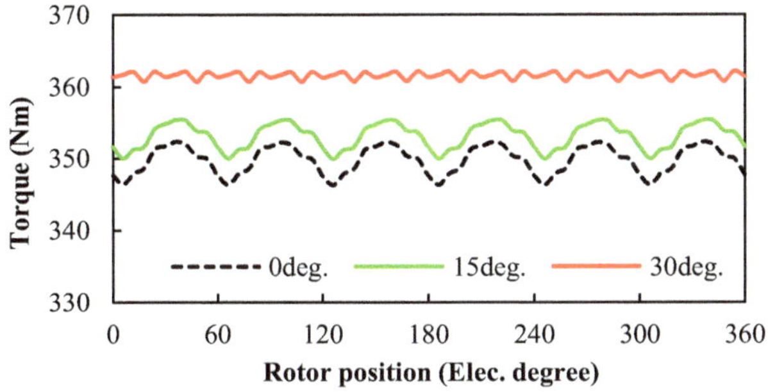

Figure 4.7 Electromagnetic torque performances for 24-slot/22-pole DTP PMSMs with different angle displacements.

Accordingly, the torque waveforms for the DTP 24-slot/22-pole machine with different angle displacements are shown in Figure 4.7. The 30° configuration has the largest torque output, followed by the 15° and 0° configurations. It can be easily explained that the 30° configuration has the highest winding factor and back-EMF amplitude (see Figure 4.5). Meanwhile, the torque ripple for the 30° is also the smallest, mainly the 12th harmonic. In comparison, the torque ripple for the 15° configuration is slightly smaller than the 0° counterpart. The differences in torque ripple for the three angle displacements can also be explained by the differences in stator MMF harmonics as explained above.

4.3 Fractional-slot DTP PMSMs with coil pitch of two slot pitches

4.3.1 Feasible angle displacements

In this section, all feasible angle displacements for DTP PMSMs with a coil pitch of two slot pitches and various slot/pole number combinations are summarized in Table 4.3 which includes all possible angle displacements (β) between the two sets of windings, as expressed in (4.1).

Similar to the case of a coil pitch of one slot pitch in Section 2.2, it can be observed from Table 4.3 that with the increase of slot number N_s ($N_s = 6k$, k is an integer number) for the DTP machines with a coil pitch of two slot pitches, in addition to the conventional angle displacements of 0° and 30° [8,12], there are other possible angle displacements. By way of example, for the 24-slot DTP PMSMs with 2-, 10-, 14, 22, and 26-pole, the angle displacement can be 15°. Besides, the angle displacement for 36-slot DTP PMSMs with 2-, 10-, 14-, 22- and 26-pole can also be either 10° or 20° [16].

4.3.2 Winding configurations of 24-slot/10-pole DTP PMSMs

A 24-slot/10-pole DTP PMSM with coil pitch of two slot pitches [16] is taken as an example for analysis to demonstrate the influence of different angle displacements,

Table 4.3 Slot/pole number combinations for DTP PMSMs with coil pitch of two slot pitches and different angle displacements β (°)

N_s	$2p$	2	4	6	8	10	12	14	16	20	22	24	26	28	30
6	K_{dp}	1	0.866	**	0.866	0.5	**	0.5	0.5	0.866	0.5	**	0.5	0.866	**
	β	0	0	**	0	0	**	0	0	0	0	**	0	0	**
12	K_{dp}	0.966	1	**	0.866	0.933	**	0.933	0.866	0.5	0.5	**	0.5	0.5	**
	β	0/30	0	**	0	0/30	**	0/30	0	0	0/30	**	0/30	0	**
18	K_{dp}	0.960	0.945	1	0.945	0.945	0.866	0.902	0.945	0.945	0.902	0.866	0.735	0.617	0.5
	β	0/20	0/20	0	0/20	0/20	0	0/20	0/20	0/20	0/20	0	0/20	0/20	0
24	K_{dp}	0.958	0.966	**	1	0.925	**	0.925	0.866	0.933	0.949	**	0.949	0.933	**
	β	0/15/30	0/30	**	0	0/15/30	**	0/15/30	0	0/30	0/15/30	**	0/15/30	0/30	**
30	K_{dp}	0.957	0.951	**	0.910	1	**	0.951	0.951	0.866	0.874	**	0.936	0.951	**
	β	0/12/24	0/12/24	**	0/12/24	0	**	0/12/24	0/12/24	0	0/12/24	**	0/12/24	0/12/24	**
36	K_{dp}	0.956	0.960	0.966	0.945	0.924	1	0.898	0.945	0.945	0.783	0.866	0.867	0.902	0.933
	β	0/10/20/30	0/20	0/30	0/20	0/10/20/30	0	0/10/20/30	0/20	0/20	0/10/20/30	0	0/10/20/30	0/20	0/30

**Unfeasible slot/pole number combinations for DTP PMSMs.

Note: The grey shades indicate different shifting angles.

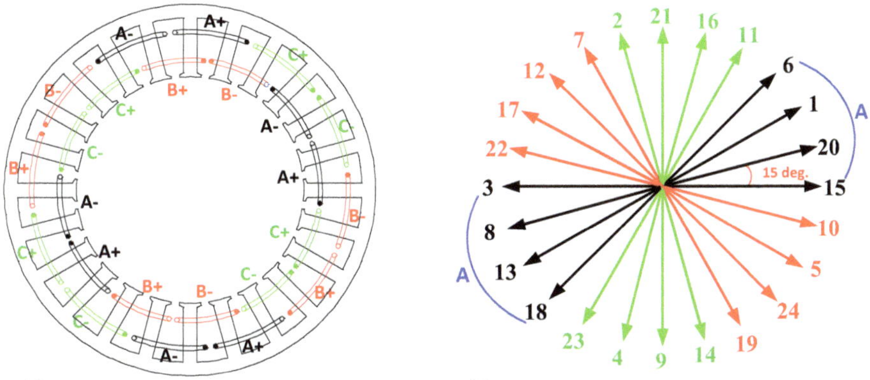

(a) (b)

Figure 4.8 Winding arrangement and EMF phasor diagram for 24-slot/10-pole STP PMSM. (a) Three-phase winding arrangement and (b) coil EMF phasor diagram.

considering the angle displacements of 0°, 15°, and 30°. In addition, it is worth noting that the selected 24-slot/10-pole PMSMs for both STP and DTP configurations have the advantages of reduced MMF harmonics and PM losses, and increased machine efficiency, etc. [9,16,17,19–23].

For the 24-slot/10-pole STP PMSM, the winding arrangement is shown in Figure 4.8(a), and the phasor diagram of coil EMF is shown in Figure 4.8(b). As can be seen, there are eight EMF phasors for each single phase, and thus, many types of coil connections are possible when splitting the eight phasors into two independent phases (e.g., phases A1 and A2) for its DTP counterpart.

Considering the balance between the two sets of three-phase windings, only three winding connections, i.e., 0°, 15°, and 30° angle displacements, are selected as shown in Figure 4.9. Their electromagnetic performance will be shown in the following section.

4.3.3 Electromagnetic performance of 24-slot/10-pole DTP PMSMs

The 24-slot/10-pole DTP SPMSM is optimized in the FE based on the genetic algorithms. The optimization target is to maximize the average torque under the constraints of the copper loss of 30W with a fixed stator out-diameter of 90 mm and stack length of 25 mm. The split ratio, stator tooth, yoke length, and PM length are selected as the optimization variables. The resultant optimized design parameters and specifications are given in Table 4.4.

4.3.3.1 Winding factor and back-EMF

According to the winding arrangement and EMF phasors in Figure 4.9, the winding factors can be obtained as follows.

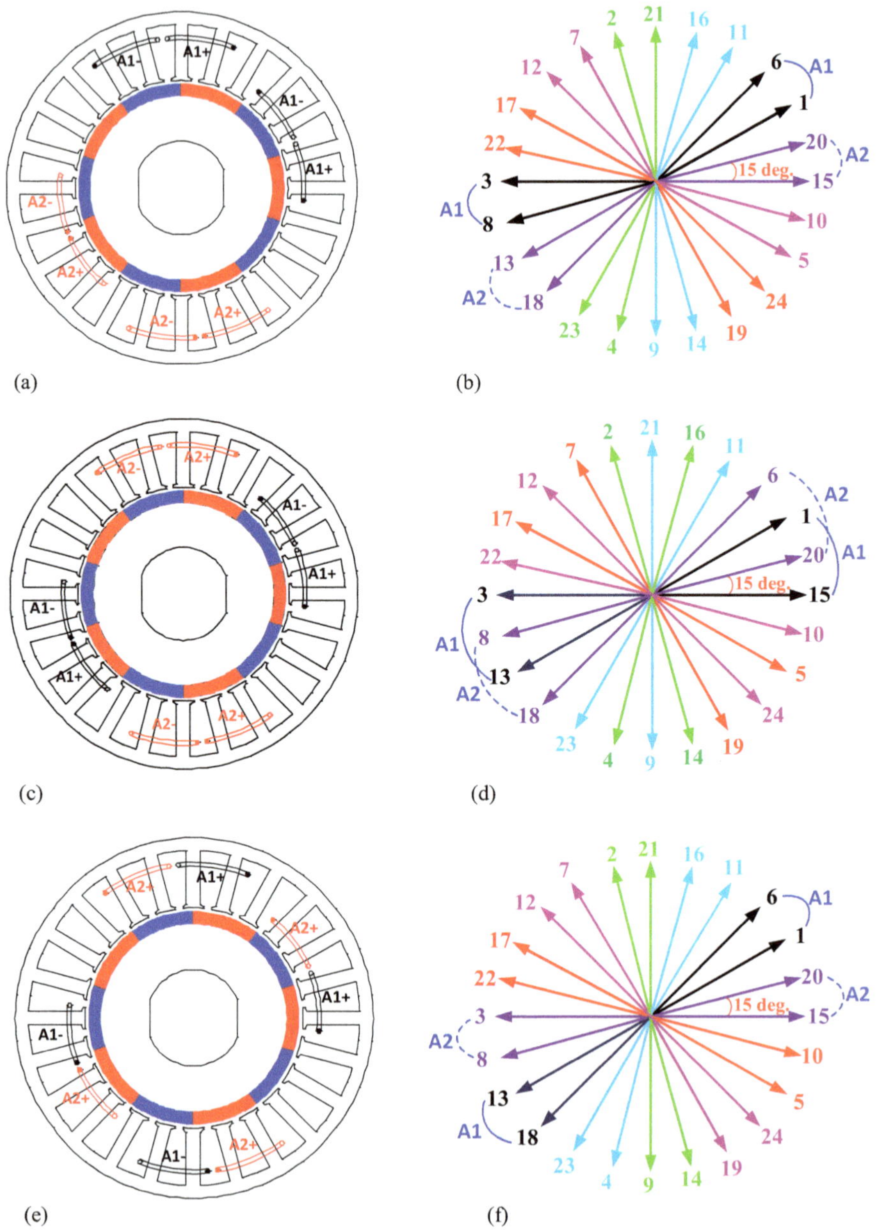

Figure 4.9 Winding arrangements and EMF phasor diagrams for 24-slot/10-pole DTP PMSM. (a) Winding arrangement, 0°. (b) Coil EMF phasor diagram, 0°. (c) Winding arrangement, 15°. (d) Coil EMF phasor diagram, 15°. (e) Winding arrangement, 30°. (f) Coil EMF phasor diagram, 30°.

Table 4.4 Main design parameters of 24-slot/10-pole PMSM

Parameter	Value	Parameter	Value
Number of poles	10	Stack length (mm)	25
Number of stator slots	24	Tooth width (mm)	3.4
Stator OD (mm)	90	Yoke thickness (mm)	3.3
Stator ID (mm)	55	PM remanence (T)	1.2
Rotor OD (mm)	53	SPM thickness (mm)	3
Air gap length (mm)	1	Turns per phase	84

For the angle displacement of 0°:

$$K_{\text{dpn}} = \sin^2\left(\frac{2n\pi}{N_s}\right)\cos\left(\frac{5n\pi}{N_s}\right)\sin\left(\frac{n\pi}{2}\right) \tag{4.7}$$

For the angle displacement of 15°:

$$K_{\text{dpn}} = \sin^2\left(\frac{n\pi}{N_s}\right)\sin\left(\frac{n\pi}{2}\right) \tag{4.8}$$

For the angle displacement of 30°:

$$K_{\text{dpn}} = \sin\left(\frac{n\pi}{N_s}\right)\cos\left(\frac{5n\pi}{2}\right)\sin\left(\frac{n\pi}{2}\right) \tag{4.9}$$

where K_{dpn} is the winding factor, n is the harmonic order in the mechanical period, N_s is the stator slot number, respectively.

According to (4.7)–(4.9), the main harmonic (i.e., the 5th) winding factor can be calculated as 0.925, 0.933, and 0.958 for the 0°, 15°, and 30° configurations, respectively. It is expected from the winding factors that the 30° configuration has the highest torque density, followed by the 15° and 0° configurations.

The waveforms and spectra of open-circuit phase- and line-back-EMFs for phases A1 and A2 with different angle displacements are calculated by the FE method and shown in Figures 4.10 and 4.11, respectively. As can be seen, the phase shifts between phases A1 and A2 are 0°, 15°, and 30° for the three angle displacements. Meanwhile, the 30° configuration has the largest 1st harmonic back-EMF amplitude due to the relatively larger winding factor, while the 15° configuration is slightly lower than the 30° configuration but slightly larger than the 0° counterpart. Besides, for the line back-EMF in Figure 4.11(b), the total harmonic distortion (THD%) can also be obtained as 1.65%, 2.63%, and 6.09% for 0°, 15° and 30° angle displacements, respectively.

4.3.3.2 Electromagnetic torque performance

The instantaneous torque waveforms for the 24-slot/10-pole DTP PMSMs with three-angle displacements are shown in Figure 4.12. The 30° configuration has a relatively larger torque output, followed by the 15° and 0° configurations due to its back-EMF amplitude.

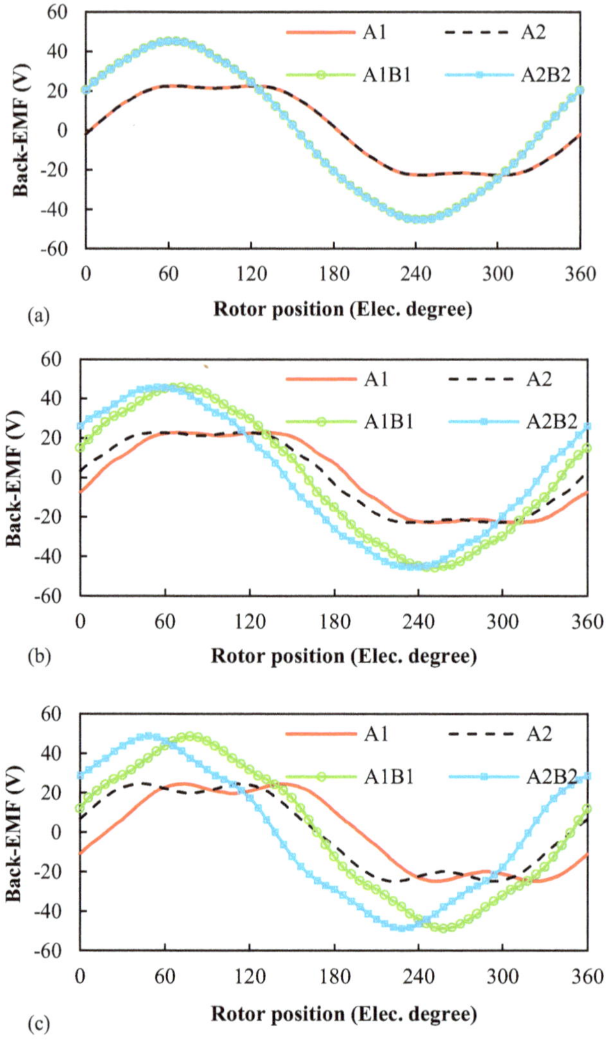

Figure 4.10 Open-circuit back-EMF for 24-slot/10-pole DTP PMSM with different angle displacements at 2,400 r/min. (a) 0°, (b) 15°, and (c) 30°.

In terms of torque ripple, the 30° is the lowest compared to the other two cases, since it has the lowest 5th and 7th stator MMF harmonics, which contribute to large torque ripple [24], as described by:

$$\frac{T_{\text{ripple}}}{T_{\text{avg}}} = \sum_{\substack{k=(6h\mp 1)p \\ h=1,2,3\ldots}} k\frac{F_{sk}}{F_{sp}}\frac{F_{rk}}{F_{rp}}\sin\left[(k\pm p)\omega_r t \pm \frac{\pi}{2}\right] \tag{4.10}$$

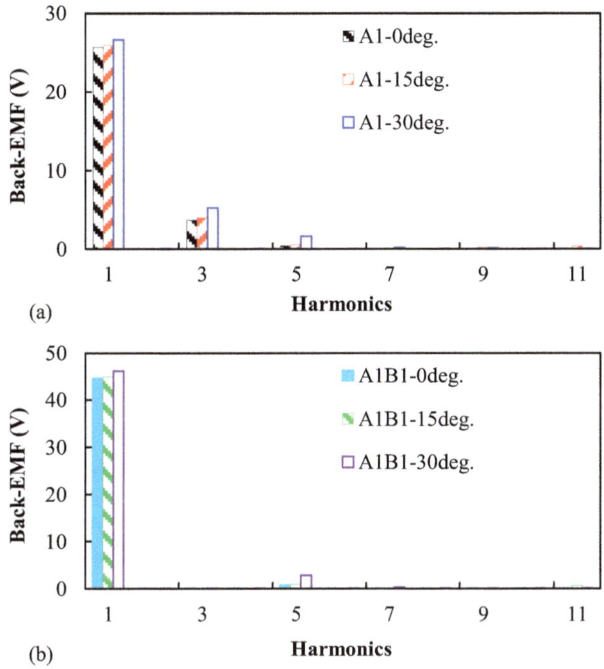

Figure 4.11 Phase- and line-back-EMF spectra for 24-slot/10-pole DTP PMSM with different angle displacements at 2,400 r/min. (a) Phase back-EMF and (b) line back-EMF.

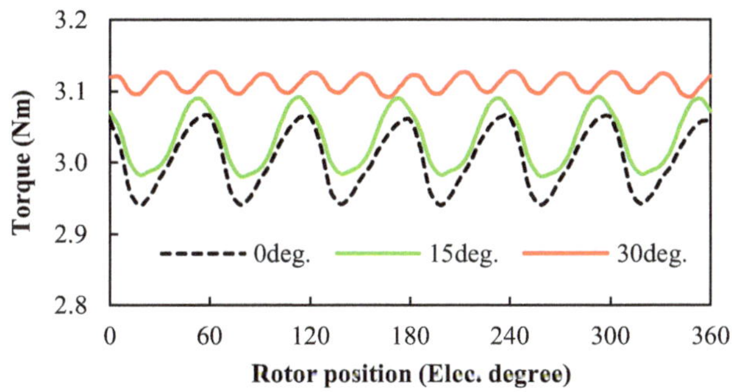

Figure 4.12 Electromagnetic torque waveforms for 24-slot/10-pole DTP PMSMs with different winding configurations under copper loss of 30 W.

where F_{sk} and F_{rk} are the kth stator and rotor MMF harmonics. F_{sp} and F_{rp} are the main harmonics of the stator and rotor MMFs, respectively. It can be observed from (4.10) that the torque ripple depends on both the stator and rotor MMF harmonic distributions, and the rotor MMF F_r for the PMSM can be expressed as [24]:

$$F_r = \sum_{\nu} F_{r\nu} \cos\left[\nu p(\alpha - \omega_r t)\right] \tag{4.11}$$

where $F_{r\nu}$ is the νth harmonic rotor PM MMF amplitude. α is the angular position in the stator reference frame. p is the number of pole-pairs. ω_r is the mechanical rotor speed.

Since the rotor MMF distribution is the same for the three angle displacements, the differences in torque ripple are related to the stator MMF harmonic distributions. The total stator MMF expression for the DTP machine can be found in (4.5). Accordingly, the waveforms and spectra of stator MMFs for one set and two sets of 3-phase windings are shown in Figure 4.13. The harmonic amplitudes are listed in Table 4.5.

As can be seen, since the 25th and 35th MMF harmonics (i.e., the 5th and 7th stator MMF harmonics in the electrical period) in Table 4.5 is zero for the 30° configuration, the 6th torque pulsations are eliminated according to (4.5), (4.10), and (4.11), which results in the lowest torque ripple for the 30° angle displacement. In comparison, the 15° configuration has a slightly lower torque ripple compared to the 0°, which can be attributed to the lower 25th and 35th MMF harmonics, as listed in Table 4.5.

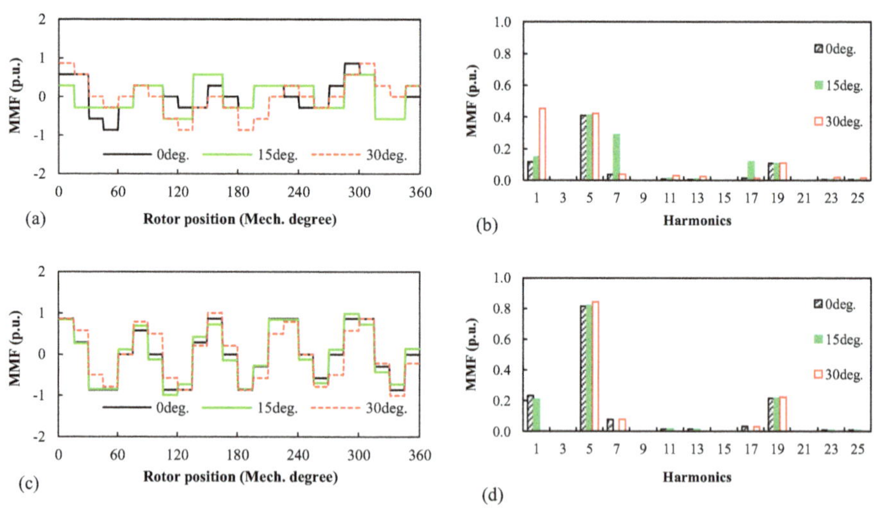

Figure 4.13 Waveforms and spectra of stator MMF distributions for 24-slot/10-pole DTP PMSM with different angle displacements. (a) Waveforms of one set of three-phase winding. (b) Spectra of one set of three-phase winding. (c) Waveform of summed two sets of three-phase winding. (d) Spectra of summed two sets of three-phase winding.

Table 4.5 MMF harmonic amplitude of DTP machine with different angle displacements in mechanical period (n = 3k is zero)

n	One set of three-phase winding			Two sets of three-phase winding		
	0°	**15°**	**30°**	**0°**	**15°**	**30°**
1	0.117	0.148	0.453	0.234	0.209	0
2	0.071	/	/	/	/	/
4	0.358	/	/	/	/	/
5	0.408	0.412	0.423	0.816	0.823	0.845
7	0.038	0.290	0.040	0.077	0	0.079
8	0.103	/	/	/	/	/
10	0.053	/	/	/	/	/
11	0.008	0.014	0.032	0.016	0.019	0
13	0.007	0.011	0.027	0.014	0.016	0
14	0.038	/	/	/	/	/
16	0.052	/	/	/	/	/
17	0.016	0.121	0.016	0.032	0	0.033
19	0.108	0.109	0.112	0.216	0.218	0.223
20	0.072	/	/	/	/	/
22	0.007	/	/	/	/	/
23	0.005	0.007	0.020	0.0103	0.009	0
25	0.005	0.006	0.018	0.0095	0.0084	0

Note: The grey shades indicate larger amplitudes.

4.4 Modular DTP PMSMs

The stators of large-size electrical machines need to be made modular to ease mechanical machining and transportation, as well as to improve fault tolerance [25,26]. When an integer-slot full-pitched single-layer winding is adopted, an example of a modular stator external rotor PMSM is shown in Figure 4.14 [27]. Although the end parts of this 144-slot/48-pole stator windings are overlapped among phases within each pole pair, there is no physical coupling between adjacent basic electrical machine units (6-slot/2-pole). Therefore, the stator can be cut into a few segments and the coils can be easily split into multi-three-phase windings.

Figure 4.15 shows another modular stator example based on the 48-slot/8-pole Toyota Prius 2010 PMSM, Figure 4.15(a), which has overlapping windings as described in Chapter 3, Figure 4.15(b) shows four 12-slot/2-pole basic electrical machine units and the single layer full pitched winding is adopted. The electromagnetic performance of the two cases will be the same, while the coils in modular stators adopted in Figure 4.15(b) can be easily split into multi-three-phase windings.

Some coils of all teeth wound windings are removed to create unwound teeth for modularity [28]. However, the removed coils must be able to construct a balanced three-phase system to maintain the balance of the modular machine. The two PMSMs with slot/pole number combinations ($N_s/2p$), i.e., $2p = N_s \pm 2$ and $2p = N_s \pm 1$, are illustrated in Figure 4.16. It is evident that more coils need to be

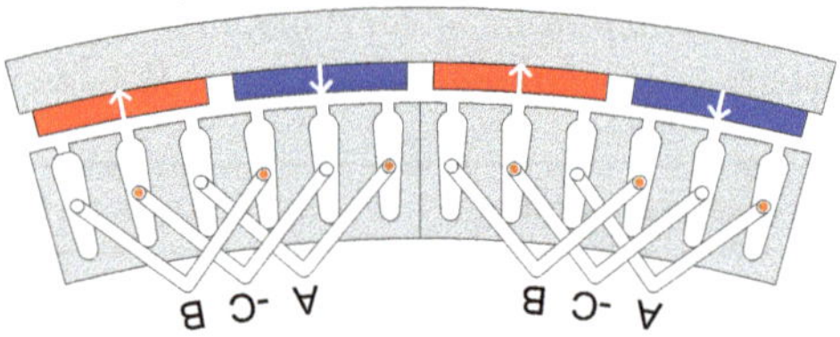

Figure 4.14 An external rotor wind power generator.

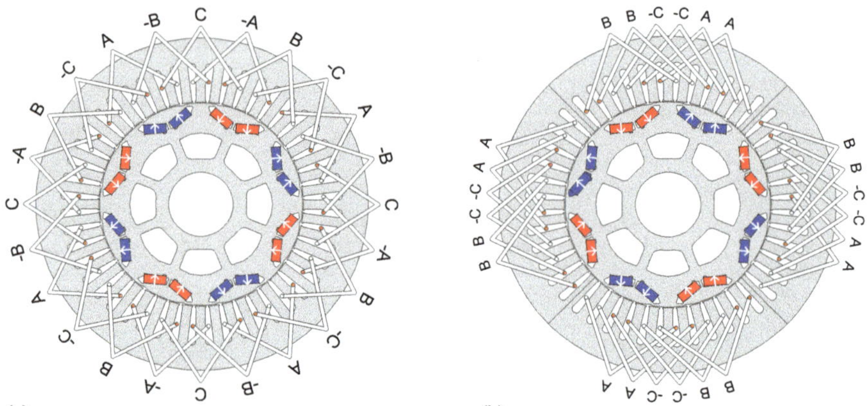

(a) (b)

Figure 4.15 Modular and non-modular 48-slot/8-pole Toyota Prius 2010 PMSM. (a) Non-modular and (b) modular.

removed for the 24-slot/22-pole PMSM, as it has two complementary winding groups for each phase. Conversely, only three coils need to be removed for the 15-slot/14-pole PMSM. Although the coils can be easily split into multi-three-phase windings, abundant harmonics pose a significant challenge for this method.

Besides, a method is proposed in [29] where intentional spaces are left between segments, allowing coils on specific teeth positions to be removed, leaving unwound teeth for stator core modularity. The resulting two modular machines are depicted in Figure 4.17. In this arrangement, each segment only contains coils belonging to one phase, and the coil pitch is assigned the same as the pole pitch to achieve a unity winding factor. The arrangement in Figure 4.17(b) can eliminate undesirable unbalanced magnetic forces existing in the PMSM with a rotating asymmetric winding layout [30], shown in Figure 4.17(a). The coils can be easily split into double-tooth pitch windings. However, the significant on-load torque ripple associated with this method may limit its application.

A similar method is proposed in [31], where each segment has a complete three-phase winding, as shown in Figure 4.18. Figure 4.18(a) shows that the original 18-slot/12-pole PMSM has all teeth wound winding. To add some redundant teeth to ease modularity, another pole pair is added and used for redundant teeth

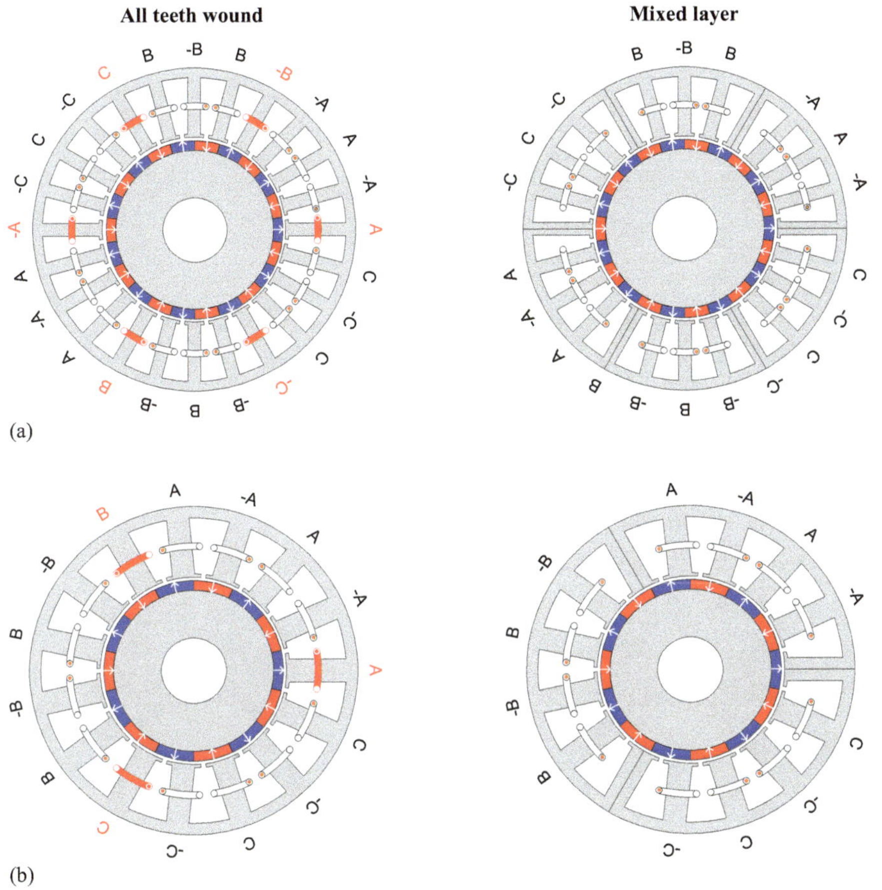

Figure 4.16 Modular machines with removed coils. (a) 24-slot/22-pole and (b) 15-slot/14-pole.

arrangement, as shown in Figure 4.18(b), where four redundant teeth are inserted to construct complementary segments. Nevertheless, since the three phases are unbalanced and the torque ripple is still large, this kind of modularity was not widely applied although the coils can also be split into multiphase windings.

In [32], two modular machines with non-overlapping and overlapping windings are shown in Figure 4.19, and they have the same rotor pole number, while the stator slot numbers are different, i.e., 42-slot/32-pole and 192-slot/32-pole. The stators of both machines can be divided into segments at the tooth center, which provides protection for coil sides located at the end parts of each segment during the transportation and assembly process. For the stators, the maximum stator segment numbers differ for the two machines, 6 and 16 for the 42-slot/32-pole and 192-slot/32-pole modular machines, respectively. For the rotors, the 192-slot/32-pole modular machine is an integer-slot machine, necessitating PM shaping to

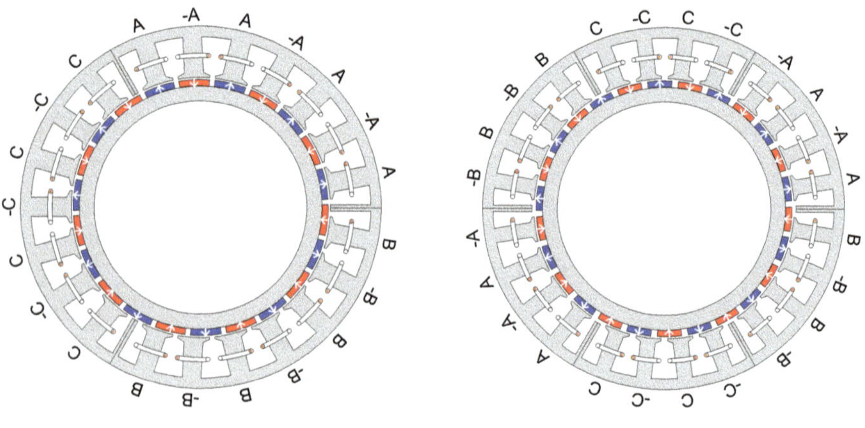

(a) (b)

Figure 4.17 Modular machines with redundant teeth. (a) 24-slot/22-pole and (b) 30-slot/26-pole.

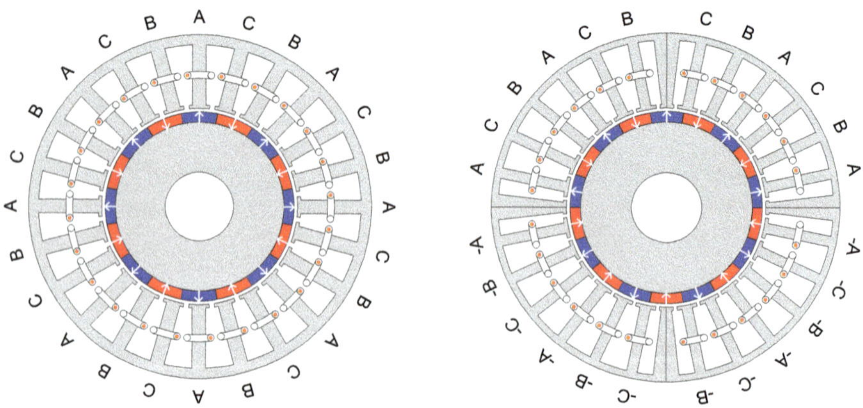

(a) (b)

Figure 4.18 Modular machines with complete winding for each segment. (a) Original 24-slot/16-pole PMSM. (b) Modified 28-slot/18-pole PMSM.

improve the EMF waveform and reduce the cogging torque. However, this is not essential for the 42-slot/32-pole modular machine, as it is a fractional-slot machine with more sinusoidal EMF waveform and low cogging torque. To arrange the 36 and 96 coils for the two machines to construct the double tooth pitch windings, the star of slots method can be used, as illustrated in Figure 4.20.

In Figure 4.20, the left column indicates that the phase shifts between adjacent coils within each segment (α_e) are 150° and 30° for the two modular machines,

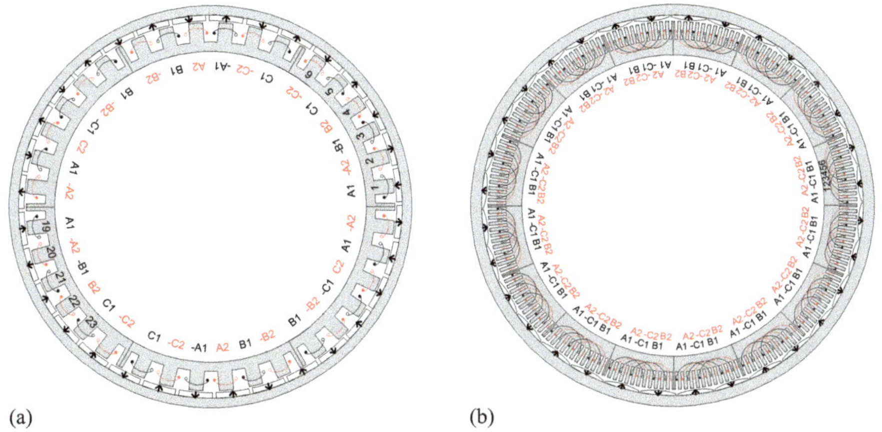

Figure 4.19 Cross-sections of the modular machines with non-overlapping/overlapping windings. (a) 42-slot/32-pole (non-overlapping windings). (b) 192-slot/32-pole (overlapping windings).

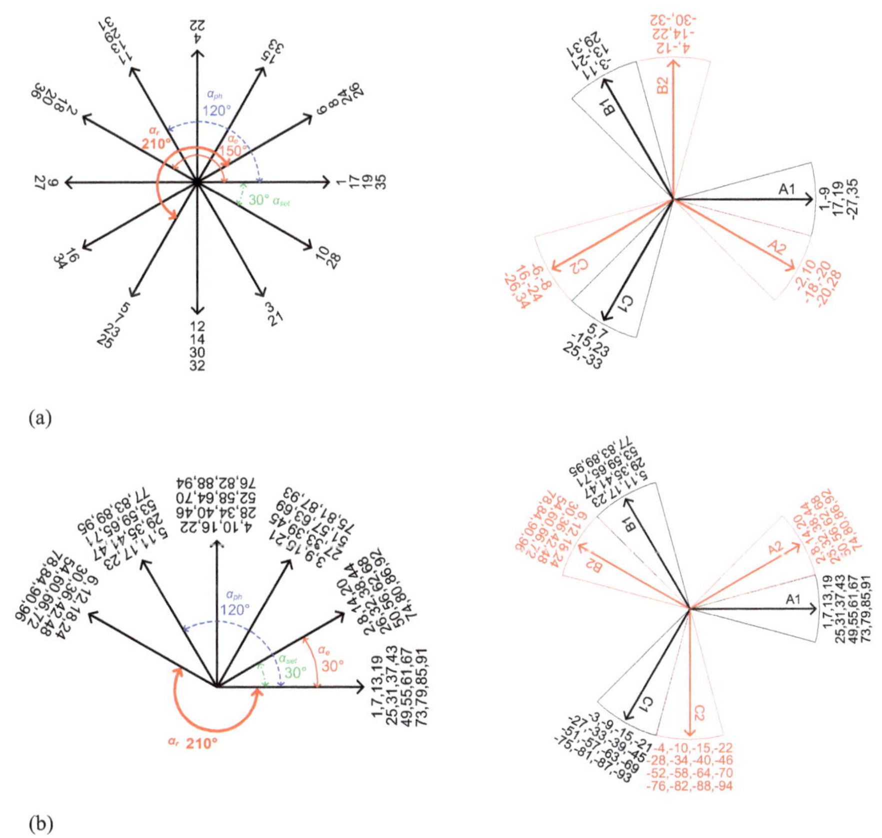

Figure 4.20 Winding arrangements of two modular machines. (a) 42-slot/32-pole (non-overlapping windings). (b) 192-slot/32-pole (overlapping windings).

respectively. The phase shift between the two adjacent end coils (α_r) is the same for the two machines, i.e., 210°. Among one set of windings, three phases have a 120° shift between each other (α_{ph}) and two sets of windings differ 30° in space (α_{set}). However, the winding set 2 is 30° ahead for the 42-slot/32-pole modular machine and vice versa for the 192-slot/32-pole one, as shown in the right column of Figure 4.20. The minus symbol before the coil number means the wound direction is opposite. The electromagnetic performance is described in [32], and the general slot/pole number combinations with feasible winding configurations for modular PMSMs can be found in [33,34].

4.5 Summary

This chapter has analyzed the fractional-slot DTP PMSMs with coil pitches equal to one and two slot pitches and different angle displacements between two sets of three-phase windings, i.e., 0°, 15°, and 30°. The 24-slot/22-pole and 24-slot/10-pole DTP PMSMs with various winding connections have been used for illustration. All theoretical analyses have been verified by 2-D FE analyses (as well as experiments that were given in [15–17]). In addition, modular stator PMSMs were also introduced, and a few examples were described briefly.

For both cases with a coil pitch equal to one and two slot pitches, it was found that under healthy conditions, i.e., three-phase OC conditions [15–17], the 30° angle displacement has the highest average torque and much lower torque ripple, followed by the 15° and the 0° configurations. However, in this case, the 15° configuration has much better over-load capability than the 0° and 30° angle displacements due to less cross-coupling saturation effects. Furthermore, the 15° angle displacement has the lowest risks of the PM irreversible demagnetization under one set three-phase OC condition. It should also be mentioned that despite the analyses shown in this chapter being based only on some specific examples, the obtained conclusions are generic and can be applied to DTP PMSMs with other winding configurations.

It is worth mentioning that DTP winding configurations generally enhance fault tolerance capabilities. On the one hand, for relatively low-power PMSMs, DTP PMSMs with non-overlapping windings may find wider applications due to their advantageous characteristics such as good torque/power capability resulting from short end-windings due to non-overlapping winding layouts. These features make them suitable for applications, e.g., power steering and aerospace, and their compatibility with automated manufacturing processes further enhances their appeal. On the other hand, for high-power PMSMs, such as wind power PMSMs, which typically experience high electric loading, particular attention should be paid to excessively high stator MMF harmonics due to fractional-slot winding configurations. These harmonics may cause increased stator and rotor iron losses, excessive rotor PM eddy current loss, as well as high acoustic noise and vibration [23]. In such cases, DTP PMSMs with overlapping distributed windings may need to be employed to mitigate these issues.

References

[1] Z. Q. Zhu and D. Howe, "Electrical machines and drives for electric, hybrid and fuel cell vehicles," *Proc. IEEE*, vol. 95, no. 4, pp. 746–765, 2007.

[2] P. B. Reddy, A. M. El-Refaie, K.-K. Huh, J. K. Tangudu, and T. M. Jahns, "Comparison of interior and surface PM machines equipped with fractional-slot concentrated windings for hybrid traction applications," *IEEE Trans. Energy Convers.*, vol. 27, no. 3, pp. 593–602, 2012.

[3] A. M. EL-Refaie, "Fractional-slot concentrated-windings synchronous permanent magnet machines: Opportunities and challenges," *IEEE Trans. Ind. Electron.*, vol. 57, no. 1, pp. 107–121, 2010.

[4] Z. Q. Zhu, "Fractional slot permanent magnet brushless machines and drives for electric and hybrid propulsion systems," *Int. J. Comput. Math. Electr. Electron. Eng. (COMPEL)*, vol. 30, no. 1, pp. 9–31, 2011.

[5] L. Alberti and N. Bianchi, "Theory and design of fractional-slot multilayer windings," *IEEE Trans. Ind. Appl.*, vol. 49, no. 2, pp. 841–849, 2013.

[6] M. Barcaro, N. Bianchi, and F. Magnussen, "Analysis and tests of a dual three-phase 12-slot 10-pole permanent-magnet motor," *IEEE Trans. Ind. Appl.*, vol. 46, no. 6, pp. 2355–2362, 2010.

[7] M. Barcaro, N. Bianchi, and F. Magnussen, "Faulty operations of a PM fractional-slot machine with a dual three-phase winding," *IEEE Trans. Ind. Electron.*, vol. 58, no. 9, pp. 3825–3832, 2011.

[8] M. Barcaro, N. Bianchi, and F. Magnussen, "Six-phase supply feasibility using a PM fractional-slot dual winding machine," *IEEE Trans. Ind. Appl.*, vol. 47, no. 5, pp. 2042–2050, 2011.

[9] P. Zheng, F. Wu, Y. Lei, Y. Sui, and B. Yu, "Investigation of a novel 24-slot/14-pole six-phase fault-tolerant modular permanent-magnet in-wheel motor for electric vehicles," *Energies*, vol. 6, no. 10, pp. 4980–5002, 2013.

[10] V. Patel, J. Wang, W. Wang, and X. Chen, "Six-phase fractional-slot-per-pole-per-phase permanent-magnet machines with low space harmonics for electric vehicle application," *IEEE Trans. Ind. Appl.*, vol. 50, no. 4, pp. 2554–2563, 2014.

[11] V. Patel, J. Wang, D. Nugraha, R. Vuletic, and J. Tousen, "Enhanced availability of drivetrain through novel multi-phase permanent magnet machine drive," *IEEE Trans. Ind. Electron.*, vol. 63, no. 1, pp. 469–480, 2016.

[12] S. Abdel-Khalik, S. Ahmed, and A. Massoud, "A six-phase 24 slot-10 pole permanent-magnet machine with low space harmonics for electric vehicle applications," *IEEE Trans. Magn.*, vol. 52, no. 6, pp. 1–10, 2016.

[13] S. Abdel-Khalik, S. Ahmed, and A. Masmoud, "Low space harmonics cancellation in double layer fractional slot winding using dual multi-phase winding," *IEEE Trans. Magn.*, vol. 51, no. 5, pp. 8104710, 2015.

[14] A. Matyas, G. Aroquiadassou, C. Martis, A. Mpanda-Mabwe, and K. Biro, "Design of six-phase synchronous and induction machines for EPS," *2010 XIX Int. Conf. Electr. Mach. (ICEM)*, Italy, 2010, pp. 1–6.

[15] P. L. Xu, Z. Q. Zhu, B. Shao, *et al.*, "Analysis of dual 3-phase fractional-slot non-overlapping winding pm synchronous machines with different angle displacements," *IEEE Energy Convers. Cong. Expo. (ECCE)*, Baltimore, MD, USA, 2019, pp. 5616–5623.

[16] P. L. Xu, J. H. Feng, S. Y. Guo, *et al.*, "Analysis of dual three-phase permanent-magnet synchronous machines with different angle displacements," *IEEE Trans. Ind. Electron.*, vol. 65, no. 3, pp. 1941–1954, 2018.

[17] J. H. Feng, P. L. Xu, Z. Q. Zhu, *et al.*, "Comparative study of dual 3-phase permanent magnet machines with coil span of two slot-pitches," *IET Electr. Power Appl.* pp. 1–13, 2021.

[18] Oak Ridge National Laboratory, "Evaluation of the 2010 Toyota Prius hybrid synergy drive system," 2011.

[19] G. Dajaku and D. Gerling, "A novel 24-slots/10-poles winding topology for electric machines," *Proc. IEEE Int. Electr. Mach. Drives Conf. (IEMDC)*, 2011, pp. 65–70.

[20] M. Harke., 'Fractional slot windings with a coil span of two slots and less content of low order harmonics', *Proc. Int. Symp. Power Electron., Elect. Drives, Automat. Motion (SPEEDAM)*, June. 2018, pp. 1309–1314.

[21] S. Zhu, T. Cox, Z. Xu, and C. Gerada, "Novel 24-slots14-poles fractional-slot concentrated winding topology with low-space harmonics for electrical machine," *J. Eng.*, vol. 2019, no. 17, pp. 3784–3788, 2019.

[22] H. Y. Sun and K. Wang, "Space harmonics elimination for fractional-slot windings with two-slot coil pitch," *IEEE Access*, vol. 7, pp. 106961–106972, 2019.

[23] S. Zhu, W. Zhao, G. Liu, Y. Mao, and Y. Sun, "Effect of phase shift angle on radial force and vibration behaviour in dual three-phase PMSM," *IEEE Trans. Ind. Electron.*, vol. 68, no. 4, pp. 2988–2998, 2021.

[24] P. B. Reddy, K.-K. Huh, and A. M. EL-Refaie, "Generalized approach of stator shifting in interior permanent-magnet machines equipped with fractional-slot concentrated windings," *IEEE Trans. Ind. Electron.*, vol. 61, no. 9, pp. 5035–5046, 2014.

[25] Y. X. Li, "*Novel modular dual 3-phase permanent magnet machines for wind power application*," PhD thesis, University of Sheffield, 2018.

[26] Z. Q. Zhu and Y. X. Li, "Modularity techniques in high performance permanent magnet machines and applications," *CES Trans. Electr. Mach. Syst.*, vol. 2, no. 1, pp. 93–103, 2018.

[27] J. Y. Chen, C. V. Nayar, and L. Y. Xu, "Design and finite-element analysis of an outer-rotor permanent-magnet generator for directly coupled wind turbines," *IEEE Trans. Magn.*, vol. 36, no. 5, pp. 3802–3809, 2000.

[28] E. Fornasiero, L. Alberti, N. Bianchi, and S. Bolognani, "Considerations on selecting fractional-slot nonoverlapped coil windings," *IEEE Trans. Ind. Appl.*, vol. 49, no. 3, pp. 1316–1324, 2013.

[29] Y. G. Chen, Z. M. Du, W. G. Zhong, and L. B. Kong, "Modular stator structure permanent magnet synchronous machine," *World Automation Congress (WAC 2008)*, 2008, pp. 1–5.

[30] Z. Q. Zhu, M. L. Mohd Jamil, and L. J. Wu, "Influence of slot and pole number combinations on unbalanced magnetic force in PM machines with diametrically asymmetric windings," *IEEE Trans. Ind. Appl.*, vol. 49, no. 1, pp. 19–30, 2013.

[31] N. J. Baker, D. J. B. Smith, M. C. Kulan, and S. Turvey, "Design and performance of a segmented stator permanent magnet alternator for aerospace," *IEEE Trans. Energy Convers.*, vol. 33, no. 1, pp. 40–48, 2018.

[32] Y. X. Li, Z. Q. Zhu, A. Thomas, Y. Wu, and X. M. Wu, "Comparison of modular dual 3-phase PM machines with overlapping/non-overlapping windings," *IEEE Trans. Ind. Appl.*, vol. 55, no. 4, pp. 3566–3576, 2019.

[33] Y. X. Li, Z. Q. Zhu, A. S. Thomas, Z. Y. Wu, and X. M. Wu, "Novel modular fractional slot permanent magnet machines with redundant teeth," *IEEE Trans. Magn.*, vol. 55, no. 9, pp. 1–10, 2019.

[34] Y. X. Li, Z. Q. Zhu, and A. S. Thomas, "Generic slot and pole number combinations for novel modular permanent magnet dual 3-phase machines with redundant teeth," *IEEE Trans. Energy Convers.*, vol. 35, no. 3, pp. 1676–1687, 2020.

Chapter 5

Modeling of multiphase PMSMs

5.1 Introduction

Mathematical modeling for multiphase permanent magnet synchronous machines (PMSMs) is essential for improving electromagnetic and control performance and will be introduced in this chapter, with a particular focus on dual three-phase (DTP) PMSMs.

In multiphase PMSMs, similar to single three-phase (STP) machines, the modeling process entails coordinate transformation. The original phase model of multiphase PMSMs can be obtained by considering different phases individually and directly. According to the basic electromagnetic principle and the relative positions between the windings of different phases, the relationship between the voltages, currents, and flux linkages of phase windings can be easily determined, and thus, the phase model can be obtained. However, the obtained phase model is often complex, featuring the characteristics of high order, nonlinearity, and strong coupling. These characteristics pose significant challenges in the analysis and design of the entire drive system. Hence, selecting an appropriate coordinate system is important in the modeling of multiphase PMSMs. With an appropriate coordinate system, the original phase model can be simplified into a form that is easy to analyze. In PMSMs, the field-oriented control (vector control) is also closely related to the selection of the coordinate system, and different coordinate systems usually correspond to different vector control strategies.

In STP PMSMs, field-oriented control is the most widely used control strategy, with the *dq* model serving as the most conventional modeling method. Furthermore, the *dq* modeling method is also widely utilized in other multiphase PMSMs. Meanwhile, for DTP PMSMs, as there are two winding sets in the machines, the increased number of variables also offers greater flexibility in the selection of the transformation matrix. Generally, both the double *dq* transformation [1–3] and the vector space decomposition (VSD) transformation [4] are commonly used to establish the mathematical model of DTP PMSMs.

The arrangement of this chapter is summarized in Figure 5.1, where the modeling of STP, multiphase, and DTP PMSMs are all described. For STP and multiphase PMSMs, only *dq* models will be explained, while for DTP PMSMs, both double *dq* and VSD will be introduced in detail. Following the elucidation of the mathematical models of STP and DTP PMSMs, the separation of the

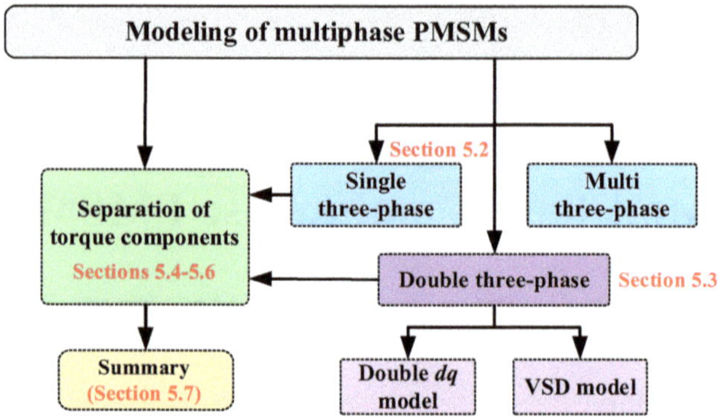

Figure 5.1 Structure of Chapter 5.

instantaneous PM and reluctance torque components in STP and DTP PMSMs will be demonstrated for further illustration, where the frozen permeability (FP) method is used.

5.2 Modeling of single three-phase and multiphase PMSMs

5.2.1 Single three-phase PMSMs

When the hysteresis and eddy current losses are ignored, the voltage equation of the STP PMSMs in stationary ABC frame can be written as

$$\begin{bmatrix} u_a \\ u_b \\ u_c \end{bmatrix} = \begin{bmatrix} R_s & 0 & 0 \\ 0 & R_s & 0 \\ 0 & 0 & R_s \end{bmatrix} \begin{bmatrix} i_a \\ i_b \\ i_c \end{bmatrix} + \frac{d}{dt}\left(\begin{bmatrix} L_{aa} & L_{ba} & L_{ca} \\ L_{ab} & L_{bb} & L_{cb} \\ L_{ac} & L_{bc} & L_{cc} \end{bmatrix} \begin{bmatrix} i_a \\ i_b \\ i_c \end{bmatrix} \right) + \frac{d}{dt} \begin{bmatrix} \psi_{fa} \\ \psi_{fb} \\ \psi_{fc} \end{bmatrix} \tag{5.1}$$

where $\begin{bmatrix} u_a & u_b & u_c \end{bmatrix}^T$, $\begin{bmatrix} i_a & i_b & i_c \end{bmatrix}^T$, and $\begin{bmatrix} \psi_{fa} & \psi_{fb} & \psi_{fc} \end{bmatrix}^T$ are the three-phase voltages, currents, and PM flux linkages, respectively. R_s is the stator phase resistance. L represents the phase self-inductance or mutual inductance between two phases.

By multiplying the Park transformation (5.2) in both sides of the equation (5.1), the voltage equation in synchronous reference dq frame can be written as (5.3).

$$\boldsymbol{C}(\theta_e) = \frac{2}{3} \begin{bmatrix} \cos\theta_r & \cos\left(\theta_r - \frac{2}{3}\pi\right) & \cos\left(\theta_r + \frac{2}{3}\pi\right) \\ -\sin\theta_r & -\sin\left(\theta_r - \frac{2}{3}\pi\right) & -\sin\left(\theta_r + \frac{2}{3}\pi\right) \end{bmatrix} \tag{5.2}$$

$$\begin{bmatrix} u_d \\ u_q \end{bmatrix} = \begin{bmatrix} R_s & 0 \\ 0 & R_s \end{bmatrix} \begin{bmatrix} i_d \\ i_q \end{bmatrix} + \frac{d}{dt} \begin{bmatrix} L_d i_d \\ L_q i_q \end{bmatrix} + \omega_e \begin{bmatrix} 0 & -1 \\ 1 & 0 \end{bmatrix} \begin{bmatrix} L_d i_d + \psi_f \\ L_q i_q \end{bmatrix} \tag{5.3}$$

where θ_r and ω_e are the electrical angle and angular speed of the rotor. $[u_d\ u_q]^T$ and $[i_d\ \ i_q]^T$ are the voltages and currents in dq-axes, respectively. L_d and L_q are the dq-axis inductances and can be expressed as $L_d = L_\sigma + L_{ad}$, $L_q = L_\sigma + L_{aq}$. L_σ is the leakage inductance. L_{ad} and L_{aq} are the armature reaction inductances of d- and q-axes. ψ_f is the PM flux linkage.

The torque equation and mechanical equation are:

$$T_e = \frac{3}{2} p \psi_f i_q + \frac{3}{2} p (L_d - L_q) i_d i_q$$

$$J \frac{d\omega_m}{dt} = T_e - T_L - B_m \omega_m \tag{5.4}$$

where p is the number of pole pairs. T_L is the load torque. ω_m is the mechanical angular speed of the rotor. B_m is the friction coefficient, and J is the inertia of the rotor.

5.2.2 Multiphase PMSMs

The winding axes of multiphase machines, including multi-three-phase machines and other multiphase machines, are shown in Figure 5.2. The multi-three-phase machine, shown in Figure 5.2(a), can be regarded as a combination of several three-phase winding sets, and thus the phase number must be the multiples of three, i.e., 6, 9, 12, etc. Two adjacent three-phase winding sets are shifted by an angle of β, where $\beta = \pi / m$. In contrast, when the phase number of other multiphase machines is more than three but not multiples of three, e.g., five-phase and seven-phase machines, as shown in Figure 5.2(b), the phase shifting angle between two adjacent phases is β but it should satisfy $\beta = 2\pi/m$.

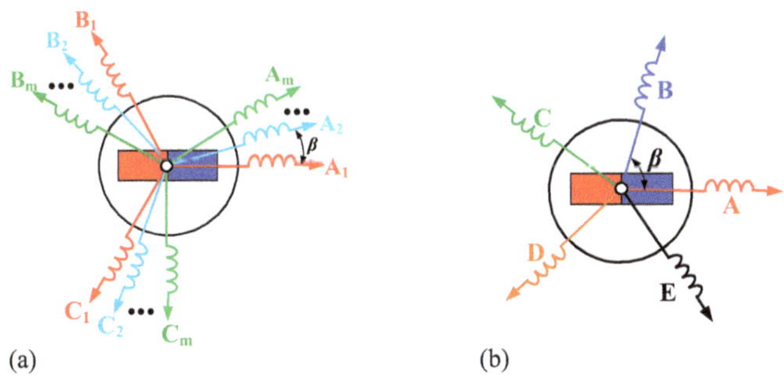

Figure 5.2 Winding axes of multiphase machines. (a) Multi-three-phase machines. (b) Other multiphase machines (a five-phase machine is shown as an example).

The VSD transformation of these multiphase machines can be written as

$$\begin{bmatrix} u_{\alpha 1} \\ u_{\beta 1} \\ u_{z1} \\ u_{z2} \\ u_{z3} \\ \cdots \\ u_0 \end{bmatrix} = \frac{2}{m} \begin{bmatrix} 1 & \cos\left(\frac{2\pi}{m}\right) & \cos\left(\frac{4\pi}{m}\right) & \cos\left(\frac{6\pi}{m}\right) & \cos\left(\frac{8\pi}{m}\right) & \cdots & \cos\left(\frac{2(m-1)\pi}{m}\right) \\ 0 & \sin\left(\frac{2\pi}{m}\right) & \sin\left(\frac{4\pi}{m}\right) & \sin\left(\frac{6\pi}{m}\right) & \sin\left(\frac{8\pi}{m}\right) & \cdots & \sin\left(\frac{2(m-1)\pi}{m}\right) \\ 1 & \cos\left(\frac{6\pi}{m}\right) & \cos\left(\frac{12\pi}{m}\right) & \cos\left(\frac{18\pi}{m}\right) & \cos\left(\frac{24\pi}{m}\right) & .. & \cos\left(\frac{6(m-1)\pi}{m}\right) \\ 0 & \sin\left(\frac{6\pi}{m}\right) & \sin\left(\frac{12\pi}{m}\right) & \sin\left(\frac{18\pi}{m}\right) & \sin\left(\frac{24\pi}{m}\right) & \cdots & \sin\left(\frac{6(m-1)\pi}{m}\right) \\ 1 & \cos\left(\frac{10\pi}{m}\right) & \cos\left(\frac{20\pi}{m}\right) & \cos\left(\frac{30\pi}{m}\right) & \cos\left(\frac{40\pi}{m}\right) & \cdots & \cos\left(\frac{10(m-1)\pi}{m}\right) \\ \cdots & \cdots & \cdots & \cdots & \cdots & \cdots & \cdots \\ \frac{1}{2} & \frac{1}{2} & \frac{1}{2} & \frac{1}{2} & \frac{1}{2} & \cdots & \frac{1}{2} \end{bmatrix} \begin{bmatrix} u_a \\ u_b \\ u_c \\ u_d \\ u_e \\ \cdots \\ u_g \end{bmatrix} \tag{5.5}$$

5.3 Modeling of dual three-phase PMSMs

5.3.1 Double dq model

The double *dq* modeling method is based on the synchronous *dq* model in STP PMSMs. Consequently, a DTP PMSM can be treated as two separate STP PMSMs, and the model is expressed under the assumptions of:

1. The windings are sinusoidally distributed.
2. The magnetic saturation, the mutual leakage inductances, and the core losses are neglected.

The STP PMSM model in synchronous reference *dq* frame can be extended to each three-phase set of the DTP PMSMs, and thus, the double *dq* model can be derived. The machine voltage equation and flux linkage equation in a stationary frame can be expressed as

$$\begin{bmatrix} u_a \\ u_b \\ u_c \\ u_x \\ u_y \\ u_z \end{bmatrix} = \begin{bmatrix} R_s & 0 & 0 & 0 & 0 & 0 \\ 0 & R_s & 0 & 0 & 0 & 0 \\ 0 & 0 & R_s & 0 & 0 & 0 \\ 0 & 0 & 0 & R_s & 0 & 0 \\ 0 & 0 & 0 & 0 & R_s & 0 \\ 0 & 0 & 0 & 0 & 0 & R_s \end{bmatrix} \begin{bmatrix} i_a \\ i_b \\ i_c \\ i_x \\ i_y \\ i_z \end{bmatrix} + \frac{d}{dt} \begin{bmatrix} \psi_a \\ \psi_b \\ \psi_c \\ \psi_x \\ \psi_y \\ \psi_z \end{bmatrix} \tag{5.6}$$

$$\begin{bmatrix} \psi_a \\ \psi_b \\ \psi_c \\ \psi_x \\ \psi_y \\ \psi_z \end{bmatrix} = \begin{bmatrix} L_{aa} & L_{ba} & L_{ca} & L_{xa} & L_{ya} & L_{za} \\ L_{ab} & L_{bb} & L_{cb} & L_{xb} & L_{yb} & L_{zb} \\ L_{ac} & L_{bc} & L_{cc} & L_{xc} & L_{yc} & L_{zc} \\ L_{ax} & L_{bx} & L_{cx} & L_{xx} & L_{yx} & L_{zx} \\ L_{ay} & L_{by} & L_{cy} & L_{xy} & L_{yy} & L_{zy} \\ L_{az} & L_{bz} & L_{cz} & L_{xz} & L_{yz} & L_{zz} \end{bmatrix} \begin{bmatrix} i_a \\ i_b \\ i_c \\ i_x \\ i_y \\ i_z \end{bmatrix} + \begin{bmatrix} \psi_{fa} \\ \psi_{fb} \\ \psi_{fc} \\ \psi_{fx} \\ \psi_{fy} \\ \psi_{fz} \end{bmatrix} \tag{5.7}$$

where ψ represents the stator flux linkage variable, and ψ_f represents the PM flux linkage variable. L means the self- or mutual inductances.

The double dq transformation is given by

$$\begin{bmatrix} u_{\alpha 1} \\ u_{\beta 1} \\ u_{\alpha 2} \\ u_{\beta 2} \end{bmatrix} = \frac{2}{3} \begin{bmatrix} 1 & \cos\left(\frac{2\pi}{3}\right) & \cos\left(-\frac{2\pi}{3}\right) & 0 & 0 & 0 \\ 0 & \sin\left(\frac{2\pi}{3}\right) & \sin\left(-\frac{2\pi}{3}\right) & 0 & 0 & 0 \\ 0 & 0 & 0 & \cos\left(\frac{\pi}{6}\right) & \cos\left(\frac{5\pi}{6}\right) & \cos\left(-\frac{\pi}{2}\right) \\ 0 & 0 & 0 & \sin\left(\frac{\pi}{6}\right) & \sin\left(\frac{5\pi}{6}\right) & \sin\left(-\frac{\pi}{2}\right) \end{bmatrix} \begin{bmatrix} u_a \\ u_b \\ u_c \\ u_x \\ u_y \\ u_z \end{bmatrix} \tag{5.8}$$

$$\begin{bmatrix} u_{d1} \\ u_{q1} \\ u_{d2} \\ u_{q2} \end{bmatrix} = \begin{bmatrix} \cos(\theta) & \sin(\theta) & 0 & 0 \\ -\sin(\theta) & \cos(\theta) & 0 & 0 \\ 0 & 0 & \cos(\theta) & \sin(\theta) \\ 0 & 0 & -\sin(\theta) & \cos(\theta) \end{bmatrix} \begin{bmatrix} u_{\alpha 1} \\ u_{\beta 1} \\ u_{\alpha 2} \\ u_{\beta 2} \end{bmatrix} \tag{5.9}$$

where “1” and “2” in the subscript represent the variables of the first and second sets, respectively.

Then, the voltage equation in the original phase coordinates can be described as follows:

$$\begin{aligned} \begin{bmatrix} u_{d1} \\ u_{q1} \end{bmatrix} &= \begin{bmatrix} R_s + L_{d1}p & 0 \\ 0 & R_s + L_{q1}p \end{bmatrix} \cdot \begin{bmatrix} i_{d1} \\ i_{q1} \end{bmatrix} + \omega_e \begin{bmatrix} -L_{q1} i_{q1} \\ L_{d1} i_{d1} \end{bmatrix} + \omega_e \begin{bmatrix} 0 \\ \psi_{fd} \end{bmatrix} \\ &+ \begin{bmatrix} M_d p & 0 \\ 0 & M_q p \end{bmatrix} \cdot \begin{bmatrix} i_{d2} \\ i_{q2} \end{bmatrix} + \omega_e \begin{bmatrix} -M_q \cdot i_{q2} \\ M_d \cdot i_{d2} \end{bmatrix} \end{aligned} \tag{5.10}$$

$$\begin{aligned} \begin{bmatrix} u_{d2} \\ u_{q2} \end{bmatrix} &= \begin{bmatrix} R_s + L_{d2}p & 0 \\ 0 & R_s + L_{q2}p \end{bmatrix} \cdot \begin{bmatrix} i_{d2} \\ i_{q2} \end{bmatrix} + \omega_e \begin{bmatrix} -L_{q2} i_{q2} \\ L_{d2} i_{d2} \end{bmatrix} + \omega_e \begin{bmatrix} 0 \\ \psi_{fd} \end{bmatrix} \\ &+ \begin{bmatrix} M_d p & 0 \\ 0 & M_q p \end{bmatrix} \cdot \begin{bmatrix} i_{d1} \\ i_{q1} \end{bmatrix} + \omega_e \begin{bmatrix} -M_q \cdot i_{q1} \\ M_d \cdot i_{d1} \end{bmatrix} \end{aligned} \tag{5.11}$$

where p represents the derivative. u_{di}, u_{qi}, i_{di}, and i_{qi} with $\{i = 1 \text{ and } 2\}$ represent the dq-axis voltages and currents of the first and second winding sets. $L_{d1,2}$ and $L_{q1,2}$ are the dq-axis inductances. M_d and M_q are the mutual inductance between the two three-phase winding sets in dq-axes and satisfy

$$\begin{cases} L_d = M_d + L_\sigma \\ L_q = M_q + L_\sigma \end{cases} \tag{5.12}$$

Overall, (5.10)–(5.12) represent the DTP model in dq-axes and can be depicted in Figure 5.3. Compared to the model of STP machines, the double dq model of the

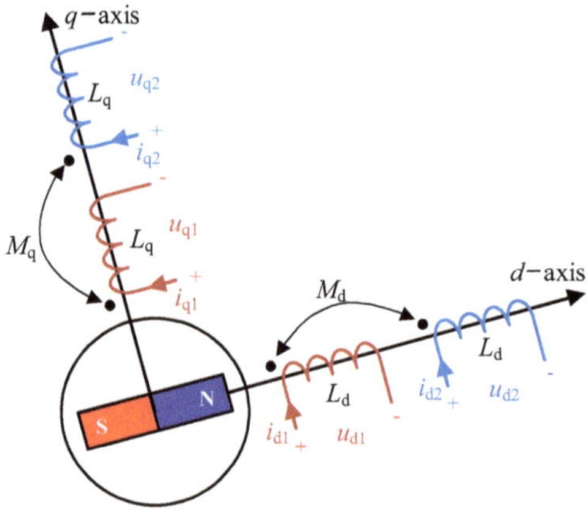

Figure 5.3 Representation of modeling of DTP PMSMs in dq-axes.

DTP machines includes more terms with M_d and M_q in the voltage equations. These terms are introduced by the magnetic coupling between the two three-phase winding sets, which means the currents in one set will generate flux linkage in the other set, leading to the induced electromotive force and motion electromotive force in the voltage equations.

Thus, when a DTP PMSM is analyzed by using a double *dq* model, the resultant electromagnetic torque can be expressed as

$$T_e = \frac{3}{2}p\left(\psi_{fd}\left(i_{q1} + i_{q2}\right) + \left(L_d - L_q\right)\left(i_{d1}i_{q1} + i_{d2}i_{q2}\right) + \left(M_d - M_q\right)\left(i_{d1}i_{q2} + i_{d2}i_{q1}\right)\right) \tag{5.13}$$

It can be seen that the double *dq* model maps the variables in two sets of three-phase windings into two *dq* coordinates, facilitating the applications of current control strategies from STP to DTP PMSMs. However, the mutual coupling between two *dq* coordinates may significantly deteriorate the control performance.

5.3.2 Vector space decomposition model

5.3.2.1 Vector space decomposition transformation

The VSD is a set of mathematical transformations to separate the components with different phase angle relationships. Since the phase angle relationship of the harmonics is different from the fundamental, VSD can be used to separate the components of voltages, currents, and flux linkages in the DTP PMSMs [4], where all variables in a DTP PMSM can be transformed into three orthogonal subspaces.

Actually, the *ABC*-$\alpha\beta z$ transformation in the general three-phase system is shown below, which can be regarded as one of the VSD transformations.

$$\begin{bmatrix} u_\alpha \\ u_\beta \\ u_z \end{bmatrix} = \frac{2}{3}\begin{bmatrix} 1 & -\frac{1}{2} & -\frac{1}{2} \\ 0 & \frac{\sqrt{3}}{2} & -\frac{\sqrt{3}}{2} \\ \frac{1}{2} & \frac{1}{2} & \frac{1}{2} \end{bmatrix}\begin{bmatrix} u_a \\ u_b \\ u_c \end{bmatrix} \tag{5.14}$$

where "u" represents the variables in the corresponding axis.

The axis definition of the *ABC-αβz* transformation is shown in Figure 5.4. The α- and z-axes are aligned to the axis of phase a, and β-axis is delayed by 90 degrees with respect to α the axis. It is known that the zero-sequence components, e.g., the 3rd and the 9th, can be decomposed from the phase variables into z-axis using the *ABC-αβz* transformation (5.14), while the other components, e.g., the fundamental, 5th, and 7th harmonics, are decomposed in α- and β-axes.

Since the multi-three-phase machines can be regarded as the composition of multiple three-phase sets, the VSD transformation can also be written in the format of three-phase sets as follows:

$$\begin{bmatrix} u_{\alpha\beta} \\ u_{z1z2} \\ u_{z3z4} \\ \cdots \\ u_{z\left(\frac{2m}{3}-3\right)z\left(\frac{2m}{3}-2\right)} \\ u_{o1} \\ u_{o2} \\ \cdots \\ u_{o\frac{m}{3}} \end{bmatrix} = \frac{3}{m}\begin{bmatrix} T(0) & T\left(\frac{\pi}{m}\right) & T\left(\frac{2\pi}{m}\right) & \cdots & T\left(\frac{(m-1)\pi}{m}\right) \\ T(0) & T\left(\frac{-5\pi}{m}\right) & T\left(\frac{-10\pi}{m}\right) & \cdots & T\left(\frac{-5(m-1)\pi}{m}\right) \\ T(0) & T\left(\frac{7\pi}{m}\right) & T\left(\frac{14\pi}{m}\right) & \cdots & T\left(\frac{7(m-1)\pi}{m}\right) \\ \cdots & \cdots & \cdots & \cdots & \cdots \\ T(0) & T\left(\frac{n\pi}{m}\right) & T\left(\frac{2n\pi}{m}\right) & \cdots & T\left(\frac{n(m-1)\pi}{m}\right) \\ T_0 & 0 & 0 & \cdots & 0 \\ 0 & T_0 & 0 & \cdots & 0 \\ \cdots & \cdots & \cdots & \cdots & \cdots \\ 0 & 0 & 0 & \cdots & T_0 \end{bmatrix}\begin{bmatrix} u_{abc1} \\ u_{abc2} \\ u_{abc3} \\ \cdots \\ u_{abc\frac{m}{3}} \end{bmatrix} \tag{5.15}$$

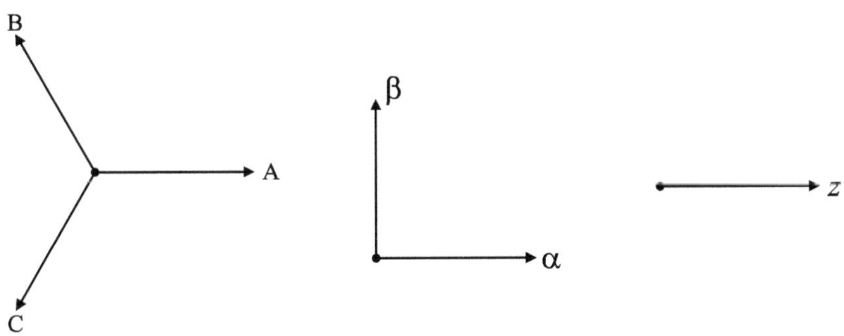

(a) (b) (c)

Figure 5.4 Definition of axes in ABC-αβz transformation. (a) abc frame, (b) αβ frame, and (c) z frame.

where $n = 1, -5, 7, -11, 13, \ldots$ and the sub-matrices can be expressed as

$$u_{abc} = \begin{bmatrix} u_{a1} & u_{b1} & u_{c1} \end{bmatrix}^T$$
$$u_{\alpha\beta} = \begin{bmatrix} u_{\alpha} & u_{\beta} \end{bmatrix}^T$$
$$u_{z1z2} = \begin{bmatrix} u_{z1} & u_{z2} \end{bmatrix}^T$$
$$T(\vartheta) = \frac{2}{3}\begin{bmatrix} \cos(\vartheta) & \cos\left(\vartheta + \frac{2\pi}{3}\right) & \cos\left(\vartheta - \frac{2\pi}{3}\right) \\ \sin(\vartheta) & \sin\left(\vartheta + \frac{2\pi}{3}\right) & \sin\left(\vartheta - \frac{2\pi}{3}\right) \end{bmatrix}$$
$$T_0 = \begin{bmatrix} \frac{1}{2} & \frac{1}{2} & \frac{1}{2} \end{bmatrix} \tag{5.16}$$

Take a dual-three-phase machine as an example, m equals 6 in dual-three-phase machines and β should be $\pi/6$, i.e., 30 degrees. $abc1 = abc$ and $abc2 = xyz$.

The VSD transformation can be rewritten as

$$\begin{bmatrix} u_{\alpha\beta} \\ u_{z1z2} \\ u_{o1} \\ u_{o2} \end{bmatrix} = \frac{1}{2}\begin{bmatrix} T(0) & T\left(\frac{\pi}{6}\right) \\ T(0) & T\left(\frac{-5\pi}{6}\right) \\ T_0 & 0 \\ 0 & T_0 \end{bmatrix}\begin{bmatrix} u_{abc} \\ u_{xyz} \end{bmatrix} \tag{5.17}$$

and considering $T\left(\frac{-5\pi}{6}\right) = -T\left(\frac{\pi}{6}\right)$, (5.17) can be expressed as

$$\begin{bmatrix} u_{\alpha\beta} \\ u_{z1z2} \\ u_{o1} \\ u_{o2} \end{bmatrix} = \frac{1}{2}\begin{bmatrix} T(0) & T\left(\frac{\pi}{6}\right) \\ T(0) & -T\left(\frac{\pi}{6}\right) \\ T_0 & 0 \\ 0 & T_0 \end{bmatrix}\begin{bmatrix} u_{abc} \\ u_{xyz} \end{bmatrix} \tag{5.18}$$

It can be clearly seen from (5.18) that the components in $\alpha\beta$ subspace can be considered as the sum of vectors from two sets, while the components in z_1z_2 subspace can be considered as the difference of the vectors from two sets. The definition of the axes in the VSD transformation is shown in Figure 5.5.

The transformation for VSD in (5.18) can be expanded as follows.

$$\begin{bmatrix} u_{\alpha} \\ u_{\beta} \\ u_{z_1} \\ u_{z_2} \\ u_{o_1} \\ u_{o_2} \end{bmatrix} = \frac{1}{3}\begin{bmatrix} 1 & -1/2 & -1/2 & \sqrt{3}/2 & -\sqrt{3}/2 & 0 \\ 0 & \sqrt{3}/2 & -\sqrt{3}/2 & 1/2 & 1/2 & -1 \\ 1 & -1/2 & -1/2 & -\sqrt{3}/2 & \sqrt{3}/2 & 0 \\ 0 & -\sqrt{3}/2 & \sqrt{3}/2 & 1/2 & 1/2 & -1 \\ 1 & 1 & 1 & 0 & 0 & 0 \\ 0 & 0 & 0 & 1 & 1 & 1 \end{bmatrix}\begin{bmatrix} u_a \\ u_b \\ u_c \\ u_x \\ u_y \\ u_z \end{bmatrix} \tag{5.19}$$

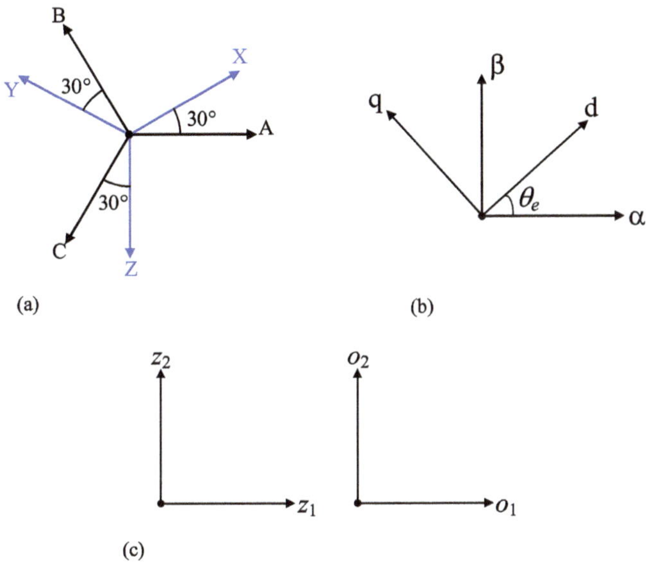

Figure 5.5 Definition of axes in VSD transformation of DTP PMSMs. (a) Natural ABC–XYZ frame. (b) Frames in αβ subspace. (c) Frames in z_1z_2 subspace and zero-sequence subspace.

and

$$\begin{bmatrix} u_d \\ u_q \\ u_{z_1} \\ u_{z_2} \\ u_{o_1} \\ u_{o_2} \end{bmatrix} = \begin{bmatrix} \cos(\theta_e) & \sin(\theta_e) & 0 & 0 & 0 & 0 \\ -\sin(\theta_e) & \cos(\theta_e) & 0 & 0 & 0 & 0 \\ 0 & 0 & 1 & 0 & 0 & 0 \\ 0 & 0 & 0 & 1 & 0 & 0 \\ 0 & 0 & 0 & 0 & 1 & 0 \\ 0 & 0 & 0 & 0 & 0 & 1 \end{bmatrix} \begin{bmatrix} u_\alpha \\ u_\beta \\ u_{z_1} \\ u_{z_2} \\ u_{o_1} \\ u_{o_2} \end{bmatrix} \tag{5.20}$$

Similar to the *abc-αβz* transformation in (5.15), all phase variables, i.e., currents, voltages, and flux linkages, can be decomposed into three orthogonal subspaces, i.e., the $\alpha\beta$ subspace (or the dq subspace), the z_1z_2 subspace, and the o_1o_2 subspace.

These subspaces exhibit the following characteristics for the fundamental and harmonics:

- The fundamental component of the machine variables and the kth order harmonics with $k = 12m \pm 1$ ($m = 1, 2, 3, \ldots$), i.e., the fundamental and the 11th, 13th, ... harmonics, are transformed into the $\alpha\beta$ subspace. The $\alpha\beta$ subspace is directly associated with the air gap flux. Hence, the variables in the $\alpha\beta$ subspace generate a substantial MMF in the machine air gap, resulting in torque production and electromechanical energy conversion.

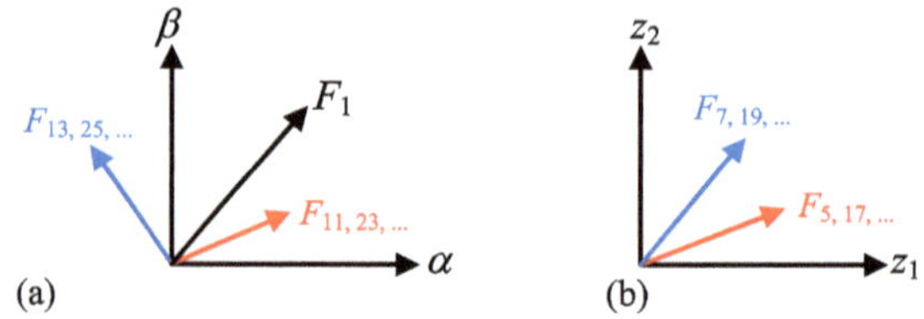

Figure 5.6 Phasors of variables. (a) $\alpha\beta$ subspace (fundamental and 11th, 13th,... harmonics). (b) z_1z_2 subspace (5th, 7th, 17th, 19th,... harmonics).

- Harmonics with $k = 6m \pm 1$ ($m = 1, 3, 5, \ldots$), i.e., the 5th, 7th, 17th, 19th,... harmonics, are mapped into the z_1z_2 subspace. As the z_1z_2 subspace is orthogonal to the $\alpha\beta$ subspace, the variables in this subspace are not related to the generation of any MMF in the air gap and do not contribute to any torque or torque ripples.
- Zero-sequence $k = 3m$ ($m = 1, 2, 3, \ldots$) harmonics, which are also non-electromechanical energy conversion related, are mapped into the o_1o_2 subspace to form the conventional zero sequence components. When two sets of three-phase windings have two isolated neutral points, no zero-sequence current path will exist in the o_1o_2 subspace. Hence, the machine model can be simplified from a six-dimensional system to a four-dimensional system for isolated neutral point topology.

Therefore, by using the VSD modeling method, the coupling between two sets of three-phase windings for DTP-PMSMs can be fully coupled to ease the control.

Figure 5.6 further illustrates the phasors of variables in the two subspaces. The fundamental and the 11th, 13th,... harmonics in the $\alpha\beta$ subspace are depicted in Figure 5.6 (a). The 5th, 7th, 17th, 19th,... harmonics in the z_1z_2 subspace are shown in Figure 5.6 (b).

Taking the current variables as an example, Figure 5.7 shows the simulation results of the current waveforms and the spectra in a DTP machine. As can be seen, there are current harmonic components in the phase currents with the major current harmonics of the 5th, 7th, 11th, and 13th. Using the VSD transformation (5.18), the current harmonics are mapped into different subspaces, as shown in Figure 5.7(b) and (c). The fundamental component together with the 11th and 13th components are decoupled into the $\alpha\beta$ subspace, while the sum of the 5th and 7th is decomposed from the phase currents and mapped into the z_1z_2 subspace. This harmonic mapping is consistent with the description above, and the currents in Figure 5.7 can be changed to other variables in DTP machines, such as voltage and flux linkage, etc.

5.3.2.2 Vector space decomposition model

Neglecting the zero-sequence subspace, the VSD model of the DTP PMSM can be established as below through left multiplying both sides of the equations (5.6) and

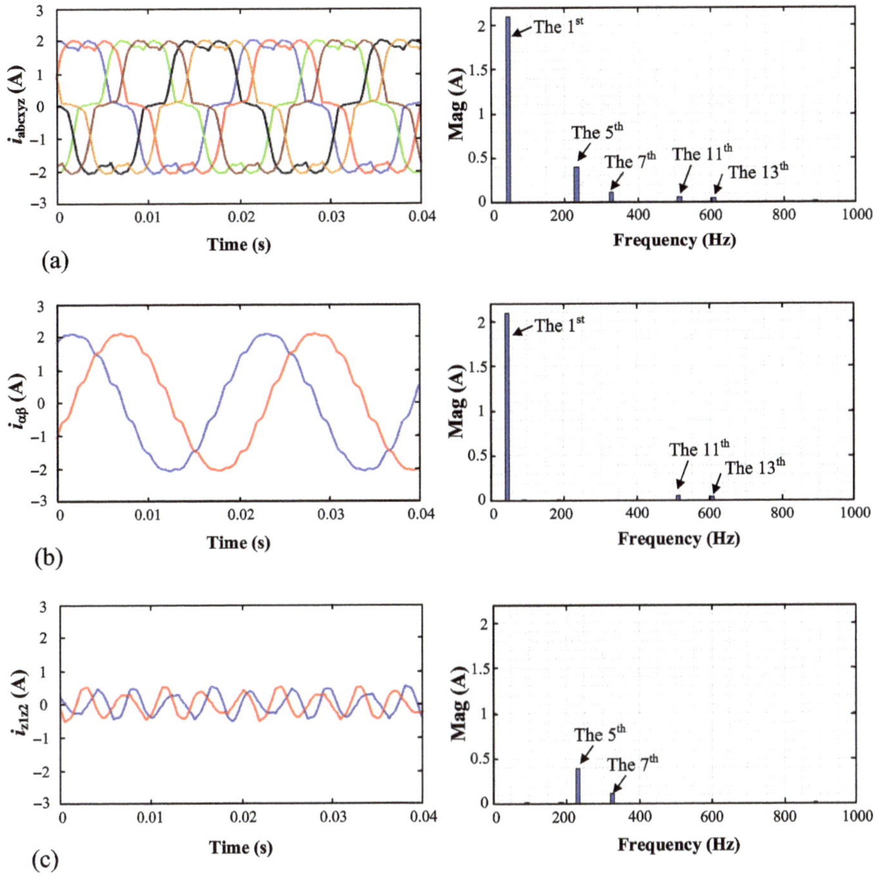

Figure 5.7 *Current harmonics in phase, αβ, and z_1z_2 subspaces after VSD transformation of DTP PMSMs. (a) Phase currents, (b) αβ subspace, and (c) z_1z_2 subspace.*

(5.7) by the transformation matrix of (5.14).

$$\begin{bmatrix} u_\alpha \\ u_\beta \\ u_{z1} \\ u_{z2} \end{bmatrix} = \begin{bmatrix} R_s & 0 & 0 & 0 \\ 0 & R_s & 0 & 0 \\ 0 & 0 & R_s & 0 \\ 0 & 0 & 0 & R_s \end{bmatrix} \begin{bmatrix} i_\alpha \\ i_\beta \\ i_{z1} \\ i_{z2} \end{bmatrix} + \frac{d}{dt} \begin{bmatrix} \psi_\alpha \\ \psi_\beta \\ \psi_{z1} \\ \psi_{z2} \end{bmatrix} \tag{5.21}$$

$$\begin{bmatrix} \psi_\alpha \\ \psi_\beta \\ \psi_{z1} \\ \psi_{z2} \end{bmatrix} = \begin{bmatrix} L_\alpha & 0 & 0 & 0 \\ 0 & L_\beta & 0 & 0 \\ 0 & 0 & L_\sigma & 0 \\ 0 & 0 & 0 & L_\sigma \end{bmatrix} \begin{bmatrix} i_\alpha \\ i_\beta \\ i_{z1} \\ i_{z2} \end{bmatrix} + \begin{bmatrix} \psi_{f\alpha} \\ \psi_{f\beta} \\ 0 \\ 0 \end{bmatrix} \tag{5.22}$$

where L_α and L_β the α- and β-axis inductances, respectively. L_σ is the leakage inductance.

Since the $\alpha\beta$ and z_1z_2 subspaces are orthogonal, the machine variables in one subspace are decoupled and isolated to those in the other subspace, which results in the diagonal resistance and inductance matrices. The fundamental components of machine variables are mapped in the $\alpha\beta$ subspace, and the torque and energy conversion occur exclusively in the $\alpha\beta$ subspace.

5.3.2.3 VSD electrical model of DTP PMSMs

To simplify the machine model and further provide the guideline for the design of the current/torque regulator, the machine model in the $\alpha\beta$ subspace is further transformed to a synchronous reference dq frame as

$$\begin{bmatrix} u_d \\ u_q \end{bmatrix} = \begin{bmatrix} R_s & 0 \\ 0 & R_s \end{bmatrix}\begin{bmatrix} i_d \\ i_q \end{bmatrix} + \frac{d}{dt}\begin{bmatrix} L_D i_d \\ L_Q i_q \end{bmatrix} + \omega_e \begin{bmatrix} 0 & -1 \\ 1 & 0 \end{bmatrix}\begin{bmatrix} L_D i_d + \psi_{fd} \\ L_Q i_q \end{bmatrix} \tag{5.23}$$

where

$$\begin{bmatrix} u_d \\ u_q \end{bmatrix} = \begin{bmatrix} \cos(\theta_r) & \sin(\theta_r) \\ -\sin(\theta_r) & \cos(\theta_r) \end{bmatrix}\begin{bmatrix} u_\alpha \\ u_\beta \end{bmatrix}$$

$$\begin{bmatrix} i_d \\ i_q \end{bmatrix} = \begin{bmatrix} \cos(\theta_r) & \sin(\theta_r) \\ -\sin(\theta_r) & \cos(\theta_r) \end{bmatrix}\begin{bmatrix} i_\alpha \\ i_\beta \end{bmatrix}$$

$$\begin{bmatrix} L_D & 0 \\ 0 & L_Q \end{bmatrix} = \begin{bmatrix} \cos(\theta_r) & \sin(\theta_r) \\ -\sin(\theta_r) & \cos(\theta_r) \end{bmatrix}\begin{bmatrix} L_\alpha & 0 \\ 0 & L_\beta \end{bmatrix}\begin{bmatrix} \cos(\theta_r) & \sin(\theta_r) \\ -\sin(\theta_r) & \cos(\theta_r) \end{bmatrix}^T \tag{5.24}$$

The phase angle relationship between the dq-axes and the $\alpha\beta$-axes is shown in Figure 5.5. By comparing (5.20) with the STP PMSM model in (5.3), it is evident that the DTP PMSM model in $\alpha\beta$ subspace has the same format as the STP PMSM model. This implies that the current/torque regulator in this subspace can be designed in the same manner as in STP PMSM systems.

The machine model in z_1z_2 subspace remains as

$$\begin{bmatrix} u_{z1} \\ u_{z2} \end{bmatrix} = \begin{bmatrix} R_s & 0 \\ 0 & R_s \end{bmatrix}\begin{bmatrix} i_{z1} \\ i_{z2} \end{bmatrix} + \frac{d}{dt}\begin{bmatrix} L_\sigma i_{z1} \\ L_\sigma i_{z2} \end{bmatrix} \tag{5.25}$$

Therefore, both (5.23) and (5.25) constitute the VSD electrical model of DTP PMSMs. The main harmonics, e.g., the 5th and 7th, are mapped in z_1z_2 subspace, and it can be clearly seen from (5.25) that the current in this subspace is only restricted by the resistance and the leakage inductance, which are usually very small. This characteristic implies that DTP PMSMs will suffer a more pronounced effect from the current harmonics compared to STP PMSMs if these current harmonics are not compensated effectively.

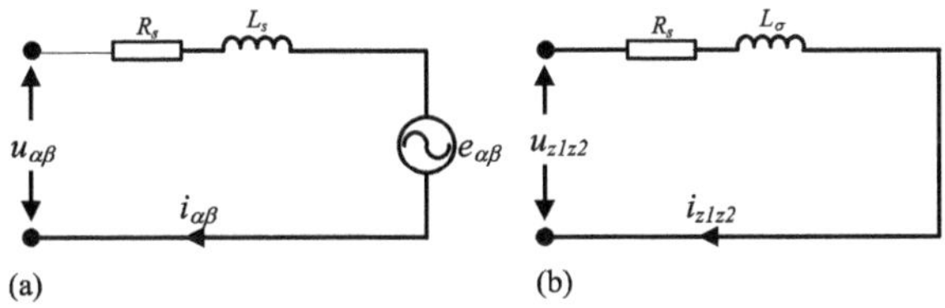

Figure 5.8 Equivalent circuits of DTP PMSMs. (a) $\alpha\beta$ subspace (fundamental and 11th, 13th,... harmonics). (b) z_1z_2 subspace (5th, 7th, 17th, 19th,... harmonics).

5.3.2.4 Equivalent electrical circuits of DTP PMSMs

Based on the electrical model of DTP PMSMs, the equivalent electrical circuits in two subspaces are shown in Figure 5.8. In the z_1z_2 subspace, the current harmonics from two sets of three-phase winding counterbalance with each other and do not generate any flux in the air gap. Consequently, the armature reaction inductance should be zero and the current harmonics are only limited by the leakage inductance in the z_1z_2 subspace, as illustrated by the equivalent circuit in Figure 5.8. Therefore, it can be expected that in DTP PMSMs, the 5th, 7th, 17th, 19th,... harmonic currents will inherently be large, and their minimization or elimination is a major concern.

5.3.2.5 Torque model of DTP PMSMs

In terms of the torque model, since the average torque and torque ripples are only relevant to the variables in $\alpha\beta$ subspace, the electromagnetic torque can be expressed by the cross-multiplication of the flux linkages and the currents in the stationary frame of $\alpha\beta$ subspace as

$$T_e = 3p\left(\psi_\alpha i_\beta - \psi_\beta i_\alpha\right) \tag{5.26}$$

Alternatively, it can also be expressed as the cross-multiple of the flux linkages and the currents in the synchronous reference *dq* frame of $\alpha\beta$ subspace as

$$T_e = 3p\left[\left(L_d - L_q\right)i_d i_q + \psi_{fd} i_q\right] \tag{5.27}$$

The main current harmonics, i.e., the 5th and 7th, are not mapped in $\alpha\beta$ subspace. Consequently, there are no 6th harmonics in i_d, i_q, and ψ_{fd}. Therefore, even if significant 5th and 7th current harmonics exist in phase current, the 6th torque ripple does not exist in DTP PMSMs as it does in STP PMSMs. The elimination of the 6th torque ripple is one of the key advantages of DTP PMSMs.

Compared with the torque equation (5.13) in the double three-phase *dq* model, the torque equation (5.26) in the VSD model is greatly simplified and is more practical in the torque estimation of DTP PMSMs.

5.3.2.6 Comparison of double *dq* and VSD models

Accordingly, the equivalent circuits for two modeling methods, i.e., the double *dq* model and the VSD model, are illustrated in Figures 5.9 and 5.10.

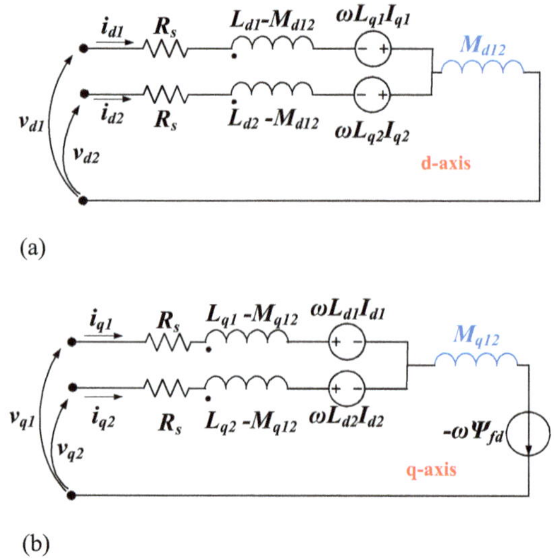

Figure 5.9 Equivalent circuits in double dq model of DTP PMSM. (a) d-axis and (b) q-axis.

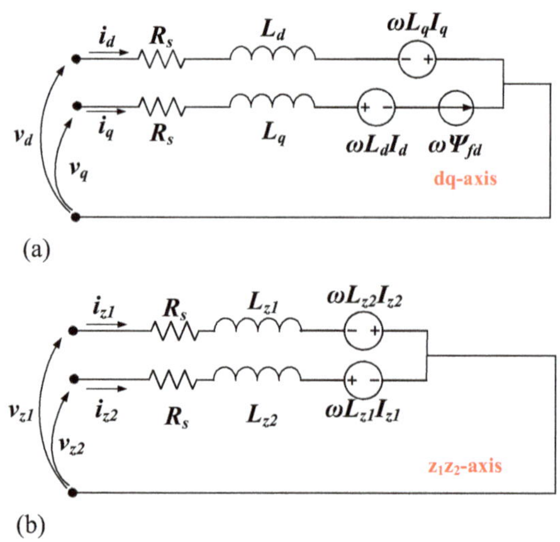

Figure 5.10 Equivalent circuits in the VSD model of DTP PMSM. (a) αβ-axis and (b) z_1z_2-axis.

The characteristics of the two modeling methods are summarized in Table 5.1. In brief, the double *dq* modeling treats the DTP PMSMs as two separate three-phase machines, where the mutual coupling between two sets of three-phase windings is ignored. It is highly practical in industrial applications as it can

Table 5.1 Comparison of double dq and VSD modeling methods

	Merits	Demerits
Double *dq* modeling	• Simple structure; • Easy duplication of vector control from STP to DTP PMSMs; • Inherently suppresses the current unbalance resulting from asymmetries between two sets of three-phase windings.	• Inferior dynamic torque performance; • Instability issues for two-individual current controls (when mutual inductance between two sets of windings is high).
VSD modeling	• Fully decoupled two sets of three-phase windings; • Excellent dynamic torque performance without the influence of coupling voltages between two sets of windings.	• Complex current control for variables mapped into different subspaces; • Asymmetric issues are complicated to control.

duplicate the space vector modeling of STP to DTP PMSMs and inherently suppresses the current unbalance resulting from asymmetries between the two sets of three-phase windings. However, it suffers from several issues caused by coupling between two sets of three-phase windings, such as potential instability in the current loops [3].

In comparison, VSD modeling techniques address this issue by decoupling all variables into three different orthogonal subspaces [4]. However, it is ineffective in decoupling the current unbalance resulting from asymmetries between the two sets of three-phase windings, and it also has a complex current control structure.

5.4 Separation of torque components in STP PMSMs

In general, the electromagnetic torque of a PMSM can be separated as PM and reluctance torque components. The PM torque component is produced by the interaction between the armature field and the PM field, and the reluctance torque component is generated by the interaction between the armature field and rotor saliency. Thus, utilizing reluctance torque components appropriately can effectively improve the torque performance of PMSMs with even reduced PM consumption. Further, an accurate torque component separation method for PMSMs can be of great importance in the design of PMSMs, which will be described in this and the following section.

In literature, several methods have been presented to separate torque components in PMSMs, such as by assuming that q-axis flux linkages only vary with q-axis current [5] or using the frozen permeability (FP) method [6–11]. In [6], the FP method was first introduced into PMSMs to consider magnetic saturation effects. Then, the FP method is utilized to find out the proportion of total flux-linkage attributable to the PMs and phase currents in [7]. Similarly, the non-linear saturation effects of d- and q-axis currents on d- and q-axis flux linkages are analyzed with the FP method in [8]. Then, the average torque separation methods with the FP method and Maxwell stress tensor/virtual work principle are compared in [9]. Based on the foregoing analyses, a method to separate and reproduce the instantaneous torque components in PMSMs under any load condition is proposed in [10], and then, a comprehensive review of the FP technique and its applications in various electrical machines is provided in [11]. It shows that the FP method can separate on-load magnetic fields, flux linkages, and inductance components accurately in all types of electrical machines and the accurate on-load parameters can further be used to improve the machine model and control.

In this section, the instantaneous torque separation method for STP PMSMs [10] is introduced in detail, and then it is extended into DTP PMSMs [12] in Section 5.5.

5.4.1 Analytical model

Accounting for the cross-coupling effects, the d- and q-axis flux linkages, ψ_d and ψ_q, in STP PMSMs can be calculated as

$$\begin{bmatrix} \psi_d \\ \psi_q \end{bmatrix} = \begin{bmatrix} \psi_{fd} \\ \psi_{fq} \end{bmatrix} + \begin{bmatrix} L_d & M_{dq} \\ M_{qd} & L_q \end{bmatrix} \begin{bmatrix} i_d \\ i_q \end{bmatrix} \tag{5.28}$$

where L is the self-inductance, M is the mutual inductance, and i is the current. ψ_{fd} and ψ_{fq} are the PM flux linkages in the dq-axes.

Thus, the d- and q-axis flux linkages in STP PMSMs can be calculated as

$$\begin{cases} \psi_d = \psi_{fd} + L_d i_d + M_{dq} i_q \\ \psi_q = \psi_{fq} + M_{dq} i_d + L_q i_q \end{cases} \tag{5.29}$$

The total electromagnetic torque produced by STP PMSMs can also be estimated from the sum of the cross products of stator current vectors and stator flux linkage vectors, as

$$T_{em} = 1.5p\left(\psi_d i_q - \psi_q i_d\right) \tag{5.30}$$

However, (5.30) cannot be directly applied to the FP process, since it does not include the torque produced due to variation of magnetic co-energy. To separate torque components accurately in the FP process, it is modified as:

$$T_{total} = 1.5p\left(\psi_{fd} i_q - \psi_{fq1} i_{d1}\right) + T_{mr} \tag{5.31}$$

where T_{mr} represents the torque component produced by rotor magnets and uneven reluctance in the stator core.

In (5.31), the total torque component of the STP PMSMs can further be separated as

$$T_{total} = 1.5p\left[\psi_{fd} i_q - \psi_{fq} i_d + \left(L_d - L_q\right) i_d i_q + M_{dq} i_q^2 - M_{qd} i_d^2\right] + T_{mr} \tag{5.32}$$

In (5.32), the PM torque component is produced by the PM flux linkages, while the reluctance torque component is caused by the interaction between the armature field and the rotor saliency. Therefore, the PM and reluctance torque components in STP PMSMs can be obtained as

$$T_{PM} = 1.5p\left(\psi_{fd}i_q - \psi_{fq}i_d\right) \tag{5.33}$$

$$T_{rel} = 1.5p\left[\left(L_d - L_q\right)i_d i_q + M_{dq}i_q^2 - M_{qd}i_d^2\right] \tag{5.34}$$

Overall, with the analytical model accounting for all cross-coupling effects in STP PMSMs, the total output torque can be separated into PM torque component (T_{PM}), reluctance torque component (T_{rel}), and magnet-iron torque component (T_{mr}).

5.4.2 Torque separation method

5.4.2.1 Calculation of relative permeability distribution

To employ the FP method, the relative permeability distribution of the machine under a specific load condition is first calculated step by step. The calculations should be operated over a whole electric cycle, and the results are stored for further analyses.

5.4.2.2 Excitation configurations of different cases

In the second step, the excitation configuration of the STP PMSM should be changed according to Table 5.2, where "$i_{d1} = i_{d1}$, $i_{q1} = 0$" denotes that in the first winding set, the d-axis current is not changed, but the q-axis current is changed to 0. "*Magnets*" denotes that the rotor magnets are not changed, but "*No magnets*" means that all the rotor magnets are removed. Since the relative permeability distributions are obtained from the specific on-load condition, all the corresponding calculated results have considered cross-saturation effects.

5.4.2.3 Flux linkage components of different cases

In the third step, the components in the resultant d- and q-axis flux linkages of the STP PMSM under different cases can be obtained, as shown in Table 5.3. It can be

Table 5.2 Excitation configurations of different cases in torque separation method for STP PMSMs

Case	Excitation configurations	
	Currents	**Magnet**
A	$i_d = i_d$, $i_q = i_q$	–
B	$i_d = i_d$, $i_q = 0$	*With*
C	$i_d = i_d$, $i_q = 0$	–
D	$i_q = i_q$, $i_d = 0$	*With*
E	$i_q = i_q$, $i_d = 0$	–
F	–	*With*

Table 5.3 Flux-linkage components of different cases in torque separation method for STP PMSMs

Case	Flux linkage components	
	ψ_d	ψ_q
A	$L_d i_d + M_{dq} i_q$	$M_{qd} i_d + L_q i_q$
B	$\psi_{md} + L_d i_d$	$\psi_{mq} + M_{qd} i_d$
C	$L_d i_d$	$M_{dq} i_d$
D	$\psi_{md} + M_{dq} i_q$	$\psi_{mq} + L_q i_q$
E	$M_{dq} i_q$	$L_q i_q$
F	ψ_{md}	ψ_{mq}

Note: The grey shades indicates that flux linkages only varying with one current.

Table 5.4 Torque components of different cases in torque separation method for STP PMSMs

Case	Torque components
A	$1.5p\left[(L_d - L_q)i_d i_q + M_{dq} i_q^2 - M_{qd} i_d^2\right]$
B	$1.5p(-\psi_{fq} i_d - M_{qd} i_d^2) + T_{mr}$
C	$-1.5pM_{qd} i_d^2$
D	$1.5p\left(\psi_{fd} i_q + M_{dq} i_q^2\right) + T_{mr}$
E	$1.5pM_{dq} i_q^2$
F	T_{mr}

seen that under Cases C and E, the *d*- and *q*-axis flux linkages only vary with one current component. Hence, for STP PMSMs, the whole inductance matrix can be obtained from Cases C and E.

5.4.2.4 Torque separation and reproduction

In the fourth step, the electromagnetic torque produced under these conditions can be calculated by (5.31). In addition, under the cases with rotor magnets (Cases B, D, and F), besides electromagnetic torque, T_{mr} also exists in the resultant total output torque. Hence, the torque components of the STP PMSM under different cases can be summarized in Table 5.4. Then, the instantaneous torque waveforms of different torque components can be separated based on the instantaneous torque waveforms of the STP PMSM under these cases.

According to (5.33), (5.34), and Table 5.4, the torque components of STP PMSMs can be separated, and the total output torque of STP PMSMs can be calculated as

$$T_{total} = T_{rel} + T_{pm} + T_{mr} = T_A + T_B - T_C + T_D - T_E - T_F \tag{5.35}$$

5.5 Separation of torque components in DTP PMSMs

Since there are two three-phase winding sets in DTP PMSMs, and the resultant armature fields are produced by the two winding sets simultaneously, the torque components in DTP PMSMs are different from those in STP PMSMs. Hence, the torque separation method [10] introduced in Section 5.4 cannot be directly applied to DTP PMSMs. In this section, the torque separation method will be extended to DTP PMSMs, where the coupling effects between the two winding sets on the resultant PM and reluctance torque components are all considered.

5.5.1 Analytical model

Since the armature windings of DTP PMSMs are composed of two separate winding sets, each winding set has its own *d*- and *q*-axes. In DTP PMSMs, the magnetic cross-coupling effect exists not only between the *d*- and *q*-axes within the same winding set, but also between the *d*- and *q*-axes of the other winding set. Therefore, to account for all the cross-coupling effects in DTP PMSMs, besides the mutual inductances between the *d*- and *q*-axes within one winding set, the mutual inductances between the *d*- and *q*-axes of the two winding sets also should be included. The *d*- and *q*-axis flux linkages of the two winding sets can be calculated as [13,14]:

$$\begin{bmatrix} \psi_{d1} \\ \psi_{q1} \\ \psi_{d2} \\ \psi_{q2} \end{bmatrix} = \begin{bmatrix} \psi_{fd1} \\ \psi_{fq1} \\ \psi_{fd2} \\ \psi_{fq2} \end{bmatrix} + \begin{bmatrix} L_{d1} & M_{d1q1} & M_{d1d2} & M_{d1q2} \\ M_{q1d1} & L_{q1} & M_{q1d2} & M_{q1q2} \\ M_{d2d1} & M_{d2q1} & L_{d2} & M_{d2q2} \\ M_{q2d1} & M_{q2q1} & M_{q2d2} & L_{q2} \end{bmatrix} \begin{bmatrix} i_{d1} \\ i_{q1} \\ i_{d2} \\ i_{q2} \end{bmatrix} \tag{5.36}$$

where ψ is the flux linkage, L is the self-inductance, M is the mutual inductance, and i is the current. The subscripts d_1, q_1, d_2, and q_2 refer to the *d*- and *q*-axis components of the first and second winding sets, respectively. It can be seen that all the mutual inductances between the *d*- and *q*-axes in one winding set, and the mutual inductances between the *d*- and *q*-axes of the two winding sets are considered in the inductance matrix in (5.36). In addition, since the mutual inductance between $d1$-axis and $q1$-axis equals that between $q1$-axis and $d1$-axis, it can be assumed that $M_{d1q1} = M_{q1d1} = M_1$. Similarly, other mutual inductances in (5.36) can be simplified as $M_{d1d2} = M_{d2d1} = M_2$, $M_{d1q2} = M_{q2d1} = M_3$, $M_{q1d2} = M_{d2q1} = M_4$, $M_{q1q2} = M_{q2q1} = M_5$, and $M_{d2q2} = M_{q2d2} = M_6$. Thus, the *d*- and *q*-axis flux linkages in DTP PMSMs can be calculated by

$$\begin{cases} \psi_{d1} = \psi_{fd1} + L_{d1}i_{d1} + M_1 i_{q1} + M_2 i_{d2} + M_3 i_{q2} \\ \psi_{q1} = \psi_{fq1} + M_1 i_{d1} + L_{q1} i_{q1} + M_4 i_{d2} + M_5 i_{q2} \\ \psi_{d2} = \psi_{fd2} + M_2 i_{d1} + M_4 i_{q1} + L_{d2} i_{d2} + M_6 i_{q2} \\ \psi_{q2} = \psi_{fq2} + M_3 i_{d1} + M_5 i_{q1} + M_6 i_{d2} + L_{q2} i_{q2} \end{cases} \tag{5.37}$$

Similar to STP PMSMs, the total electromagnetic torque produced by DTP PMSMs can also be estimated from the sum of the cross products of stator current vectors and stator flux linkage vectors in the two winding sets [13], as

$$T_{em} = 1.5p\left(\psi_{d1}i_{q1} - \psi_{q1}i_{d1} + \psi_{d2}i_{q2} - \psi_{d2}i_{q2}\right) \tag{5.38}$$

Based on (5.37) and (5.38), the total electromagnetic torque of DTP PMSMs can be obtained as

$$\begin{aligned} T_{em} = 1.5p\big[&\psi_{fd1}i_{q1} - \psi_{fq1}i_{d1} + \psi_{fd2}i_{q2} - \psi_{fq2}i_{d2} + \left(L_{d1} - L_{q1}\right)i_{d1}i_{q1} \\ &+\left(L_{d2} - L_{q2}\right)i_{d2}i_{q2} + M_1\left(i_{q1}i_{q1} - i_{d1}i_{d1}\right) + M_6\left(i_{q2}i_{q2} - i_{d2}i_{d2}\right) \\ &+(M_2 - M_5)\left(i_{d1}i_{q2} + i_{d2}i_{q1}\right) + (M_3 + M_4)\left(i_{q1}i_{q2} - i_{d1}i_{d2}\right)\big] \end{aligned} \tag{5.39}$$

From (5.39), since PM torque components are produced by the interaction between the armature field and the PM field, the PM torque components produced by the two winding sets and the rotor, can be separated as

$$T_{pm1} = 1.5p\left(\psi_{md1}i_{q1} - \psi_{mq1}i_{d1}\right) \tag{5.40}$$

$$T_{pm2} = 1.5p\left(\psi_{md2}i_{q2} - \psi_{mq2}i_{d2}\right) \tag{5.41}$$

$$T_{pm} = T_{pm1} + T_{pm2} \tag{5.42}$$

where T_{pm1} represents the PM torque component produced by the first winding set and the rotor, and T_{pm2} represents that produced by the second winding set and the rotor, respectively. T_{pm} represents the total PM torque component of the whole DTP PMSM.

Similarly, since reluctance torque components T_{rel} are generated by the interaction between the armature field and the rotor saliency, it can be separated into three components as

$$T_{rel1} = 1.5p\left[\left(L_{d1} - L_{q1}\right)i_{d1}i_{q1} + M_1\left(i_{q1}i_{q1} - i_{d1}i_{d1}\right)\right] \tag{5.43}$$

$$T_{rel2} = 1.5p\left[\left(L_{d2} - L_{q2}\right)i_{d2}i_{q2} + M_6\left(i_{q2}i_{q2} - i_{d2}i_{d2}\right)\right] \tag{5.44}$$

$$T_{relm} = 1.5p\left[(M_2 - M_5)\left(i_{d1}i_{q2} + i_{d2}i_{q1}\right) + (M_3 + M_4)\left(i_{q1}i_{q2} - i_{d1}i_{d2}\right)\right] \tag{5.45}$$

$$T_{rel} = T_{rel1} + T_{rel2} + T_{relm} \tag{5.46}$$

Considering that $M_1 = M_{d1q1} = M_{q1d1}$, and $M_2 = M_{d2q2} = M_{q2d2}$, T_{rel1} and T_{rel2} indicate the reluctance torque components, which are only influenced by the first and second winding sets respectively. All the inductances in (5.45), M_2 (M_{d1d2}, M_{d2d1}), M_3 (M_{d1q2}, M_{q2d1}), M_4 (M_{q1d2}, M_{d2q1}), and M_5 (M_{q1q2}, M_{q2q1}), are the mutual inductances between the two winding sets, which indicate the interactions between the two winding sets. Hence, T_{relm} represents the reluctance torque component that is produced due to the differences in the mutual inductances between the two winding sets. T_{rel} represents the total reluctance torque component of the DTP PMSM.

In (5.45), it can be inferred that when the two winding sets in DTP PMSMs are physically isolated, all the mutual inductances between the two winding sets (M_2, M_3, M_4, and M_5) will be zero, and T_{relm} will also be zero. Only under this condition can the total reluctance torque of DTP PMSMs be obtained as the sum of T_{rel1} and T_{rel2}. Otherwise, T_{relm} must be considered in the calculation of the total reluctance torque in DTP PMSMs.

Besides PM and reluctance torque components, as introduced and explained in [10], the interaction between rotor magnets and un-even reluctance in stator iron core can also contribute to the resultant output torque in DTP PMSMs. Thus, the total output torque of DTP PMSMs can be expressed as

$$T_{total} = T_{em} + T_{mr} = T_{pm1} + T_{pm2} + T_{rel1} + T_{rel2} + T_{relm} + T_{mr} \tag{5.47}$$

Overall, as shown in (5.47), with the analytical model accounting for all cross-coupling effects in DTP PMSMs, the total output torque of DTP PMSMs can be separated into two PM torque components (T_{pm1} and T_{pm2}), three reluctance torque components (T_{rel1}, T_{rel2}, and T_{relm}), and one torque component produced by rotor magnets and uneven reluctance in stator core component (T_{mr}).

5.5.2 Torque separation method

5.5.2.1 Calculation of relative permeability distribution

To employ the FP method in DTP PMSMs, the relative permeability distribution under a specific load condition is first calculated step by step. The calculations should be operated over a whole electric cycle, and the results are stored for further analyses.

5.5.2.2 Excitation configurations of different cases

The excitation configuration of the DTP PMSM should be changed according to Table 5.5, where "$i_{d1} = i_{d1}$, $i_{q1} = 0$" denotes that in the first winding set, the d-axis current is not changed, but the q-axis current is changed to 0. "*Magnets*" denotes

Table 5.5 Excitation configurations of different cases in torque separation method for DTP PMSMs

Case	Excitation configurations	
	Currents	**Magnet**
A	$i_{d1} = i_{d1}, i_{q1} = i_{q1}, i_{d2} = i_{d2}, i_{q2} = i_{q2}$	–
A1	$i_{d1} = i_{d1}, i_{q1} = i_{q1}, i_{d2} = i_{q2} = 0$	–
A2	$i_{d1} = i_{q1} = 0, i_{d2} = i_{d2}, i_{q2} = i_{q2}$	–
B1	$i_{d1} = i_{d1}, i_{q1} = i_{d2} = i_{q2} = 0$	With
C1	$i_{d1} = i_{d1}, i_{q1} = i_{d2} = i_{q2} = 0$	–
D1	$i_{q1} = i_{q1}, i_{d1} = i_{d2} = i_{q2} = 0$	With
E1	$i_{q1} = i_{q1}, i_{d1} = i_{d2} = i_{q2} = 0$	–
B2	$i_{d2} = i_{d2}, i_{d1} = i_{q1} = i_{q2} = 0$	With
C2	$i_{d2} = i_{d2}, i_{d1} = i_{q1} = i_{q2} = 0$	–
D2	$i_{q2} = i_{q2}, i_{d1} = i_{q1} = i_{d2} = 0$	With
E2	$i_{q2} = i_{q2}, i_{d1} = i_{q1} = i_{d2} = 0$	–
F	–	With

that the rotor magnets are not changed, but "*No magnets*" means that all the rotor magnets are removed.

5.5.2.3 Flux linkage components of different cases

The components in the resultant *d*- and *q*-axis flux linkages of the DTP PMSM under different cases can be obtained, as shown in Table 5.6. Under Cases C1, E1, C2, and E2, the *d*- and *q*-axis flux linkages only vary with one current component. Hence, for DTP PMSMs, the whole inductance matrix can be obtained from Cases C1, E1, C2, and E2.

5.5.2.4 Torque separation and reproduction

As the *d*- and *q*-axis flux linkages under different cases have been provided in Table 5.6, the electromagnetic torque produced under these conditions can be calculated by (5.47). In addition, under the cases with rotor magnets (Cases B1, D1, B2, D2, and F), besides electromagnetic torque, T_{mr} also exists in the resultant total output torque. Hence, the torque components of the researched DTP PMSM under different cases can be summarized in Table 5.7. Then, the instantaneous torque waveforms of different torque components can be separated based on the instantaneous torque waveforms of the researched DTP PMSM under these cases.

According to (5.40), (5.41), (5.42), and Table 5.7, the PM torque components in DTP PMSMs can be obtained as

$$T_{pm1} = T_{B1} - T_{C1} + T_{D1} - T_{E1} - 2T_F \quad (5.48)$$

Table 5.6 Flux-linkage components of different cases in torque separation method for DTP PMSMs

Case	Flux linkage components			
	ψ_{d1}	ψ_{q1}	ψ_{d2}	ψ_{q2}
A	$L_{d1}i_{d1} + M_1 i_{q1} + M_2 i_{d2} + M_3 i_{q2}$	$M_1 i_{d1} + L_{q1} i_{q1} + M_4 i_{d2} + M_5 i_{q2}$	$M_2 i_{d1} + M_4 i_{q1} + L_{d2} i_{d2} + M_6 i_{q2}$	$M_3 i_{d1} + M_5 i_{q1} + M_6 i_{d2} + L_{q2} i_{q2}$
A1	$L_{d1}i_{d1} + M_1 i_{q1}$	$M_1 i_{d1} + L_{q1} i_{q1}$	$M_2 i_{d1} + M_4 i_{q1}$	$M_3 i_{d1} + M_5 i_{q1}$
A2	$M_2 i_{d2} + M_3 i_{q2}$	$M_4 i_{d2} + M_5 i_{q2}$	$L_{d2} i_{d2} + M_6 i_{q2}$	$M_6 i_{d2} + L_{q2} i_{q2}$
B1	$\psi_{fd1} + L_{d1} i_{d1}$	$\psi_{fq1} + M_1 i_{d1}$	$\psi_{fd2} + M_2 i_{d1}$	$\psi_{fq2} + M_3 i_{d1}$
C1	$L_{d1} i_{d1}$	$M_1 i_{d1}$	$M_2 i_{d1}$	$M_3 i_{d1}$
D1	$\psi_{fd1} + M_1 i_{q1}$	$\psi_{fq1} + L_{q1} i_{q1}$	$\psi_{fd2} + M_4 i_{q1}$	$\psi_{fq2} + M_5 i_{q1}$
E1	$M_1 i_{q1}$	$L_{q1} i_{q1}$	$M_4 i_{q1}$	$M_5 i_{q1}$
B2	$\psi_{fd1} + M_2 i_{d2}$	$\psi_{fq1} + M_4 i_{d2}$	$\psi_{fd2} + L_{d2} i_{d2}$	$\psi_{fq2} + M_6 i_{d2}$
C2	$M_2 i_{d2}$	$M_4 i_{d2}$	$L_{d2} i_{d2}$	$M_6 i_{d2}$
D2	$\psi_{fd1} + M_3 i_{q2}$	$\psi_{fq1} + M_5 i_{q2}$	$\psi_{fd2} + M_6 i_{q2}$	$\psi_{fq2} + L_{q2} i_{q2}$
E2	$M_3 i_{q2}$	$M_5 i_{q2}$	$M_6 i_{q2}$	$L_{q2} i_{q2}$
F	ψ_{fd1}	ψ_{fq1}	ψ_{fq1}	ψ_{fq2}

Note: The grey shades indicate the flux-linkage components due to one current only.

Table 5.7 Torque components of different cases in torque separation method for DTP PMSMs

Case	Torque components
A	$1.5p\big[(L_{d1}-L_{q1})i_{d1}i_{q1}+(L_{d2}-L_{q2})i_{d2}i_{q2}+M_1(i_{q1}i_{q1}-i_{d1}i_{d1})+$ $M_6(i_{q2}i_{q2}-i_{d2}i_{d2})+(M_2-M_5)(i_{d1}i_{q2}+i_{d2}i_{q1})+(M_3+M_4)(i_{q1}i_{q2}-i_{d1}i_{d2})\big]$
A1	$1.5p\big[(L_{d1}-L_{q1})i_{d1}i_{q1}+M_1(i_{q1}i_{q1}-i_{d1}i_{d1})\big]$
A2	$1.5p\big[(L_{d2}-L_{q2})i_{d2}i_{q2}+M_6(i_{q2}i_{q2}-i_{d2}i_{d2})\big]$
B1	$1.5p(-\psi_{fq1}i_{d1}-M_1i_{d1}i_{d1})+T_{mr}$
C1	$-1.5pM_1i_{d1}i_{d1}$
D1	$1.5p(\psi_{fd1}i_{q1}+M_1i_{q1}i_{q1})+T_{mr}$
E1	$1.5pM_1i_{q1}i_{q1}$
B2	$1.5p(-\psi_{fq2}i_{d2}-M_6i_{d2}i_{d2})+T_{mr}$
C2	$-1.5pM_6i_{d2}i_{d2}$
D2	$1.5p(\psi_{fd2}i_{q2}+M_6i_{q2}i_{q2})+T_{mr}$
E2	$1.5pM_6i_{q2}i_{q2}$
F	T_{mr}

$$T_{pm2}=T_{B2}-T_{C2}+T_{D2}-T_{E2}-2T_F \tag{5.49}$$

$$T_{pm}=T_{pm1}+T_{pm2}=T_{B1}-T_{C1}+T_{D1}-T_{E1}+T_{B2}-T_{C2}+T_{D2}-T_{E2}-4T_F \tag{5.50}$$

For reluctance torque components, from (5.43), (5.44), (5.45), (5.46), and Table 5.7, the reluctance torque components in DTP PMSMs can be obtained as

$$T_{rel}=T_A \tag{5.51}$$

$$T_{rel1}=T_{A1} \tag{5.52}$$

$$T_{rel2}=T_{A2} \tag{5.53}$$

$$T_{relm}=T_{rel}-T_{rel1}-T_{rel2}=T_A-T_{A1}-T_{A2} \tag{5.54}$$

It is worth noting that T_{rel}, T_{rel1}, and T_{rel2} can be obtained directly from the instantaneous torque waveforms under Cases A, A1, and A2. However, T_{relm}, which represents the reluctance torque component produced due to the differences in mutual inductances between the two winding sets, can only be calculated as the difference between total reluctance torque (T_{rel}) and the sum of reluctance torque components produced by the two winding sets separately (T_{rel1} and T_{rel2}). With the separated torque components, the total output torque of DTP PMSMs can be calculated as

$$\begin{aligned}T_{total}&=T_{rel}+T_{pm}+T_{mr}\\&=T_A+T_{B1}-T_{C1}+T_{D1}-T_{E1}+T_{B2}-T_{C2}+T_{D2}-T_{E2}-3T_F\end{aligned} \tag{5.55}$$

5.6 Comparison of torque components in STP and DTP PMSMs

In this section, the torque separation methods introduced in Sections 5.4 and 5.5 will be applied to the Toyota Prius 2010 machine, the calculations are obtained with the finite element (FE) method. Then, the torque components in STP and DTP PMSMs are compared.

5.6.1 Machine configuration

To illustrate the procedure of the torque separation method for DTP PMSMs and also to demonstrate the differences in PM and reluctance torque components between STP and DTP PMSMs, as well as those between different types of DTP configurations, three PMSMs in Section 3.2 (STP, DTP-SL, and DTP-DL PMSMs) are chosen as benchmarks in this chapter, as shown in Figure 5.11. The detailed design parameters of three PMSMs and the specifications of different winding configurations can be found in Section 3.2 and are also presented in Tables 5.8 and 5.9. The cross-sections of the benchmark STP and DTP PMSMs are shown in Figure 5.11(a)–(c), respectively.

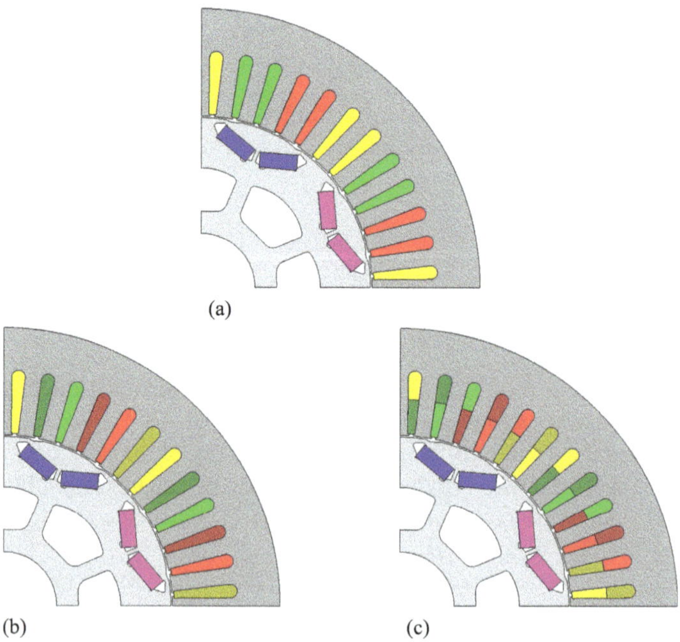

Figure 5.11 Cross-sections of benchmark STP and DTP PMSMs. (a) STP, (b) DTP-SL, and (c) DTP-DL.

Table 5.8 Geometric specifications of benchmark STP and DTP PMSMs

Parameter	Value	Parameter	Value
Number of stator slots	48	Number of rotor poles	8
Stator OD, mm	264	Rotor OD, mm	160.44
Stator ID, mm	161.9	Rotor ID, mm	51
Stator stack length, mm	50.8	Lamination thickness, mm	0.305
Slot depth, mm	30.9	Slot opening, mm	1.88
Phase resistance, Ω	0.077	Magnet dimensions, mm	49.3 × 17.88 × 7.16

Table 5.9 Specifications of STP, DTP-SL, and DTP-DL winding configurations

	STP	DTP-SL	DTP-DL
Number of phases	3	6	6
Wires in parallel per turn	12	6	6
Turns in series per coil	11	22	11
Coils in series per phase	8	4	8
Coil pitch number	5	6	5

5.6.2 *Torque separation method for DTP-SL PMSM*

In this sub-section, the torque separation method introduced earlier will be applied to the DTP-SL PMSM under the condition, i.e., the phase current amplitude I_A of 118 A and the current advancing angle β_a of 50°.

5.6.2.1 Relative permeability distribution

The relative permeability distribution of the DTP-SL machine under the load condition is calculated first. The calculations are performed over a complete electric cycle (60 steps), and the results are stored for further analyses. The relative permeability distribution of the DTP-SL machine at Step 1 is shown in Figure 5.12.

5.6.2.2 Instantaneous inductances

Under the on-load condition ($I_A = 118$ A, $\beta_a = 50°$), the d- and q-axis flux linkages under Cases C1, E1, C2, and E2 are calculated with the stored relative permeability distributions, and then, the self and mutual inductances under these cases can be obtained. The variations of these inductances with rotor position are shown in Figure 5.13.

As shown in Figure 5.13, the average inductance matrix can be obtained, as

$$\begin{bmatrix} 1.84 & 0.36 & 1.24 & 0.48 \\ 0.36 & 2.82 & 0.47 & 1.83 \\ 1.24 & 0.47 & 1.84 & 0.36 \\ 0.48 & 1.83 & 0.36 & 2.83 \end{bmatrix} (mH) \tag{5.56}$$

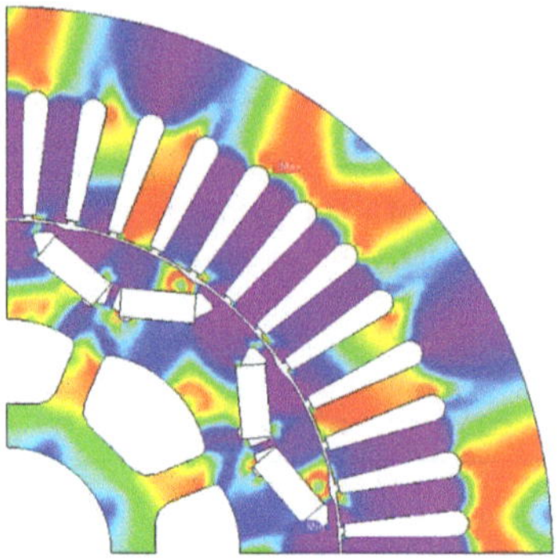

Figure 5.12 Relative permeability distribution of DTP-SL PMSM at Step 1 under on-load condition (I_A = 118 A, β_a = 50°).

From Figure 5.13 and (5.56), it can be observed that the inductances in one winding set (L_{d1}, L_{q1}, and M_1 in the first winding set) are almost the same as those in the other winding set (L_{d2}, L_{q2}, and M_6 in the second winding set). Furthermore, it can be seen that M_2 (M_{d1d2}, M_{d2d1}, 1.24 mH) and M_5 (M_{q1q2}, M_{q2q1}, 1.83 mH) are only slightly lower than L_{d1}, L_{d2} (1.84 mH) and L_{q1}, L_{q2} (2.82 mH), which suggests that the cross-coupling effect between the two winding sets is significant in the DTP PMSM.

5.6.2.3 Torque separation and reproduction

Under the same condition (I_A = 118 A, β_a = 50°), the instantaneous torque waveforms of the PM torque components are shown in Figure 5.14. It can be seen that the PM torque components caused by the two winding sets are shifted by 30°.

It is worth noting that T_{rel}, T_{rel1}, and T_{rel2} can be obtained directly from the instantaneous torque waveforms under Cases A, A1, and A2. However, T_{relm}, which represents the reluctance torque component produced due to the differences in mutual inductances between the two winding sets, can only be calculated as the difference between the total reluctance torque (T_{rel}) and the sum of reluctance torque components produced by the two winding sets separately (T_{rel1} and T_{rel2}). In the DTP-SL PMSM, the instantaneous torque waveforms of the reluctance torque components are shown in Figure 5.15.

Based on the above analyses, in the DTP-SL PMSM, the instantaneous torque waveforms of T_{total}, T_{pm}, T_{rel}, and T_{mr}, as well as the output torque waveform, calculated using the FE method, are shown in Figure 5.16. It can be seen that the reproduced total output torque (T_{total}) agrees well with the direct FE results (T_{direct}).

5.6.3 Comparison of torque components in STP and DTP PMSMs

Since the described torque separation method is applicable to all DTP PMSMs, the torque components in DTP-DL PMSM can also be separated. To ensure a fair

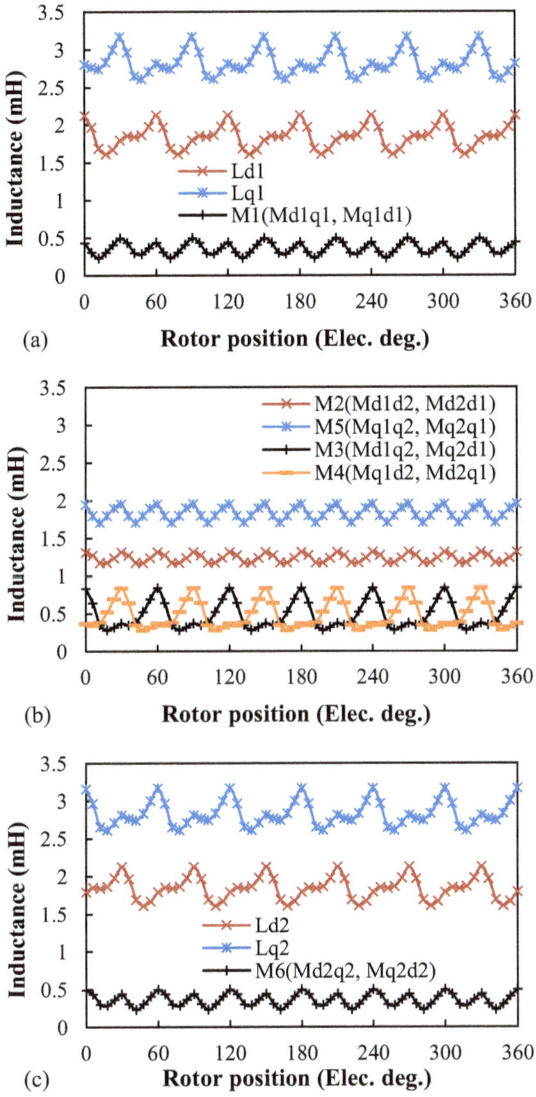

Figure 5.13 Variations of different inductances with rotor position. (a) L_{d1}, L_{q1}, and M_1, (b) M_2, M_5, M_3, and M_4, and (c) L_{d2}, L_{q2}, and M_6.

comparison, the torque components in the DTP-DL PMSM are separated under the same load condition (I_A = 118A, β_a = 50°), where the instantaneous torque waveforms of the PM and reluctance torque components are calculated and shown in Figures 5.17 and 5.18, respectively.

Based on Figures 5.17 and 5.18, the total torque of the DTP-DL PMSM can be reproduced by taking PM, reluctance, and magnet-iron torque components into consideration. The variations of these torque components with rotor position are

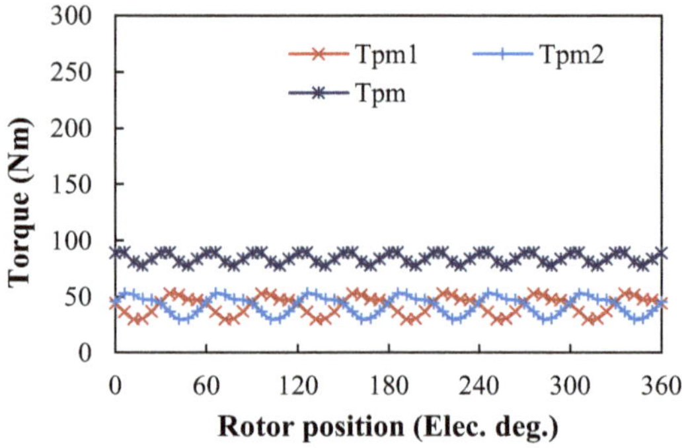

Figure 5.14 Variations of PM torque components with rotor position in DTP-SL PMSM under on-load condition (I_A = 118 A, β_a = 50°).

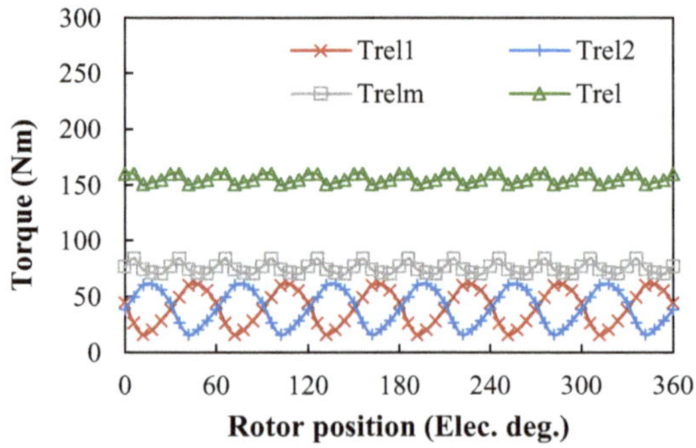

Figure 5.15 Variations of reluctance torque components with rotor position in DTP-SL PMSM under on-load condition (I_A = 118 A, β_a = 50°).

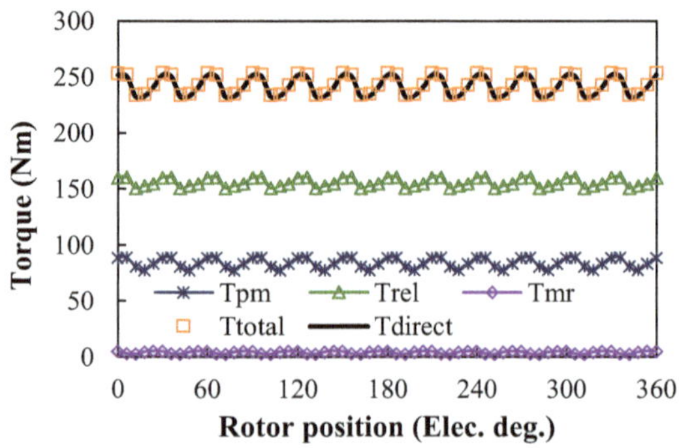

Figure 5.16 Variations of different torque components with rotor position in DTP-SL PMSM under on-load condition (I_A = 118 A, β_a = 50°).

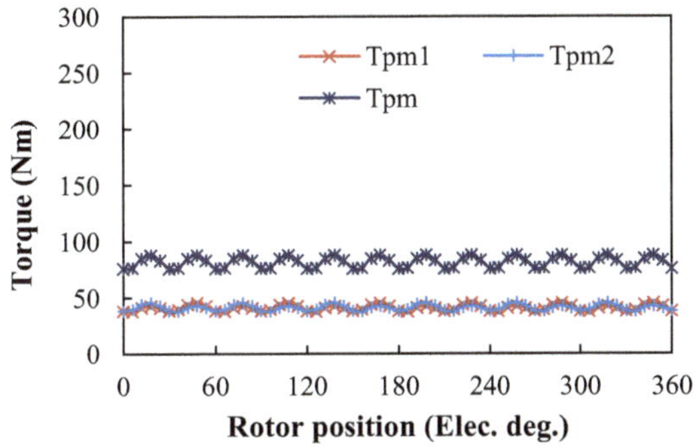

Figure 5.17 Variations of PM torque components with rotor position in DTP-DL PMSM under on-load condition (I_A = 118 A, β_a = 50°).

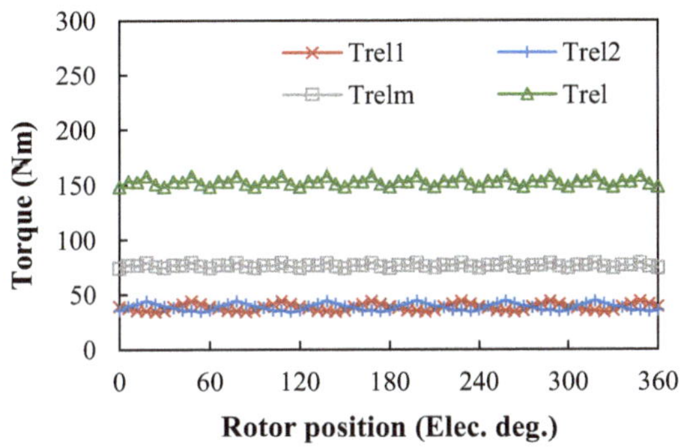

Figure 5.18 Variations of reluctance torque components with rotor position in DTP-DL PMSM under on-load condition (I_A = 118 A, β_a = 50°).

shown in Figure 5.19 and compared with the FE results. It can be found that the reproduced total torque still agrees well with the FE directly calculated total torque.

Moreover, the different torque components in the benchmark STP PMSM are separated under the specific on-load condition (I_A = 236 A, β_a = 50°). It should be mentioned that due to different winding configurations in STP and DTP PMSMs, the phase current amplitude in STP PMSMs is doubled from those in DTP PMSMs for a fair comparison. In the benchmark STP PMSM under the on-load condition (I_A = 236 A, β_a = 50°), the variations of different torque components with rotor position are shown in Figure 5.20. Similarly, the instantaneous torque waveform obtained directly by the FE method is also presented in Figure 5.20 to show the effectiveness of the torque separation method in STP PMSMs.

From Figures 5.16, 5.19, and 5.20, the different torque components in the benchmark STP, DTP-SL, and DTP-DL PMSMs can be summarized in Table 5.10. It should be noted that the total average torques in Table 5.10 are obtained as the sum of the separated torque components by (5.35), which are slightly different from the average torques obtained directly from FE analyses (shown in Section 3.3).

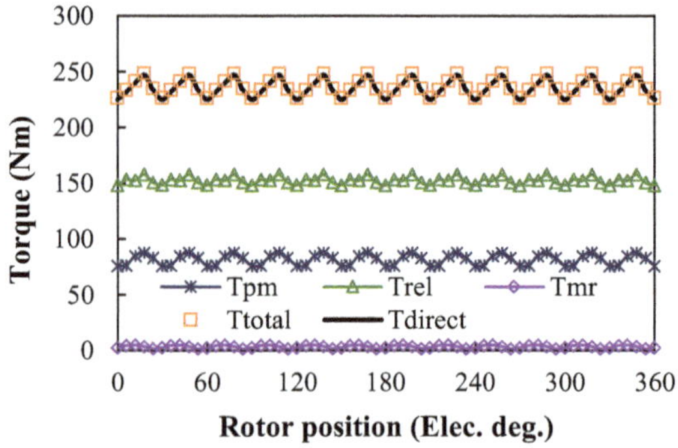

Figure 5.19 Variations of different torque components with rotor position in DTP-DL PMSM under on-load condition (I_A = 118 A, β_a = 50°).

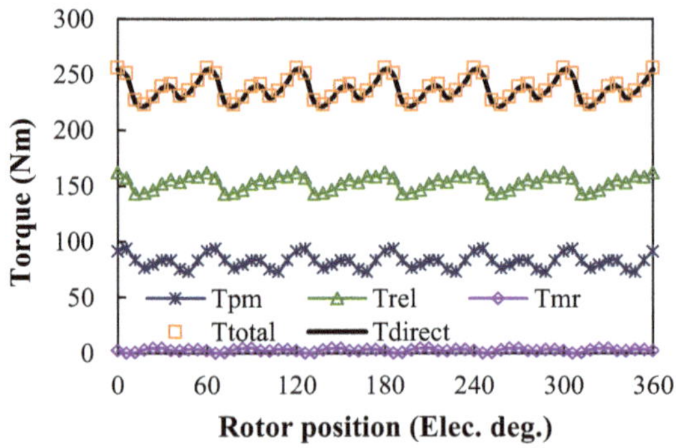

Figure 5.20 Variations of different torque components with rotor position in STP PMSM under on-load condition (I_A = 236 A, β_a = 50°).

Table 5.10 Comparison of torque components in STP, DTP-SL, and DTP-DL PMSMs

	STP		**DTP-SL**		**DTP-DL**	
	Average torque (Nm)	**Ratio (%)**	**Average torque (Nm)**	**Ratio (%)**	**Average torque (Nm)**	**Ratio (%)**
T_{pm1}	–	–	42.08	17.26	40.81	17.20
T_{pm2}	–	–	42.09	17.26	40.81	17.20
T_{pm}	82.35	34.56	84.17	34.52	81.61	34.39
T_{rel1}	–	–	40.19	16.48	38.14	16.07
T_{rel2}	–	–	40.09	16.44	38.08	16.05
T_{relm}	–	–	75.48	30.96	76.20	32.11
T_{rel}	153.16	64.28	155.76	63.89	152.43	64.24
T_{mr}	2.76	1.16	3.88	1.59	3.24	1.37
T_{total}	238.27	100.00	243.81	100.00	237.29	100.00

From Table 5.10, it can be found that although the detailed torque components in STP, DTP-SL, and DTP-DL PMSMs are different, the ratios of PM, reluctance, and magnet-iron torque components in the total output torque of the three machines are very similar. In other words, the DTP winding configuration cannot change the contributions of PM and reluctance torque components in total output torque in STP PMSMs. In addition, it can be found that in DTP-SL and DTP-DL PMSMs, the reluctance torque component produced by the mutual inductances between the two winding sets, T_{relm}, is the largest component among T_{pm1}, T_{pm2}, T_{rel1}, T_{rel2}, and T_{relm}. Hence, when separating torque components in DTP PMSMs, the cross-coupling effect between the two winding sets cannot be ignored.

5.7 Summary

In this chapter, the mathematical modeling methods for STP, multiphase, and DTP PMSMs were described. For DTP PMSMs, both double *dq* and VSD models were compared. The double *dq* model mapped the variables in two sets of three-phase windings into two *dq* coordinates, and the current control strategies of STP PMSMs can be easily applied. However, the mutual coupling between two *dq* coordinates deteriorates the control performance. In the VSD modeling method, all phase variables, i.e., the currents, voltages, and flux linkages, can be decomposed into three orthogonal subspaces, i.e. the $\alpha\beta$ subspace (or the *dq* subspace, the fundamental component of the machine variables and the *k*th order harmonics with $k = 12m \pm 1$ (m =1, 2, 3,...), i.e., the fundamental and the 11th, 13th,... harmonics), the z_1z_2 subspace (harmonics with $k = 6m\pm1$ (m = 1, 3, 5,...), i.e., the 5th, 7th, 17th, 19th,... harmonics,), and the o_1o_2 subspace (zero-sequence related components, $k = 3m$ (m = 1, 2, 3,...)). Thus, the coupling between two sets of three-phase windings for DTP-PMSMs can be fully coupled to ease the control.

The precise torque separation results can provide more insights into the design and control of DTP PMSMs. With the *dq* model of STP PMSMs and the double *dq* model of DTP PMSMs, the instantaneous torque separation methods for STP and DTP PMSMs were also introduced in this chapter. To further illustrate the torque separation methods in Sections 5.4 and 5.5, an example based on Toyota Prius 2010 machine was given in Section 5.6 for illustration.

The output torques of STP and DTP PMSMs are analyzed analytically, accounting for the cross-coupling effect within and between different winding sets. It was found that the total output torque of DTP PMSMs can be separated into two PM torque components (T_{pm1} and T_{pm2}), three reluctance torque components (T_{rel1}, T_{rel2}, and T_{relm}), and one magnet-iron torque component (T_{mr}). Then, the different torque components in DTP PMSMs are separated by setting different excitation configurations in FE models with the help of the FP method. It was found that the reluctance torque produced by the mutual inductance between the two winding sets contributes the most to the resultant output torque. Hence, to separate torque components in DTP PMSMs, the cross-coupling effects between different winding sets cannot be ignored.

It also should be mentioned that the described torque separation method can be further extended to other PMSMs with multiple winding sets.

References

[1] R. Bojoi, M. Lazzari, F. Profumo, and A. Tenconi, "Digital field-oriented control for dual three-phase induction motor drives," *IEEE Trans. Ind. Appl.*, vol. 39, no. 3, pp. 752–760, May/Jun. 2003.

[2] G. K. Singh, K. Nam, and S. K. Lim, "A simple indirect field-oriented control scheme for multiphase induction machine," *IEEE Trans. Ind. Electron.*, vol. 52, no. 4, pp. 1177–1184, 2005.

[3] Y. Hu, Z.Q. Zhu, and M. Odavic, "Comparison of two-individual current control and vector space decomposition control for dual three-phase PMSM," *IEEE Trans. Ind. Appl.*, vol. 53, no. 5, pp. 4483–4492, 2017.

[4] Y. Zhao and T. A. Lipo, "Space vector PWM control of dual three-phase induction machine using vector space decomposition," *IEEE Trans. Ind. Appl.*, vol. 31, no. 5, pp. 1100–1109, 1995.

[5] G. Qi, J. T. Chen, Z. Q. Zhu, D. Howe, L. B. Zhou, and C. L. Gu, "Influence of skew and cross-coupling on flux-weakening performance of permanent-magnet brushless AC machines," *IEEE Trans. Magn.*, vol. 45, no. 5, pp. 2110–2117, 2009.

[6] N. Bianchi and S. Bolognani, "Magnetic models of saturated interior permanent magnet motors based on finite element analysis," *Proc. 33rd IAS Annu. Meeting*, vol. 1, Oct. 12–15, 1998, pp. 27–34.

[7] J. A. Walker, D. G. Dorrell, and C. Cossar "Flux-linkage calculation in permanent-magnet motors using the frozen permeabilities method," *IEEE Trans. Magn.*, vol. 41, no. 10, pp. 3946–3948, 2013.

[8] S. Y. Kwak, J. K. Kim, and H. K. Jung, "Characteristic analysis of multilayer-buried magnet synchronous motor using fixed permeability method," *IEEE Trans. Energy Convers.*, vol. 20, no. 3, pp. 549–555, 2005.

[9] W. Q. Chu and Z. Q. Zhu, "Average torque separation in permanent magnet synchronous machines using frozen permeability," *IEEE Trans. Magn.*, vol. 49, no. 3, pp. 1202–1210, 2013.

[10] X. Chen, J. Wang, V. I. Patel, P. Lazari, L. Chen, and P. Lombard, "Reluctance torque evaluation for interior permanent magnet machines using frozen permeability," *Int. Conf. Power Electr. Mach. & Drives (PEMD)*, Manchester, UK, 2014.

[11] Z. Q. Zhu, and W. Q. Chu, "Advanced frozen permeability technique and applications in developing high performance electrical machines," *Trans. of China Electrotechnical Society*, vol. 31, no. 20, pp.13–29, 2016.

[12] S. Wang, "Investigation of torque and torque ripple in dual three-phase permanent magnet synchronous machines for electric vehicles," *PhD dissertation*, University of Sheffield, 2023.

[13] J. Karttunen, S. Kallio, P. Peltoniemi, P. Silventoinen, and O. Pyrhönenet, "Dual three-phase permanent magnet synchronous machine supplied by two independent voltage source inverters," *Proc. Int. Symp. Power Electron. Elect. Drives*, Autom. Motion, Sorrento, Italy, pp. 741–747, 2012.

[14] J. Karttunen, S. Kallio, P. Peltoniemi, P. Silventoinen, and O. Pyrhönen. "Decoupled vector control scheme for dual three-phase permanent magnet synchronous machines," *IEEE Trans. Ind. Electron.*, vol. 61, no. 5, pp. 2486–2494, 2014.

Chapter 6

Design optimization and comparison of multiphase PMSMs

6.1 Introduction

This chapter compares multiphase interior permanent magnet synchronous machine (IPMSM) systems for electric vehicle applications [1]. Amongst various multiphase candidates, six-phase, also known as dual three-phase (DTP) machines, are commonly used [2]. In DTP machines, two sets of three-phase windings offer advantages of fault tolerance, power sharing, torque ripple reduction, etc., while the conventional three-phase inverters can still be used to reduce the cost of the drive [3–5]. Five-phase machines can utilize the third harmonic magnetic field and back-electromotive force (EMF) in addition to the fundamental component, to enhance their electromagnetic performance [6–8]. Although not as popular as five- and six-phase machines, nine-phase machines have also demonstrated higher performance in terms of better efficiency and lower harmonics [9,10]. In this chapter, four IPMSMs, i.e., three-, five-, six-, and nine-phase IPMSMs, are designed and optimized by the finite element (FE) method for maximizing output torque and minimizing torque ripple and PM volume, all within the same space envelope as the baseline single three-phase (STP) IPMSM used in commercial electric vehicles. The key performance, including torque and power capabilities, torque ripples, PM volume, torque per PM volume, losses, and efficiencies, are compared. Additionally, the effects of phase number on both the machine side [1] and the drive side are analyzed accounting for the costs of machines and drive systems. Additionally, this analysis also considers the requirements of power switches, peripherals, and DC-link capacitors.

The structure of this chapter is shown in Figure 6.1. Section 6.2 describes the parametrization of the baseline STP IPMSM and the multiphase machines. The possible slot/pole number combinations for multiphase machines are also investigated since a specific slot/pole number combination may not be generally suitable for all of these machines. Section 6.3 discusses the optimized results and compares the electromagnetic performance of optimized machines, including back-EMFs, output torque, losses, and efficiency maps. Section 6.4 focuses on the effect of using multiphase machines on the drive side by investigating the number of power switches and the value of DC-link capacitance that is required for the same level of performance. Section 6.5 analyses the corresponding costs of each component in the multiphase machine and the inverter.

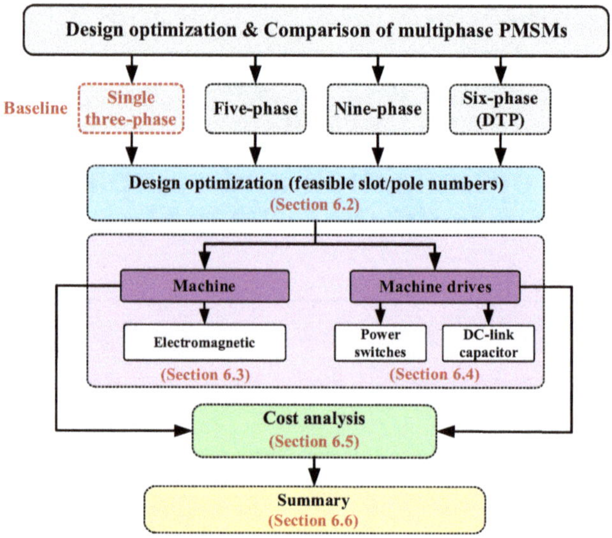

Figure 6.1 Structure of Chapter 6.

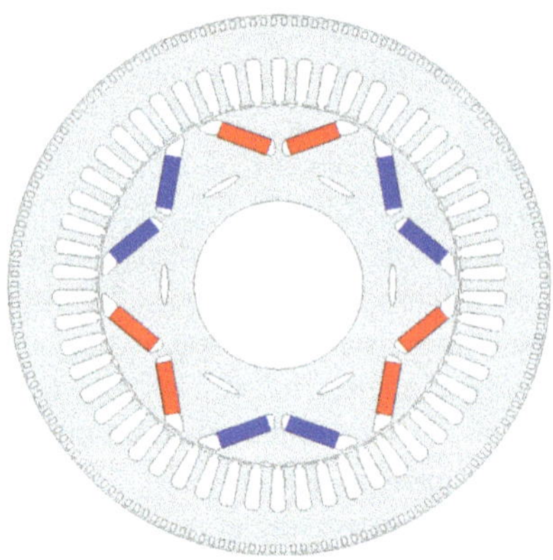

Figure 6.2 Cross-section of baseline STP IPMSM.

6.2 Design optimization of three-, five-, six- (DTP), and nine-phase PMSMs

The baseline STP IPMSM is a commercial traction machine for electric vehicles [11]. Its cross-section is presented in Figure 6.2 and specifications are given in Table 6.1, which are used in the optimization of all four machines.

The rotor lamination is optimized with two notches at the outer surface for each pole and also features arc notches at the iron bridges to reduce the leakage flux, increase the output torque, and reduce the back-EMF harmonics and torque ripple [12]. NdFeB magnets with a typical V-shape design and three-step skewing are used in the rotor.

Table 6.1 Specifications of baseline STP IPMSM

Parameter	Value	Parameter	Value
Torque (Nm)/power (kW)	439/230	DC-link voltage, (V)	350
Peak current (A)	1,100	Maximum speed (r/min)	18,000
Air gap (mm)	0.7	Lamination length (mm)	134
Stator outer diameter (mm)	225	Stator inner diameter (mm)	151.3
PM dimension (mm)	33 × 21 × 6.5	PM remanence (T)	1.44
Number of stator slots	54	Number of poles	6

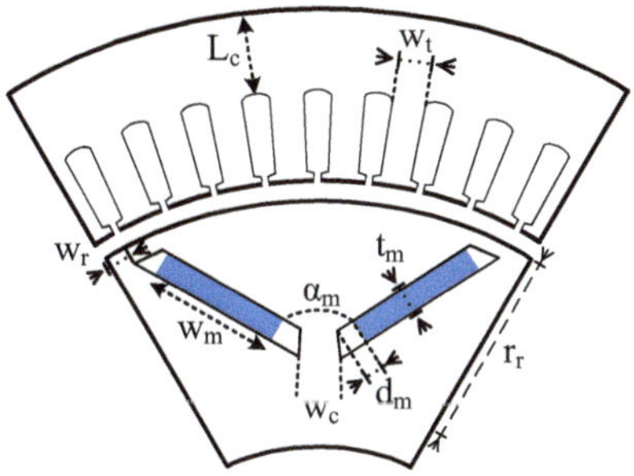

Figure 6.3 Parametrized model of IPMSM.

6.2.1 Parametrization of IPMSM

Based on the baseline IPMSM, the machine model is parametrized. The cross-section with the parameters used for the global optimization is shown in Figure 6.3.

The stator is optimized by considering the split ratio γ, the tooth width w_t, and the back-iron thickness L_c. The split ratio γ is defined as:

$$\gamma = \frac{r_r}{r_s} \tag{6.1}$$

where r_r and r_s are the rotor and stator outer radii, respectively.

For the stator, the upper limit of the back-iron thickness is constrained by various design factors, including the split ratio. As the split ratio becomes lower, the maximum value of the back-iron thickness also decreases to eliminate infeasible solutions from the optimization search space. This adjustment aims to filter out impractical solutions from the optimization process.

The rotor configuration is determined by several variables, including the split ratio (γ), the rib width (w_r), the center post width (w_c), the angle between magnets (α_m), the magnet thickness (t_m), the magnet width (w_m), and the distance of magnets from the center iron rib (d_m). The maximum magnet width is influenced by these parameters.

The width of the iron bridge remains fixed at 1.5 mm as the minimum iron bridge in the baseline IPMSM due to constraints related to mechanical stress induced by centrifugal force. Furthermore, an additional parameter, the phase advance angle of the stator current (β), is introduced into the optimization process since the optimum current angle for maximum torque may be different due to the changed rotor structure and saliency. The upper and lower boundaries of the design variables are outlined in Table 6.2.

6.2.2 Design constants and limits

Another crucial aspect of the optimization process involves the design constants, as presented in Table 6.3. Constants such as the outer diameter of the stator, the air-gap length, the slot opening width, and the inner diameter of the rotor remain fixed due to mechanical constraints. Airgap length, current density in the coils, and fill factor values are set as the same as the baseline design to have a fair comparison between all designs. Moreover, the maximum phase current is limited by the capabilities of the battery used in the EV. Thus, the maximum phase current is limited to the phase current of the baseline design. Copper loss is the biggest contributor to the machine loss, and the total loss is limited by the capabilities of the cooling system. Since the baseline design has a maximum copper loss of 20 kW at maximum phase current, all designs are limited by the same copper loss limit to ensure the optimum machines do not operate above thermal limitations. The current density of the reference design at the peak operating condition has been used as a limit. The number of stator slots is maintained constant for the STP IPMSM but can vary for multiphase machines.

Table 6.2 Lower and upper limits of design variables

Variables	Lower limit	Upper limit
Tooth width, w_t (mm)	3	4.75
Back-iron thickness, L_c (mm)	10	35
Magnet angle, α_m (°)	130	170
Split ratio, γ	0.55	0.75
Rib width, w_r (mm)	1	10
Magnet distance, d_m (mm)	0	18
Magnet width, w_m (mm)	10	40
Magnet thickness, t_m (mm)	2	8
Center iron rib width, w_c (mm)	3	10
Current phase advance, α_{adv} (°)	40	60

Table 6.3 Design constants used in optimization

Parameters	Value	Parameters	Value
Stator outer diameter (mm)	225	Rotor inner diameter (mm)	70
Axial length (mm)	134	Number of stator slots	54
Slot opening width (mm)	2.5	Airgap (mm)	0.7
Current density (A/mm^2)	37	Fill factor	0.4
Phase current (A)	<1,100	Copper loss (kW)	<20

Additionally, the current density in the coils and the fill factor in the slot are set as constants to ensure a fair comparison between designs. The phase current is constrained by the following equation:

$$I_m = \frac{3I_{p,\max}}{m} \tag{6.2}$$

The phase current limit for multiphase machines (I_m) is determined, where $I_{p,\max}$ represents the maximum phase current of the STP machine, set at 1,100 A, and m represents the number of phases in the multiphase machines. This current limit serves to restrict the total current drawn from the battery; another design constant constrained by the battery architecture. Furthermore, an essential parameter is the total copper loss of the machine, governed by thermal design considerations. Therefore, a copper loss limit is also incorporated to ensure a fair comparison between designs.

To calculate the phase resistance, it is necessary to first determine the mean length per turn of the winding and the end winding length for each machine. These values can be calculated by using (6.3) and (6.4), respectively.

$$MLT = 2L_{motor} + 2L_{end} \tag{6.3}$$

$$L_{end} = \pi\, p_{slot} \tag{6.4}$$

where MLT is the mean length per turn, L_{motor} and L_{end} are the machine's axial length and the end winding length, respectively. The end winding shape is assumed to be semi-circular, and the radius of the coil pitch p_{slot} is calculated by

$$p_{slot} = \frac{2\pi\, r_{slot}\, t_{coil}}{N_s} \tag{6.5}$$

where r_{slot} represents the radius of the middle of the slot, t_{coil} denotes the throw of the coil, and N_s stands for the number of stator slots. The resistivity of copper varies with the temperature of the coils, as described in (6.6).

$$\rho_{cu,s} = \rho_{cu,0}[1 + 0.0039(\vartheta_s - \vartheta_0)] \tag{6.6}$$

where $\rho_{cu,s}$ representing the resistivity of copper at the selected temperature (ϑ_s), ρ_{cu} is the resistivity of copper at the reference temperature ϑ_0. In this case, T_s and T_0 are set to 80 °C and 20 °C, respectively. The operating temperature of the machine is set at 80 °C. Adjustments in the remanence and magnetic coercivity of NdFeB magnets are made accordingly. Finally, the phase resistance, R_{phase}, is calculated by

$$R_{phase} = \frac{\rho_{cu,s}\, MLT\, N_{turn}}{A_{coil}} \tag{6.7}$$

where N_{turn} represents the number of series turns in one phase, and A_{coil} denotes the cross-sectional area of copper. The copper cross-sectional area is determined by the slot area and the copper fill factor. The phase resistance obtained from these parameters is subsequently used to compute the maximum phase current, considering a predefined copper loss limit. This ensures that the optimization algorithm remains constrained by both the maximum phase current and the maximum copper loss of the machine.

6.2.3 Possible slot/pole number combinations for multiphase machines

To ensure a fair comparison between the baseline design and the optimal designs, both the STP and nine-phase machines utilize the same number of stator slots. For the nine-phase machine, this slot/pole combination is the only feasible option. In contrast, for five- and DTP machines, multiple slot/pole number combinations are explored to identify the specific configuration that offers the best performance. Table 6.4 lists the winding factors of five-, DTP, nine-phase machines with various slot/pole number combinations.

For the five-phase machines, the 60-slot/6-pole combination exhibits the highest winding factor compared to others and it is selected as the optimal slot/pole combination. Notably, it also displays a higher winding factor for the third harmonic, which could potentially enhance the average output torque through the third harmonic injection method. However, the effectiveness of the third harmonic injection depends on both the winding configuration and the distribution of air-gap flux density. Despite the suitability of the winding configuration in the five-phase machine for achieving a high third harmonic injection due to its higher winding factor, the increase in the third harmonic at the air-gap flux density is constrained by the limited pole arc in this IPMSM. This limitation results in the cancelation of the third harmonic at the air-gap flux density. Consequently, the third harmonic current injection method is not used in the optimization of the five-phase IPMSM, as it does not provide significant benefits due to the restricted pole arc.

For the DTP machines, despite the 36-slot/6-pole machine exhibiting a higher winding factor, it is not selected in this study. The primary reason is that this configuration has a unity winding factor for all harmonics, and its cogging torque frequency is 12. This presents an issue with a DTP machine because the torque ripple induced by the electromagnetic torque operates at the same frequency. Thus, selecting the 36-slot/6-pole configuration would lead to significantly high torque ripples in the output torque, making it unsuitable for the intended application. Thus, the 72-slot/6-pole configuration is chosen.

The winding configurations of one pole pair and the resultant magnetomotive force (MMF) distributions and spectra of phase A windings are presented in Figure 6.4. As the number of phases increases the magnitude of the MMF decreases

Table 6.4 Winding factors of different winding configurations for different harmonic orders

Machine	Number of stator slots	Fundamental	Third	Fifth	Seventh
STP	54-slot/6-pole	0.960	0.667	0.218	0.177
Five-phase	40-slot/6-pole	0.973	0.772	0.462	0.176
	45-slot/6-pole	0.980	0.830	0.577	0.300
	50-slot/6-pole	0.982	0.848	0.616	0.344
	60-slot/6-pole	0.988	0.891	0.707	0.454
	70-slot/6-pole	0.980	0.830	0.578	0.303
DTP	36-slot/6-pole	1	1	1	1
	72-slot/6-pole	0.991	0.924	0.793	0.609
Nine-phase	54-slot/6-pole	1	1	1	1

Note: The grey shades indicate final selected for comparison.

since the slots are divided into multiple windings. The angle difference between two three-phase winding sets of the DTP machine is chosen as 30 degrees.

The optimization of the machines is conducted using NSGA-II within ANSYS Maxwell. The optimization results for a 54-slot/6-pole STP machine, a 60-slot/6-pole five-phase machine, a 72-slot/6-pole DTP machine, and a 54-slot/6-pole nine-phase machine are depicted in Figure 6.5, where the purple dots signify the selected optimal design. Additionally, cross-sections of all machines are shown in Figure 6.6, together

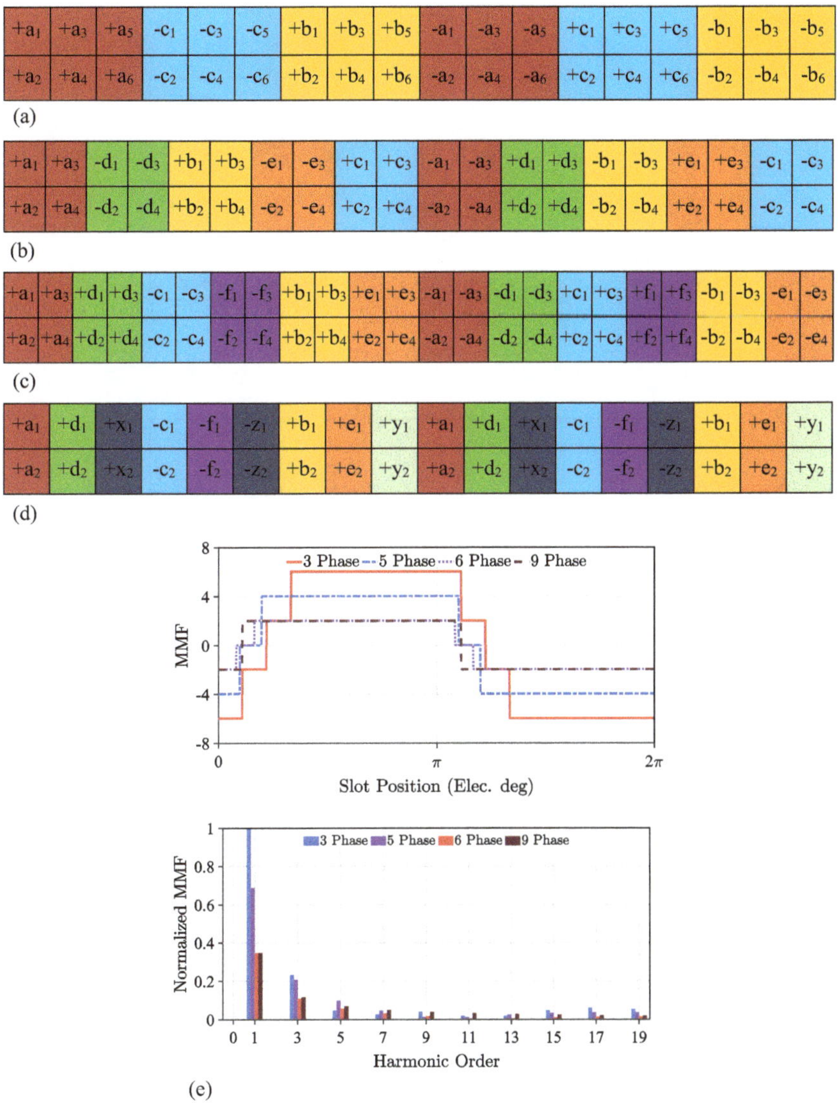

Figure 6.4 Winding configurations and MMFs of selected machines. (a) 54-slot/6-pole STP machine. (b) 60-slot/6-pole five-phase machine. (c) 72-slot/6-pole DTP machine. (d) 54-slot/6-pole nine-phase machine. (e) MMF distributions and spectra of phase A winding of each machine.

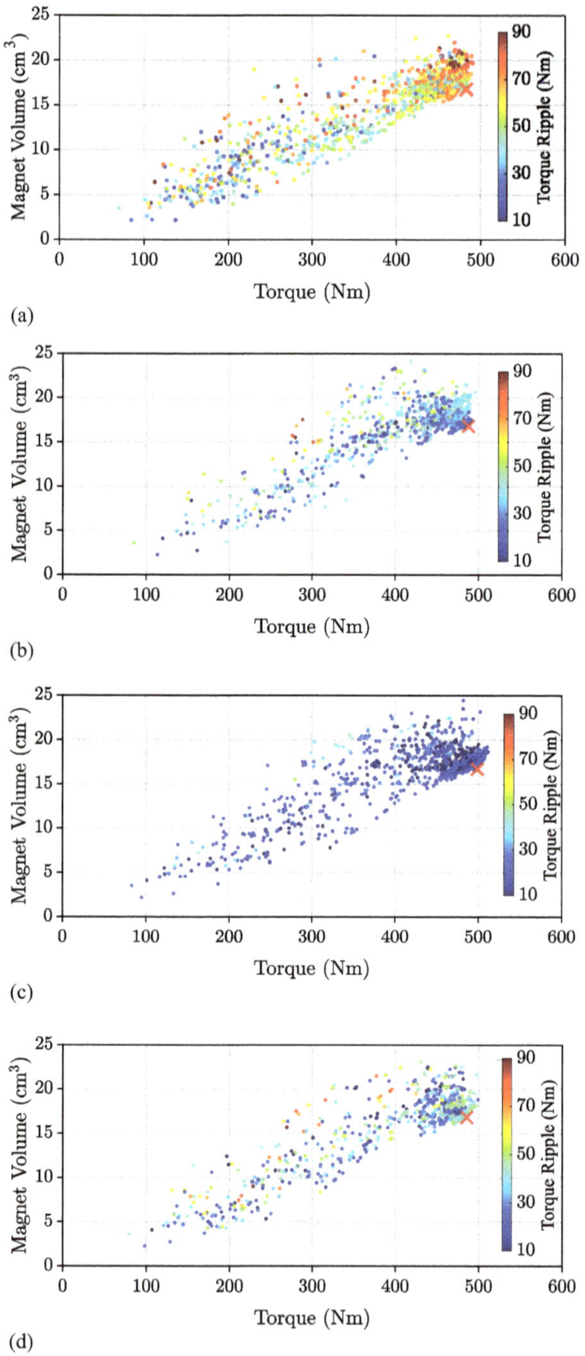

Figure 6.5 Results of multi-objective optimization. (a) 54-slot/6-pole STP machine. (b) 60-slot/6-pole five-phase machine. (c) 72-slot/6-pole DTP machine. (d) 54-slot/6-pole nine-phase machine.

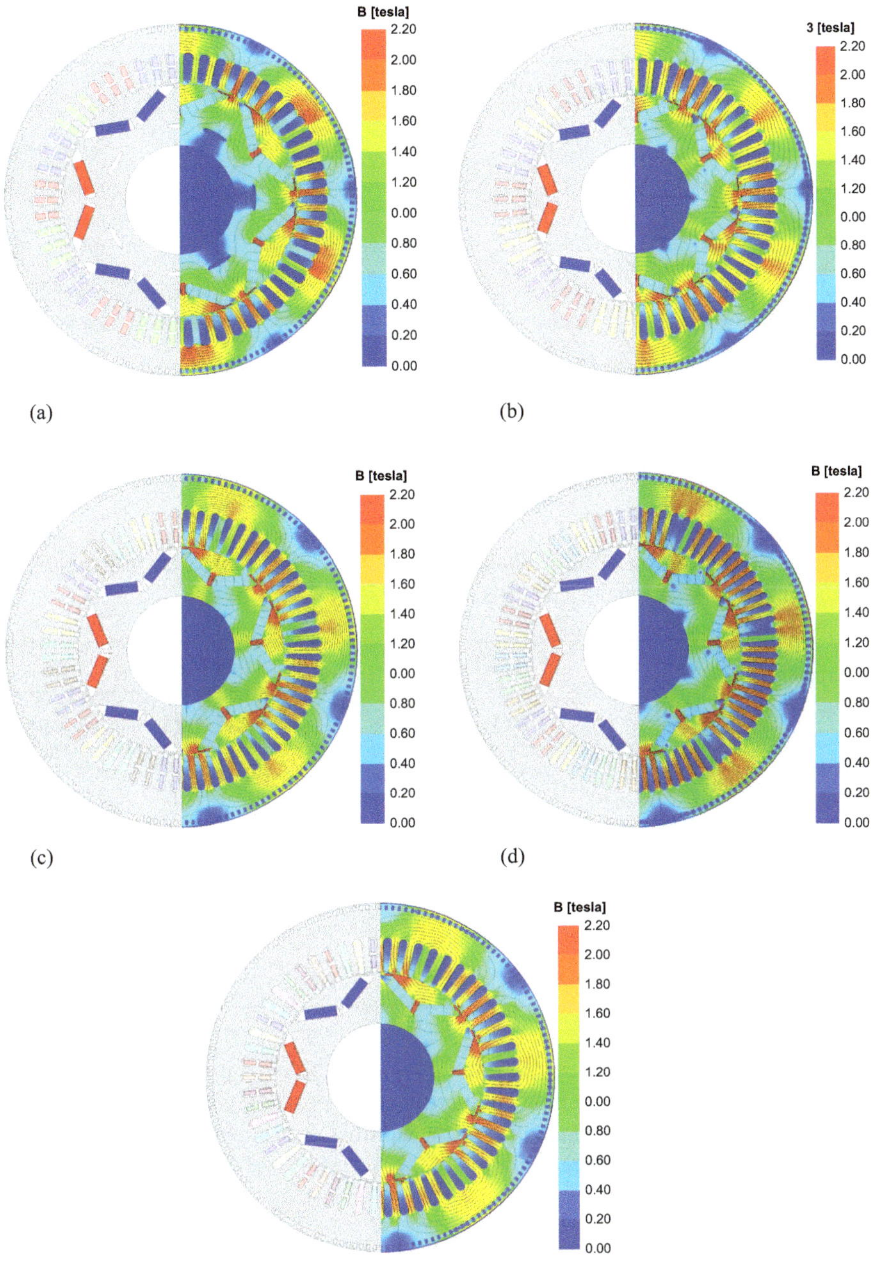

Figure 6.6 Cross-sections, flux density, and flux contour distributions of optimized machines. (a) Baseline, (b) optimized STP, (c) optimized five-phase, (d) optimized DTP, and (e) optimized nine-phase.

Table 6.5 Design parameters after optimization

Variables	Baseline	Optimized			
	STP	STP	Five-phase	DTP	Nine-phase
Slot/pole combination	54/6	54/6	60/6	72/6	54/6
Split ratio	0.666	0.59	0.59	0.6	0.6
Tooth width (mm)	4.35	4.64	4.43	3.49	4.65
Back-iron thickness (mm)	18	21.9	22	21	22.3
Rib width (mm)	7.6	9.73	9.9	9.97	9.43
Magnet width (mm)	21.6	18.8	20.38	20.8	20.5
Magnet thickness (mm)	6.5	6.66	6.15	6	6.11
Center iron rib width (mm)	3.2	3.0	3.26	3.26	3.0
Magnet angle (°)	140	139.3	131.8	135.9	135.4
Magnet distance (mm)	1.27	0.62	0.43	0.19	0.4
Current phase advance (°)	54	55	56	58	54.7

Table 6.6 Comparison of multiphase machines at 5,000 r/min and maximum phase currents by ANSYS Maxwell

Variables	Baseline	Optimized			
	STP	STP	Five-phase	DTP	Nine-phase
Output torque (Nm)	439	474	488	499	485
Output power (kW)	230	248	255	261	254
Torque ripple (Nm/%)	16.2/3.7	55/11.6	18/3.7	24.4/4.9	43/8.9
PM volume (cm^3)	18.8	16.8	16.8	16.73	16.83
Torque per PM volume (Nm/cm^3)	23.4	28.2	29.1	29.8	28.8
RMS phase current (A)	1100	1100	635	547	355
Phase advance (°)	54	55	56	58	54.7
Iron losses (kW)	0.87	1.04	1.11	1.05	1.11
Efficiency (%)	91.7	92.2	92.37	92.56	92.33

with equal potential distributions and flux contours. Table 6.5 presents the alterations in the design variables throughout the optimization process, while Table 6.6 shows the electromagnetic performance etc. of the optimized machines which will be discussed in detail in the next section.

6.3 Comparison of electromagnetic performance of optimized machines

The optimized machines are analyzed by using ANSYS Maxwell under both open circuit and maximum load conditions, operating at 5,000 r/min. The flux density distributions of the machines under full load conditions are depicted in Figure 6.6.

Notably, the DTP machine exhibits a slightly higher tooth flux density compared to the other machines. This is attributed to its higher number of stator slots, which consequently leads to a reduced tooth width. For reference, Table 6.5 presents the machine parameters, together with both the baseline STP and optimized STP machines.

6.3.1 Comparison of three-phase machines

The FE predicted results are presented in Table 6.6. It is important to highlight that the baseline STP IPMSM employs three-step skewing, whereas the other machines do not utilize any skewing. It shows that all of the designed machines, including the optimized STP machine, exhibit higher average output torque with a reduced amount of PM volume compared to the baseline design. Notably, the optimized STP machine shows an 8% increase in average torque and a 10.7% reduction in PM volume, resulting in a 20% enhancement in torque per PM volume. This performance improvement is primarily attributed to changes in the split ratio. The aim of increasing the average electromagnetic torque leads to an increase in the magnetic flux density in the air gap. Consequently, both the tooth width and the stator back-iron thickness are increased to handle the higher flux density and avoid magnetic saturation. Since the copper loss of the designed machines is limited, the stator bore diameter is reduced to increase the slot area and reduce phase resistance. Thus, the optimal split ratio of the machine has been decreased since the magnetic loading is increased [13]. However, it should be noted that the torque ripple is higher than that of the reference design due to the lack of skewing or rotor core shaping. Another factor contributing to the increased average output torque is the higher winding factor achievable with multiphase windings. As shown in Table 6.4, the multiphase machines have a higher fundamental winding factor compared to the STP machine, which increases the average output torque.

6.3.2 Comparison of multiphase machines

The machines are evaluated at 5,000 r/min under open-circuit conditions. The resulting back-EMF and spectra are illustrated in Figure 6.7. It is observed that the multiphase machines exhibit higher harmonics, particularly the fifth and seventh harmonics, in comparison to the baseline design. This increase in higher-order harmonics contributes to higher core losses, as highlighted in Table 6.6. Interestingly, the baseline machine demonstrates the lowest winding factors at high-order harmonics, which subsequently reduces the iron losses.

Figure 6.8 illustrates the output torque waveforms and spectra. Notably, all of the designed machines exhibit higher average output torque compared to the baseline IPMSM, with the best performance for the DTP machine due to higher magnetic loading and flux linkage compared to other designs. Additionally, both the five-phase and nine-phase machines show enhanced torque performance compared to the optimized STP machine, primarily due to improved winding factors.

The torque ripples of optimized multiphase machines are lower than the optimized STP machine, but they are still higher than the baseline STP IPMSM due to

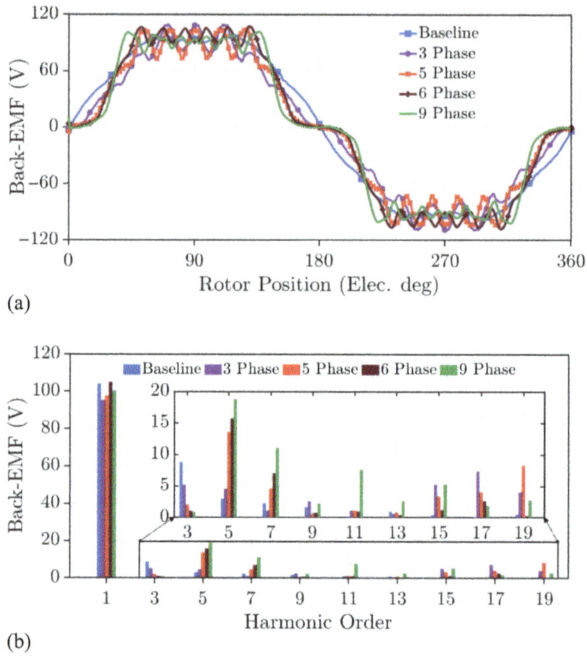

Figure 6.7 Back-EMFs at 5,000 r/min. (a) Waveforms and (b) spectra.

the absence of PM skewing and rotor core shaping methods. Implementing these techniques in addition to multiphase winding configurations can further reduce torque ripples, ensuring smoother operation. However, it is worth noting that the nine-phase machine exhibits a higher torque ripple compared to other multiphase machines. This discrepancy can be attributed to the chosen slot/pole number combination for the nine-phase machine. The cogging torque frequency of a PM machine, f_{cog}, can be calculated by

$$f_{cog} = \frac{2N_s f}{GCD(N_s, 2p)} \tag{6.8}$$

where N_s is the number of stator slots, f is the electrical frequency, $GCD(N_s, 2p)$ is the greatest common divisor of the number of stator slots, N_s, and the number of poles, $2p$. The cogging frequency of a 54-slot/6-pole machine is the 18th harmonic of the electrical frequency. Similarly, the torque ripple created by the electromagnetic torque of a nine-phase machine is at the same frequency which creates an even higher torque ripple when combined with the cogging torque.

Figure 6.9 shows the variation of output torque with copper loss. Notably, all designed machines outperform the baseline design in terms of torque performance. Specifically, the DTP machine exhibits the highest performance, while all multiphase machines demonstrate slightly superior torque performance compared to the

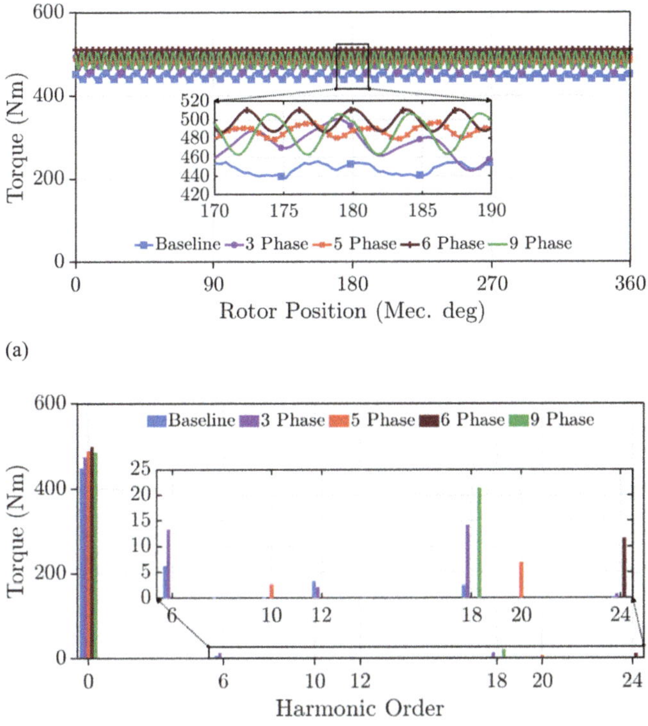

(a)

(b)

Figure 6.8 Torque performance. (a) Waveforms and (b) spectra.

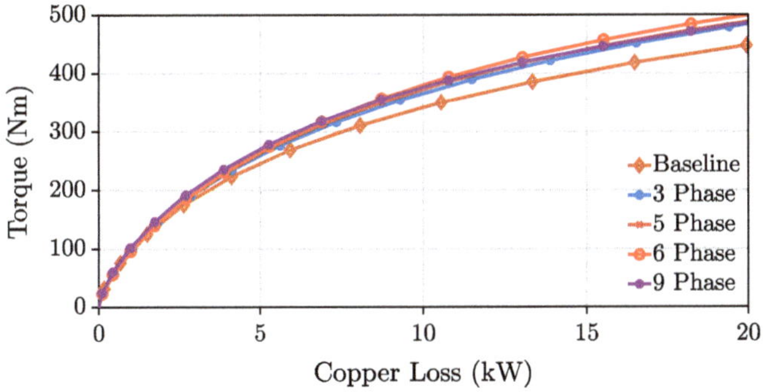

Figure 6.9 Variation of average output torque with copper loss.

optimized STP machine. Although all of the designed machines have the same copper loss, the designed multiphase machines show better average torque per copper loss.

The torque-speed characteristics of the machines are investigated using the maximum torque per ampere (MTPA) control algorithm with a 350 V DC-link voltage. The maximum rotational speed of the baseline IPMSM is 18,000 r/min.

The torque and voltage are calculated from the flux linkage and current by

$$T = \frac{M}{2} p\left(\psi_d I_q - \psi_q I_d\right) \tag{6.9}$$

$$V_a = \sqrt{\left(R_a I_q + \omega\psi_d\right)^2 + \left(R_a I_d - \omega\psi_q\right)^2} \tag{6.10}$$

where T is the average torque, M is the number of phases, p is the number of pole pairs, ψ_d and ψ_q are the d- and q-axis flux linkages, I_d and I_q are the d- and q-axis currents, R_a is the phase resistance, ω is the rotational speed, V_a is the phase voltage. The maximum voltage is limited by the DC-link voltage of 350 V and the SVPWM method, while the maximum d- and q-axis currents are limited by the maximum phase current by

$$I_a = \sqrt{I_q^2 + I_d^2} \tag{6.11}$$

where I_a is the phase current.

The torque-speed and power-speed characteristics of the machines are depicted in Figure 6.10. In the constant torque region, the DTP machine demonstrates the best

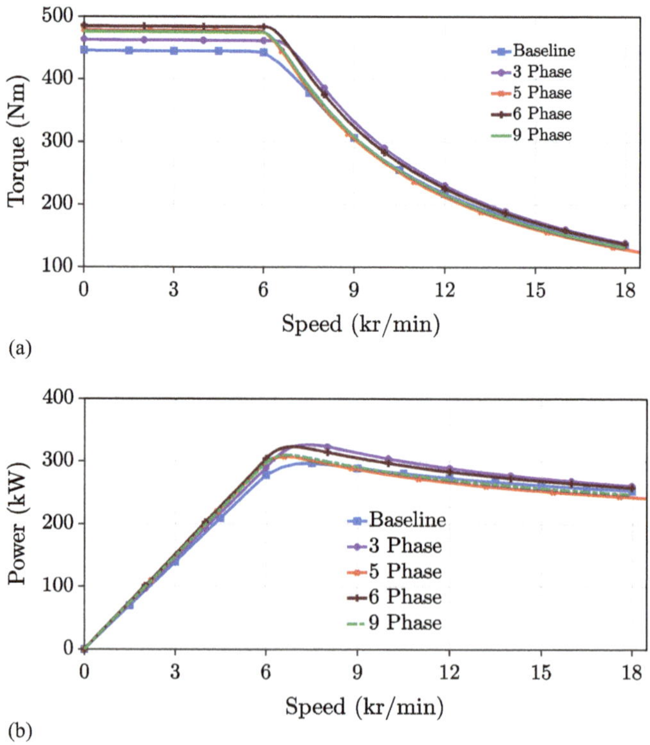

Figure 6.10 Torque/power-speed characteristics in the whole speed range. (a) Torque-speed and (b) power-speed.

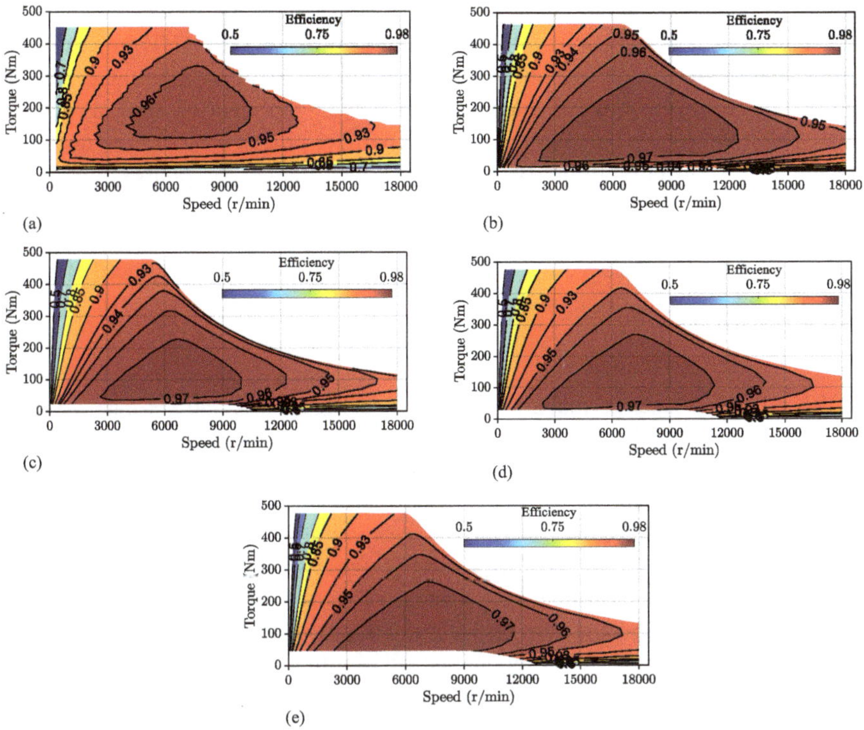

Figure 6.11 Efficiency maps of baseline and optimized IPMSMs. (a) Baseline, (b) optimized STP, (c) optimized five-phase, (d) optimized DTP, and (e) optimized nine-phase.

performance. Conversely, the optimized STP machine exhibits the highest base speed due to its lowest back-EMF, as shown in Figure 6.7. Since the machines are driven using the DC-link voltage and control algorithm the machine with lower back-EMF has a higher base speed as the base speed is limited by the voltage ellipse. Consequently, the optimized STP machine achieves the highest power output when comparing the power-speed curves. Efficiency maps of the machines are presented in Figure 6.11. It is important to note that mechanical losses are not considered in this analysis. The higher iron losses of the designed machines result in lower efficiency in the high-speed regions as the frequency increases. Nonetheless, the efficiencies of the multiphase machines surpass that of the baseline design overall, attributable to the increased output power achieved by raising the average output torque.

6.4 Comparison of drives for multiphase machines

The multiphase machines offer significant advantages over their STP counterparts considering the drive and control. The first major advantage is the lowered phase current. The phase current of the multiphase machines becomes lower as the number of phases is increased following the equation (6.2). This is important for designs that require high currents since oftentimes the phase current cannot be supplied by a single power switch in high-power applications such as electric vehicles or wind power generators. Multiple parallel power switches are necessary to achieve the current rating. By increasing the number of phases, the power switches with lower current ratings can be used and the cost can be decreased. Another important component to consider is the DC-link capacitor which takes up a big portion of the drive's volume. This section investigates the effects of multi-phase machines on the drive components.

6.4.1 Power switches

The drive for the baseline machine uses 24 SiC MOSFETs and six gate drives in total [11]. A two-level three-phase inverter has been used in the drive with four MOSFETs in parallel in each leg to achieve the required phase current rating [14]. The MOSFET used in the inverter has the part number SCTW100N65G2AG which has a 650 V maximum drain-source voltage, and an absolute maximum current rating of 280 A [15]. If the same MOSFET is assumed to be used with the multi-phase machine's drives, the required number of power switches can be calculated for each drive assuming the same peak current passes through in each case. The comparison of the number of components used in the drives is given in Table 6.7.

It can be seen from Table 6.7 that the number of power switches used in the drives is not directly proportional to the number of phases. The current flowing through each leg of the inverter is reduced as the number of phases is increased. This, in turn, reduces the number of parallel power switches required to be used in each leg. By doing so, the total number of power switches can remain constant even if the number of phases is increased. This phenomenon can be observed for DTP machines where the same number of power switches is used as the baseline design. However, the number of power switches used in the five-phase and nine-phase machines is higher than the baseline STP design. The number of gate drives is higher since a higher number of half-bridge legs are required to drive multiphase machines.

Table 6.7 Comparison of the number of components used in drives

Parameter	**Baseline**	**Optimized**			
	STP	**STP**	**Five-phase**	**DTP**	**Nine-phase**
Peak phase current (A)	1,100	1,100	635	547	355
Number of legs	3	3	5	6	9
Number of parallel power switches	4	4	3	2	2
Number of gate drives	6	6	10	12	18
Total number of power switches	24	24	30	24	36

Moreover, the fault tolerance of the drives can be increased by utilizing a multiphase configuration [16]. Thus, the DTP machine can replace the baseline machine with the same number of MOSFETs while having better electromagnetic performance and efficiency with lower PM volume, and higher fault tolerance. The drawback of increasing the number of phases on the drive side is the increased number of peripheral components such as gate drives, sensors, etc. Also, the control algorithm required to drive the multiphase machines is more complicated than the conventional STP machines [2].

6.4.2 DC-link capacitor

Another important component to be considered while comparing the drives for multiphase machines is the DC-link capacitor. The DC-link capacitor is the biggest component in the drive board by volume, and due to the size and the requirements of the capacitor, it is an expensive part as well [17]. On top of these, it is the least reliable part of the drive since it is more prone to failure than the other components [18]. Combining these factors, reducing the size and ratings of the DC-link capacitor would increase the reliability of the drive, and reduce the size and cost.

The required capacitance values of the DC-link capacitors for multiphase machines are investigated in this section by assuming a 5% voltage ripple present on the capacitor voltage. The two-level inverter design for a generic drive for the machine with an "N" number of phases is shown in Figure 6.12.

The voltage ripple $V_{\%\text{ripple}}$ can be calculated by

$$V_{\%\text{ripple}} = \frac{\Delta V_{\text{c}}}{V_{\text{dc}}} = \frac{\Delta Q_{\text{c}}}{C V_{\text{dc}}} \tag{6.12}$$

where ΔV_{c} is the change in capacitor voltage. V_{dc} is the DC-link voltage, which is 350 V in the baseline design. ΔQ_{c} is the change in the capacitor charge. C is the required capacitance value. The change in the capacitor charge can be found by

$$Q(t) = \int_0^t I_{\text{c}}(t)\text{d}t \tag{6.13}$$

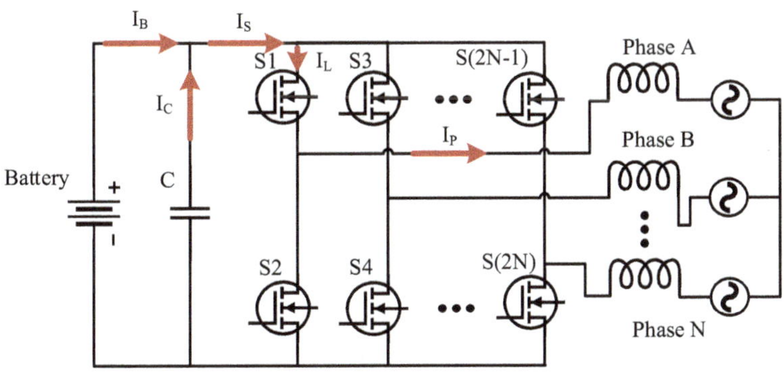

Figure 6.12 Drive circuit schematic for N phase machine.

where I_c is the capacitor current. The capacitor charge is the integral of the capacitor current over an electrical frequency (see Figure 6.10). The capacitor current is designed as:

$$Q(t) = I_s(t) - \frac{1}{t}\int_0^t I_s(t)\mathrm{d}t \tag{6.14}$$

where I_s is the total current supplied to the drive. It is assumed that the battery supplies a constant current while the capacitor supplies the pulsed AC. The capacitor current can be found by subtracting the average out of the total current. The total inverter input current is the summation of currents on each leg, I_L, where they are defined by the phase currents of the machine by

$$I_{P-N}(t) = \widehat{I}_P \cos(\omega t - \beta - (N-1)\alpha) \tag{6.15}$$

where I_{P-N} is the phase current of the N-phase winding of the machine. $\widehat{I}_P$ is the peak current. ω is the frequency. β is the current advance angle. α is the electrical angle difference of different phases. The change in the DC-link capacitor charge can be calculated using this method [19]. The maximum normalized DC-link charge changes for multiphase machines [20] are presented in Table 6.8.

Using capacitor charge change values for STP, five-phase, DTP, and nine-phase machines along with their phase current values, it is possible to estimate the required DC-link capacitance. As the DC-link voltage and the maximum voltage ripple allowed at the DC-link are the same for all cases as shown by (6.12), the change in the capacitor charge is directly proportional to the required DC-link capacitance value. By multiplying the normalized charge change with the phase current value for each machine, it is possible to estimate the required capacitance value. Table 6.8 presents the per unit DC-link capacitance for each case assuming a 5% DC-link voltage ripple and 350 V DC supply voltage.

It is clear from Table 6.8 that increasing the number of phases from three to five or six has a major advantage in terms of smaller capacitance as it is possible to use a ~50% smaller DC-link capacitor which is a major advantage over three-phase systems. Reducing the size of the DC-link capacitor helps increase the power density of the inverter as the DC-link capacitor is the biggest component in terms of

Table 6.8 Comparison of DC-link capacitances of multiphase machines

Parameter	Baseline	Optimized			
	STP	STP	Five-phase	DTP	Nine-phase
Maximum phase current (A)	1,100	1,100	635	547	355
Maximum normalized DC-link charge change, ΔQ_c, [21] (μC/A)	21.70	21.70	19.10	21.65	29.90
DC-link capacitance (p.u.)	1	1	0.51	0.50	0.45
Maximum phase current (A)	1,100	1,100	635	547	355

volume. The RMS capacitor current value can be decreased by utilizing the multiphase inverter and machine design which in turn reduces the DC-link capacitor charge change and reduces the required capacitance value. The reduction in the capacitor current also helps to increase the reliability of the DC-link capacitance since the thermal stress and temperature change in the capacitor gets higher as the capacitor current is increased [18]. However, increasing the number of phases further does not show a significant improvement. Nevertheless, multiphase machines have enabled designs with a much lower DC-link capacitor.

6.5 Cost analysis and discussions

This section compares the costs of electric powertrains with various multiphase IPMSM systems.

6.5.1 Electrical machines

The relative costs of the machines are estimated at high volume production at the scale of 100,000 units per year. The per unit cost of the baseline STP IPMSM has been reported to be 624 $ per machine [22]. If the cost of the N48 grade NdFeB magnet is assumed to be 70 $/kg [21], then the PM cost can be estimated from the total weight of PMs used. Another important factor to consider while comparing the multiphase machines is the production complexity of the winding. As the number of phases increases, the time to manufacture the winding and the tooling costs increase in proportion. To account for the change in both of these important cost factors, the cost distribution of each component used in the baseline IPMSM at high volume production (>100,000 units/year) is presented in Figure 6.13, where "Material" represents materials used in the machine such as PMs, copper, electrical sheets, etc., "Purchased Parts" are the parts purchased from suppliers such as bearing, cooling pump, etc., "Equity" is the cost of investment in the production line, "Labor" is the cost of the human labor, "Assembly" is the cost of final assembly of all components, "Tools & Area" is the costs associated with the tooling cost and the area used by the tools, and "Maintenance" is the cost to maintain the

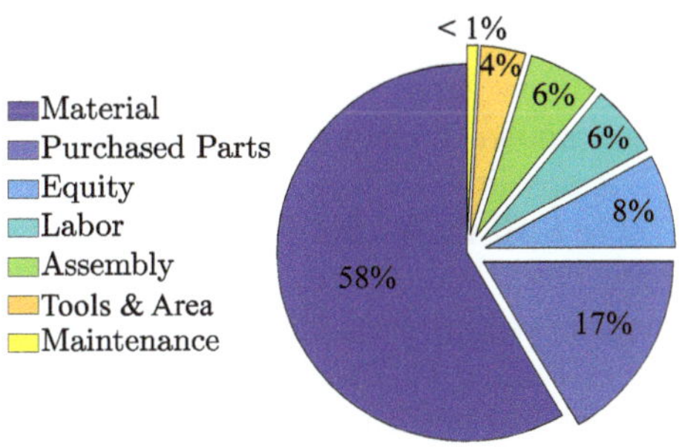

Figure 6.13 Distribution of baseline STP IPMSM costs at high-volume production.

Table 6.9 Estimated cost of designed machines at high-volume production

Parameter	Baseline	Optimized			
	STP	STP	Five-phase	DTP	Nine-phase
Output power (kW)	230	248	255	261	254
Total PM weight (kg)	1.692	1.512	1.512	1.506	1.515
Estimated PM cost ($)	137.3	105.8	105.8	105.4	106.1
Estimated tooling cost ($)	24.96	24.96	41.6	49.92	74.88
Estimated total cost of machine ($)	624	592.5	609.1	617.5	642.4
Power per USD (W/$)	369	418.6	418.7	422.7	395.4

production line [22]. Most of the costs at this high-volume production level come from the materials. However, tooling costs still account for 4% of the total costs of the machine. The tooling cost associated with the winding is calculated in proportion to the cost distribution and the costs of tooling for multiphase machines are estimated from the three-phase winding. The estimated total costs of the machines are shown in Table 6.9.

The tooling cost of multiphase machines increases proportionally to the number of phases as the complexity of the winding increases. PM cost is reduced compared to the baseline design. Although the optimized STP machine has the lowest cost among all designs, it has lower electromagnetic performance compared to the multiphase machines. Thus, the ratio of generated output power per cost values has been included in Table 6.9, which shows that the DTP machine has slightly higher power per cost value than other designs. The nine-phase machine has the lowest power per cost value among optimized machines since the number of phases is high, the power improvement gained by increasing the number of phases is not enough to account for the cost of the tooling.

6.5.2 Drives

The other part affected by the changes is the inverter of the powertrain. The inverter for the baseline IPMSM is estimated to cost around 522 $ per unit [23]. The distribution of the components' costs of a three-phase inverter with SiC switches is shown in Figure 6.14 [24,25]. 66% of the cost associated with the inverter comes from the SiC switches. The DC-link capacitor accounts for 3% of the total cost. Using the known total cost of the baseline's drive and the distribution of the costs, it is possible to estimate the cost of each component used in the baseline design. Table 6.10 shows the estimated costs of each component of the three-phase inverter.

This data is then used to calculate the change in the price tag of the multiphase inverters which is given in Table 6.11. The results show that increasing the number of switches has a significant cost burden. The inverter of the nine-phase machine is 33% more expensive than the baseline design due to the higher number of switches used in the driver. A DTP machine results in a slightly cheaper inverter cost since

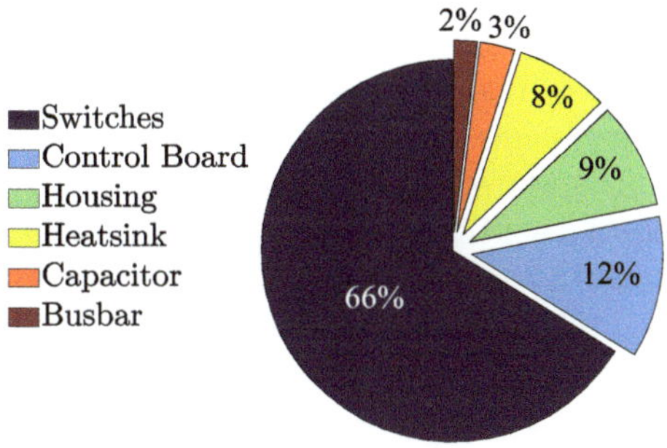

Figure 6.14 Estimated distribution of inverter costs for baseline IPMSM.

Table 6.10 Cost estimation of each component used in the inverter of the baseline design

Parameter	Total price ($)	Quantity	Unit price ($)
SiC switches	344.5	24	14.36
Control board	62.6	1	62.6
Housing	46.8	1	46.8
Heatsink	41.5	1	41.5
Capacitor	15.6	1	15.6
Busbar	10.2	1	10.2
Total cost of inverter ($)	522	1	522

Table 6.11 Costs of multiphase machine inverters

Parameter	Baseline	Optimized			
	STP	STP	Five-phase	DTP	Nine-phase
Output power (kW)	230	248	255	261	254
Total number of switches	24	24	30	24	36
Cost of switches	344.5	344.5	430.8	344.5	516.9
DC-link capacitance (p.u.)	1	1	0.51	0.50	0.45
Cost of capacitor	15.6	15.6	7.96	7.8	7.02
Estimated total cost of inverter ($)	522	522	615.9	521.8	693.5
Power per USD (W/$)	440.6	475.1	414.1	500.2	366.3

the same number of switches has been used considering paralleled switches in STP machines while the DC-link capacitor is smaller than that in the baseline. The power per cost value is also added in Table 6.11, this shows that the DTP machine has the best performance/price ratio compared to others, while the nine-phase machine has the worst performance due to the increased number of switches.

6.5.3 Discussions

The overall performance of machine and drive systems can be compared to see the effect of multiphase systems. In terms of the electromagnetic performance, all the multiphase machines show better performance than the baseline design. Reasons for the improvement include a higher winding factor that can be achievable by using multiphase winding and split ratio optimization of multiphase machines. As a result of the higher output power, the multiphase machines show higher efficiency in all torque-speed ranges which is crucial in ever-demanding electric vehicle powertrains. The effect of the multiphase winding is not limited to the electrical machine. The change in the inverter topology is discussed in two crucial components: switches and the DC-link capacitor. The number of switches used in the inverter is of vital importance since the cost of these SiC switches is most of the total cost of the inverter. The DTP machine has been shown to use the same number of switches as the baseline design thanks to the lower phase current that can be achieved by utilizing the DTP configuration. On the other hand, five-phase and nine-phase machines have resulted in a higher number of switches. Although the DC-link capacitor is not the most expensive part of the driver circuitry, it is the biggest in terms of volume and it is the least reliable part since it is more prone to failures. Thus, a reduction in the required DC-link capacitor would not only improve the power density of the overall system but also improve the reliability of the powertrain. The multiphase machines show around 50% lower required DC-link capacitance compared to the STP baseline which is a significant improvement to the inverter performance. Lastly, the overall effect of the multiphase systems on the cost of the powertrain is shown in Table 6.12. The data shows that the total cost of the powertrain does not necessarily increase when multiphase systems are utilized, contrary to what is often stated in the literature. While five- and nine-phase machines have higher costs, the DTP machine has lower total costs and higher performance/price ratios when compared to all other designs. The improvement in the performance comes from various factors: using lower PM volume, using the same number of switches while requiring less DC-link capacitance, and having higher electromagnetic performance.

Table 6.12 Performance and cost comparison of all machines

Parameter	Baseline	Optimized			
	STP	STP	Five-phase	DTP	Nine-phase
Output power (kW)	230	248	255	261	254
Cost of machine ($)	624	592.5	609.1	617.5	642.4
Cost of inverter ($)	522	522	615.9	521.8	693.5
Total cost ($)	1,146	1,114.5	1,225	1,139.3	1,335.9
Power per USD (W/$)	200.7	222.5	208.2	229.1	190.1

6.6 Summary

A comparative study of three-, five-, six-, and nine-phase IPMSM systems for electric vehicles has been presented in this chapter, with a reference to a commercial STP IPMSM in terms of torque and power capabilities, torque ripples, PM volume, torque per PM volume, losses, and efficiencies, as well as the requirements of power switches, peripherals, and DC-link capacitor, and the costs for machine and inverter.

The results indicate that the average output torque of the optimized STP machine has a 20% increase in torque per PM volume compared to the baseline IPMSM. Despite higher iron losses, multiphase machines exhibit a 3%–6% higher average torque per PM volume. Additionally, they have inherently low torque ripples without additional rotor modifications. It is shown that increasing the number of phases does not necessarily increase the number of MOSFETs used in the drive considering paralleled switches in the STP machine since the phase current decreases as the number of phases is increased.

Furthermore, it is also shown that the DC-link capacitance value can be decreased by ~50% by using multiphase machines, which not only increases the power density of the inverter due to reduced volume but also improves the reliability of the powertrain. The cost analysis of the multiphase machines shows that although the multiphase machines result in lower PM usage, the tooling costs, due to the more complex winding, increase. The number of switches used in the inverter creates the majority of the cost of the inverters. A DTP machine has a 14.2% higher performance/price ratio compared to other investigated electric powertrains since it has better electromagnetic performance while having the same number of SiC switches considering paralleled switches in STP machines and smaller DC-link capacitors.

The above conclusions have been derived for electric vehicles and should be applicable for wind power generation etc.

References

[1] F. Tokgoz, Z.Q. Zhu, X. Chen, *et al.*, "Comparative study of multiphase IPMSMs for EV applications," *The 26th Int. Conf. Electr. Mach. Syst. (ICEMS)*, Zhuhai, China, 2023, pp. 383–388.

[2] Z. Q. Zhu, S. Wang, B. Shao, L. Yan, P. L. Xu, and Y. Ren, "Advances in dual-three-phase permanent magnet synchronous machines and control techniques," *Energies*, vol. 14, no. 22, pp. 1–46, 2021.

[3] S. Wang, Z. Q. Zhu, A. Pride, J. Shi, R. Deodhar, and C. Umemura, "Comparison of different winding configurations for dual three-phase interior PM machines in electric vehicles," *World Electr. Veh. J.*, vol. 13, no. 3, pp. 1–34, 2022.

[4] J. K. Park, C. Babetto, G. Berardi, J. Hur, and N. Bianchi, "Comparison of fault characteristics according to winding configurations for dual three-phase

synchronous reluctance motor," *IEEE Trans. Ind. Appl.*, vol. 57, no. 3, pp. 2398–2406, 2021.

[5] S. V. Nair, K. Layek, and K. Hatua, "An unequal split dual three-phase PMSM with extended torque-speed characteristics for automotive application," *IEEE Trans. Power Electron.*, vol. 37, no. 10, pp. 12437–12449, 2022.

[6] K. Wang, Z. Y. Gu, C. Liu, and Z. Q. Zhu, "Design and analysis of a five-phase SPM machine considering third harmonic current injection," *IEEE Trans. Energy Convers.*, vol. 33, no. 3, pp. 1108–1117, 2018.

[7] J. Huang, Y. Sui, Z. Yin, S. Yang, and P. Zheng, "Five-phase hybrid single-/double-layer fractional slot-winding PMSM for torque improvement under third harmonic current injection condition," *IEEE Trans. Magn.*, vol. 58, no. 8, pp. 1–6, 2022.

[8] S. Sadeghi, L. Guo, H. A. Toliyat, and L. Parsa, "Wide operational speed range of five-phase permanent magnet machines by using different stator winding configurations," *IEEE Trans. Ind. Electron.*, vol. 59, no. 6, pp. 2621–2631, 2012.

[9] X. Chen, J. Wang, V. I. Patel, and P. Lazari, "A nine-phase 18-slot 14-pole interior permanent magnet machine with low space harmonics for electric vehicle applications," *IEEE Trans. Energy Convers.*, vol. 31, no. 3, pp. 860–871, 2016.

[10] Y. Demir, E. Yolacan, A. M. EL-Refaie, and M. Aydin, "Investigation of different winding configurations and displacements of a nine-phase permanent-magnet-synchronous motor with unbalanced AC winding structure," *IEEE Trans. Ind. Appl.*, vol. 55, no. 4, pp. 3660–3670, 2019.

[11] S. Lingang, "Tesla Model 3 machine disassembly," *Inventchip Technology*, Aug. 2018, Available: https://www.inventchip.com.cn/uploads/files/20200919/1600488691370280.pdf.

[12] F. Tokgoz, Z.Q. Zhu, X. Chen, *et al.*, "Asymmetric and symmetrical rotor core shaping techniques of IPMSMs for EVs," *2023 26th Int. Conf. Electr. Mach. Syst. (ICEMS)*, Zhuhai, China, 2023, pp. 377–382.

[13] Y. Pang, Z. Q. Zhu, and D. Howe, "Analytical determination of optimal split ratio for permanent magnet brushless motors," *IET Proc.*, vol. 153, no. 1, p. 7, 2006.

[14] R. Kroeze, C. Campbell, and N. R. Kalayjian, "Fast switching for power inverter," United States Patent 2012/0195079, 2, 2012.

[15] STMicroelectronics, "Automotive-grade silicon carbide Power MOSFET 650 V, 20 mΩ typ., 100 A in an HiP247 package," SCTW100N65G2AG datasheet, September 2021.

[16] M. A. Frikha, J. Croonen, K. Deepak, Y. Benômar, M. E. Baghdadi, and O. Hegazy, "Multiphase motors and drive systems for electric vehicle powertrains: state of the art analysis and future trends," *Energies*, vol. 16, no. 2, p. 768, 2023.

[17] H. Wen, W. Xiao, X. Wen, and P. Armstrong, "Analysis and evaluation of DC-link capacitors for high-power-density electric vehicle drive systems," *IEEE Trans. Veh. Tech.*, vol. 61, no. 7, pp. 2950–2964, 2012.

[18] S. Yang, D. Xiang, A. Bryant, P. Mawby, L. Ran, and P. Tavner, "Condition monitoring for device reliability in power electronic converters: a review," *IEEE Trans. Power Electron.*, vol. 25, no. 11, pp. 2734–2752, 2010.

[19] Z. Nie and N. Schofield, "Multi-phase VSI DC-link capacitor considerations," *IET Electric Power Appl.*, vol. 13, no. 11, pp. 1804–1811, 2019.

[20] Z. Nie, "Multiphase power electronic converters for electric vehicle machine drive systems," *Ph.D. dissertation*, Electrical and Computer Engineering, McMaster University, Canada, Jun. 15, 2018.

[21] J. Cui, J. Ormerod, D. Parker, *et al.*, "Manufacturing processes for permanent magnets: Part I—Sintering and casting," *J. Minerals*, vol. 74, no. 4, pp. 1279–1295, 2022.

[22] J. Hemsen, N. Nowak, and L. Eckstein, "Production cost modeling for permanent magnet synchronous machines for electric vehicles," *Automot. Engine Technol.*, vol. 8, no. 2, pp. 109–126, 2023.

[23] G. Rogers, V. Nair, and S. Pillai, "48 V and battery electric vehicles costs for revised 2023 and later model year light-duty vehicle greenhouse gas emission standards," Roush Industries, Inc., 2021. Available: https://downloads.regulations.gov/EPA-HQ-OAR-2021-0208-0210/attachment_3.pdf.

[24] S. Yu, J. Wang, X. Zhang, Y. Liu, N. Jiang, and W. Wang, "The potential impact of using traction inverters with SiC MOSFETs for electric buses," *IEEE Access*, vol. 9, pp. 51561–51572, 2021.

[25] R. Burkart and J. W. Kolar, "Component cost models for multi-objective optimizations of switched-mode power converters," *2013 IEEE Energy Convers. Cong. Expos. (ECCE)*, Denver, CO, USA, 2013, pp. 2139–2146.

Chapter 7
Modulation strategies for multiphase PMSMs

7.1 Introduction

The modulation strategy plays an important role in motor control, aiming to generate the desired pulse-width-modulated (PWM) signals to the inverter switches (such as MOSFET and IGBT) for the single three-phase (STP) or multiphase machines. Amongst various PWM methods, the space vector (SV) modulation (SVM) strategy is the most popular, due to several advantages including fixed switching frequency, low harmonic content, and higher DC bus voltage utilization. However, in some specific control algorithms, such as direct torque control (DTC) and model predictive control (MPC), voltage vectors are generated in the absence of an SVM modulator, and thus, the SVM strategy may not be required.

In this chapter, various modulation strategies employed in different control algorithms for multiphase machines will be introduced, including modulation strategies for field-oriented control (FOC), switching-table (ST), and SVM-based DTC (ST- and SVM-DTC). This chapter will describe the principles behind these modulation strategies, and experimental results are also presented for illustrative purposes. The structure of this chapter is summarized in Figure 7.1.

7.2 Control algorithms of single three-phase PMSMs

In controlling PMSMs, the modulation strategies employed in FOC and DTC systems differ in their approach. On the one hand, the modulation strategy in an FOC system utilizes proper voltage vectors (VVs) to generate a rotating magnetic field. On the other hand, the modulation strategy in a DTC system utilizes proper voltage vectors to generate a pulsating magnetic field. Therefore, to clearly illustrate modulation strategies for multiphase machine drives, the principles of FOC and DTC and their modulation strategies in STP PMSMs are first presented and discussed in this section.

7.2.1 Field-oriented control (FOC)

The concept of indirect FOC was first proposed by Hasse [1,2] in 1968. Later in 1971, direct FOC was developed within Siemens by Blaschke [3]. The two methods

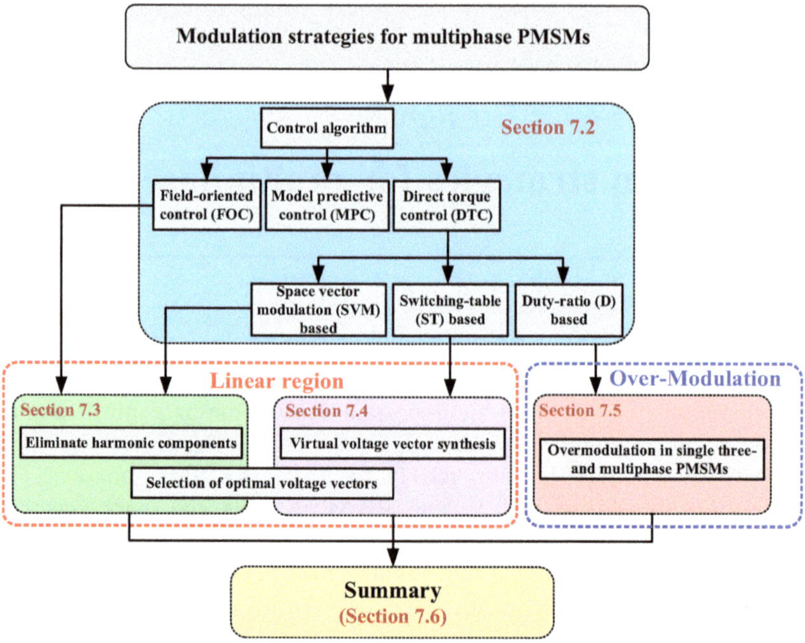

Figure 7.1 Structure of Chapter 7.

differ in the determination of the rotor angle. In direct FOC, the angle is obtained by the terminal voltages and currents, while in indirect FOC, the angle is obtained by using rotor position measurement or estimation. In terms of industrial applications, *Toshiba* was the pioneer in implementing the vector-controlled AC drive in 1979, comprising a voltage-source inverter (VSI) and an induction machine. In 1980, *Werner Leonhard* at the Technical University of Braunschweig contributed to the advancement of FOC techniques, thereby expanding opportunities for AC drives [4,5]. The utilization of the FOC accelerated with the development of microprocessors. It was not long before several manufacturers began offering microprocessor-based vector control solutions over a wide range of power options. Soon the term “vector control” became synonymous with high-performance drives. Later, a less ambiguous and more technically precise term was proposed: “Field Oriented Control.”

7.2.1.1 Model of STP PMSMs in rotation reference frame

AC machines are ideally fed by sinusoidal voltages. Since the machine windings are distributed in space as the phases, the electrical quantities such as currents, are distributed in time and space. Hence, it is quite simple to represent the electrical quantities using space vectors. However, the general frame of reference, also known as the “*abc*” reference frame, as shown in Figure 7.2, describes all the electrical quantities along the three axes.

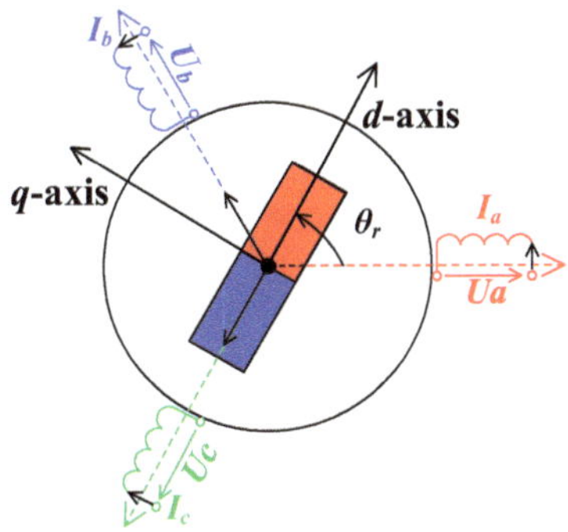

Figure 7.2 Schematic representation of stator winding in abc and dq reference frames.

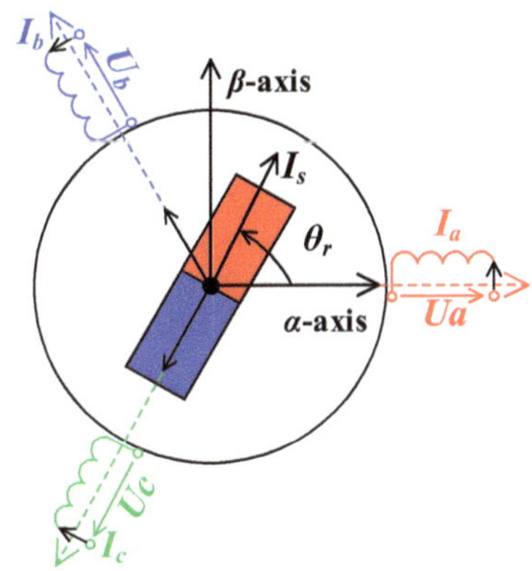

Figure 7.3 Stator current space vectors in stationary αβ-reference frame.

The stator current vector in the *abc* reference frame can be defined as:

$$i_s = i_A + i_B e^{j\cdot 2\pi/3} + i_c e^{j\cdot 4\pi/3} \tag{7.1}$$

To simplify the calculations and reduce the number of manipulated variables, two transformation methods, i.e., Clark transformation and Park transformation, are widely utilized.

Clark Transformation: The Clark transformation can transform three-phase quantities, such as voltages or currents, that vary in time along the axes a, b, and c into two-phase quantities that vary in time along the axes α and β, which is known as the Clarke transformation, as described in (7.2) and shown in Figure 7.3. In this case, a factor 2/3 is applied to scale the length of the space vector to the amplitude of the phase values. If the active and reactive powers are required to preserve

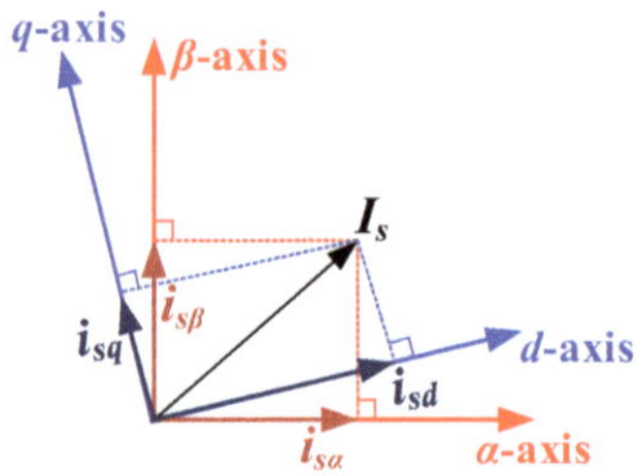

Figure 7.4 Stator current space vectors in rotating dq-reference frame.

before and after the transformation, the factor becomes $\sqrt{2/3}$.

$$\begin{bmatrix} u_\alpha \\ u_\beta \end{bmatrix} = \frac{2}{3} T_{\alpha\beta} \begin{bmatrix} u_A \\ u_B \\ u_C \end{bmatrix} = \frac{2}{3} \begin{bmatrix} 1 & -\frac{1}{2} & -\frac{1}{2} \\ 0 & \frac{\sqrt{3}}{2} & -\frac{\sqrt{3}}{2} \end{bmatrix} \begin{bmatrix} u_A \\ u_B \\ u_C \end{bmatrix} \tag{7.2}$$

Park Transformation: The Park transformation converts vectors in the $\alpha\beta$ reference frame (stationary reference frame) to the *dq* reference frame (rotating reference frame). The *d*-axis is aligned with the rotor flux, and the *q*-axis is orthogonal to it, as shown in Figure 7.4. It should be noted that after the Clark transformation, the variables in the $\alpha\beta$ reference frame still depend on the time and speed. In comparison, after the Park transformation, if the rotor flux position is precise, the *dq* components are constant values and independent of time. The Park transformation matrix is described as:

$$\begin{bmatrix} u_d \\ u_q \end{bmatrix} = \begin{bmatrix} \cos\theta_e & \sin\theta_e \\ -\sin\theta_e & \cos\theta_e \end{bmatrix} \begin{bmatrix} u_\alpha \\ u_\beta \end{bmatrix} \tag{7.3}$$

The Park transformation is the fundament of FOC, by which the stator current space vector is decomposed into a flux component (*d*-axis) and a torque-component (*q*-axis). To derive the machine model, the following assumptions are made.

- The windings are distributed sinusoidally.
- The magnetic saturation, the mutual leakage inductances, and the core losses are neglected.

The STP PMSM model in the rotating reference frame can be expressed as:

$$\begin{bmatrix} u_d \\ u_q \end{bmatrix} = \begin{bmatrix} R_s + L_d p & 0 \\ 0 & R_s + L_q p \end{bmatrix} \cdot \begin{bmatrix} i_d \\ i_q \end{bmatrix} + \omega \begin{bmatrix} -L_q i_q \\ L_d i_d \end{bmatrix} + \omega \begin{bmatrix} 0 \\ \psi_f \end{bmatrix} \tag{7.4}$$

$$\begin{bmatrix} \psi_d \\ \psi_q \end{bmatrix} = \begin{bmatrix} L_d & 0 \\ 0 & L_q \end{bmatrix} \cdot \begin{bmatrix} i_d \\ i_q \end{bmatrix} + \begin{bmatrix} 0 \\ \psi_f \end{bmatrix} \tag{7.5}$$

$$T_e = \frac{3}{2} p \left[\left(L_d - L_q \right) i_d i_q + \psi_f i_q \right] \tag{7.6}$$

It should be noted that in (7.4), p is the operator for d/dt, while in (7.6), p is the number of pole pairs.

7.2.1.2 Principle of FOC

FOC is implemented by current control employing a PI controller, which generates PWM signals for the inverter by employing an SVM technique. This method facilitates motor control in a manner similar to the control of a DC motor with independent excitation. The control process is described in Figure 7.5.

As can be seen, FOC can be used to control either synchronous or induction machines by simply changing the flux reference and obtaining the rotor flux position accordingly. In the case of PMSMs, where the rotor flux is predetermined by the magnets and remains constant, the flux reference (i_d^{ref}) can be set to zero if there is no rotor saliency. In comparison, induction machines require rotor flux to operate, thus necessitating a non-zero flux reference. This adaptability addresses a significant limitation of traditional control structures, allowing for a transition from asynchronous to synchronous drives. The outputs of the current regulators, the dq-axis reference voltages, are then fed into the inverse Park transformation. This transformation represents the components of the stator vector voltage in the stationary orthogonal $\alpha\beta$ reference frame. These values serve as inputs for the SVPWM, generating signals to drive the inverter. In addition, it is worth noting that both the Park and inverse Park transformations rely on knowing the rotor flux position. Determining this position depends on the type of AC machine being utilized, whether synchronous or asynchronous.

7.2.1.3 Space vector modulation strategy

A typical 2-level three-phase inverter fed by a DC power supply, as shown in Figure 7.6, can be connected to a three-phase machine via a series of switches. S_A, S_B, and S_C represent the switch states of the three legs (A, B, C) of the inverter. When the upper transistor is activated, i.e., the switch A+, B+, or C+ is switched on, the switch state is set as "1". When the lower transistor is activated, i.e., the switch A-, B-, or C- is switched on, the switch state is set as "0". It should be noted that the switches in the same leg are required to avoid being turned on simultaneously; otherwise, the DC supply would be short-circuited.

The relationship between the switching states [S_A, S_B, S_C] and the line-to-line voltage vector [V_{AB}, V_{BC}, V_{CA}] is given by (7.7) as follows.

$$\begin{bmatrix} V_{AB} \\ V_{BC} \\ V_{CA} \end{bmatrix} = U_{dc} \begin{bmatrix} 1 & -1 & 0 \\ 0 & 1 & -1 \\ -1 & 0 & 1 \end{bmatrix} \begin{bmatrix} S_A \\ S_B \\ S_C \end{bmatrix} \tag{7.7}$$

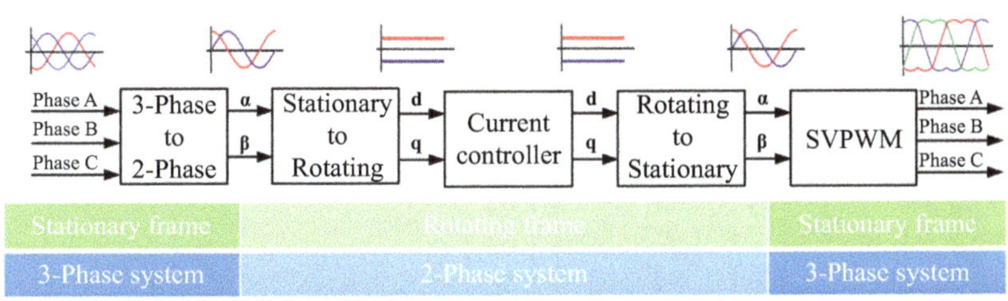

Figure 7.5 FOC control process diagram for STP PMSMs.

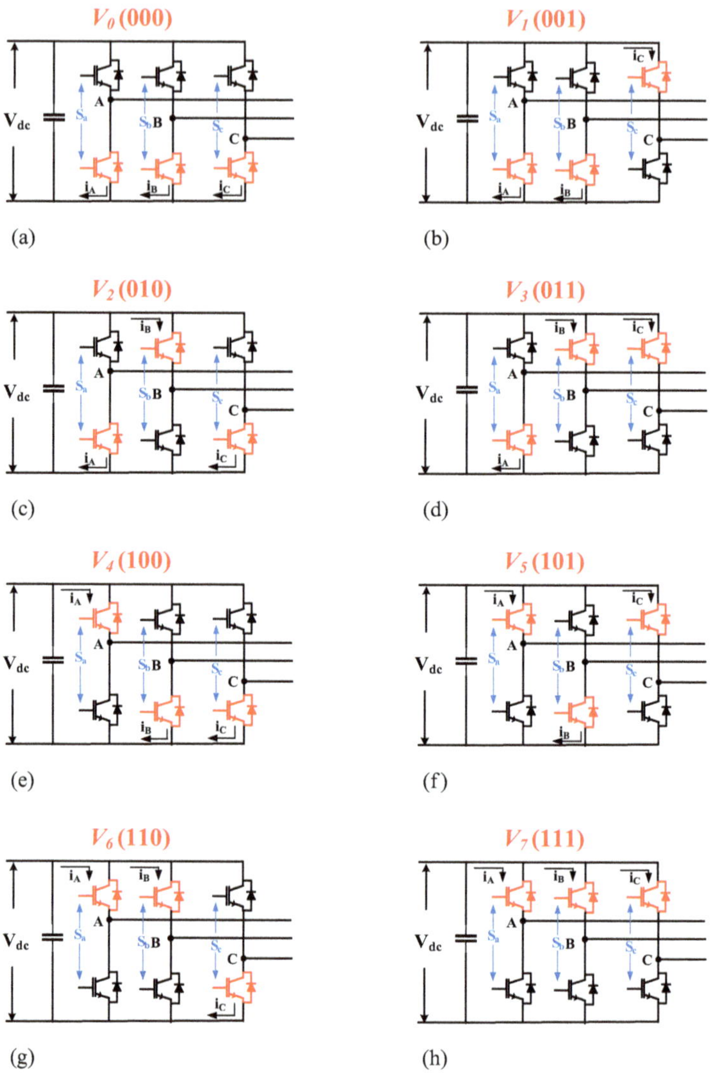

Figure 7.6 Eight voltage vectors (V_0-V_7) of a three-phase inverter. (a) V_0 (000). (b) V_1 (001). (c) V_2 (010). (d) V_3 (011). (e) V_4 (100). (f) V_5 (101). (g) V_6 (110). (h) V_7 (111).

Also, the relationship between the switching states [S_A, S_B, S_C] and the phase voltage vector [V_A, V_B, V_C] is given by (7.8) as follows.

$$\begin{bmatrix} V_A \\ V_B \\ V_C \end{bmatrix} = \frac{U_{dc}}{3} \begin{bmatrix} 2 & -1 & -1 \\ -1 & 2 & -1 \\ -1 & -1 & 2 \end{bmatrix} \begin{bmatrix} S_A \\ S_B \\ S_C \end{bmatrix} \tag{7.8}$$

There are in total $2^3 = 8$ possible switch patterns (V_0 to V_7) in a 2-level three-phase inverter. According to (7.7) and (7.8), the output line-to-neutral voltage (phase voltage) and the output line-to-line voltages in terms of DC-link U_{dc} are given in Table 7.1.

The SVPWM technique aims to approximate the reference voltage vector V_{ref} using the eight switching patterns. One simple method of approximation is to ensure that the generated average output of the inverter over a small period T_s is the same as the reference voltage vector, as shown in Figure 7.7, where the reference voltage V_{ref} located in sector I is used for illustration.

In summary, the SVPWM implementation can be summarized as three steps:

1. Determine the located sector.
2. Determine the time durations for the adjacent voltage vectors.
3. Determine the output switch signals to 6 switches, i.e., the PWM switch signals, and the time durations for the adjacent voltage vectors can be calculated as

$$T_1 = \frac{|V_{\text{ref}}|}{2/3U_{\text{dc}}} \cdot \frac{\sin(\pi/3 - \alpha)}{\sin(\pi/3)} T_s \tag{7.9}$$

Table 7.1 Switching vectors, phase voltages, and line-to-line voltages

Voltage vectors	Switch vector			Phase voltage			Line-to-line voltage			Status
	S_A	S_B	S_C	V_A	V_B	V_C	V_{AB}	V_{BC}	V_{CA}	
V_0	0	0	0	0	0	0	0	0	0	Zero
V_1	1	0	0	2/3	−1/3	−1/3	1	0	−1	Active
V_2	0	1	0	−1/3	2/3	−1/3	−1	1	0	Active
V_3	1	1	0	1/3	1/3	−2/3	0	1	−1	Active
V_4	0	0	1	−1/3	−1/3	2/3	0	−1	1	Active
V_5	1	0	1	1/3	−2/3	1/3	1	−1	0	Active
V_6	0	1	1	−2/3	1/3	1/3	−1	0	1	Active
V_7	1	1	1	0	0	0	0	0	0	Zero

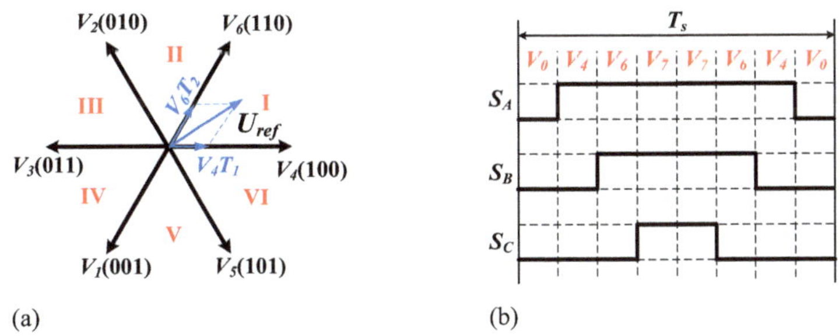

Figure 7.7 Example of SVPWM strategy when V_{ref} is located in Sector I. (a) Space vector hexagon. (b) Switching state sequence.

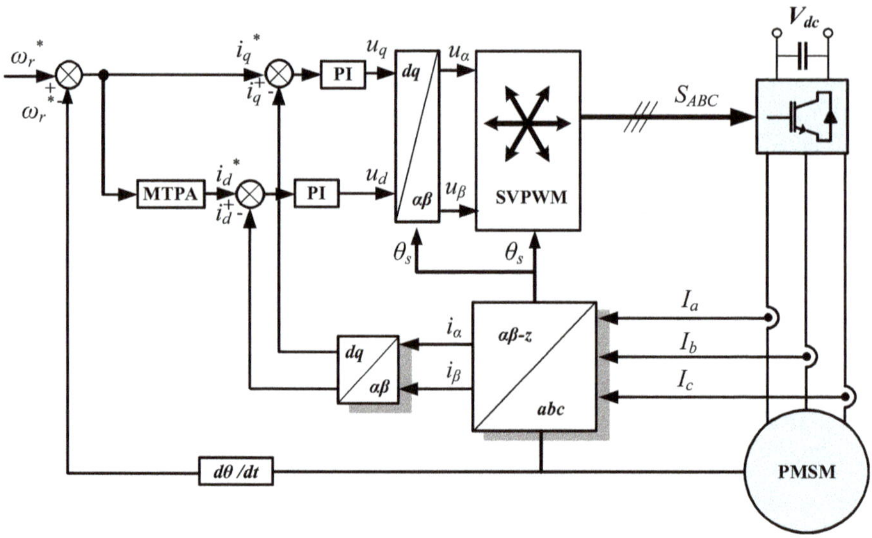

Figure 7.8 Basic structure of FOC for STP PMSM.

$$T_2 = \frac{|V_{\mathrm{ref}}|}{2/3V_{\mathrm{dc}}} \cdot \frac{\sin(\alpha)}{\sin(\pi/3)} T_s \tag{7.10}$$

$$T_2 = T_s - (T_1 + T_2) \tag{7.11}$$

where T_0, T_1, and T_2 are the time durations of the zero voltage vector, the voltage vectors V_4 and V_6, respectively. α is the angle between the reference voltage vector V_{ref} and the voltage vector V_4. T_s is the sampling period.

In summary, the FOC structure is presented in Figure 7.8. The FOC uses the output of current control with a linear PI controller and then generates a PWM signal using an SVM technique.

7.2.2 Direct torque control (DTC)

In FOC, the rotor-oriented coordinate transformation is necessary, and the PI-tuned parameters are strongly dependent on the machine parameters. Therefore, the DTC is proposed by I. Takahashi and M. Depenbrock in the late 1990s to reduce the complexity of control algorithms and improve the control performance [6–11]. The design of DTC is characterized by its independence from machine parameters and avoidance of coordinate transformations. Unlike FOC, which aims to mimic the electromagnetic behavior of a DC motor, DTC focuses on fully exploiting the flux and torque-producing capabilities of the machine. By selectively choosing the inverter switching states, DTC allows the direct controlling of flux and torque.

7.2.2.1 Model of STP PMSMs in stationary reference frame

In DTC, the rotor-oriented coordinate transformation is unnecessary. As a consequence, the Clarke transformation is solely employed in formulating the mathematical model of the machine for DTC, where the three-phase system can be referenced to the stationary reference frame ($\alpha\beta$ coordinate system):

$$V_{s\alpha\beta} = R_s I_{s\alpha\beta} + p\psi_{s\alpha\beta} \tag{7.12}$$

$$\psi_{s\alpha\beta} = L_s I_{s\alpha\beta} + \psi_{r\alpha\beta} \tag{7.13}$$

$$\psi_{r\alpha\beta} = \psi_f e^{j\theta_r} \tag{7.14}$$

$$T_e = 3/2p\left(\psi_{s\alpha} i_{s\beta} - \psi_{s\beta} i_{s\alpha}\right) \tag{7.15}$$

where R_s is the stator resistance. ψ_f is the permanent magnet flux. L_s is the stator inductance. $V_{s\alpha\beta}$ and $I_{s\alpha\beta}$, are the $\alpha\beta$ stator voltages and currents, respectively. $\psi_{s\alpha\beta}$ is the stator fluxes; θ_r is the electrical rotor angle, and p is the number of pole pairs.

7.2.2.2 Principle of DTC

The stator flux vector for STP PMSM can be calculated as:

$$\psi_s = \int_0^t (V_s - R_s I_s)dt + \psi_{s|t=0} \tag{7.16}$$

where V_s and I_s are the stator voltage and the current space vectors, respectively.

When the voltage drop on the stator resistance R_s is neglected, the stator flux linkage vector is expressed as

$$\Delta\psi_s = V_s T_{on} \tag{7.17}$$

Each of the six possible active voltage vectors for an STP inverter has a component tangential to the stator flux linkage vector for a switch-on time T_{on} of the voltage vector V_{s}, as illustrated in Figure 7.9.

Therefore, it can be concluded that the trajectory of the space flux ψ_s in the $\alpha\beta$ frame can be determined by the application of a sequence of voltage space vectors.

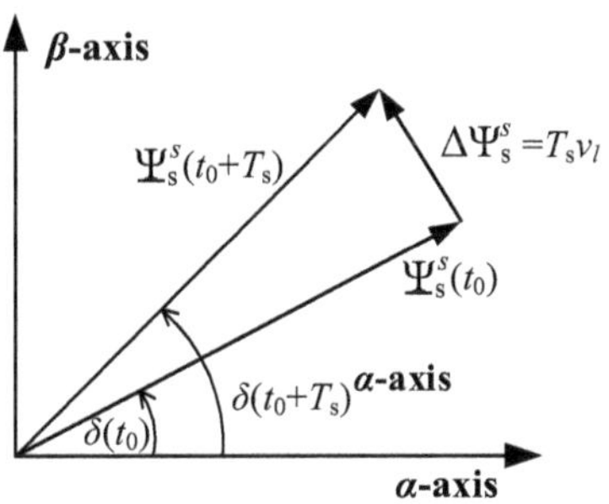

Figure 7.9 Effect of space vector on the variation of stator flux linkage ψ_s.

Neglecting cogging torque, the steady-state electromagnetic torque T_e of an STP PMSM can be written as:

$$T_e = \frac{3}{4}\frac{p|\psi_s|}{L_d L_q}\left[2|\psi_f|L_q \sin(\delta) - |\psi_s|(L_q - L_d)\sin(2\delta)\right] \tag{7.18}$$

where δ is the load angle and is defined as the angle between the stator flux linkage vector ψ_s and the permanent magnet flux linkage vector ψ_f. For PMSMs without saliency, i.e., SPMSMs, L_d is equal to L_q, and (7.18) can be simplified as:

$$T_e = \frac{3}{2}\frac{p}{L_d}|\psi_s||\psi_f|\sin(\delta) \tag{7.19}$$

7.2.2.3 Formulation of control concept

As illustrated in Figure 7.9, during the time period T_s, the stator flux can be affected in a variety of ways by the active space voltage vectors of the inverter. Irrespective of the flux position, three of these voltage vectors contribute to an increase in the flux magnitude, while others lead to a decrease. In addition, three of the space vectors include an anticlockwise rotation, thereby increasing the relative angle, whereas the remaining space vectors tend to decrease the angle due to a clockwise rotation.

To independently control the stator flux and torque, it is necessary to ensure that only appropriate voltage vectors are used to control the magnitude of the flux. As depicted in Figure 7.10, the voltage vector components can be divided into two different components, i.e., direct component and indirect component. The direct voltage vector component solely affects the stator flux magnitude, whereas the indirect voltage vector component affects torque angles, consequently altering the torque.

Figure 7.11 further illustrates the contributions of voltage vectors to flux and torque control. $\psi_s(t_0)$ is the initial flux vector positioned in the first quadrant of the $\alpha\beta$ frame. In this case, the stator flux magnitude increases by using vectors V_2 and V_6, since both the vectors have a positive direct component that positively

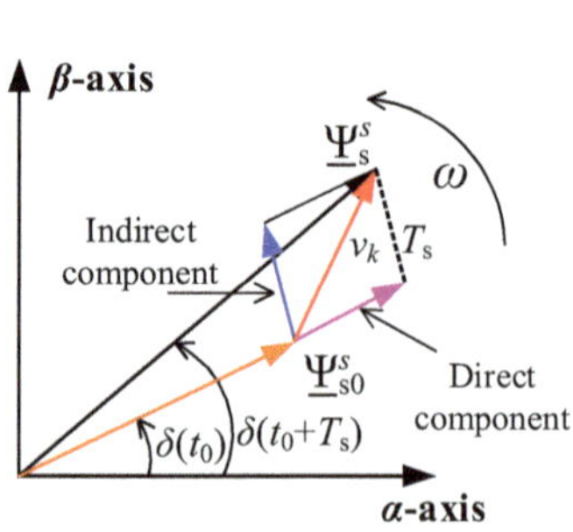

Figure 7.10 Direct and indirect components of stator flux produced by voltage vectors.

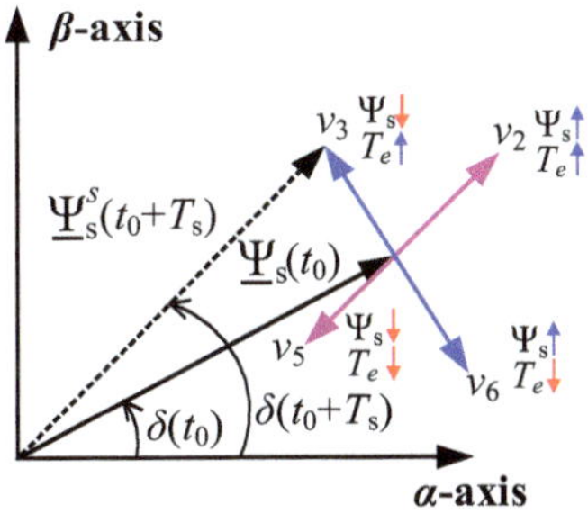

Figure 7.11 Stator flux ψ_s and torque T_e control using voltage vectors.

Table 7.2 Switching look-up table for DTC of STP PMSM

ψ_s	T_e	I	II	III	IV	V	VI
1	1	V_6	V_2	V_3	V_1	V_5	V_4
	0	V_0, V_7	V_0, V_7	V_0, V_7	V_0, V_7	V_0, V_7	V_0, V_7
	−1	V_5	V_4	V_6	V_2	V_3	V_1
−1	1	V_2	V_3	V_1	V_5	V_4	V_6
	0	V_0, V_7	V_0, V_7	V_0, V_7	V_0, V_7	V_0, V_7	V_0, V_7
	−1	V_1	V_5	V_4	V_6	V_2	V_3

influences flux augmentation. In contrast, the vectors V_3 and V_5 can be used to decrease the flux magnitude due to their negative direct component.

Moreover, the indirect components of vectors V_2 and V_3 align in the same direction, contributing to an increase in torque, whereas the vectors V_5 and V_6 can be utilized to decrease the torque. Consequently, the vector V_2 can be used to increase both flux magnitude and torque, and the vector V_5 can be used to decrease both stator flux and torque. Vector V_3 decreases stator flux but increases torque, whereas vector V_6 increases stator flux and decreases torque.

A lookup table, as listed in Table 7.2, summarizes the relationship between the voltage space vectors and their effects on the flux and torque within each other.

The DTC structure for an STP PMSM is shown in Figure 7.12. As can be seen, it utilizes the outputs of stator flux and torque hysteresis regulators according to the estimated flux and torque, and then the voltage vector can be determined by a prefixed switching look-up table.

The torque and stator flux estimation equations are expressed as:

$$\begin{cases} \psi_{s\alpha} = \int (u_{s\alpha} - R_s i_{s\alpha}) dt \\ \psi_{s\beta} = \int (u_{s\beta} - R_s i_{s\beta}) dt \end{cases} \tag{7.20}$$

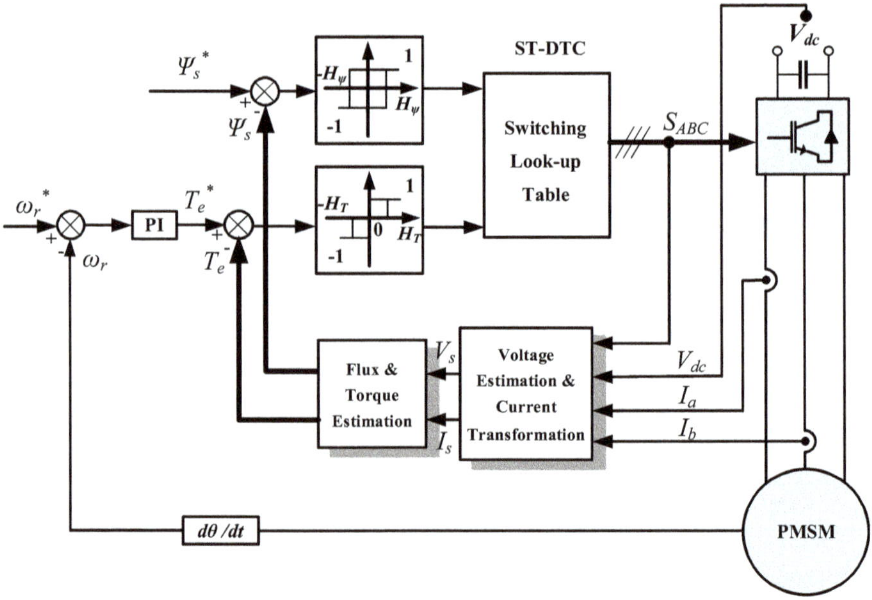

Figure 7.12 Basic structure of DTC for STP PMSM.

$$\begin{cases} \psi_s = \sqrt{\psi_{s\alpha}^2 + \psi_{s\beta}^2} \\ \gamma_s = \arctan\left(\dfrac{\psi_{s\beta}}{\psi_{s\alpha}}\right) \end{cases} \tag{7.21}$$

$$T_e = \frac{3}{2}p(\psi_{s\alpha}i_{s\beta} - \psi_{s\beta}i_{s\alpha}) \tag{7.22}$$

In general, the stator flux is controlled using a 2-level hysteresis regulator, while the torque is controlled using a 3-level hysteresis regulator, as shown in Figure 7.13. The outputs of the torque/flux hysteresis regulator, along with sector flux information, are used to index the lookup table, by which the appropriate voltage vectors can be selected to control both the stator flux and the torque simultaneously.

The output of the torque hysteresis regulator can take the values -1, 0, and 1, indicating a decrease, maintenance, or increase in torque, respectively. Similarly, the flux hysteresis regulator outputs can be -1 and 1, indicating a decrease and increase in flux, respectively. It should be noted that since the response time for flux is much faster than the torque, there is no value 0 output for the flux regulator. This type of control is commonly known as a *bang-bang* control.

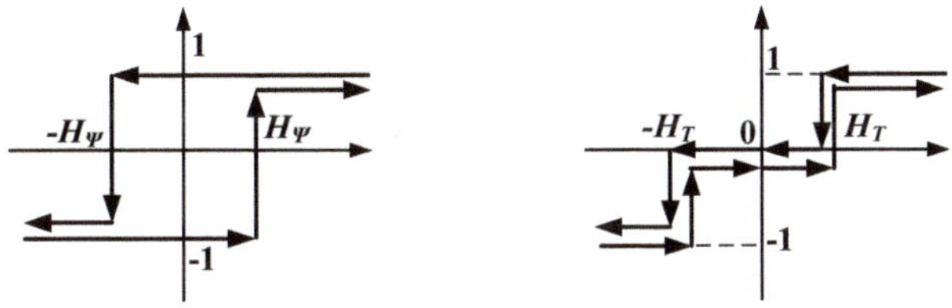

(a) (b)

Figure 7.13 Hysteresis regulators in DTC for STP PMSM. (a) 2-level flux regulator. (b) 3-level torque regulator.

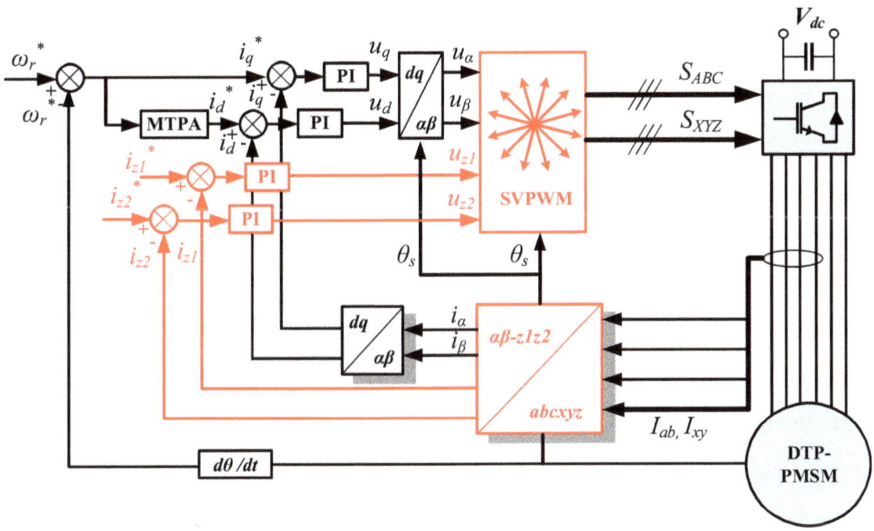

Figure 7.14 Structure diagram of FOC for DTP machines.

7.3 Modulation strategies of FOC and SVM-DTC for multiphase PMSMs

The inverters for multiphase machines have a large number of possible voltage vectors compared to those for the STP machines. Consequently, there are more potential modulation strategies involving different voltage vector combinations. In this section, the modulation strategy for FOC and SVM-DTC are investigated in DTP machines, as well as other multiphase machine topologies.

7.3.1 Overview of FOC and SVM-DTC for DTP PMSMs

The structure of FOC for DTP machines is shown in Figure 7.14. It should be noted that the synchronous PI controllers are employed to ensure the modulation strategy operates normally.

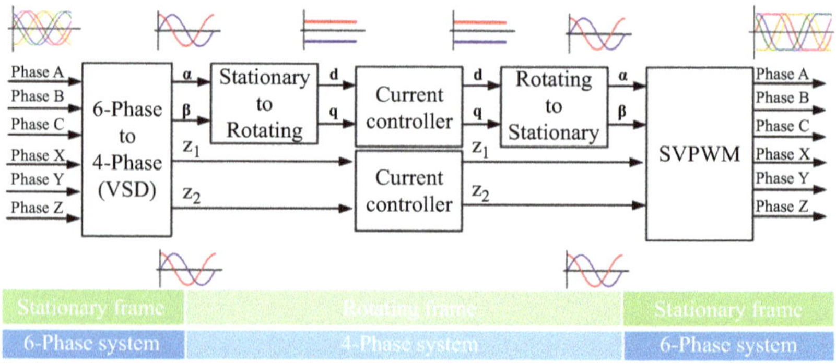

Figure 7.15 FOC control process diagram for DTP machines.

It should be noted that the FOC for DTP machines has the same basic structure as the FOC for STP machines. However, the FOC for DTP machines has two additional current controllers in the z_1z_2 subspaces to control the current harmonics. Meanwhile, the SVPWM modulator is more complex due to the increased number of possible switch states, and there is a different transformation from a 6-phase system into a 4-coordinate (dq, z_1z_2) system. The process diagram of the FOC for DTP PMSMs is summarized in Figure 7.15. It should be noted that the *Park* transformation is only required in the $\alpha\beta$ subspace since only the components in the $\alpha\beta$ subspace are related to the torque production and excitation field. In addition, more details of the FOC for DTP machines can be found in Section 8.2 in Chapter 8.

The ST-DTC has the demerits of variable switching frequency and undesired torque and current ripples. To address these issues, a direct torque control (DTC) combined with space vector modulation (SVM), i.e., SVM-DTC, is proposed in [12], where a continuous rotating voltage vector is synergized to regulate the stator flux and torque more accurately instead of relying on the switching table to select the voltage vector. The crucial aspect of this strategy lies in determining the command voltage vector. One of the most commonly used SVM-DTCs is the deadbeat-based SVM-DTC, as shown in Figure 7.16. This method does not require regulator tuning but needs to solve a quadratic equation online. Hence, the precise machine parameters are required.

Another widely used SVM-DTC scheme is to employ the torque and flux PI controllers, alongside SVM to replace the hysteresis regulators and the look-up table in ST-DTC to obtain the command voltage vector, as shown in Figure 7.17. The PI-based SVM-DTC provides an excellent capability for flux and torque ripple reduction during steady state, excellent dynamic performance, and high robustness during load and speed variations. Further, the parameters for the PI controller can be easily tuned in industrial applications, and the PI controller provides good bandwidth in motor control.

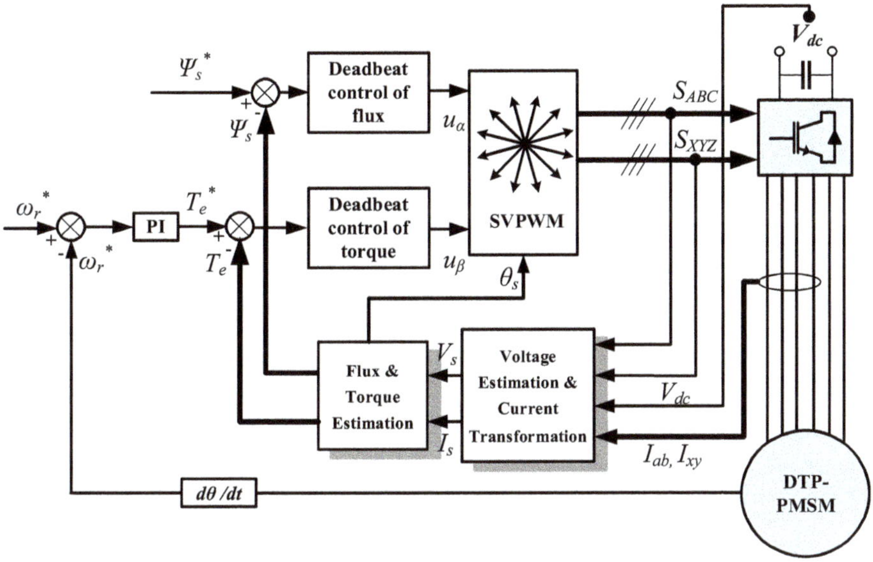

Figure 7.16 Deadbeat controller-based SVM-DTC diagram for DTP PMSM.

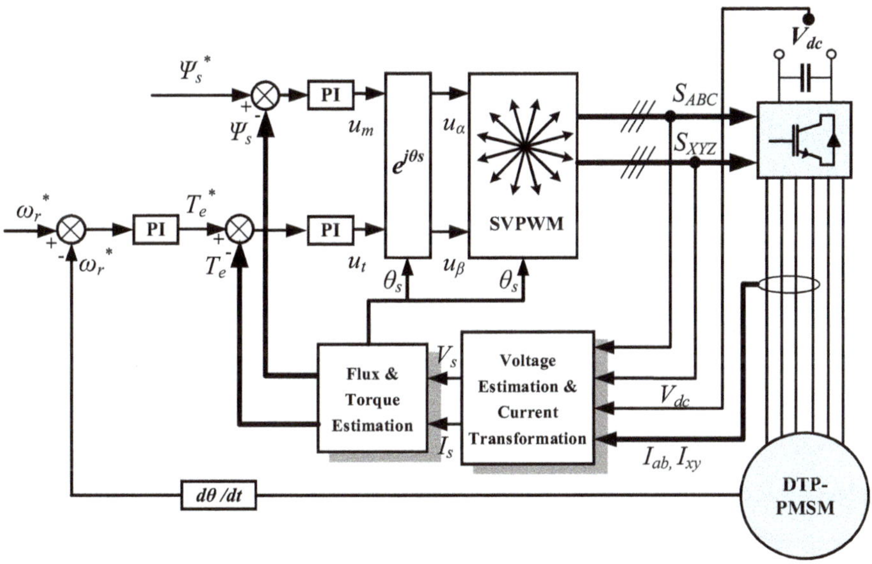

Figure 7.17 PI controller-based SVM-DTC diagram for DTP machines.

7.3.2 Modulation strategies of FOC for DTP PMSMs

In the DTP system, there are $2^6 = 64$ voltage vectors. By defining the switching function as $S = [S_A, S_B, S_C, S_X, S_Y, S_Z]$, the phase voltage can be described as

$$\begin{bmatrix} V_a \\ V_b \\ V_c \\ V_x \\ V_y \\ V_z \end{bmatrix} = \frac{V_{dc}}{3} \begin{bmatrix} 2 & -1 & -1 & 0 & 0 & 0 \\ -1 & 2 & -1 & 0 & 0 & 0 \\ -1 & -1 & 2 & 0 & 0 & 0 \\ 0 & 0 & 0 & 2 & -1 & -1 \\ 0 & 0 & 0 & -1 & 2 & -1 \\ 0 & 0 & 0 & -1 & -1 & 2 \end{bmatrix} \begin{bmatrix} S_A \\ S_B \\ S_C \\ S_X \\ S_Y \\ S_Z \end{bmatrix} \tag{7.23}$$

where S_i represents the switching state of the corresponding leg and the value of the component S_i. "1" means the switch is "on" and "0" means the switch is "off".

According to the VSD approach [13], all voltage vectors can be transferred into the $\alpha\beta$ and z_1z_2 subspaces. By defining voltage vectors as the binary value $V_i = S_Z S_Y S_X S_C S_B S_A$, the voltage vectors are shown in Figure 7.18. It is worth noting that the groups of voltage vectors from the largest to the smallest magnitude in the $\alpha\beta$ subspace are defined as L (D_1), ML (D_2), M (D_3), S (D_4), and this $\alpha\beta$ subspace magnitude based definition has no physical meaning, as listed in Table 7.3, where U_{dc} is the amplitude of the DC-link voltage.

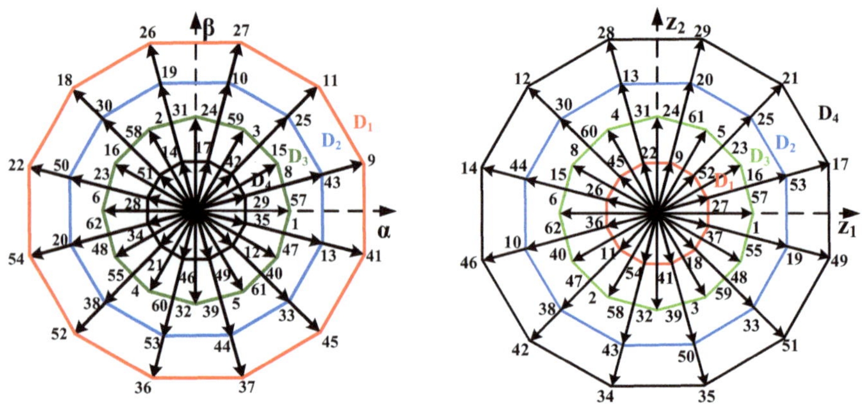

(a) (b)

Figure 7.18 DTP machine drives space voltage vectors. (a) αβ subspace. (b) z_1z_2 subspace.

Table 7.3 Magnitudes of voltage vectors in different subspaces

VVs group	Magnitude	
	$\alpha\beta$ subspace	**z_1z_2 subspace**
L (D_1)	$\frac{\sqrt{6}+\sqrt{2}}{6}U_{dc}$	$\frac{\sqrt{6}-\sqrt{2}}{6}U_{dc}$
ML (D_2)	$\frac{\sqrt{2}}{3}U_{dc}$	$\frac{\sqrt{2}}{3}U_{dc}$
M (D_3)	$\frac{1}{3}U_{dc}$	$\frac{1}{3}U_{dc}$
S (D_4)	$\frac{\sqrt{6}-\sqrt{2}}{6}U_{dc}$	$\frac{\sqrt{6}+\sqrt{2}}{6}U_{dc}$

7.3.2.1 Experimental results of different combinations of voltage vectors in SVM-based FOC

It can be easily seen from Figure 7.18 and Table 7.3 that there are many possible combinations for the voltage vector selection in SVM. The combinations of voltage vectors in DTP machines can be summarized as follows according to the employed voltage vectors.

FOC with 2L-SVM [5]

The FOC with 2*L*-SVM method utilizes the voltage vectors in *L* (D_1) groups only, since the voltage vectors in *L* (D_1) group have the largest amplitudes to provide higher voltage utilization. The issue with this method is that the uncontrolled z_1z_2 voltage components would introduce significant current harmonics, and thereby, increase the system losses and decrease system efficiency. The 2*L*-SVM for this method is shown in Figure 7.19. The 2*L* means only two adjacent voltage vectors of the located sector are utilized in one sampling period. Since only the voltage vectors in the *L* (D_1) group are selected, the harmonic components in the z_1z_2 subspace cannot be eliminated, as verified by experimental results.

FOC with 4L-SVM [14]

The FOC with 4*L*-SVM method utilizes the voltage vectors in the *L* (D_1) groups. The 4*L*-SVM is shown in Figure 7.20. The 4*L* means four adjacent voltage vectors of the located sector are utilized in one sampling period. For example, if the

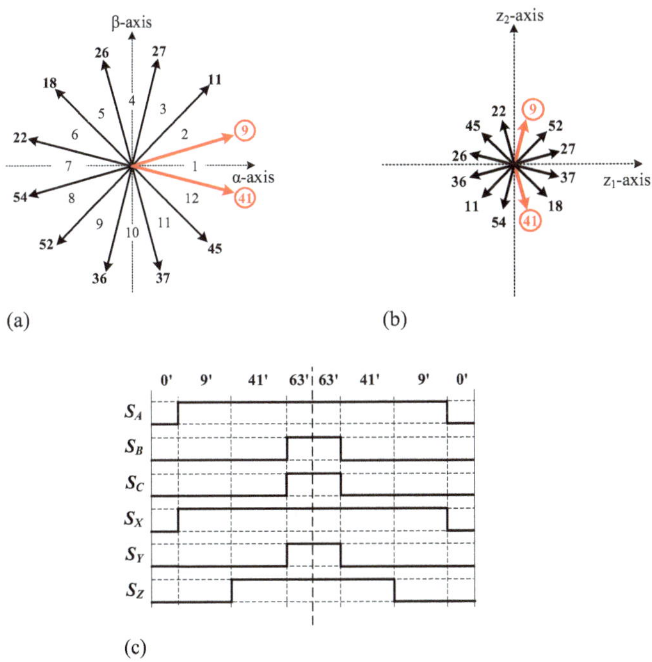

Figure 7.19 Modulation strategy of the 2L-SVM [5]. (a) Voltage vector selection in αβ subspace. (b) Voltage vector selection in z_1z_2 subspace. (c) Corresponding switching patterns.

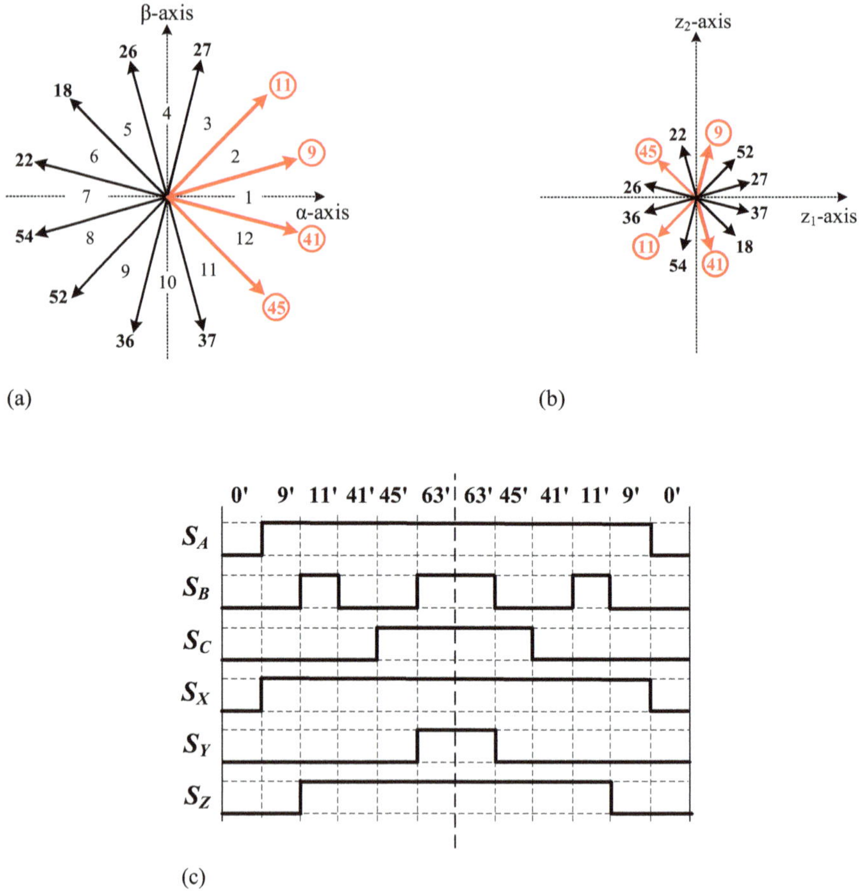

Figure 7.20 Modulation strategy of 4L-SVM [14]. (a) Voltage vector selection in $\alpha\beta$ subspace. (b) Voltage vector selection in z_1z_2 subspace. (c) Corresponding switching patterns.

reference voltage is located in sector 1 in the $\alpha\beta$ subspace, the voltage vectors V_9, V_{41}, V_{11}, and V_{45} are utilized. Then, in the z_1z_2 subspace, these voltage vectors can be controlled to zero by using the voltage-second technique.

FOC with L+M-SVM [15]

The FOC with L+M-SVM method utilizes the voltage vectors in the L (D_1) and M (D_3) groups, and this selection method can also be utilized to zero the components in the z_1z_2 subspace to reduce the current harmonics. The modulation strategy of L+M-SVM is shown in Figure 7.21. The L+M means voltage vectors are selected from the L (D_1) and M (D_3) groups. For example, if the reference voltage is located in sector 1 in the $\alpha\beta$ subspace, the voltage vectors V_9, V_{41}, V_{11}, and V_{15} are utilized. Using this method, the low-order harmonics (especially the 5th and 7th) can also be reduced significantly.

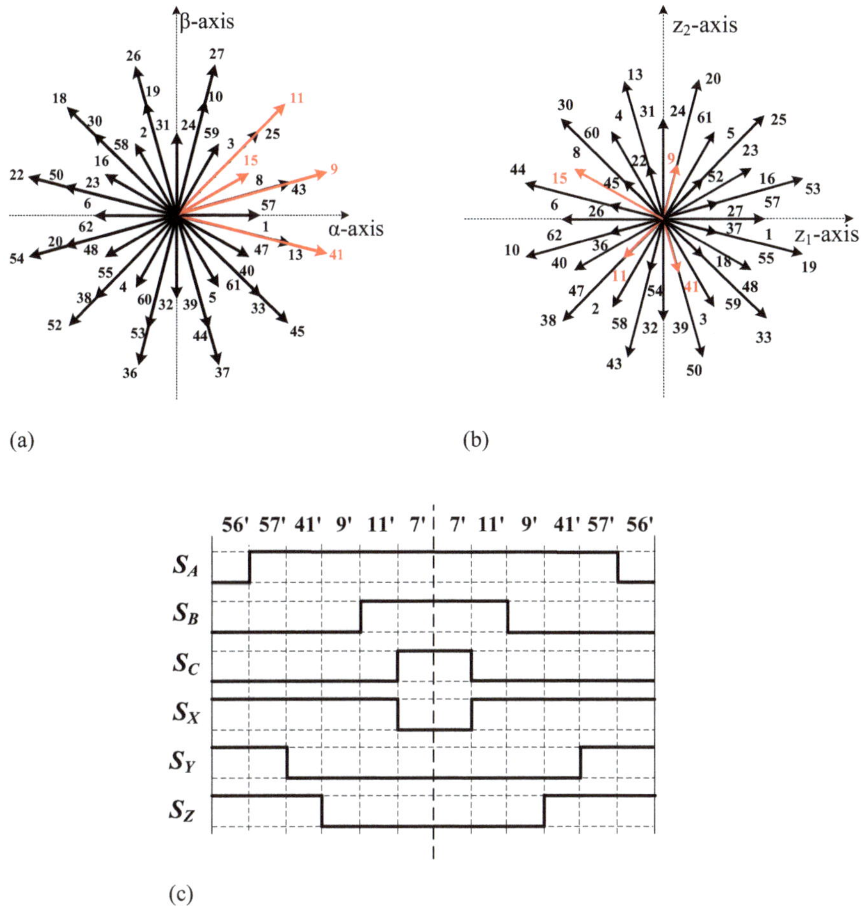

Figure 7.21 Modulation strategy of L+M-SVM [15]. (a) Voltage vector selection in $\alpha\beta$ subspace. (b) Voltage vector selection in z_1z_2 subspace. (c) Corresponding switching patterns.

FOC with L+ML+M-SVM [16,17]

The FOC with *L*+*ML*+*M*-SVM method utilizes the voltage vectors in the *L* (D_1), *ML* (D_2), and *M* (D_3) groups. In this case, the optimum voltage vectors can be selected from different sectors [16,17]. For instance, if the reference voltage is located in sector 1 in the $\alpha\beta$ subspace, the selected voltage vectors in [16] are V_9, V_1, V_{11}, and V_{43}, as shown in Figure 7.22 (I), while they are V_9, V_{41}, V_8, and V_{43} in [17], as shown in Figure 7.22 (II). As can be seen, the latter method can decrease the switch losses in PWM waveforms.

The steady-state and dynamic experimental results for different SVM-based FOC are presented in Figures 7.23 and 7.24, respectively. The steady-state experimental results include the phase current, current spectra, torque, and stator

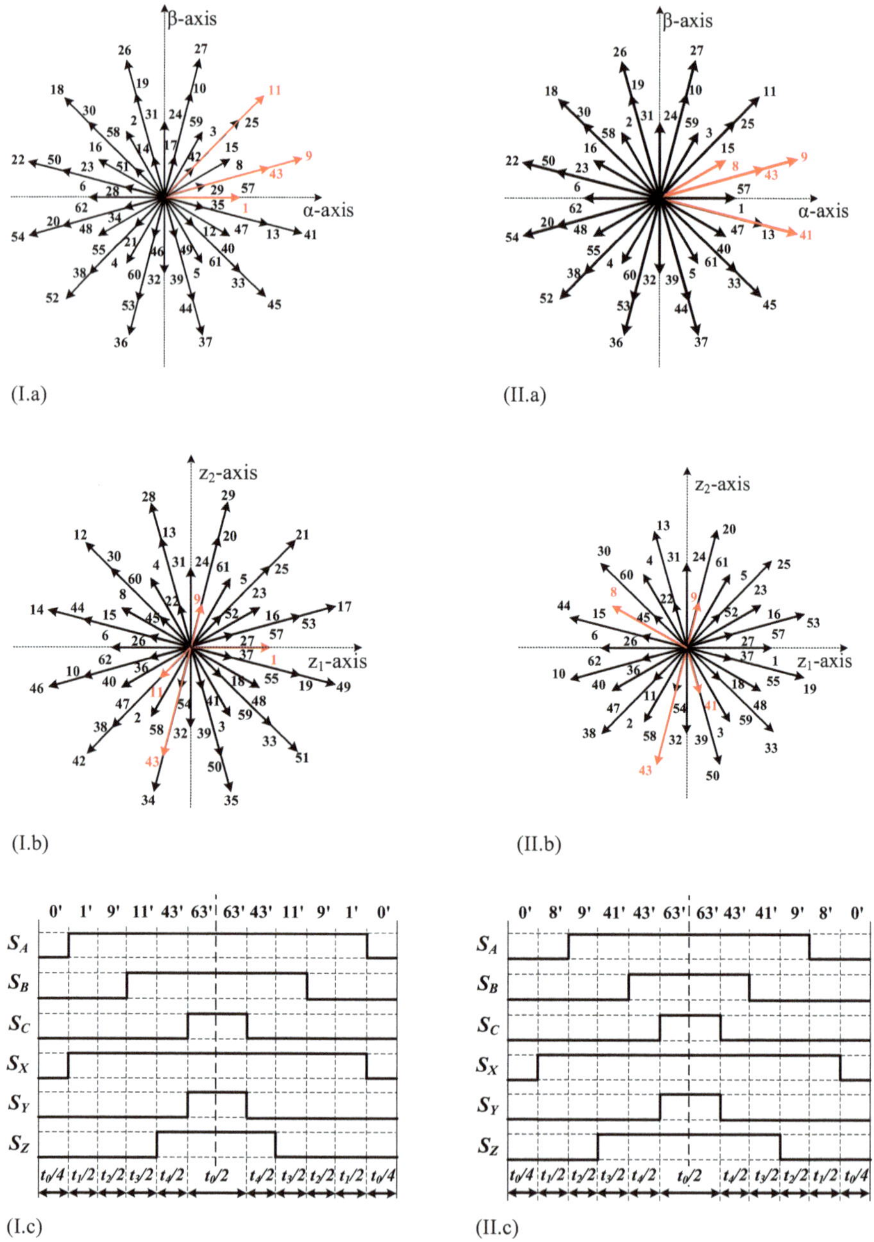

Figure 7.22 Two modulation strategies of L+ML+M-SVM. (I.a) Voltage vector selection in $\alpha\beta$ subspace. (I.b) Voltage vector selection in z_1z_2 subspace. (I.c) Switching patterns. (II.a) Voltage vector selection in $\alpha\beta$ subspace. (II.b) Voltage vector selection in z_1z_2 subspace. (II.c) Switching patterns.

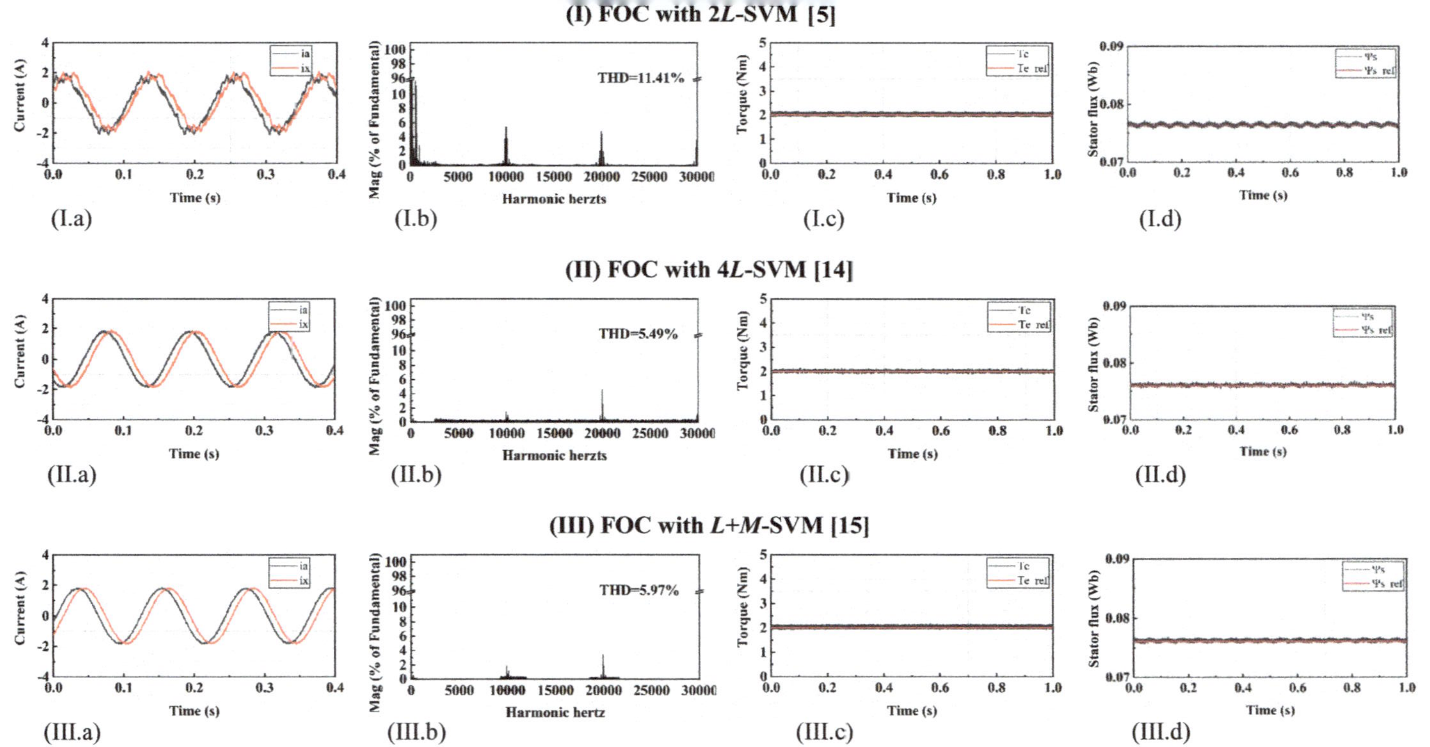

Figure 7.23 *Steady-state experimental results for all FOC with different SVM strategies. (I) FOC with 2L-SVM [5]. (I.a) Phase currents. (I.b) Current spectrum. (I.c) Torque. (I.d) Stator flux. (II) FOC with 4L-SVM [14]. (II.a) Phase currents. (II.b) Current spectrum. (II.c) Torque. (II.d) Stator flux. (III) FOC with L+M-SVM [15]. (III.a) Phase currents. (III.b) Current spectrum. (III.c) Torque. (III.d) Stator flux. (IV) FOC with L+ML+M-SVM [16]. (IV.a) Phase currents. (IV.b) Current spectrum. (IV.c) Torque. (IV.d) Stator flux. (V) FOC with L+ML+M-SVM [17]. (V.a) Phase currents. (V.b) Current spectrum. (V.c) Torque. (V.d) Stator flux.*

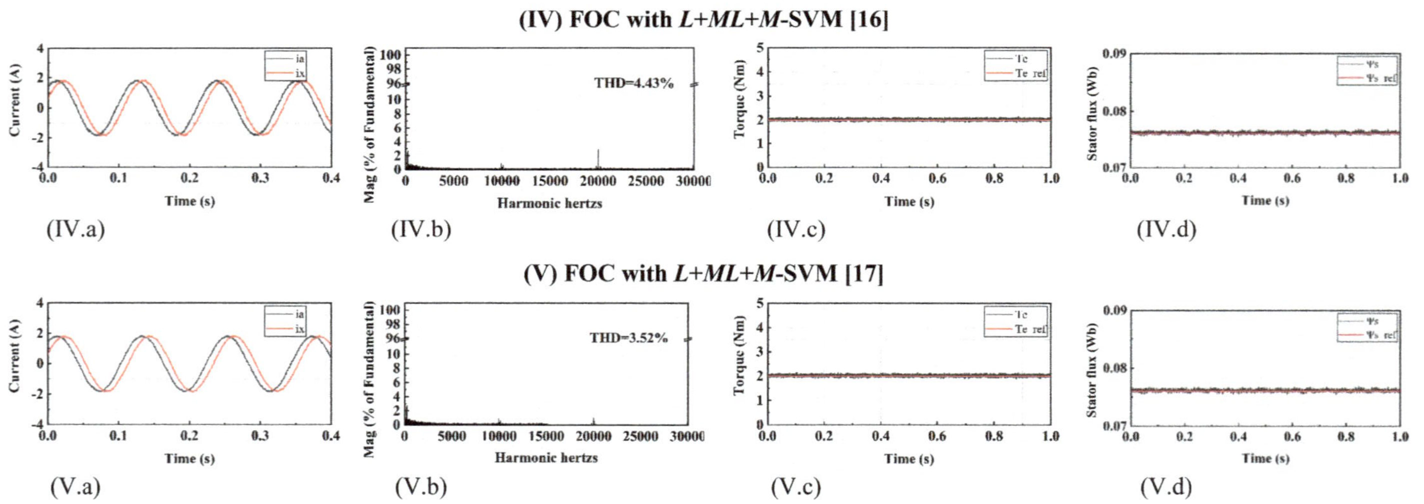

Figure 7.23 (Continued)

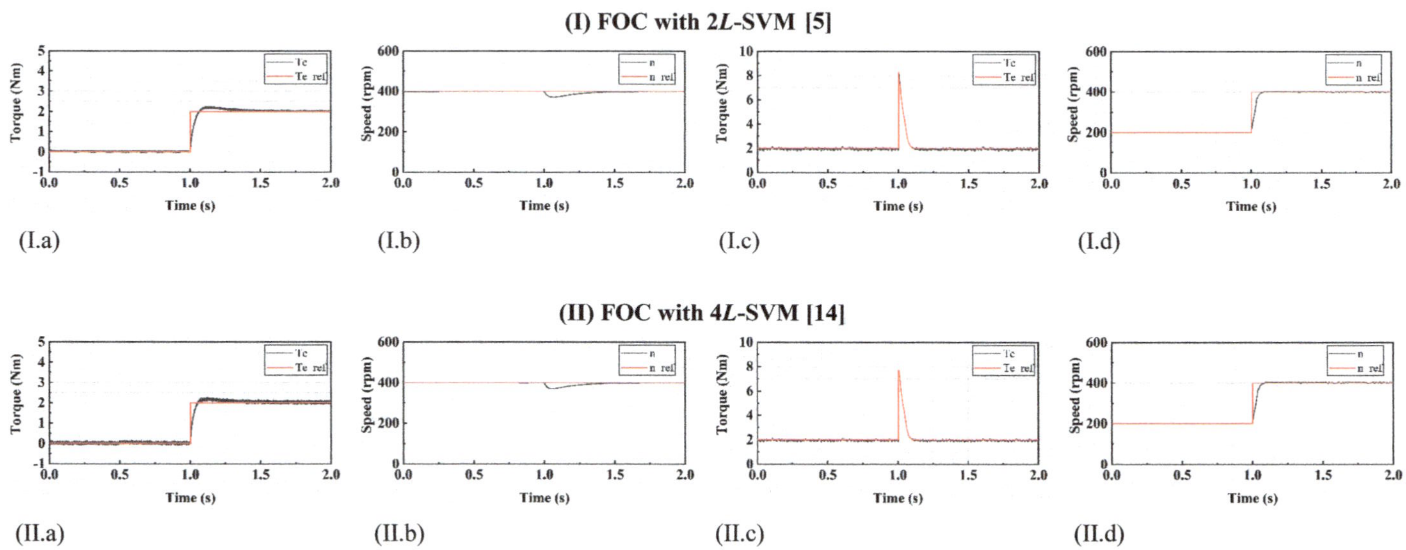

Figure 7.24 *Dynamic experimental results for all FOC with different SVM strategies. (I) FOC with 2L-SVM [5]. (I.a) Torque response with step change of load. (I.b) Speed response with step change of load. (I.c) Torque response with a step change of speed. (I.d) Speed response with a step change of speed. (II) FOC with 4L-SVM [14]. (II.a) Torque response with step change of load. (II.b) Speed response with step change of load. (II.c) Torque response with a step change of speed. (II.d) Speed response with a step change of speed. (III) FOC with L+M-SVM [15]. (III.a) Torque response with step change of load. (III.b) Speed response with step change of load. (III.c) Torque response with a step change of speed. (III.d) Speed response with a step change of speed. (IV) FOC with L+ML+M-SVM [16]. (IV.a) Torque response with step change of load. (IV.b) Speed response with step change of load. (IV.c) Torque response with a step change of speed. (IV.d) Speed response with a step change of speed. (V) FOC with L+ML+M-SVM [17]. (V.a) Torque response with step change of load. (V.b) Speed response with step change of load. (V.c) Torque response with a step change of speed. (V.d) Speed response with a step change of speed.*

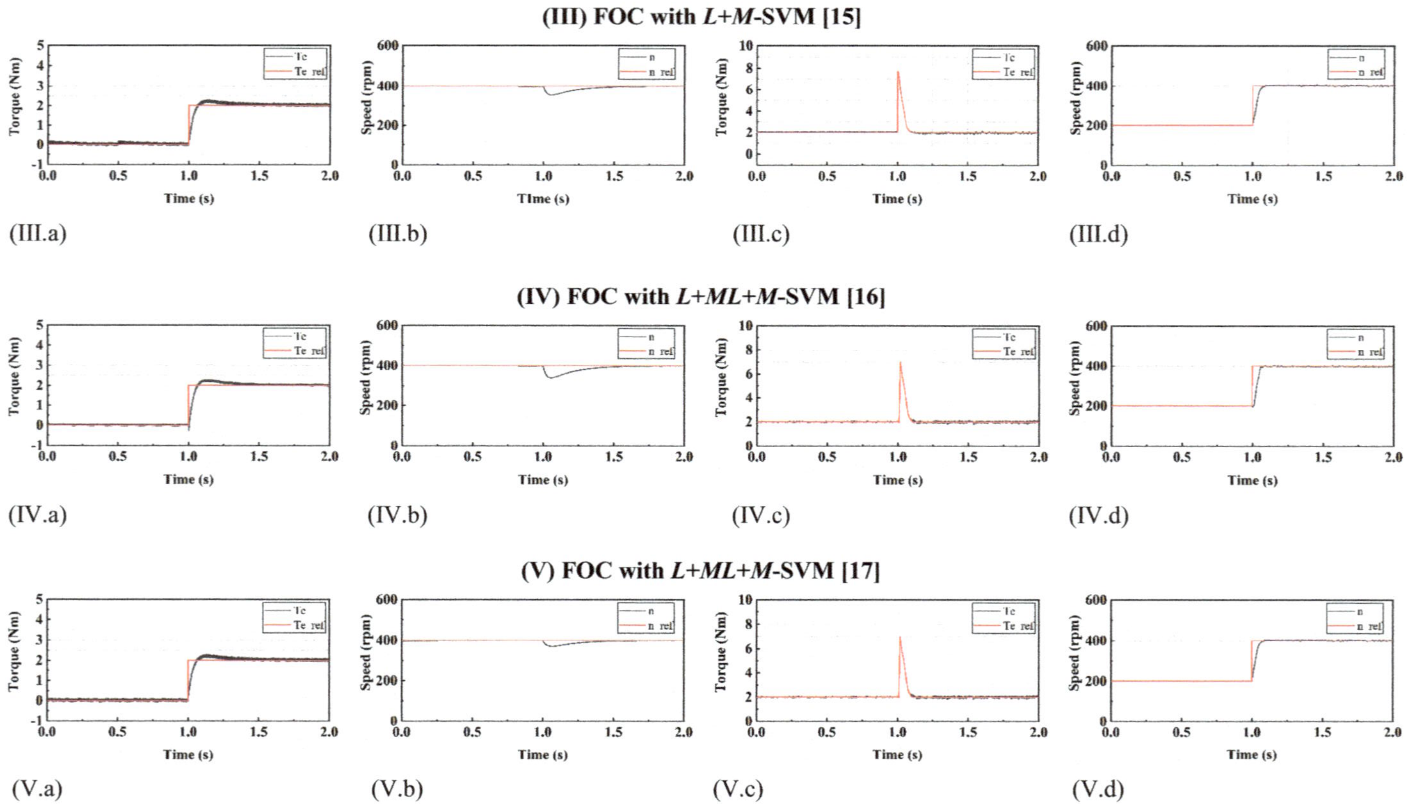

Figure 7.24 (Continued)

flux, in which the motor is operating at 200 r/min, and load torque is 2 Nm. The dynamic experimental results are conducted based on the step changes in load and speed at instant $t = 1.0$ s.

7.3.2.2 Summary and evaluation

Based on the experimental results of these SVM strategies, it can be concluded that the improved *L*+*ML*+*M*-SVM has the best steady-state performance. It should be noted that the FOC structure for the comparison is kept the same, and only the SVM strategy is different. Then, the results for these FOCs with different SVM strategies are compared in Table 7.4, where their features are also summarized accounting for the reductions of current harmonics and torque ripples, the implementation complexity, and the computation burden.

7.3.3 Modulation strategies of SVM-DTC for DTP PMSMs

On the basis of the previous analyses, in this section, some descriptions for SVM-DTCs with different SVM strategies are introduced in detail. Since the 4*L*-SVM, *L*+*M*-SVM, *L*+*ML*+*M*-SVM, and improved *L*+*ML*+*M*-SVM have similar steady-state performance as concluded in the FOC section, the SVM-DTC with improved *L*+*ML*+*M* based SVM is only presented to compare with SVM-DTC with 2*L* based SVM, and FOC, as shown in Figures 7.25 and 7.26, respectively.

Table 7.4 Experimental results and characteristics for all FOC with different SVM strategies

	2*L*-SVM	**4*L*-SVM**	***L*+*M*-SVM**	***L*+*ML*+*M*-SVM**	**Improved *L*+*ML*+*M*-SVM**
	Experimental results				
THD_I_{5th} (%)	27.10	8.39	2.85	3.72	2.87
THD_I_{7th} (%)	3.75	2.10	1.84	2.27	1.36
THD_I_a (%)	45.72	15.82	7.29	10.57	6.33
Torque ripple (Nm)	0.2015	0.1897	0.1904	0.0975	0.0255
Modulation index	1.12	1	0.894	1 (1 and 0.577)	1 (1 and 0.577 and 0.289)
DC voltage utilization	0.644 U_{dc}	0.577 U_{dc}	0.512 U_{dc}	0.577 U_{dc}	0.577 U_{dc}
	Characteristics				
Reduction of low-order current harmonics	Worse	Good	Good	Good	Good
Reduction of switching-frequency-order current harmonics	Worse	Good	Good	Better	Best
Implementation complexity	Simplest	Simple	Normal	Normal	Complex
Dynamic performance	Low	Low	Low	Low	Low

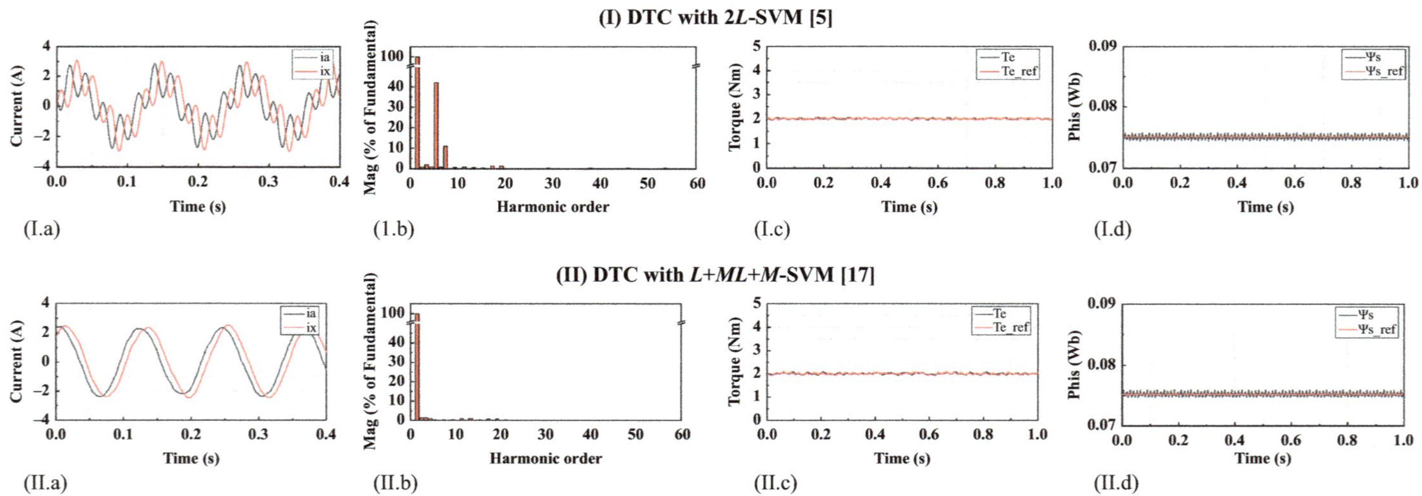

Figure 7.25 Steady-state experimental results for DTC with different SVM strategies. (I) DTC with 2L-SVM [5]. (I.a) Phase currents. (I.b) Current spectrum. (I.c) Torque. (I.d) Stator flux. (II) DTC with L+ML+M-SVM [17]. (II.a) Phase currents. (II.b) Current spectrum. (II.c) Torque. (II.d) Stator flux.

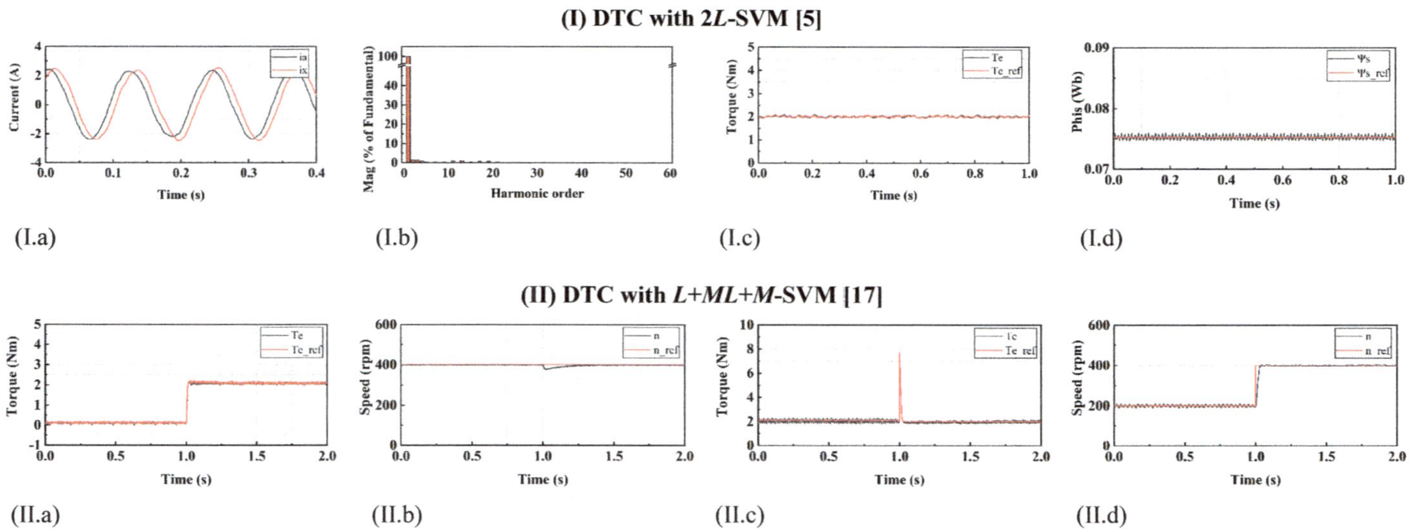

Figure 7.26 *Dynamic experimental results for DTC with different SVM strategies. (I) DTC with 2L-SVM [5]. (I.a) Torque response with step change of load. (I.b) Speed response with step change of load. (I.c) Torque response with a step change of speed. (I.d) Speed response with a step change of speed. (II) DTC with L+ML+M-SVM [17]. (II.a) Torque response with step change of load. (II.b) Speed response with step change of load. (II.c) Torque response with a step change of speed. (II.d) Speed response with a step change of speed.*

Table 7.5 Experimental results and characteristics for DTC with different SVM strategies

	2*L*-SVM (DTC)	**2*L*-SVM (FOC)**	**Improved *L*+*ML*+*M*-SVM (DTC)**	**Improved *L*+*ML*+*M*-SVM (FOC)**
	Experimental results			
THD_I_{5th} (%)	27.10	8.39	2.85	3.72
THD_I_{7th} (%)	3.75	2.10	1.84	2.27
THD_I_a (%)	45.72	15.82	7.29	10.57
Torque ripple (Nm)	0.2015	0.1897	0.1904	0.0975
Modulation index	1.12	1	0.894	1 (1 & 0.577)
DC voltage utilization	0.644 U_{dc}	0.577 U_{dc}	0.512 U_{dc}	0.577 U_{dc}
	Characteristics			
Reduction of low-order current harmonics	Worse	Worse	Good	Good
Reduction of switching-frequency-order current harmonics	Worse	Worse	Best	Best
Implementation complexity	Best	Low	Best	Low
Dynamic performance	Simplest	Simplest	Normal	Normal

The results for DTCs with different SVM strategies are summarized and compared in Table 7.5. As can be seen, they have almost the same steady-state performance, particularly in phase currents and amplitudes of the 5th and 7th harmonics. The differences between SVM-DTC and FOC are more on the dynamic performance, i.e., the torque and speed response times for SVM-DTC are much smaller than those for the FOC, no matter which SVM strategy is employed. Therefore, the dynamic performance for SVM-DTC is much better than the FOC, as demonstrated by the experimental results.

However, the SVM-DTC exhibits drawbacks when operating in low-speed conditions. As the SVM-DTC can be considered one form of sensorless control, its low-speed performance will deteriorate, particularly in the steady-state performance. In comparison, the FOC can achieve excellent steady-state performance across the entire speed region.

7.3.4 Modulation strategies of FOC and SVM-DTC for other multiphase PMSMs

In n-phase multiphase machine systems, there are 2^n switching states for a 2-level n-phase inverter. Each switching state generates a voltage vector that is mapped into $(n - 1)/2$ different orthogonal planes according to the general Clarke's transformation (7.24), where $\alpha = 2\pi/n$.

$$C_n = \frac{2}{n}\begin{bmatrix} 1 & \cos(\alpha) & \cos(2\alpha) & \cos(3\alpha) & \cdots & \cos[(n-1)\alpha] \\ 0 & \sin(\alpha) & \sin(2\alpha) & \sin(3\alpha) & \cdots & \sin[(n-1)\alpha] \\ 1 & \cos(2\alpha) & \cos(4\alpha) & \cos(6\alpha) & \cdots & \cos[2(n-1)\alpha] \\ 0 & \sin(2\alpha) & \sin(4\alpha) & \sin(6\alpha) & \cdots & \sin 2(n-1)\alpha \\ \vdots & \vdots & \vdots & \vdots & \vdots & \vdots \\ 1 & \cos\left(\frac{n-1}{2}\alpha\right) & \cos\left(2\frac{n-1}{2}\alpha\right) & \cos\left(3\frac{n-1}{2}\alpha\right) & \cdots & \cos\left[\frac{(n-1)^2}{2}\alpha\right] \\ 0 & \sin\left(\frac{n-1}{2}\alpha\right) & \sin\left(2\frac{n-1}{2}\alpha\right) & \sin\left(3\frac{n-1}{2}\alpha\right) & \cdots & \sin\left[\frac{(n-1)^2}{2}\alpha\right] \\ \frac{1}{2} & \frac{1}{2} & \frac{1}{2} & \frac{1}{2} & \cdots & \frac{1}{2} \end{bmatrix} \tag{7.24}$$

This transformation maps harmonic components into different planes according to the following rules:

1. The $2nk \pm 1$ ($k = 0, 1, 2, 3 \ldots$) harmonics including the fundamental frequency into the $\alpha\beta$ subspace.
2. The $2nk \pm (2m - 1)$ ($1 \leq m \leq (n-1)/2$, $k = 0, 1, 2, 3 \ldots$) harmonics into the xy or z_1z_2 subspace and nk harmonics into the zero-sequence subspace.

In this case, the SVPWM for multiphase machines has more possible voltage vector selections compared to the STP machines. With the increase of phases, more possible voltage vectors are provided and complicate the SVM strategy.

Taking a 5-phase machine as an example for illustration, the voltage vectors in the $\alpha\beta$ and z_1z_2 (d_2q_2) subspaces are shown in Figure 7.27 (a) and (b), respectively.

By using the simplest SVPWM of $2L$ voltage vectors in [18–20], the components in the harmonic-related subspace are ignored, and thus, the phase currents contain large low-order harmonics, as shown in Figure 7.28 (I.a) and (I.b).

Furthermore, when the reference voltage is located in sector 1, the voltage vectors V_{25}, V_{16}, V_{24}, and V_{29} are selected in the 2L+2M-SVM method, while the voltage vectors V_{17}, V_{24}, V_{25}, and V_{28} are selected in $4L$-SVM method.

As can be seen, these two methods can both eliminate the components in the harmonic-related subspaces, thereby reducing the 3rd current harmonics, the main order harmonic in 5-phase machines. The experimental results are shown in Figure 7.28 (II) and (III), respectively. It can be easily seen from the experimental results that both the $2L$+$2M$ and $4L$-based SVMs can significantly reduce the current harmonics, while $2L$-based SVM suffers from large current harmonics.

The conclusions drawn regarding SVM strategies in other multiphase machines, such as 7-phase, 9-phase, and 12-phase machine configurations, are

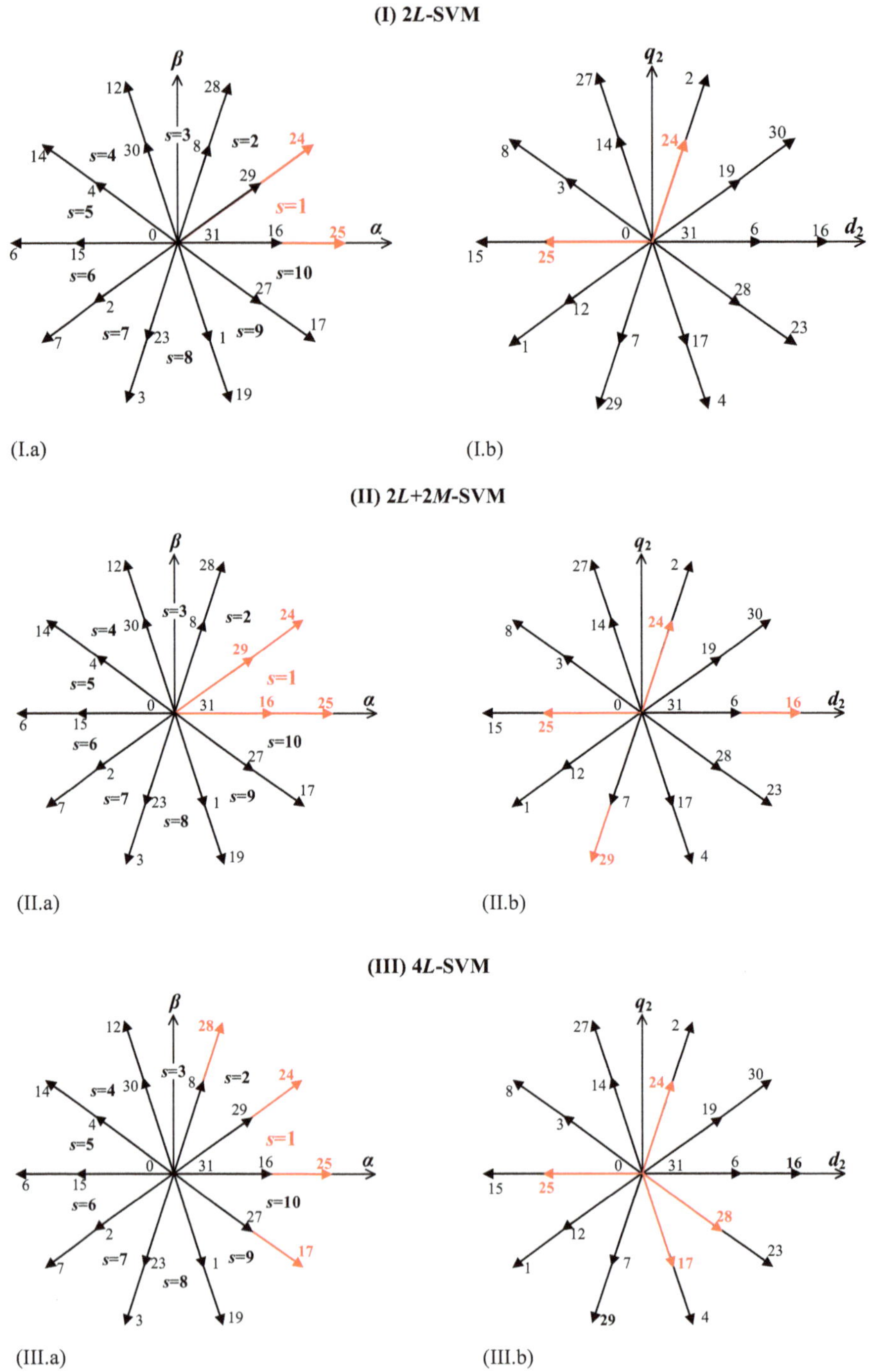

Figure 7.27 Voltage vector selections in αβ and z_1z_2 (d_2q_2) subspace for 5-phase machines. (I) 2L-SVM. (I.a) αβ subspace. (I.b) z_1z_2 (d_2q_2) subspace. (II) 2L+2M-SVM. (II.a) αβ subspace. (II.b) z_1z_2 (d_2q_2) subspace. (III) 4L-SVM. (III.a) αβ subspace. (III.b) z_1z_2 (d_2q_2) subspace.

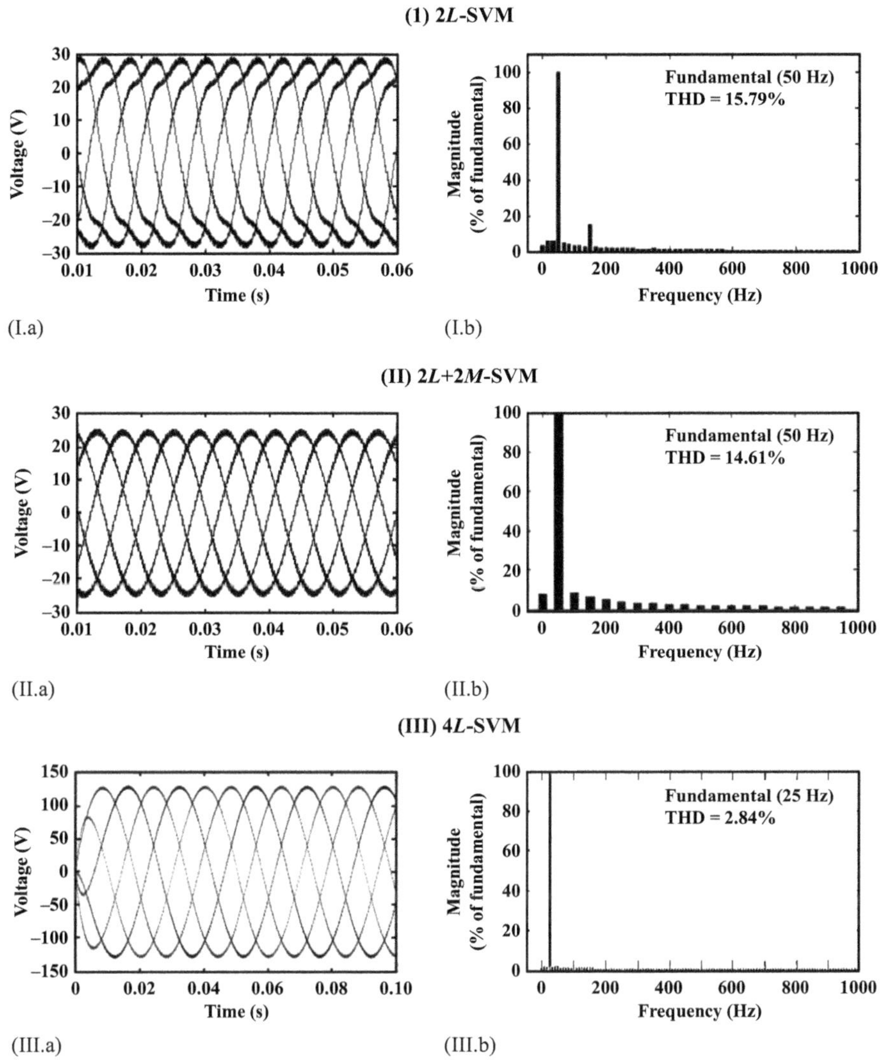

Figure 7.28 Experimental results of phase currents and spectra for 5-phase machines. (I) 2L-SVM. (I.a) Phase currents. (I.b) Spectrum. (II) 2L+2M-SVM. (II.a) Phase currents. (II.b) Spectrum. (III) 4L-SVM. (III.a) Phase currents. (III.b) Spectrum.

similar. SVM strategies that consider the elimination of components in the harmonic-related subspaces tend to offer improved steady-state performance, particularly in reducing low-order current harmonics. Therefore, the SVM strategy for other multiphase machines will not be discussed in detail in this book.

7.4 Modulation strategies of ST-DTC for multiphase PMSMs

7.4.1 ST-DTC for DTP PMSMs

The ST-DTC, i.e., the classical DTC, employs only flux and torque hysteresis regulators and a look-up table to regulate machine flux and torque. It has the merits of fast response and dynamic performance. For DTP machines, the main structure for ST-DTC is shown in Figure 7.29, which is similar to that for the STP PMSM in Figure 7.12. In addition, the red parts are the modified parts based on the ST-DTC for STP PMSMs.

The voltage estimation and current transformations for DTP machines are described in Section 7.3. The flux estimation is the same as that for the STP machines, as shown in (7.16) and (7.17). The torque equation is expressed as

$$T_e = 3p\left(\psi_{s\alpha} i_{s\beta} - \psi_{s\beta} i_{s\alpha}\right) \tag{7.25}$$

Since the inverter voltage vectors are modified in DTC for DTP machines, the switching look-up table is required to modify for selecting the proper voltage vectors, as shown in Figure 7.30 and listed in Table 7.6. Indeed, DTP machines offer a broader selection of voltage vectors compared to STP machines. However, in classical ST-DTC for DTP machines, only the outermost voltage vectors are typically chosen and employed [21]. Meanwhile, the innermost and medium voltage vectors still hold potential for selection and utilization.

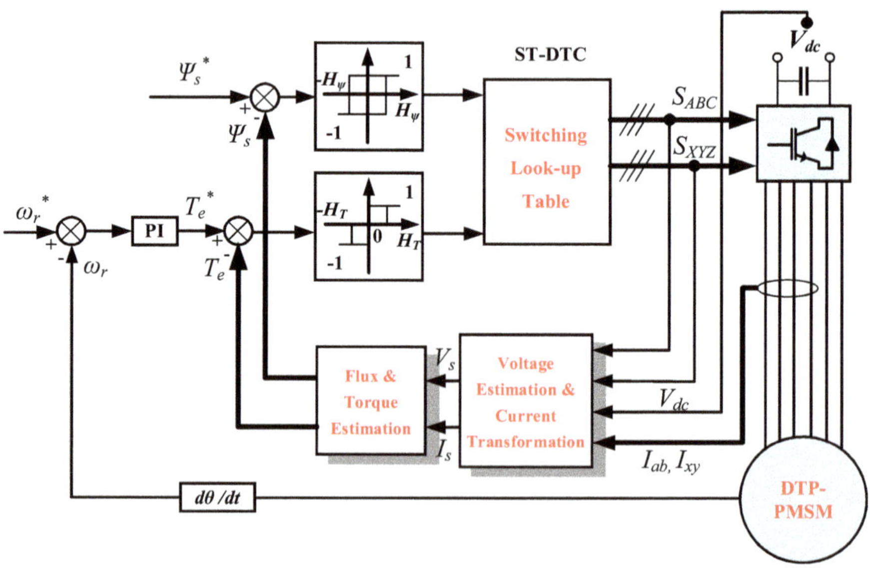

Figure 7.29 ST-DTC structure for DTP PMSM.

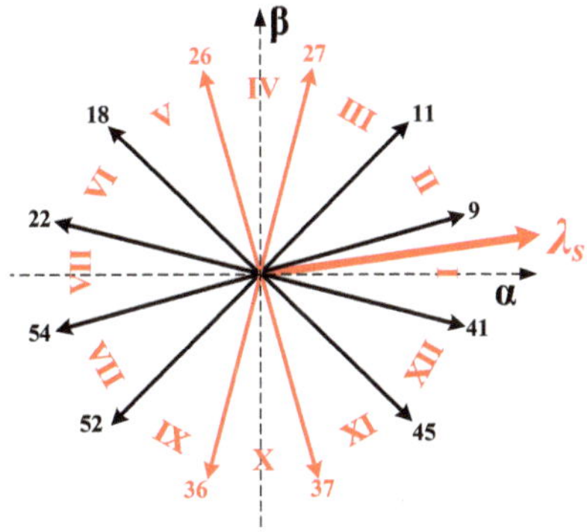

Figure 7.30 Voltage vectors and sections for DTP machines.

Table 7.6 Switching table of vector ST-DTC [21]

ψ_s	T_e	I	II	III	IV	V	VI
1	1	V_{27}	V_{26}	V_{18}	V_{22}	V_{54}	V_{52}
	0	V_0	V_0	V_0	V_0	V_0	V_0
	−1	V_{37}	V_{45}	V_{41}	V_9	V_{11}	V_{27}
−1	1	V_{26}	V_{18}	V_{22}	V_{54}	V_{52}	V_{36}
	0	V_0	V_0	V_0	V_0	V_0	V_0
	−1	V_{36}	V_{37}	V_{45}	V_{41}	V_9	V_{11}
ψ_s	T_e	**VII**	**VIII**	**IX**	**X**	**XI**	**XII**
1	1	V_{36}	V_{37}	V_{45}	V_{41}	V_9	V_{11}
	0	V_0	V_0	V_0	V_{03}	V_0	V_0
	−1	V_{26}	V_{18}	V_{22}	V_{54}	V_{52}	V_{36}
−1	1	V_{37}	V_{45}	V_{41}	V_9	V_{11}	V_{27}
	0	V_0	V_0	V_0	V_{03}	V_0	V_0
	−1	V_{27}	V_{26}	V_{18}	V_{22}	V_{54}	V_{52}

7.4.2 Virtual voltage vector synthesis techniques

As discussed in Section 7.3.2, there are $2^6 = 64$ voltage vectors in a DTP system. The phase voltage equation is written in (7.26), and the voltage vectors are shown in Figure 7.31, which is identical to the depiction in Figure 7.18 and provided here again for convenience.

$$\begin{bmatrix} V_a \\ V_b \\ V_c \\ V_x \\ V_y \\ V_z \end{bmatrix} = \frac{V_{dc}}{3} \begin{bmatrix} 2 & -1 & -1 & 0 & 0 & 0 \\ -1 & 2 & -1 & 0 & 0 & 0 \\ -1 & -1 & 2 & 0 & 0 & 0 \\ 0 & 0 & 0 & 2 & -1 & -1 \\ 0 & 0 & 0 & -1 & 2 & -1 \\ 0 & 0 & 0 & -1 & -1 & 2 \end{bmatrix} \begin{bmatrix} S_A \\ S_B \\ S_C \\ S_X \\ S_Y \\ S_Z \end{bmatrix} \tag{7.26}$$

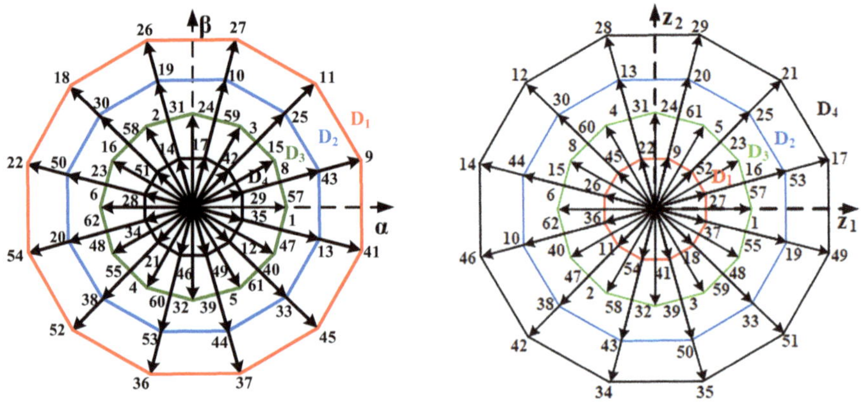

(a) (b)

Figure 7.31 DTP-PMSM drives space voltage vectors. (a) $\alpha\beta$ subspace. (b) z_1z_2 subspace.

where S_i represents the switching state of the corresponding leg, and the value of the component S_i "1" means the switch is "on" and "0" means the switch is "off".

It is evident that the voltage vectors in D_1 groups have the largest magnitudes. As previously mentioned, the torque production in DTP PMSMs is only associated with the fundamental components, i.e., the components in the $\alpha\beta$ subspace. Considering the computation burden, the D_1 groups of voltage vectors are widely employed in the classic ST-DTC schemes due to their having the largest amplitude [21]. However, by employing voltage vectors from the D_1 groups, the components in the z_1z_2 subspace remain uncontrolled, resulting in the generation of significant current harmonics in DTP-PMSM systems.

Due to the mapping of the components in the z_1z_2 subspace being primarily associated with increased losses without contributing electromechanical energy conversion, eliminating the amplitudes of voltage components in the z_1z_2 subspace is an effective approach to reducing current harmonics. To address this challenge, the virtual voltage vector synthesis (VVVS) technique was first proposed in [22] for DTP machines. By utilizing two groups of voltage vectors, i.e., D_1 and D_2 groups, the harmonic components in the z_1z_2 subspace can be eliminated, and the current harmonics can be effectively controlled.

The basic principle of the VVVS method in [22] is that the voltage vectors of the D_1, D_2, and D_4 groups share the same direction in the $\alpha\beta$ subspace. However, the voltage vector from D_2 have the opposite direction with the voltage vectors of D_1 and D_4 vectors in the z_1z_2 subspace, e.g., the vector V_{10} of the D_2 group has the same phase with vector V_{27} of D_1 group and V_{17} of D_4 group in the $\alpha\beta$ subspace, while the vector V_{10} of D_2 group has the opposite phase with vector V_{27} of D_1 group and V_{17} of D_4 group in the z_1z_2 subspace. It implies that the voltage vectors from the D_1, D_2, and D_4 groups have similar effects on torque production, but opposite effects on flux ψ_{sz1z2} and current harmonics generation.

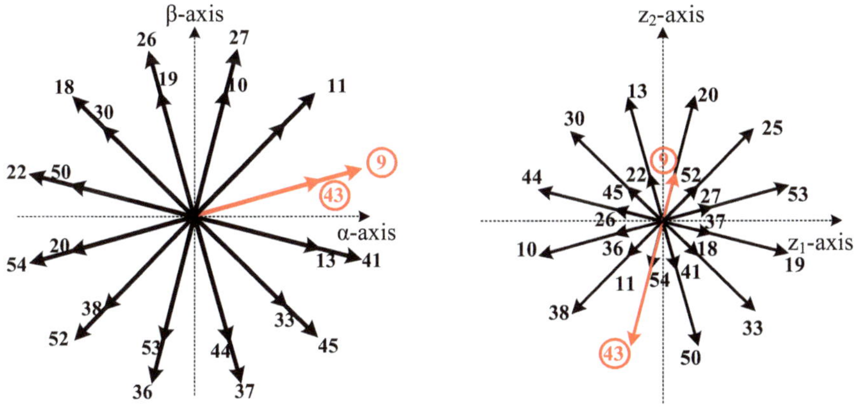

(a) (b)

Figure 7.32 Virtual voltage vector-based DTC vector selection in different subspaces. (a) $\alpha\beta$ subspace. (b) z_1z_2 subspace.

Moreover, since the voltage vectors of D_3 groups have different phases and are not aligned with any others, they are not utilized in the classic ST-DTC method. In summary, accounting for the computation burden, the combination of voltage vectors from the D_1 and D_2 groups can effectively control the components in the z_1z_2 subspace and reduce current harmonics, as shown in Figure 7.32 [22].

The amplitudes of these VVVs in the z_1z_2 subspace are forced to zero, while amplitudes in the $\alpha\beta$ subspace are calculated by

$$\left|VV_{\alpha\beta}\right| = \left|V_{1-\alpha\beta}\right|T_1 + \left|V_{2-\alpha\beta}\right|T_2 \tag{7.27}$$

By setting the magnitude synthetic space vector in z_1z_2 subspace to zero, the magnitude of stator flux ψ_{sz1z2} can be minimized and current harmonics can be suppressed. Consequently, the duty times for D_1 and D_2 vectors (T_{D1} and T_{D2}) can be defined as

$$\begin{cases} V_{D_{1z_1z_2}} \cdot T_{D_1} - V_{D_{2z_1z_2}} \cdot T_{D_2} = 0 \\ T_{D_1} + T_{D_2} = T_s \end{cases} \tag{7.28}$$

where T_s is the PWM period.

It can be concluded that the synthesis of VVVs requires adherence to the following principle [22–24]: the employed voltage vectors must have opposite directions in the z_1z_2 subspace, which ensures that the amplitudes of the synthesized voltage vector in the z_1z_2 subspace are zero. In particular, since the vectors in the D_3 groups are unaligned with those in the D_1, D_2, and D_4 groups, as shown in Figure 7.33, they are not used for synthesizing [22,24,26]. Following this principle, a maximum of two groups of VVs with different amplitudes are chosen from vectors in the V-D_1, V-D_2, and V-D_4 groups.

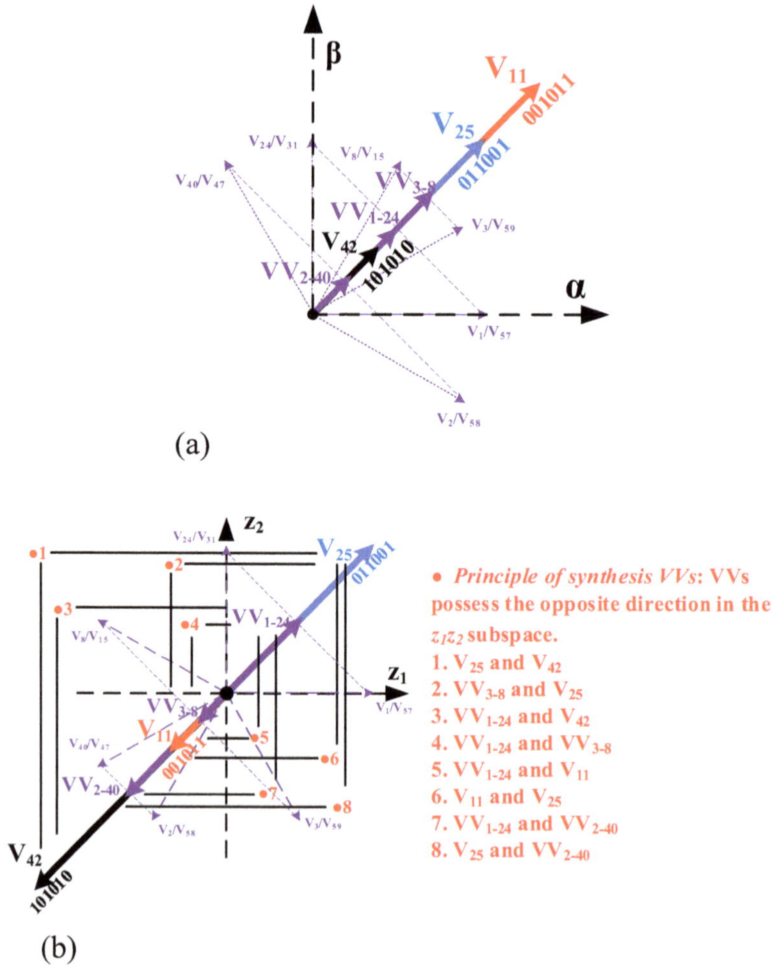

Figure 7.33 All possible virtual voltage vector synthesis in DTP machines. (a) $\alpha\beta$ subspace. (b) z_1z_2 subspace.

Then, according to (7.29) and (7.30), these eight groups of VVs can be divided into six groups according to their amplitudes in the $\alpha\beta$ subspace, i.e. the largest VVVs (*VVV-L*), the medium to large VVVs (*VVV-ML*), the medium to slightly large VVVs (*VVV-MSL*), the medium to slightly small VVVs (*VVV-MSS*), the medium to small VVVs (*VVV-MS*) and the smallest VVVs (*VVV-S*), as concluded in [25] and summarized in Table 7.7.

Using the same definitions of active voltage vectors, as previously mentioned in Table 7.1, this magnitude-based definition has no physical meaning as well.

In this way, the VVVs-based ST-DTC for DTP machines in the existing literature can be summarized as follows:

Table 7.7 Amplitudes of all possible VVVs for DTP machines

VVVs	**Synthesis components (example)**	**Magnitude in the $\alpha\beta$ subspace**
VVV-L	V_{25} and V_{11} $VV_{1\text{-}24}$ and V_{11}	$\frac{6\sqrt{2}-2\sqrt{6}}{6}U_{dc} \approx 0.598U_{dc}$
VVV-ML	V_{25} and V_{42} V_{25} and $VV_{3\text{-}8}$	$\frac{\sqrt{6}-\sqrt{2}}{3}U_{dc} \approx 0.345U_{dc}$
VVV-MSL	$VV_{3\text{-}8}$ and $VV_{1\text{-}24}$	$\frac{3\sqrt{2}-\sqrt{6}}{6}U_{dc} \approx 0.299U_{dc}$
VVV-MSS	$VV_{1\text{-}24}$ and $VV_{2\text{-}40}$	$\frac{5\sqrt{6}-3\sqrt{2}}{33}U_{dc} \approx 0.243U_{dc}$
VVV-MS	V_{42} and $VV_{1\text{-}24}$	$\frac{2\sqrt{6}-3\sqrt{2}}{3}U_{dc} \approx 0.219U_{dc}$
VVV-S	$VV_{1\text{-}24}$ and $VV_{2\text{-}40}$	$\frac{\sqrt{6}-\sqrt{2}}{6}U_{dc} \approx 0.173U_{dc}$

1. One VVV-based ST-DTC [21]: Voltage vectors only in the D_1 group are utilized to generate the VVV, i.e., the Classical ST-DTC.
2. Two VVVs-based ST-DTC [22]: Voltage vectors in D_1 and D_2 groups are utilized to generate the VVVs.
3. Three VVVs-based ST-DTC [23]: Voltage vectors in D_1, D_2, and D_3 groups are utilized to generate the VVVs. The synthesis of VVVs is based on the three groups of voltage vectors together.
4. Four VVVs-based ST-DTC [24]: Voltage vectors in D_1, D_2, and D_4 groups are utilized to generate the VVVs. The synthesis of VVVs is a pairwise combination based on three groups of voltage vectors. Hence, it can be renamed as 2VVVs+2VVVs method.
5. Four VVVs-based ST-DTC [25]: Voltage vectors in D_1, D_2, D_3, and D_4 groups are utilized to generate the VVVs.

Although employing more VVVs in ST-DTC could further improve the control performance by reducing torque ripple and suppressing current harmonic, it would also increase the structure's complexity and computation burden. Hence, the utilization of VVVs-based ST-DTCs in industrial applications depends on varying requirements and hardware constraints. In the following section, several experimental results are presented and compared to illustrate the characteristics of different VVVs-based ST-DTC for DTP machines.

7.4.3 Modulation strategies for ST-DTC of DTP PMSMs

7.4.3.1 Modulation strategies of different VVV-based ST-DTC

One VVV-based ST-DTC [21]

One VVV-based ST-DTC utilizes the voltage vectors in D_1 groups only, since the voltage vectors in the D_1 group have the largest amplitudes to provide higher voltage utilization. However, the uncontrolled z_1z_2 voltage components would introduce significant current harmonics, thereby increasing the system losses and decreasing the system's efficiency. The corresponding VVV selection strategy is shown in Figure 7.34. As can be seen, only the voltage vectors in the D_1 group are selected, but the z_1z_2 voltage components cannot be eliminated.

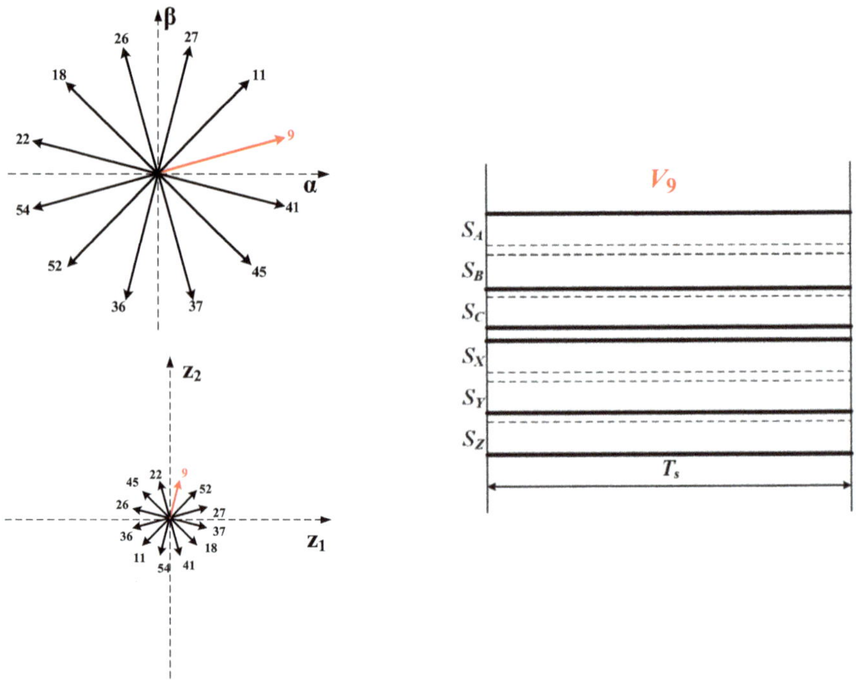

(a) (b)

Figure 7.34 Modulation strategy of one VVV-based ST-DTC. (a) Voltage vector selection in αβ and z_1z_2 subspace. (b) Corresponding switching patterns.

Two VVVs-based ST-DTC [22]

Two VVV-based ST-DTC utilizes the voltage vectors in the D_1 and D_2 groups and synthesizes the virtual voltage vector based on the aforementioned VVVS method. The issues in the one VVV-based ST-DTC could be addressed by controlling and eliminating z_1z_2 voltage components. The corresponding VVV selection strategy is shown in Figure 7.35(b1).

Three VVVs-based ST-DTC [23]

Three VVV-based ST-DTC utilizes the voltage vectors in the D_1, D_2 and D_3 groups and synthesizes the VVV. The VVV selection strategy is shown in Figure 7.36(b1).

Four VVVs (2 VVVs + 2 VVVs)-based ST-DTC [24]

The four VVVs (2 VVVs + 2 VVVs) method also utilizes the voltage vectors in the D_1, D_2 and D_4 groups and synthesizes the VVV. The VVV selection strategy is shown in Figure 7.37(b1).

Four VVVs-based ST-DTC [25]

This method utilizes the voltage vectors in the D_1, D_2, D_3 and D_4 groups and synthesizes the virtual voltage vector based on the VVVS method. The VVV selection strategy is shown in Figure 7.38(b1).

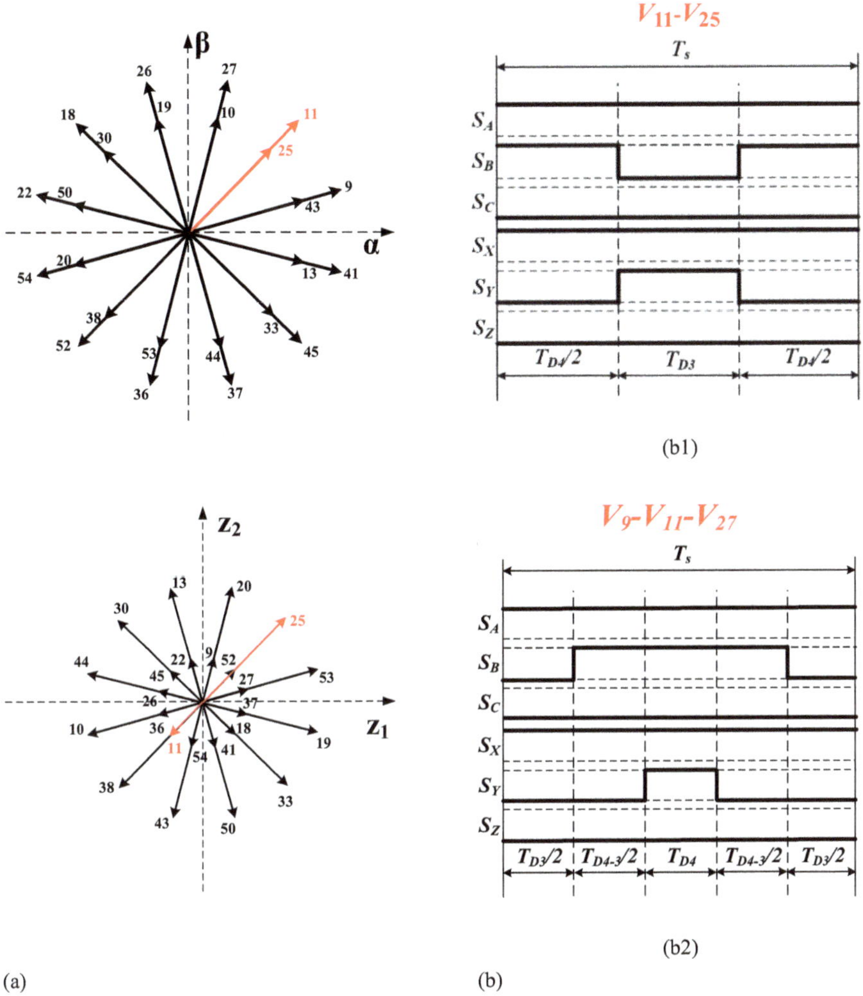

Figure 7.35 Modulation strategy of two VVV-based ST-DTC. (a) Voltage vector selection in $\alpha\beta$ and z_1z_2 subspace. (b) Corresponding switching patterns.

It is worth noting that based on the VVVS method, the virtual voltage vectors are synthesized with different numbers of VVV in Figures 7.34(a)–7.38(a), and the corresponding VVV switching patterns are shown in Figures 7.34(b) and 7.35(b1)–(b1). To reduce the current harmonics and the number of switches to minimize switching losses, the switching patterns have been modified, as shown in Figures 7.35(b2)–7.38(b2). In this case, other voltage vectors have also been used. By selecting the appropriate duration time of each vector, it ensures the actual synthesized voltage vectors are equivalent.

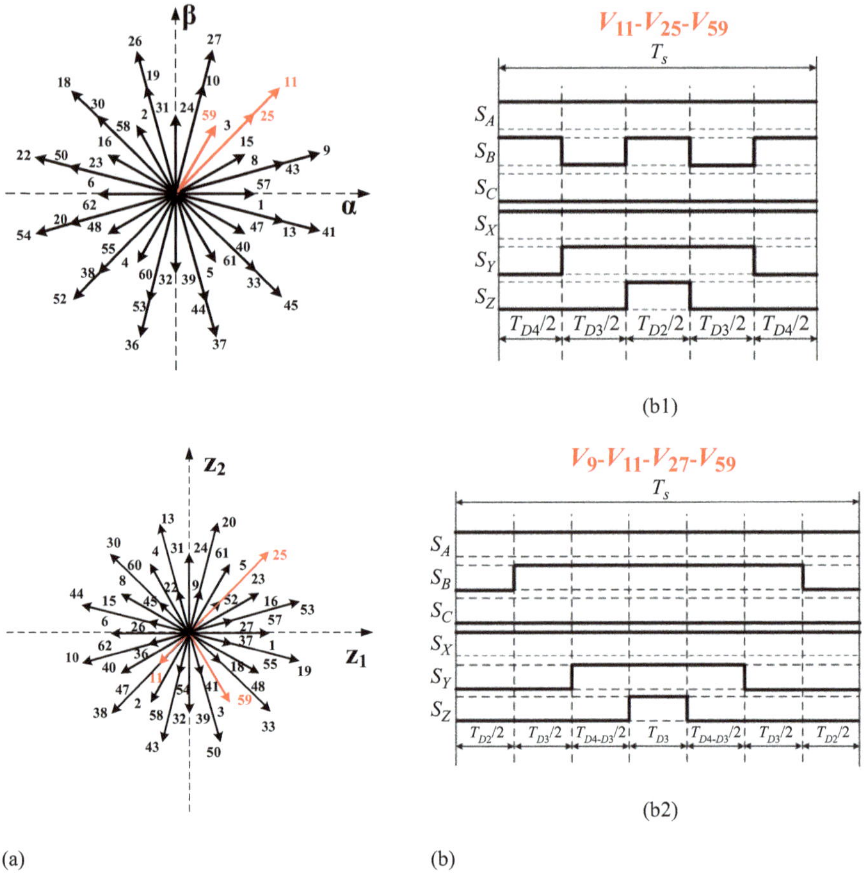

Figure 7.36 Modulation strategy of three VVV-based ST-DTC. (a) Voltage vector selection in $\alpha\beta$ and z_1z_2 subspace. (b) Corresponding switching patterns.

7.4.3.2 Experimental results of different VVVs-based ST-DTC

Figures 7.39 and 7.40 show the steady-state and dynamic experimental results of different VVVs-based ST-DTC [21–25] based on Machine 3 (see Appendix A). The steady-state tests are carried out under the same rotor speed of 200 r/min under the same load torque of 2 Nm, and the presented results include the phase current, current spectrum, torque, stator flux, and duty cycle of phase A. In the dynamic tests, the responses with step changes of load and speed at instant $t = 1.0$ s are presented.

7.4.3.3 Summary and evaluation

For more details, the results for these DTCs are illustrated in Table 7.8, where the torque and flux ripples are calculated by

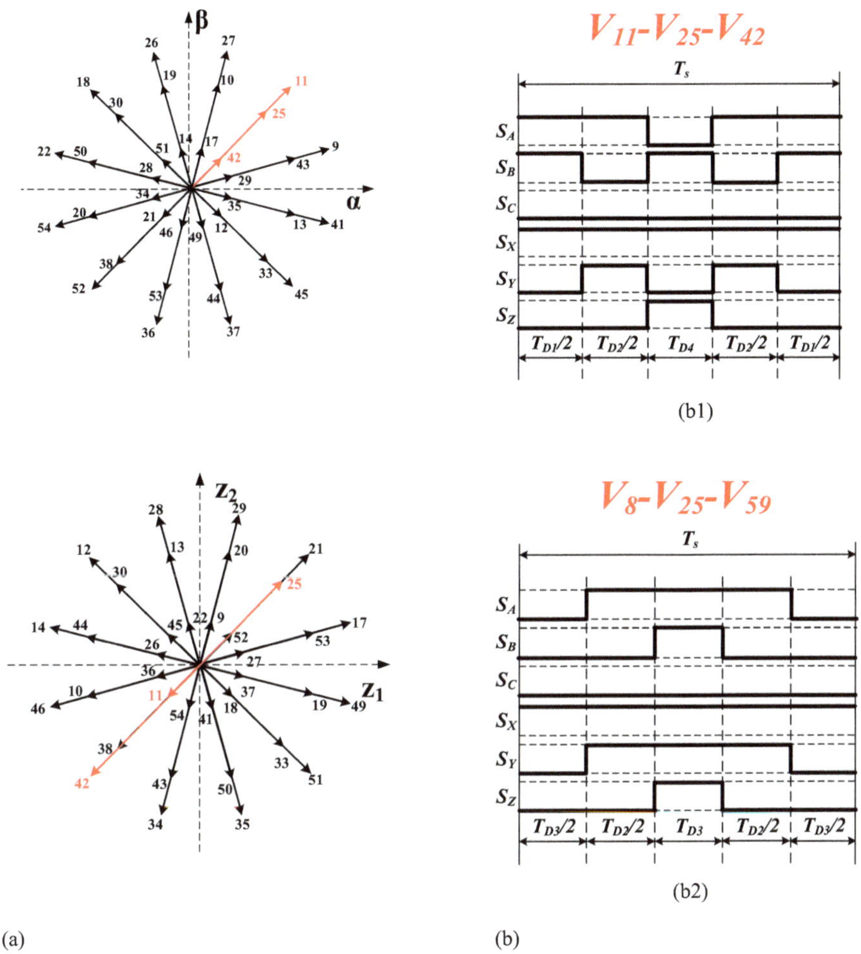

Figure 7.37 Modulation strategy of four VVVs (2VVVs + 2VVVs) based ST-DTC. (a) Voltage vector selection in αβ and z_1z_2 subspace. (b) Corresponding switching patterns.

$$\psi_{s\_ripple} = \sqrt{\frac{1}{n}\sum\nolimits_{i=1}^{n}\left(\psi_{si} - \psi_{sav}\right)^2} \tag{7.29}$$

$$T_{e\_ripple} = \sqrt{\frac{1}{n}\sum\nolimits_{i=1}^{n}\left(T_{ei} - T_{eav}\right)^2} \tag{7.30}$$

where ψ_{si} and T_{ei} are the instant values of the stator flux and torque, respectively; ψ_{sav} and T_{eav} are the average values of the stator flux and torque, respectively.

The modulation index (m) is defined as the ratio of the fundamental component of the phase voltage ($U_{\text{ref,phase}}$) to the fundamental magnitude of the phase voltage

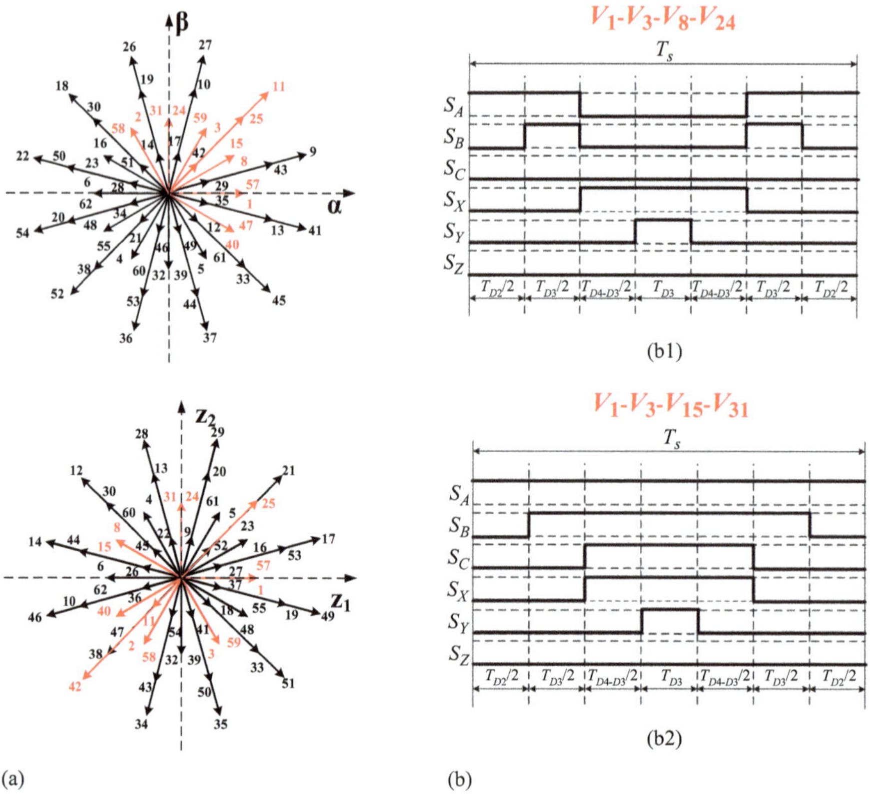

Figure 7.38 Modulation strategy of four VVV-based ST-DTC. (a) Voltage vector selection in $\alpha\beta$ and z_1z_2 subspace. (b) Corresponding switching patterns.

in six-step mode ($2U_{dc}/\pi$), i.e.,

$$m = \frac{\pi}{2}\frac{U_{\text{ref,phase}}}{U_{dc}} \tag{7.31}$$

The utilization of DC-link voltage (α) is defined as the ratio of the fundamental component magnitude of the line voltage $U_{\text{ref,line}}$ to the DC-link voltage U_{dc}, i.e.

$$\alpha = \frac{U_{\text{ref,line}}}{U_{dc}} \tag{7.32}$$

For instance, the modulation index for SPWM in STP machines is 0.785, and the maximum utilization of DC-link voltage is 0.866. For SVPWM, the modulation index in STP machines is 0.906, and the maximum utilization of DC-link voltage is 1.0. Finally, the characteristics of these VVVs-based ST-DTCs can be summarized

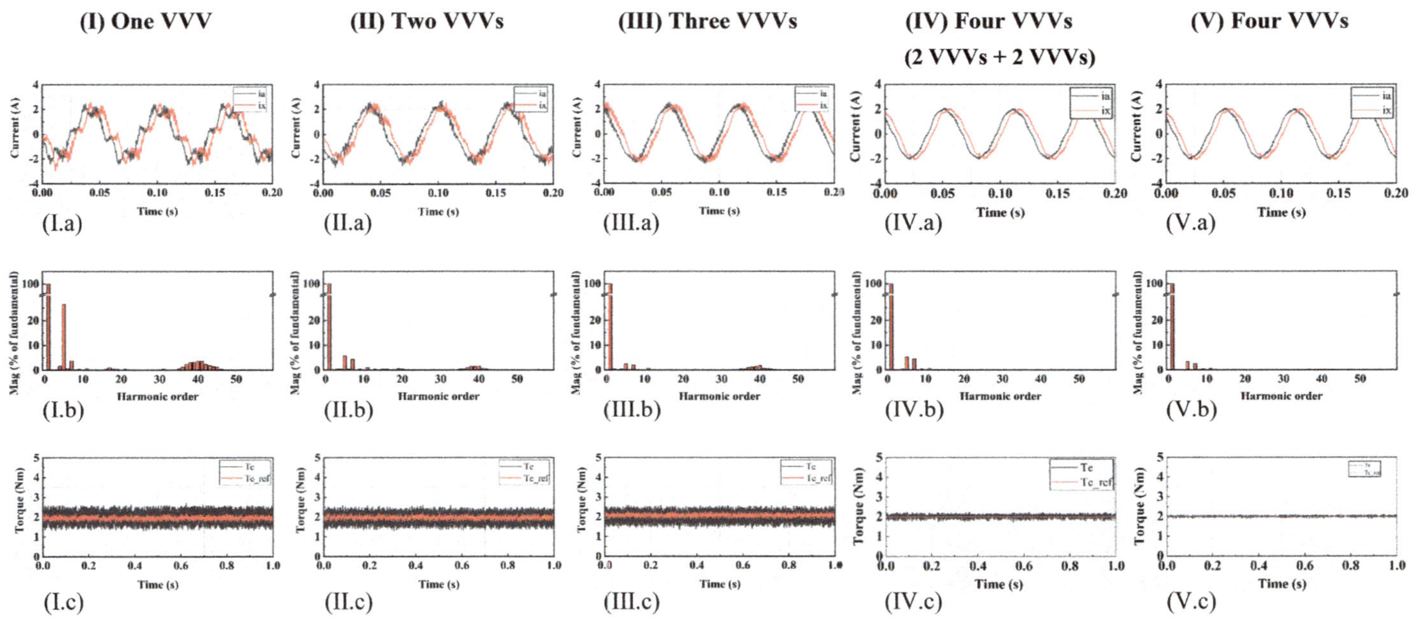

Figure 7.39 *Steady-state experimental results of different VVVs based ST-DTC. (I) One VVV. (I.a) Phase current. (I.b) Current spectrum. (I.c) Torque. (I.d) Stator flux. (I.e) Duty cycle (Phase A). (II) Two VVVs. (II.a) Phase current. (II.b) Current spectrum. (II.c) Torque. (II.d) Stator flux. (II.e) Duty cycle (Phase A). (III) Three VVVs. (III.a) Phase current. (III.b) Current spectrum. (III.c) Torque. (III.d) Stator flux. (III.e) Duty cycle (Phase A). (IV) Four VVVs (2 VVVs+ 2 VVVs). (IV.a) Phase current. (IV.b) Current spectrum. (IV.c) Torque. (IV.d) Stator flux. (IV.e) Duty cycle (Phase A). (V) Four VVVs. (V.a) Phase current. (V.b) Current spectrum. (V.c) Torque. (V.d) Stator flux. (V.e) Duty cycle (Phase A).*

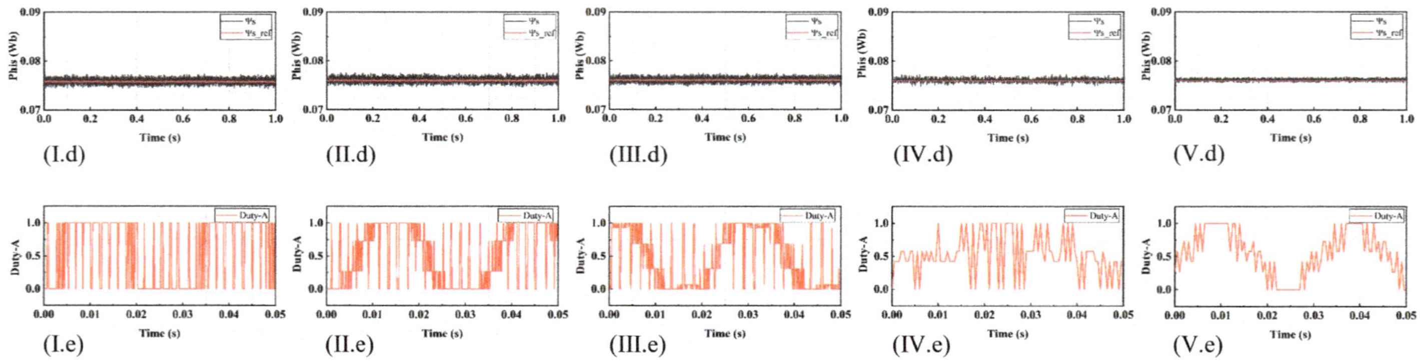

Figure 7.39 (Continued)

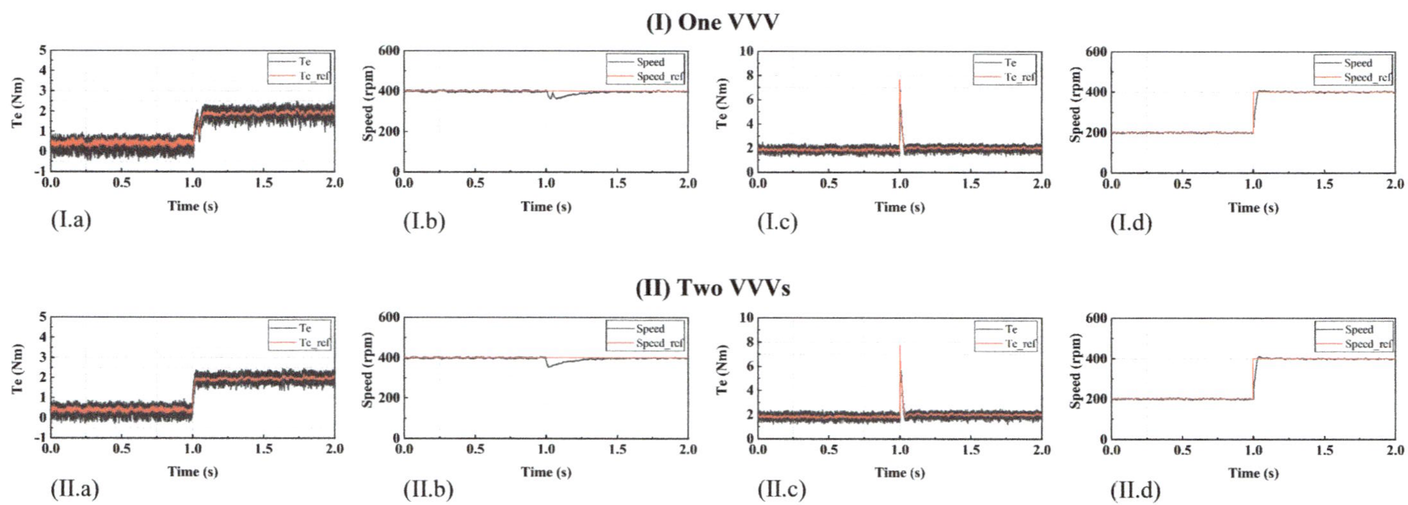

Figure 7.40 *Dynamic experimental results of different VVVs-based ST-DTC. (I) One VVV. (I.a) Torque response with step change of load. (I.b) Speed response with step change of load. (I.c) Torque response with a step change of speed. (I.d) Speed response with a step change of speed. (II) Two VVVs. (II.a) Torque response with step change of load. (II.b) Speed response with step change of load. (II.c) Torque response with a step change of speed. (II.d) Speed response with a step change of speed. (III) Three VVVs. (III.a) Torque response with step change of load. (III.b) Speed response with step change of load. (III.c) Torque response with a step change of speed. (III.d) Speed response with a step change of speed. (IV) Four VVVs (2 + 2 VVs). (IV.a) Torque response with step change of load. (IV.b) Speed response with step change of load. (IV.c) Torque response with a step change of speed. (IV.d) Speed response with a step change of speed. (V) Four VVVs. (V.a) Torque response with step change of load. (V.b) Speed response with step change of load. (V.c) Torque response with a step change of speed. (V.d) Speed response with a step change of speed.*

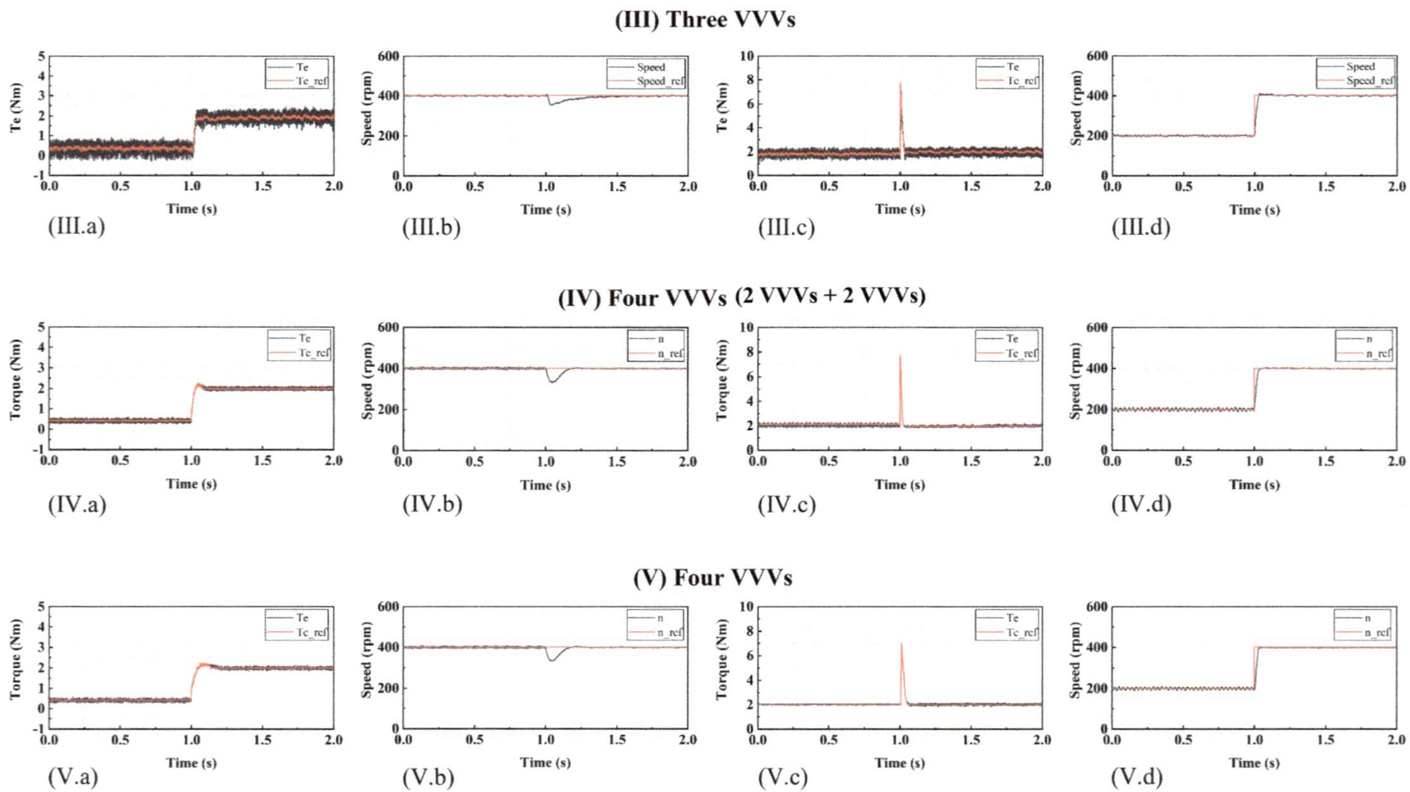

Figure 7.40 (Continued)

Table 7.8 Experimental results and characteristics for all VVVs-based ST-DTC

	One VVV	**Two VVVs**	**Three VVVs**	**Four VVVs (2 VVVs + 2 VVVs)**	**Four VVVs**
			Experimental results		
THD_I_{5th} (%)	27.10	8.39	2.85	3.72	2.87
THD_I_{7th} (%)	3.75	2.10	1.84	2.27	1.36
THD_I_a (%)	45.72	15.82	7.29	10.57	6.33
Torque ripple (Nm)	0.2015	0.1897	0.1904	0.0975	0.0255
Ψs_ripple (Wb)	9.75×10^{-5}	9.23×10^{-5}	8.77×10^{-5}	8.91×10^{-5}	8.95×10^{-5}
Maximum modulation index	1.12	1	0.894	1 (1 and 0.577)	1 (1 and 0.577 and 0.289)
DC voltage utilization	0.644	0.577	0.512	0.577	0.577
Computation time (μs)	3	5	6	8	11
			Characteristics		
Reduction of low-order current harmonics	Worse	Good	Best	Good	Best
Reduction of switching-frequency-order current harmonics	Worse	Worse	Worse	Better	Best
Implementation complexity	Simplest	Simple	Normal	Normal	Complex
Dynamic performance	Lowest	Low	High	High	Highest

in Table 7.8 with reference to the capability to reduce current harmonics, capability to reduce torque ripples, implementation complexity, and computation burden.

7.4.4 Modulation strategies of ST-DTC for other multiphase PMSMs

This section introduces the different modulation strategies of ST-DTC for other multiphase machines, where the 5-phase [27–29], 7-phase [30], and 9-phase [31] PMSMs are taken as examples for illustration.

7.4.4.1 5-phase machines

As previously mentioned, the simplest way for a modulation strategy of ST-DTC in multiphase machines is selecting a single voltage vector in each switching period. Typically, the voltage vectors with the largest amplitudes are the optimal choice. For example, it has been noted in [27] that large vectors in the fundamental subspace (*dq* in a 5-phase system) will map into small vectors in the harmonic subspace (*xy* in a 5-phase system) in a 5-phase system. Therefore, by applying large vectors, the amplitude of harmonic currents can be minimized, and thus, the switching and copper losses (due to the harmonic currents) can also be minimized.

However, this modulation strategy based on the large voltage vectors may lead to large current distortion and torque ripples, as well as low efficiency. To address these issues, the VVVs, described in Section 7.4.1, have been widely proposed and employed.

The VVVs method is first proposed based on 5-phase machines [28,2]. In 5-phase machines, similar to the VSD method in DTP machines, the fundamental component and the harmonic variables with an order of $10k \pm 1$ (*k* is an integer) are located in the d_1q_1 or *dq* subspace, while the harmonic variables with an order of $10k \pm 3$(*k* is an integer) are located in the d_3q_3 or *xy* subspace. By setting the components in the d_3q_3 or *xy* subspace to zero, the low-frequency (mainly the 3rd order) harmonic currents can be effectively eliminated. It should be noted that the voltage vectors in the largest and medium groups are selected to form the VVVs in [28,29], as shown in Figure 7.41, where the current harmonics can also be reduced significantly.

Furthermore, an improved modulation strategy for 5-phase machines is proposed in [30]. Instead of using two groups of voltage vectors in the largest and

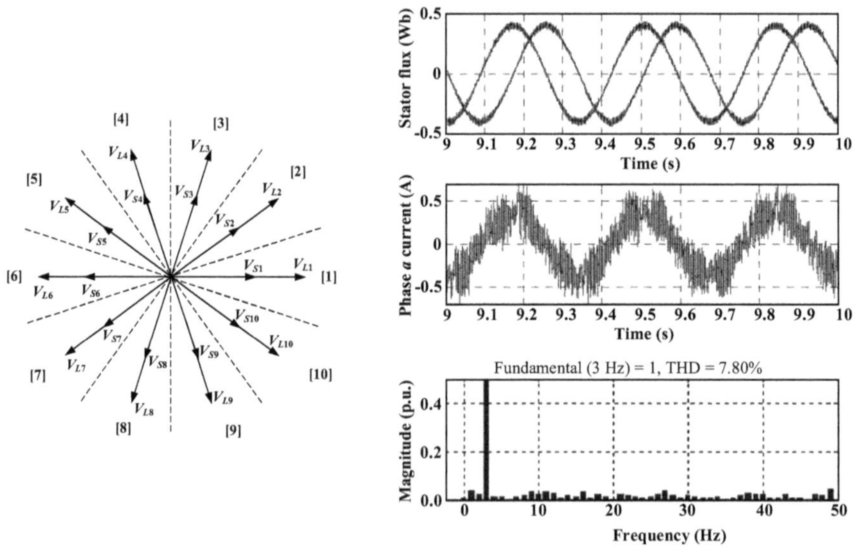

(a) (b)

Figure 7.41 Virtual voltage vector for 5-phase machines. (a) VVVs in d_1q_1 space with null d_3q_3 voltage vectors. (b) Simulation results of VVVs-based DTC.

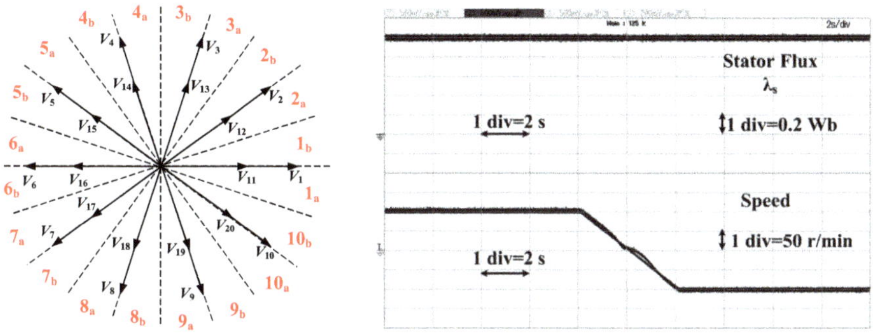

(a) (b)

Figure 7.42 Virtual voltage vector for 5-phase machines. (a) VVVs in $\alpha\beta$ plane. (b) Waveforms of stator flux and rotor speed.

medium groups, the VVVs are formed based on three groups of voltage vectors in [30], i.e., the largest, medium, and smallest groups of voltage vectors. Then, an optimal selection/modulation strategy is proposed according to the theoretical analysis. The voltage vector selection strategy in [30] is shown in Figure 7.42(a), and the proposed method is proven that it can improve the performance at low-speed operation and address the problem of the flux demagnetization, as verified by experimental results and shown in Figure 7.42(b).

7.4.4.2 7-phase machines

The 7-phase machines are also a popular topology of the multiphase machines. The inverter of a 7-phase machine has a total of 2^7=128 available voltage vectors, and it definitely provides more potential for voltage vector selection. After VSD transformation, the fundamental component and the $14k\pm1$ ($k = 0, 1, 2, \ldots$) order harmonics (i.e., 13th, 15th, 27th) are mapped into the $\alpha\beta$ subspace. Meanwhile, the $14k \pm 5$ ($k = 0, 1, 2, \ldots$) order harmonics (i.e., 5th, 9th, 19th) are mapped into the x_1y_1 subspace. The $14k\pm3$($k = 0, 1, 2, \ldots$) order harmonics (i.e., 3th, 11th, 17th) are mapped into the x_2y_2 subspace. Similarly, the components in the x_1y_1 and x_2y_2 subspaces are most responsible for producing low-order harmonics in the output current, it is therefore essential to eliminate these harmonic components. Further, the voltage vectors of a 7-phase machine can be divided into seven groups according to the amplitude in the $\alpha\beta$ subspace after the VSD transformation, which is defined as V_1, V_2, V_3, V_4, V_5, V_6, and V_7 from the smallest to largest.

Specifically, to eliminate the harmonic components, the VVVS method is extended into the 7-phase induction machines in [31]. For the simplified implementation, the voltage vectors in V_4, V_6, and V_7 groups are selected, as shown in Figure 7.43. As can be seen, the VVVS method in 7-phase machines also contributes to the reduction of current harmonics, as shown in Figure 7.43.

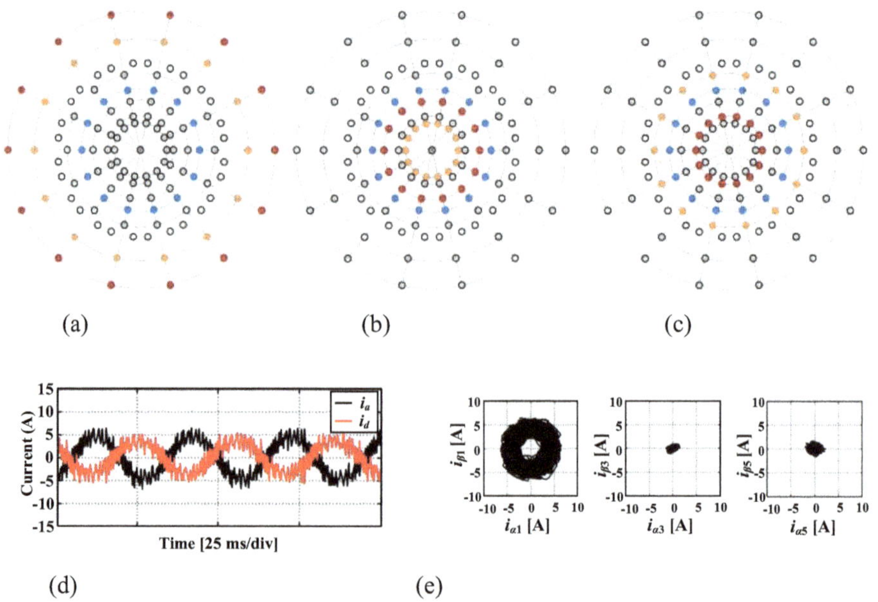

Figure 7.43 VVV$_S$ for 7-phase machines. (a)–(c) Position of voltage vector distributions. (a) $\alpha_1\beta_1$ plane. (b) $\alpha_3\beta_3$ plane. (c) $\alpha_5\beta_5$ plane. (d) Phase "a" and "d" currents. (e) Trajectories of stator currents projected on three planes.

7.4.4.3 9-phase machines

The 9-phase machines, or triple three-phase (TTP) machines, are composed of three sets of three-phase windings termed $a_1b_1c_1$, $a_2b_2c_2$, and $a_3b_3c_3$, spatially shifted 20° and with three isolated neutrals. After VSD transformation, the fundamental components and the $9k \pm 8(k = 1, 3, 5, \ldots)$ order harmonics (i.e., 17th, 19th) are mapped into the α-β subspace. Meanwhile, the $9k \pm 4$ $(k = 1, 3, 5, \ldots)$ order harmonics (i.e., 5th, 13th, 23th) are mapped into the x_1y_1 subspace. The $9k \pm 2$ $(k = 1, 3, 5, \ldots)$ order harmonics (i.e., 7th, 11th, 25th) are mapped into the x_2y_2 subspace. The $9k \pm 6$ $(k = 1, 3, 5, \ldots)$ order harmonics (i.e., zero sequences component) are mapped into the x_3y_3 subspace.

In the case of a 2-level nine-phase converter, there are 2^9=512 possible switching states. Hence, it has many more potential voltage vectors than the 5-phase and 6-phase machines. For simplifying the implementation, two groups of voltage vectors that have the largest amplitudes in the $\alpha\beta$ subspaces (i.e., O_1 and O_2 listed in Table 7.9) are selected in [32]. As can be seen from Figure 7.44, the current harmonics can be reduced significantly compared to the DTC with only one voltage vector in one switching period.

Although 9-phase machines provide more potential on voltage vector selections and modulation strategies, there is minimal literature focusing on the voltage

Table 7.9 Modulus of voltage vectors in each subspace of VSD, αβ ratio ($R_{\alpha\beta}$) and percentage of αβ contribution considering the maximum ratio as 100% [32]

Values	$\boldsymbol{O_1}$	$\boldsymbol{O_2}$	O_3	O_4	O_5	O_6	O_7	O_8	O_9	O_{10}
$\lvert v_{\alpha\beta}\rvert \cdot 10^2$	64	56	42	34	30	22	20	15	12	8
$\lvert v_{x_1y_1}\rvert \cdot 10^2$	15	20	8	42	56	22	30	12	64	34
$\lvert v_{x_2y_2}\rvert \cdot 10^2$	12	30	34	8	20	22	56	64	15	42
$R_{\alpha\beta}$	3.3	1.5	1.2	0.8	0.5	0.7	0.3	0.23	0.18	0.15
%	100	46	36	24	15	21	9	7	5	4

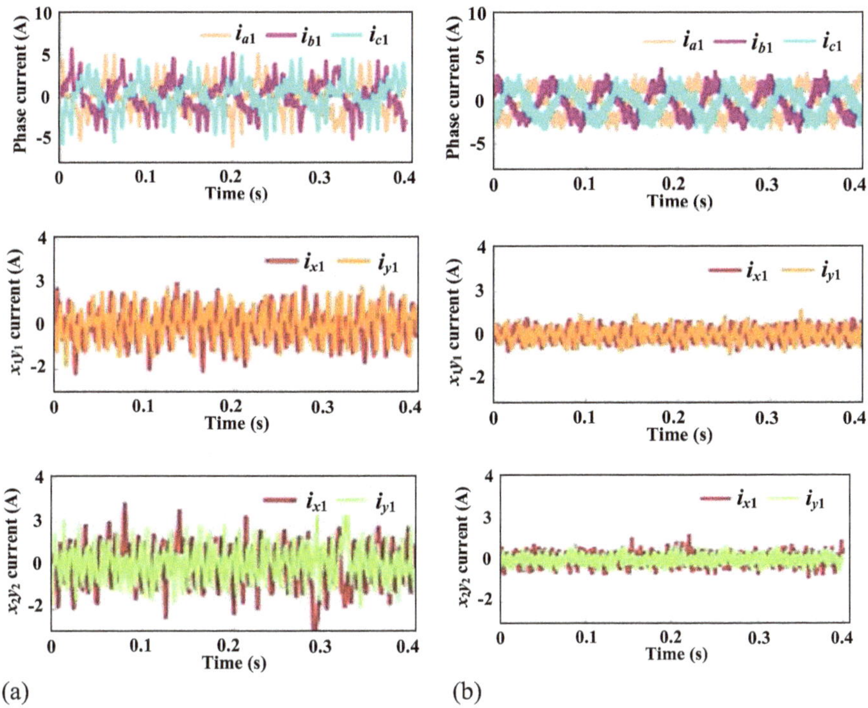

Figure 7.44 Steady-state tests for one and four VVVs-based ST-DTC for 9-phase machines. (a) One VVVs. (b) Four VVVs.

vector selection based on other VVV combinations, since when more groups of voltage vectors are utilized, the implementation process becomes even more complex.

Indeed, as the number of phases increases beyond nine in PMSMs, the number of available voltage vectors also increases. This implies that there is a greater diversity of modulation strategies in ST-DTC, such as 11-phase, 12-phase, and 15-phase. At present, there is a scarcity of literature addressing the specific issue of voltage vector selection for multiphase machines with more than nine phases, but it is predictable that these high-phase multiphase machines offer great research potential in modulation strategies in ST-DTC.

7.5 Overmodulation techniques for multiphase machine drives

7.5.1 Overmodulation techniques in STP PMSMs

The SVM utilizes the time average of a set of suitable switching state vectors to generate a sampled reference vector. By employing the Clark transformation, six voltage vectors (V_1-V_6) form a hexagon centered at the origin of the $\alpha\beta$ space. The reference voltage V_{ref} can be calculated by

$$V_{ref}T_s = V_it_i + V_kt_k + V_0t_0 \tag{7.33}$$

where T_s is the cycle period. t_i, t_k, and t_0 are the durations of the adjacent voltage vectors (V_{i} and V_{k}) and zero vector (V_0).

As defined in (7.31), the modulation indices for SPWM and SVPWM are 0.785 and 0.907, respectively. According to the modulation index, the voltage range can be divided into two modulation regions, i.e., the linear and overmodulation regions, as shown in Figure 7.45.

First, the linear region is located in the circle inscribed by the voltage vector hexagon, i.e., the modulation index is between $0 \leq m \leq 0.907$. In this region, the on-duration of the zero voltage vector is nonzero. With the modulation index m increased to 0.907, the on-duration of the zero vector reduces to zero when the reference vector reaches the boundary of the hexagon. This is the upper limit of the linear region.

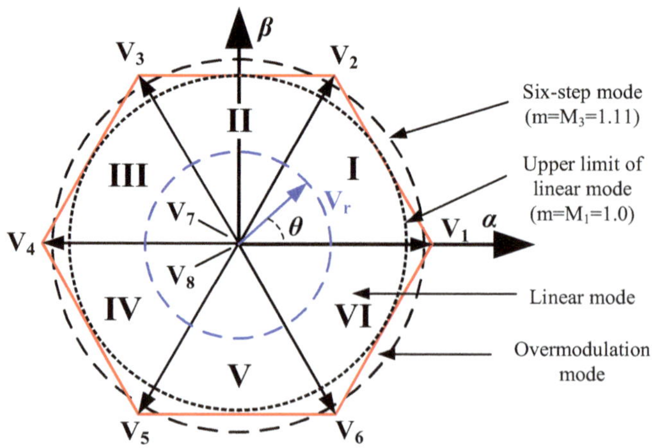

Figure 7.45 Voltage ranges with linear modulation region and overmodulation region for STP system.

Second, to achieve a higher output voltage, the reference voltage vector smoothly deviates from a sinusoidal circular trajectory reaching ultimately a discrete six-step switching sequence. This is the overmodulation region, i.e., the modulation index is between $0.907 \leq m \leq 1$. In general, to maintain continuous control over the modulation index and achieve the lowest possible distortion, the transition from sinusoidal modulation to six-step operation is divided into two different modes, referred to as overmodulation regions I and II [33].

The maximum output voltage in overmodulation regions I and II are obtained as $0.907 \leq m \leq 0.952$ and $0.952 \leq m \leq 1$, respectively. The overmodulation region I is between the circles inscribed by the voltage vector hexagon and the hexagon side. In comparison, the overmodulation region II is between the hexagon side and the circles circumscribed by the voltage vector hexagon.

7.5.1.1 In overmodulation region I

In overmodulation region I, the reference vector V_{ref} is distorted at a specific angle, and the distorted vector is allowed to move along the hexagon side until the requested modulation index is obtained. It should be noted that the on-duration action time for the zero vector is zero, as long as the reference vector remains on the hexagon side. The maximum modulation index in this region is 0.952, when the reference vector V_{ref} remains on the hexagon side. Figure 7.46 presents the phase voltage waveforms of an actual voltage reference vector (bold line) transformed in the time domain and the trajectory of three voltage vectors rotating in a complex plane, respectively.

Since the maximum output is limited to the side of the hexagon, the inverter is unable to generate an output voltage that is as large as the voltage reference vector for a voltage reference vector that extends beyond that side. The switching intervals are then adjusted as follows:

$$t_i' = \frac{t_i}{t_i + t_k} T_s \tag{7.34}$$

$$t_k' = \frac{t_k}{t_i + t_k} T_s \tag{7.35}$$

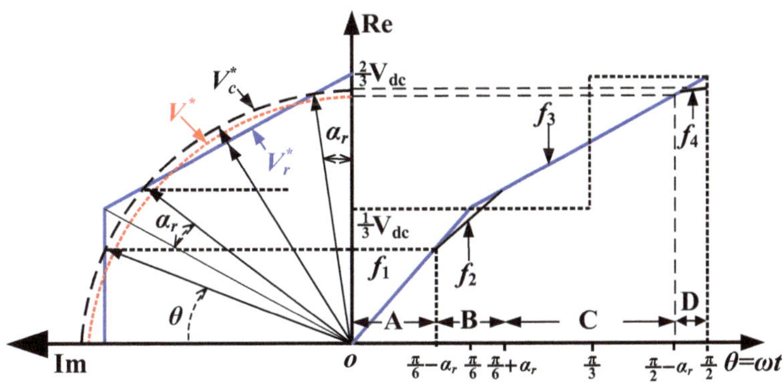

Figure 7.46 Trajectory of reference voltage vector in overmodulation region I.

$$t_0 = 0 \tag{7.36}$$

When the modulation index $m < 0.907$, the output voltages for the SVM are sinusoidal. When $m = 0.907$, the output voltages follow a circle engraved on the hexagon. When the modulation index increases, the voltage waveform of the inverter is distorted, where the magnitude becomes smaller than that of the reference voltage.

7.5.1.2 In overmodulation region II

In overmodulation region I, the angular velocities of the compensated and actual voltage reference vectors are the same and constant for each fundamental period. However, when the modulation index $m > 0.952$, the overmodulation strategy in the region I cannot generate enough voltage.

Therefore, in overmodulation region II, the actual voltage reference vector is held at a vertex for a particular time and then moves along the side of the hexagon for the rest of the switching period. Such a switching method is identified by a hold angle α_h, which controls the duration that the distorted reference remains at the vertices. The trajectory for overmodulation region II is shown in Figure 7.47.

As can be observed, the phase voltage waveform of an actual voltage reference vector (bold line) is transformed in the time domain in the right part, and the trajectory of three voltage vectors is rotating in a complex plane in the left part.

7.5.2 Overmodulation techniques in multiphase PMSMs

In DTP machines, injecting $6k\pm1$ harmonics ($k = 1, 3, 5, \ldots$) in [34] is an effective method to extend the modulation region into the overmodulation region. Since these harmonics do not contribute to the air-gap MMF due to the DTP motor configuration, the currents produced by these harmonic voltages have no impact on torque production and are limited solely by the stator impedance. The voltage waveforms with injected voltage in the overmodulation region of one phase (A) and the current waveforms and spectra are shown in Figure 7.48. It can be seen that the

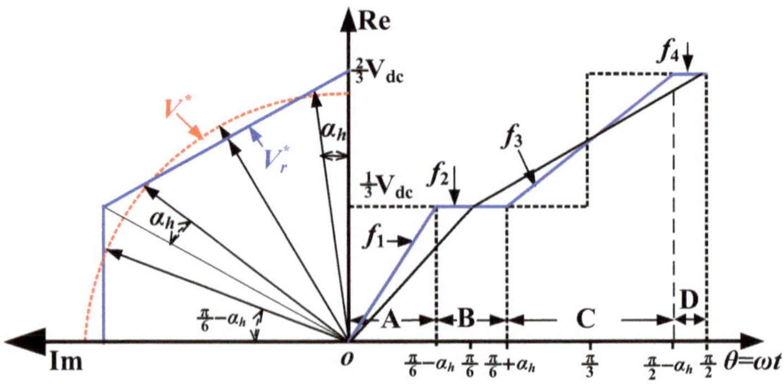

Figure 7.47 Trajectory of reference voltage vector and phase voltage waveform in overmodulation region II.

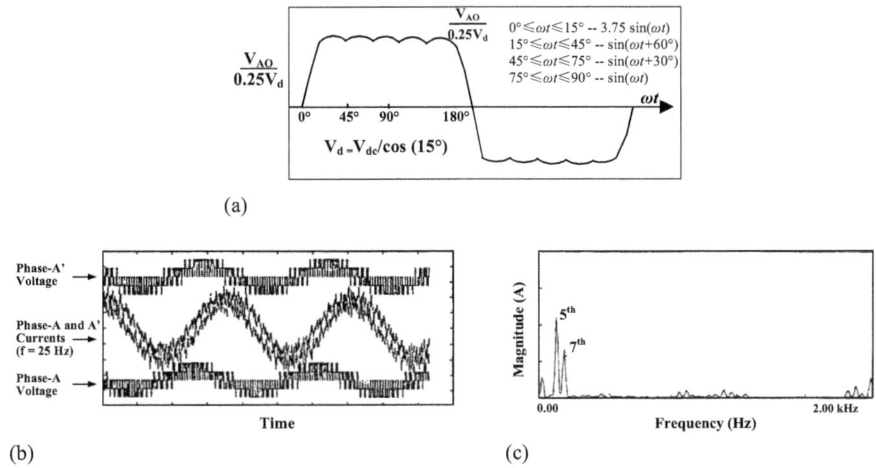

Figure 7.48 Overmodulation strategy for DTP machine. (a) Injected waveforms in phase A. (b) Waveforms of voltage and phase current. (c) Current spectra.

significant $6k\pm1$ harmonics ($k = 1, 3, 5, \ldots$) are introduced since the harmonic impendence for these order harmonics is only comprised of the stator resistance and leakage inductance.

Extending the overmodulation strategy from STP machines to DTP machines has also proven to be effective, as demonstrated in [35]. Similar to STP machines, the overmodulation region for DTP machines can also be divided into two overmodulation regions I and II.

In overmodulation region I, as with the conventional STP machines, the reference vector jumps from a circular trajectory to the hexagon side and moves along the hexagon side until the requested modulation index is achieved, as shown in Figure 7.49(a). In this mode, the dwelling time of the switching state vectors can be calculated by (7.34)–(7.36).

In modulation region II, the modulation index further increases from $m = 0.952$ to $m = 1$. The reference voltage vector V_{ref} is distorted in both magnitude and phase angle. The trajectory of the distorted reference vector jumps from a discrete six-step switching to a continuous hexagon side, as shown in Figure 7.49(b).

The experimental results in [35] are shown in Figure 7.50. As can be seen from Figure 7.50(a), the overmodulation schemes offer a voltage gain increase without the noticeable presence of low-order harmonics in a wide range of modulation. Meanwhile, it can be proved from Figure 7.50(b) that the overmodulation schemes in [35] can extend the linear characteristic of the modulator to the maximum six-step operation.

It is worth noting that the overmodulation methods presented in [34,2] ignore the low-order harmonics in DTP machines, resulting in the presence of large current harmonics. In [36], the low-order harmonics, especially in the z_1z_2 subspace,

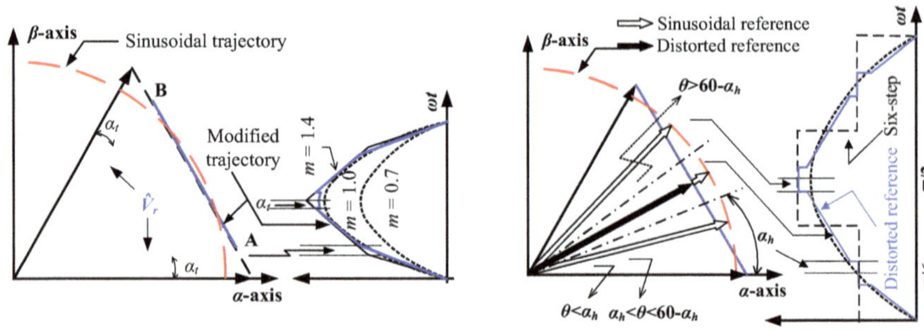

(a) (b)

Figure 7.49 Overmodulation strategy for DTP machine. (a) Overmodulation region I. (b) Overmodulation region II.

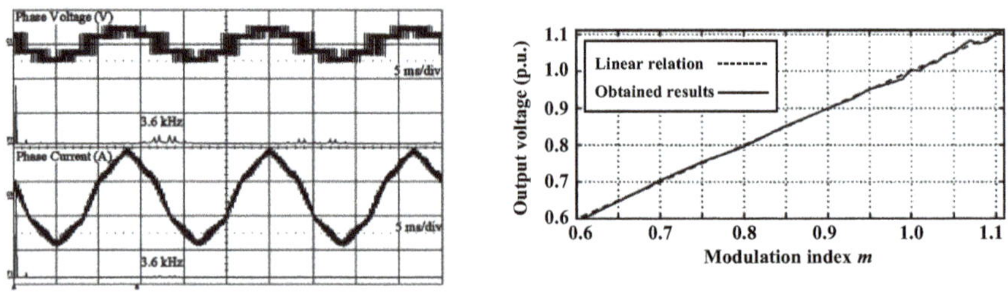

(a) (b)

Figure 7.50 Experimental results of overmodulation strategy. (a) Phase voltage and current waveforms for m = 1.04. (b) Trajectory of output voltage and modulation index.

are considered and minimized. In this case, the overmodulation scheme optimizes the dwell time of voltage vectors to minimize the voltage harmonics, and the voltages in the z_1z_2 subspaces, i.e., u_{z1}^* and u_{z2}^*, satisfy the following condition:

$$\min\left(u_{z1}^{*2} + u_{z2}^{*2}\right) \tag{7.37}$$

The results of the overmodulation scheme in [36] are presented in Figure 7.51. As can be seen, the low-order harmonics of the method in [36] are much lower than those in [35].

In terms of voltage vector selection, the overmodulation strategies in DTP machines can be divided into two types, i.e., with one group of voltage vectors [34] and with two groups of voltage vectors [35,36], as shown in Figure 7.52(a), (b), and (c), respectively. It is easy to understand that the overmodulation strategy with two groups of voltage vectors has the potential to reduce the current harmonics in the z_1z_2 subspace, which is very similar to the VVV method in DTC.

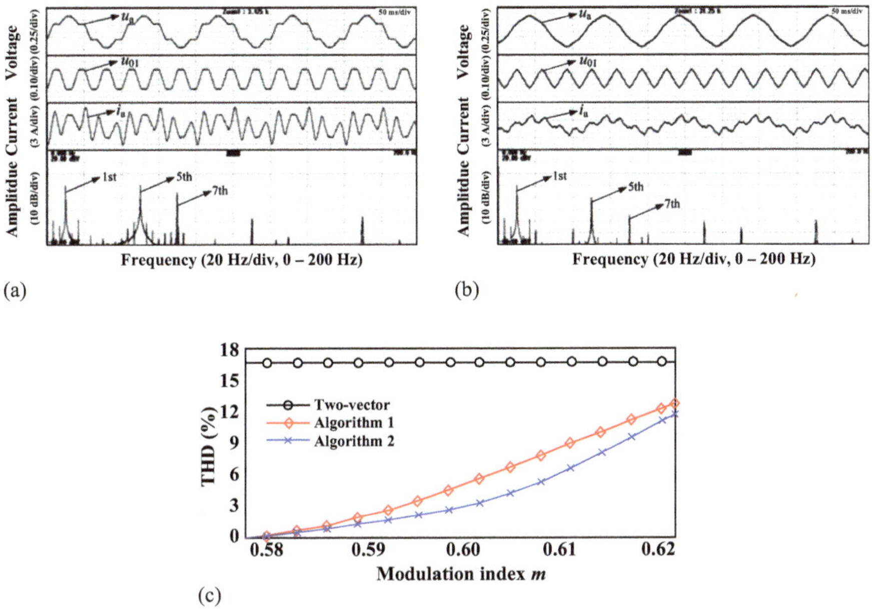

Figure 7.51 Overmodulation strategy for DTP machines. (a) Experimental results for classical overmodulation strategy. (b) Experimental results for overmodulation strategy. (c) THD of two strategies with different modulation indices.

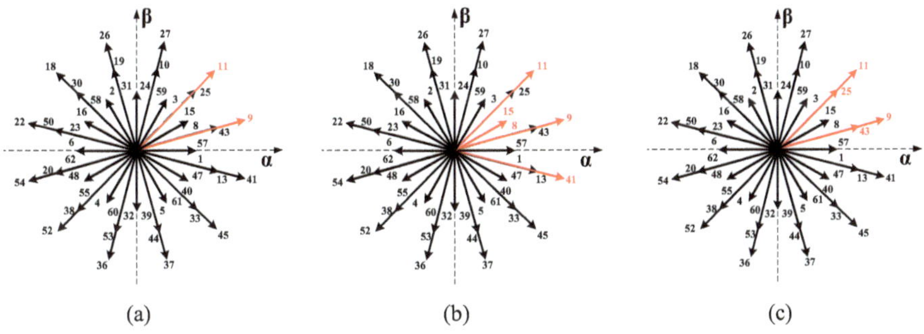

Figure 7.52 Voltage vector selection for different overmodulation strategies. (a) one group (L), (b) two groups (L+M), (c) two groups (L+ML).

For other topologies of n-phase multiphase machines, such as 5-phase and 7-phase machines, the classical overmodulation strategies are very similar to the STP machines and DTP machines. For n-phase machines, each switching state generates a voltage vector that is mapped into $(n - 1)/2$ different orthogonal planes according to Clarke's transformation. This transformation maps harmonic components into different subspaces according to the following rules: $2nk \pm 1$ harmonics

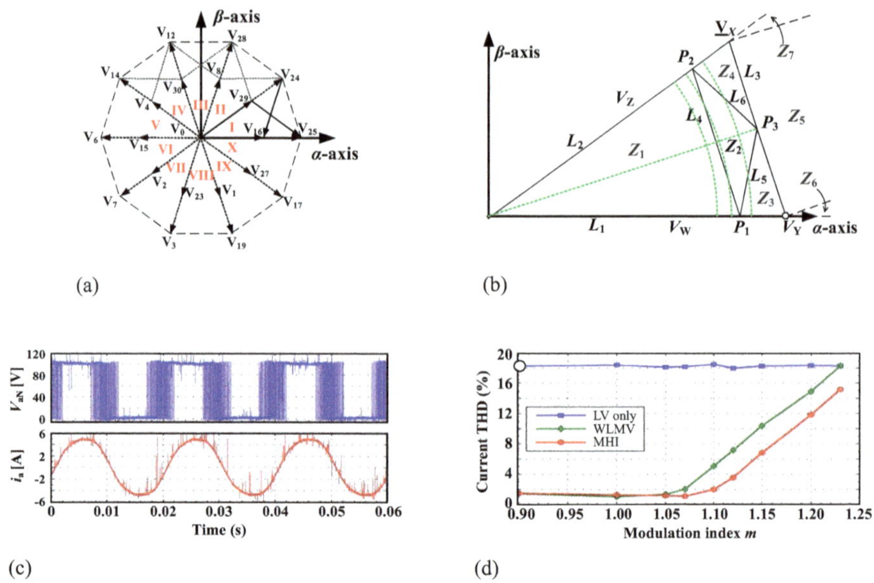

Figure 7.53 Overmodulation strategy for 5-phase machines. (a) Voltage vector selection in fundamental related subspace. (b) Voltage vector selection in harmonic-related subspace. (c) Voltage waveforms and current waveforms. (d) Current THD with different modulation indices.

including the fundamental frequency into the $\alpha\beta$ subspace, $2nk \pm (2j - 1)$ harmonics into xy subspace, and the nk harmonics into the zero-sequence subspace, with $1 \leq j \leq (n - 1)/2$ and $k = 0, 1, 2, 3\ldots$

The research of SVM algorithms to the n-phase PMSMs focuses on the $\alpha\beta$ plane, as torque is typically associated with the current components within this subspace. However, it is crucial to consider the harmonics in the xy or z_1z_2 subspace to effectively reduce the current harmonics, similar to the modulation strategies in FOC and DTC.

For instance, two groups of voltage vectors are selected in both the linear and overmodulation regions in [37], as illustrated in Figure 7.53. It can be seen that the THD for the phase current decreases significantly using the concept of the VVV method, compared to the modulation with only the largest voltage vectors in 5-phase machines.

7.6 Summary

In this chapter, modulation strategies for multiphase machines with different control algorithms, including FOC, SVM-DTC, and ST-DTC, have been presented and analyzed. For n-phase multiphase machines, the voltage vectors can be decoupled into $(n - 1)/2$ different orthogonal subspaces according to the general Clarke's transformation. Controlling the components in the harmonic-related subspaces in modulation is effective in reducing the current harmonics for the multiphase machines. In DTP machines, the components in the z_1z_2 subspace can be controlled to reduce the 5th and 7th harmonics. Similarly, in 5-phase machines, eliminating components in the d_2q_2 subspace can reduce the 3rd harmonics.

In both FOC and SVM-DTC, the primary principle is to calculate the dwelling times for the selected voltage vectors in SVM to control the harmonic components to zero. Different voltage vector selection schemes are closely related to harmonic reduction, especially the low-order harmonics and switching frequency harmonics. Detailed discussions on modulation strategies in FOC and SVM-DTC for DTP machines highlight the significant contribution of voltage vector selection to current harmonics, both low-order and high-order.

In ST-DTC, the applications of the virtual voltage vector (VVV) method in multiphase machines have also been introduced. By using VVV methods, the low-order current harmonics can be eliminated effectively, as demonstrated by extensive experimental results. It can be concluded that the steady-state control performance can be improved when more voltage vectors are considered and utilized, while the computation burden and the implementation complexity will be increased accordingly. Therefore, a trade-off between implementation and performance is usually required.

In addition, the control schemes for multiphase machines will be presented and analyzed in the next chapter.

References

[1] K. Hasse, "Zum dynamischen Verhalten der Asynchronmaschine bei Betrieb mit variabler Standerfrequenz und Standerspannung," *ETZ-A*, vol. 89, pp. 77, 1968.

[2] K. Hasse, "Feldorientierte Regelung im Ruckblick," *IEEE IAS/PELS/IES German Chapter Meeting*, Braunschweig (vol. 22), 2001.

[3] F. Blaschke, "The principle of field orientation applied to the new transvector closed-loop control system for rotating field machines," *Siemens-Rev.*, vol. 39, pp. 217–220, 1972.

[4] R. Gabriel, W. Leonhard, and C. J. Nordby, "Field-oriented control of a standard AC motor using microprocessors," *IEEE Trans. Ind. Appl.*, vol. IA-16, no. 2, pp. 186–192, 1980.

[5] W. Leonhard, "Adjustable-speech AC drives," *Proc. IEEE*, vol. 76, no. 4, pp. 455–471, 1988.

[6] T. Noguchi and I. Takahashi, "Quick torque response control of an induction motor based on a new concept," (in Japanese), *IEEJ Tech. Meet. Rotating Mach.*, vol. 84, no. 76, pp. 61–70, 1984.

[7] I. Takahashi and T. Noguchi, "A new quick-response and high efficiency control strategy of an induction machine," *IEEE Trans. Ind. Appl.*, vol. 22, pp. 820–827, 1986.

[8] I. Takahashi and Y. Ohmori, "High-performance direct torque control of an induction motor," *IEEE Trans. Ind. Appl.*, vol. 25, pp. 257–264, 1989.
[9] M. Depenbrock, "Direkte selbstregelung (DSR) für hochdynamische drehfeldantriebe mit stromrichterspeisung," (in German), *ETZ Archive*, no. 7, pp. 211–218, 1985.
[10] M. Depenbrock, "Direct self-control (DSC) of inverter-fed induction machine," *IEEE Trans. Power Electron.*, vol. 3, pp. 420–429, 1988.
[11] M. Depenbrock and A. Steimel, "High power traction drives and convertors," *in Proc. Elect. Drives Symp.'90*, Capri, 1990, pp. 1–9.
[12] Casadei. D, Serra.G, and Tani. K, "Implementation of a direct control algorithm for induction motors based on discrete space vector modulation," *IEEE Trans. Power Electron.*, vol. 15, no. 4, pp. 769–777, 2000.
[13] Y. Zhao and T. A. Lipo, "Space vector PWM control of dual three-phase induction machine using vector space decomposition," *IEEE Trans. Ind. Appl.*, vol. 31, no. 5, pp. 1100–1109, 1995.
[14] D. Hadiouche, L. Baghli, and A. Rezzoug, "Space-vector PWM techniques for dual three-phase ac machine: Analysis, performance evaluation, DSP implementation," *IEEE Trans. Ind. Appl.*, vol. 42, no. 4, pp. 1112–1122, 2006.
[15] K. Marouani, L. Baghli, D. Hadiouche, A. Kheloui, and A. Rezzoug, "A new PWM strategy based on a 24-sector vector space decomposition for a six-phase VSI-fed dual stator induction motor," *IEEE Trans. Ind. Electron.*, vol. 55, no. 5, pp. 1910–1920, 2008.
[16] S. M. Suhel and R. Maurya, "Realization of 24-sector SVPWM with new switching pattern for six-phase induction motor drive," *IEEE Trans. Power Electron.*, vol. 34, no. 6, pp. 5079–5092, 2019.
[17] B. Shao, "Direct torque control of dual three phase permanent magnet synchronous machine drives with particular reference to reduction of current harmonics and torque ripples," *Ph.D. dissertation*, Univ. Sheffield, Sheffield, UK., 2022.
[18] M. S. B. Zulkifli, W. N. B. W. A. Munim, and H. C. M. Haris, "Five phase space vector modulation voltage source inverter using large vector only," in *2012 Int. Symp. Computer Appl. Ind. Electron.* (ISCAIE), Dec 2012, pp. 5–9.
[19] K. A. Chinmaya and G. K. Singh, "Modeling and comparison of space vector PWM Schemes for a five- phase induction motor drive," *IECON 2018 – 44th Annu. Conf. IEEE Ind. Electron. Soc.*, 2018, pp. 559–564.
[20] D. Dujic, M. Jones, E. Levi, J. Prieto, and F. Barrero, "Switching ripple characteristics of space vector PWM schemes for five-phase two-level voltage source inverters—Part 1: Flux harmonic distortion factors," *IEEE Trans. Ind. Electron.*, vol. 58, no. 7, pp. 2789–2798, 2011.
[21] R. Bojoi, F. Farina, G. Griva, F. Profumo, and A. Tenconi, "Direct torque control for dual three-phase induction motor drives," *IEEE Trans. Ind. Appl.*, vol. 41, no. 6, pp. 1627–1636, 2005.
[22] Y. Ren and Z. Zhu, "Enhancement of steady-state performance in direct torque controlled dual-three phase permanent magnet synchronous machine

drives with modified switching table," *IEEE Trans. Ind. Electron.*, vol. 62, no. 6, pp. 3338–3350, 2015.

[23] B. Shao, Z. Q. Zhu, J. Feng, *et al.*, "Improved direct torque control method for dual-three-phase permanent-magnet synchronous machines with back-EMF harmonics," *IEEE Trans. Ind. Electron.*, vol. 68, no. 10, pp. 9319–9333, 2021.

[24] Y. Ren and Z. Q. Zhu, "Reduction of both harmonic current and torque ripple for dual three-phase permanent-magnet synchronous machine using modified switching-table-based direct torque control," *IEEE Trans. Ind. Electron.*, vol. 62, no. 11, pp. 6671–6683, 2015.

[25] B. Shao, Z. Q. Zhu, L. C. Yan, *et al.*, "Torque ripple reduction for direct torque control of dual three phase PMSM based on multiple virtual voltage vectors," *IEEE Trans. Energy Convers.*, vol. 38, no. 1, pp. 296–309, 2023.

[26] J. K. Pandit, M. V. Aware, R. V. Nemade, and E. Levi, "Direct torque control scheme for a six-phase induction motor with reduced torque ripple," *IEEE Trans. Power Electron.*, vol. 32, no. 9, pp. 7118–7129, 2017.

[27] L. Parsa and H. A. Toliyat, "Sensorless direct torque control of five-phase interior permanent-magnet motor drives," *IEEE Trans. Ind. Appl.*, vol. 43, no. 4, pp. 952–959, 2007.

[28] G. Liliang, J. E. Fletcher, and Z. Libo, "Low-speed control improvements for a two-level five-phase inverter-fed induction machine using classic direct torque control," *IEEE Trans. Ind. Electron.*, vol. 58, no. 7, pp. 2744–2754, 2011.

[29] L. Zheng, J. Fletcher, B. Williams, and X. He, "A novel direct torque control scheme for a sensorless five-phase induction motor drive," *IEEE Trans. Ind. Electron.*, vol. 58, no. 2, pp. 503–513, 2011.

[30] S. Payami and R. K. Behera, "An improved DTC technique for low-speed operation of a five-phase induction motor," *IEEE Trans. Ind. Electron.*, vol. 64, no. 5, pp. 3513–3523, 2017.

[31] G. Yang, J. Yang, S. Li, Y. Wang, H. Hussian, R. Deng, and L. Yan, "A sequential direct torque control scheme for seven-phase induction machines based on virtual voltage vectors," *IEEE Trans. Ind. Appl.*, vol. 57, no. 4, pp. 3722–3734, 2021.

[32] P. Garcia-Entrambasaguas, I. Zoric, I. González-Prieto, M. J. Duran, and E. Levi, "Direct torque and predictive control strategies in nine-phase electric drives using virtual voltage vectors," *IEEE Trans. Power Electron.*, vol. 34, no. 12, pp. 12106–12119, 2019.

[33] D. Lee and G. Lee, "A novel overmodulation technique for space-vector PWM inverters," *IEEE Trans. Power Electron.*, vol. 13, no. 6, pp. 1144–1151, 1998.

[34] K. Gopakumar, V. T. Ranganathan, and S. R. Bhat, "An efficient PWM technique for split-phase induction motor operation using dual voltage source inverters," *in Proc. Conf. Rec. IEEE IAS Annu. Meeting*, 1993, pp. 582–587.

[35] D. Yazdani, S. Khajehoddin, A. Bakhshai, and G. Joos, "Full utilization of the inverter in split-phase drives by means of a dual three-phase space vector

classification algorithm," *IEEE Trans. Ind. Electron.*, vol. 56, no. 1, pp. 120–129, 2009.

[36] C. Zhou, G. Yang, and J. Su, "PWM strategy with minimum harmonic distortion for dual three-phase permanent-magnet synchronous motor drives operating in the overmodulation region," *IEEE Trans. Power Electron.*, vol. 31, no. 2, pp. 1367–1380, 2016.

[37] G. Carrasco and C. A. Silva, "Space vector PWM method for five-phase two-level VSI with minimum harmonic injection in the overmodulation region," *IEEE Trans. Power Electron.*, vol. 60, no. 5, pp. 2042–2053, 2013.

Chapter 8

Control strategies for multiphase PMSMs

8.1 Introduction

The utilization of multiphase PMSMs has emerged across a wide spectrum of modern applications, and the employed control strategy plays a critical role in ensuring their reliable and efficient operation. In this chapter, the most popular control strategies for dual- and multiphase PMSMs, including field-oriented control (FOC), direct torque control (DTC), and model predictive control (MPC), are described, together with their effectiveness in minimizing current harmonics, minimizing torque ripple, maximizing efficiency, or achieving fast response. Additionally, the influences of different control strategies on machine performance are analyzed and evaluated, including implementation complexity, robust capability, and reliability. Overall, the structure of this chapter is illustrated in Figure 8.1.

8.2 Field-oriented control for multiphase PMSMs

FOC (also called vector control) was first proposed by Hasse and Blaschke in the early 1970s. Due to its simplicity, excellent steady-state performance, and accurate position control, FOC has become the most common control method in machine

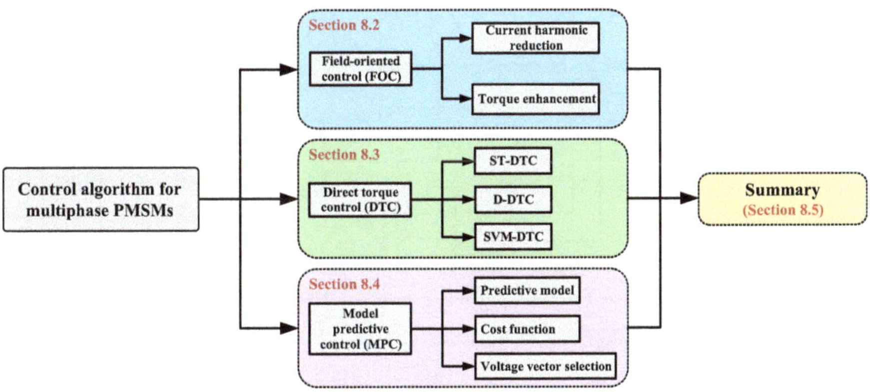

Figure 8.1 Structure of Chapter 8.

drives. FOC aims to control the AC synchronous and induction machines as a DC motor, by decomposing the stator currents into two orthogonal components, i.e., flux and torque-producing components, as detailed in Chapter 7. Specific to multiphase machines, the FOC algorithms can be developed according to different purposes, e.g., current harmonics reduction and torque enhancement, as will be discussed in this section. Since the DTP machine is one of the most common topologies of multiphase machines, its FOC will be discussed first.

8.2.1 Field-oriented control for DTP PMSMs

The basic structure of FOC for DTP machines is presented in Figure 8.2, which is comprised of inner current loops and an outer speed loop.

The differences in the FOC for STP to DTP PMSMs lay in two aspects, i.e., the coordinate transformation and the additional current controller in the harmonic-related subspace or another *dq* subspace. In this section, the main control goals of FOC for DTP can be summarized as follows.

(1) Reduce low-order current harmonics, which is an inherent drawback in DTP machine drives.
(2) Enhance the torque production by harmonics injection.

8.2.1.1 Various controllers for current harmonics reduction

As introduced in Chapter 5, the currents in DTP PMSMs contain $k = 6m \pm 1$ (m =1,2,3,...) order harmonics, i.e., the 5th, 7th, 17th, and 19th, which are mapped into the z_1z_2 subspace. The impedance of the z_1z_2 subspace is only comprised of leakage inductance and stator resistance, which is much smaller than the impedance of the $\alpha\beta$ subspace. Hence, all of the above current harmonics are significant in

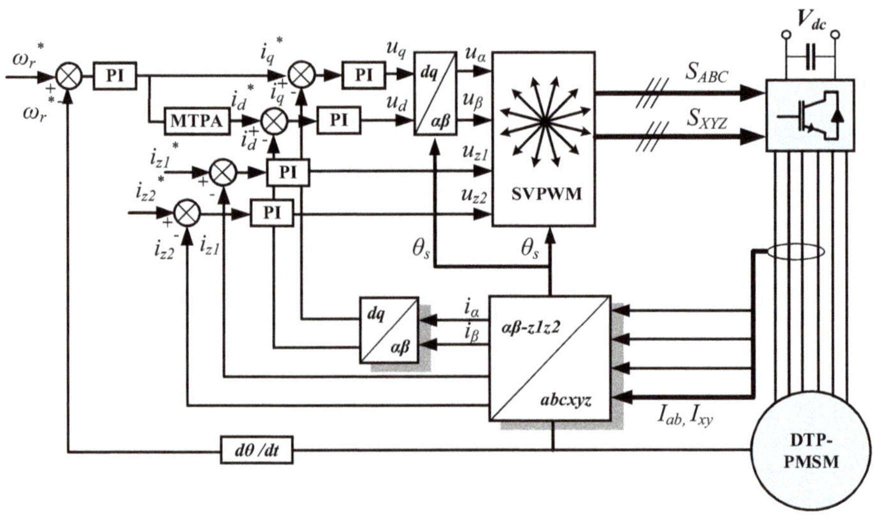

Figure 8.2 FOC diagram for DTP machines.

DTP machines, which are caused by back-EMF (electromotive force) distortions, the voltage source inverter (VSI) non-linearity, and the pulse-width-modulation (PWM) strategies.

Generally, the current harmonics can be reduced by using either passive or active methods. First, the passive methods aim to increase the impedance of the harmonic current path, e.g., with a specific machine design or external filters. However, the passive method tends to increase the cost and decrease the system's efficiency and is rarely employed. On the other hand, in active methods, the current harmonics can be reduced by controlling currents, and thus, no external modifications to the system are required. The active approach is often referred to as the current harmonic control.

In a DTP machine drive system there are several controller options, and each of these is described below.

Proportional integral (PI) controller

The proportional integral (PI) controller is widely employed due to its simplification. However, the PI controller is incapable of eliminating AC errors due to its limited bandwidth, particularly the 5th and 7th current harmonics in DTP machine drive systems. Thus, the elimination of the remaining AC errors needs either additional coordinate transformation (change AC errors to DC errors) or a specific controller.

Proportional resonant (PR) controller

The proportional resonant (PR) controller could track the sinusoidal signal reference with zero errors if the reference signal has a fixed frequency, which can eliminate AC errors in steady-state operation. The ideal resonant (R) controller can be expressed as

$$H_{\mathrm{R}}(s) = \frac{K_r s}{s^2 + \omega^2} \tag{8.1}$$

where K_r is the resonant coefficient, ω is the resonant frequency. The Bode plot of an ideal resonant controller is shown in Figure 8.3, where K_r is set to 10 and ω is set to 100 rad/s.

In the DTP machine drive system, the transfer function of the PR controller is expressed as:

$$H_{\mathrm{PR}}(s) = K_p + \frac{K_r s}{s^2 + \omega^2} \tag{8.2}$$

where K_p is the proportional gain coefficient. The Bode plot of the PR controller is shown in Figure 8.4. As can be seen, there is a peak in the magnitude of its frequency response at the resonant frequency of the system, which can be used to tune the controller to the resonant frequency and achieve optimal performance.

In the digital system, the resonate controller can be discretized with forward Euler or backward Euler functions and can be expressed as:

$$H_{\mathrm{PR}}(z) = K_p T_s + K_i T_s \frac{z^{-1} - z^{-2}}{1 + z^{-1}\left(\omega^2 T_s^2 - 2\right) + z^{-2}} \tag{8.3}$$

where T_s is the sampling time.

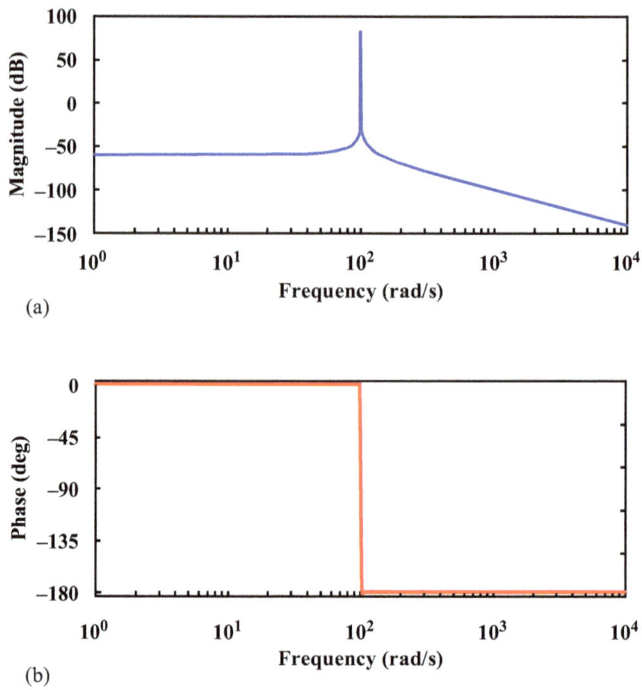

Figure 8.3 Bode diagram for an ideal resonant controller. (a) Gain margin. (b) Phase margin.

The control structure for the PR controller is shown in Figure 8.5, which is applied in [1] to suppress the 2nd (caused by asymmetry) and 6th (the 5th and 7th current harmonics after coordinate transformation) current harmonics for DTP machines. The complete control structure is shown in Figure 8.6, and an effective current harmonic reduction performance can be observed from the experimental results in Figure 8.7.

Quasi-proportional resonant (QPR) controller

Since the ideal resonant controller is not conducive in practice due to its narrow bandwidth and the limitations in the precisions of the digital system [2], a quasi-resonant controller can be adopted in the DTP control system, which is expressed as:

$$H_{\mathrm{QR}}(s) = \frac{K_r \cdot 2\omega_c s}{s^2 + 2\omega_c s + {\omega_0}^2} \tag{8.4}$$

where K_r is the resonant coefficient, ω_0 and ω_c are the resonant frequency and the cut-off frequency of the quasi-resonant controller, respectively. The Bode plot of the quasi-resonant controller is shown in Figure 8.8.

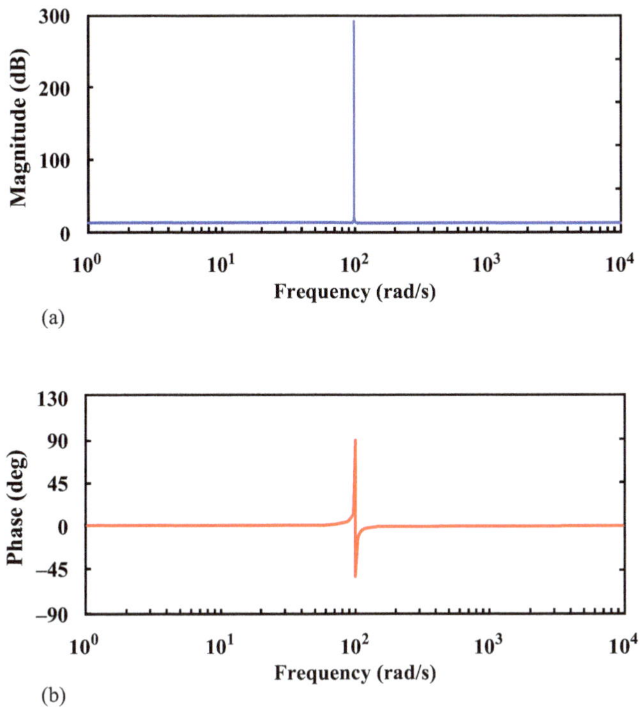

Figure 8.4 Bode diagram for a PR controller. (a) Gain margin. (b) Phase margin.

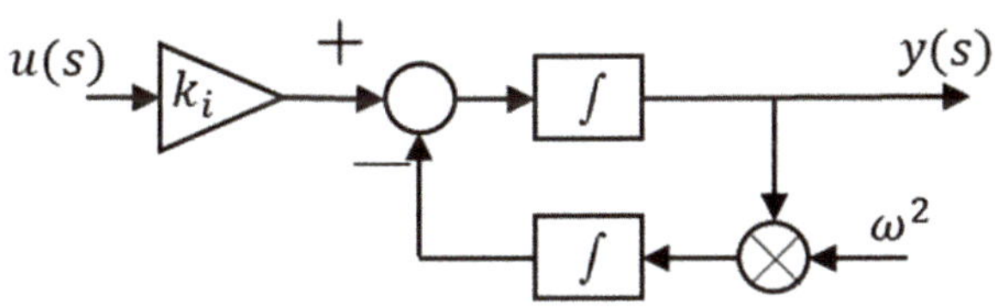

Figure 8.5 Control structure of a PR controller.

In a DTP machine drive system, the transfer function of the QPR controller can be written as:

$$H_{\mathrm{QPR}}(s) = K_p + \frac{K_r \cdot 2\omega_c s}{s^2 + 2\omega_c s + {\omega_0}^2} \tag{8.5}$$

The Bode plot of the QPR controller is shown in Figure 8.9. As can be seen, the response bandwidth of the QPR controller is obviously higher than that of the PR controller. However, it has one more parameter required to tune, i.e., the cut-off frequency ω_c, which is more complicated than a PR controller.

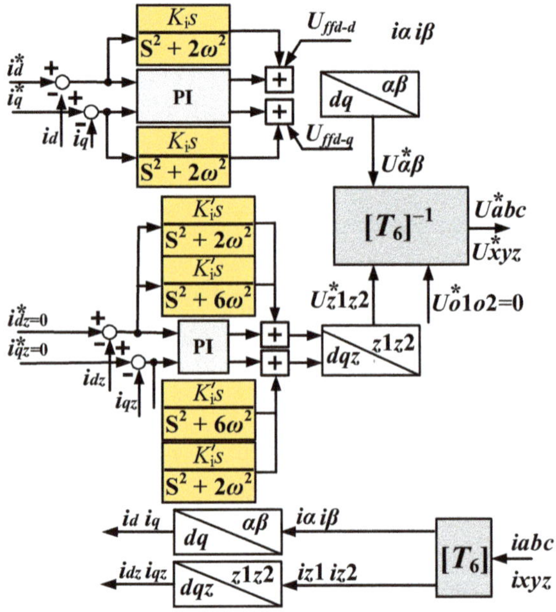

Figure 8.6 Control structure with a PR controller to reduce current harmonics.

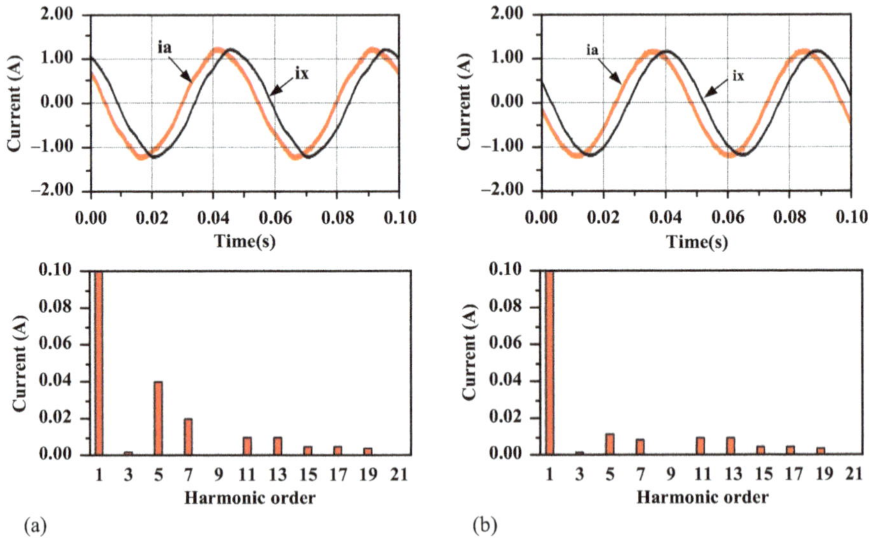

Figure 8.7 Measured currents and spectra. (a) Phase currents and spectra without PR control. (b) Phase currents and spectra with PR control.

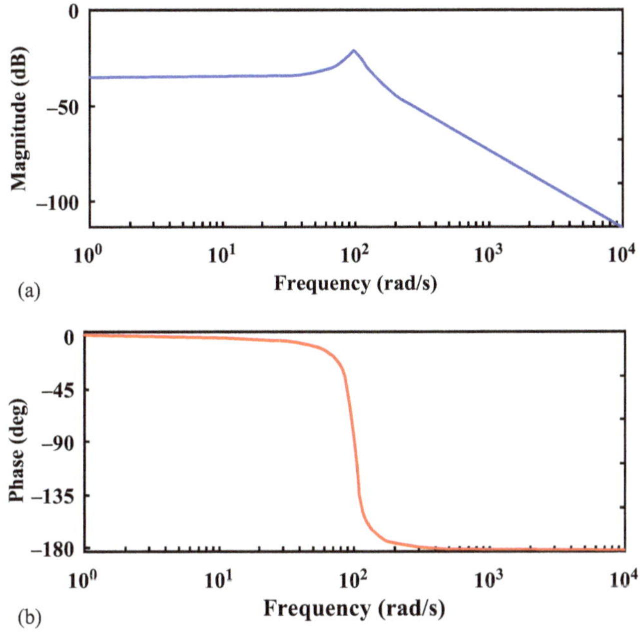

Figure 8.8 Bode diagram for the quasi-resonant controller. (a) Gain margin. (b) Phase margin.

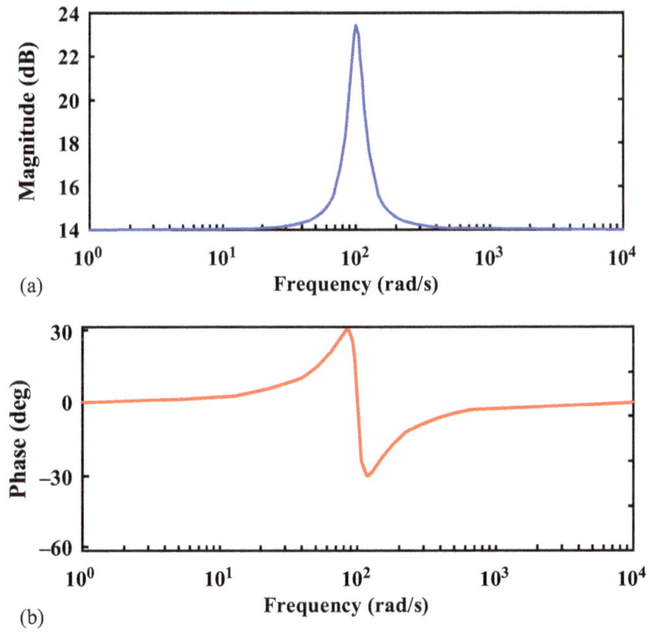

Figure 8.9 Bode diagram for the QPR controller. (a) Gain margin. (b) Phase margin.

Similarly, to be implemented into a dSpace/DSP system, the QPR controller needs to be discretized, expressed as:

$$H_{\mathrm{QPR}}(z) = K_pT_s + K_iT_s\frac{2\omega_c(z^{-1} - z^{-2})}{1 + z^{-1}\left(\omega_0^2T_s^2 - 2 + 2\omega_cT_s\right) + (1 - 2\omega_cT_s)z^{-2}} \tag{8.6}$$

The control structure is shown in Figure 8.10. For example, the QPR controller is employed in [2] to reduce the current harmonics in the z_1z_2 subspace for DTP machines based on the VSD modeling method. It is worth noting that due to the high bandwidth characteristics of the QPR controller, the sufficiently large gain in a wide frequency range easily achieves error-free steady-state operation. The experimental results are presented in Figure 8.11. The results show that the QPR

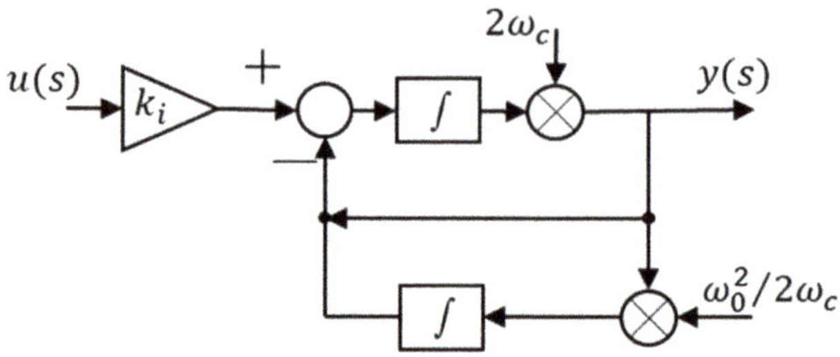

Figure 8.10 Block diagrams of a QPR controller.

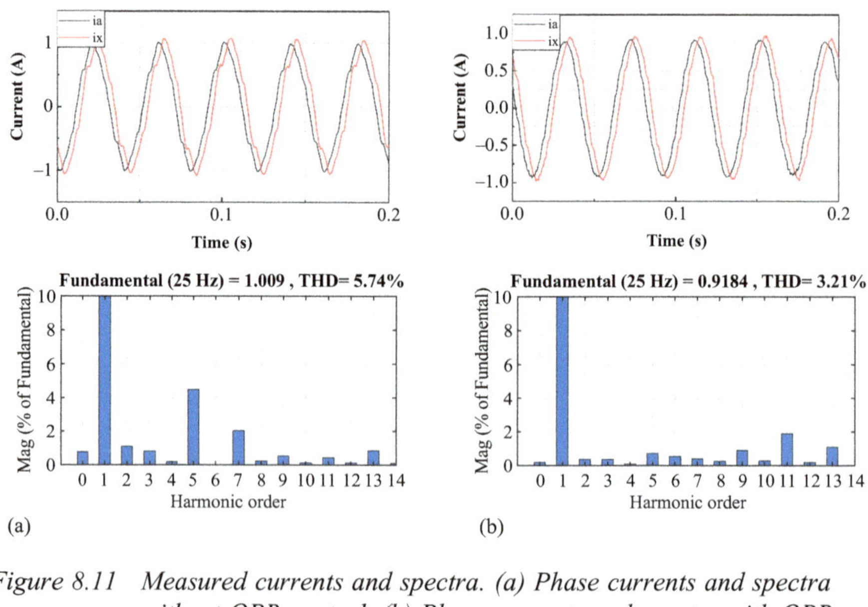

Figure 8.11 Measured currents and spectra. (a) Phase currents and spectra without QPR control. (b) Phase currents and spectra with QPR control.

controller can suppress the 5th and 7th harmonics effectively, compared to the method without harmonic component control.

Synchronous reference frame PI (SRF-PI) control

As mentioned above, the PI controller has the inherent drawbacks of a limited control capability to the AC components, and hence, it is insufficient to regulate sinusoidal signals such as the 5th and 7th current harmonics in the z_1z_2 subspace. Consequently, the 5th and 7th current harmonics are required to be transferred to DC components by coordinate transformation, if a PI controller is applied to achieve good static control performance with no static error. Since this method is to control each frequency component (mainly the 5th and 7th) in its own synchronous reference frame, it is called the "Synchronous reference frame PI (SRF-PI) control".

Figure 8.12 shows the overall control diagram based on an SRF-PI control. The scheme provides control for the fundamental component and the 5th and 7th harmonics. The current control is performed using the VSD technique. Also, the conventional PI controller is utilized as the current controller. The synchronously rotating transformation matrix $T_{5\text{-}dqz}$ and $T_{7\text{-}dqz}$ are expressed as:

$$T_{5-dqz} = \begin{bmatrix} \cos 5\theta & \sin 5\theta \\ -\sin 5\theta & \cos 5\theta \end{bmatrix} \tag{8.7}$$

$$T_{7-dqz} = \begin{bmatrix} \cos 7\theta & -\sin 7\theta \\ -\sin 7\theta & -\cos 7\theta \end{bmatrix} \tag{8.8}$$

where θ is the flux position of the permanent magnet in the electrical angle.

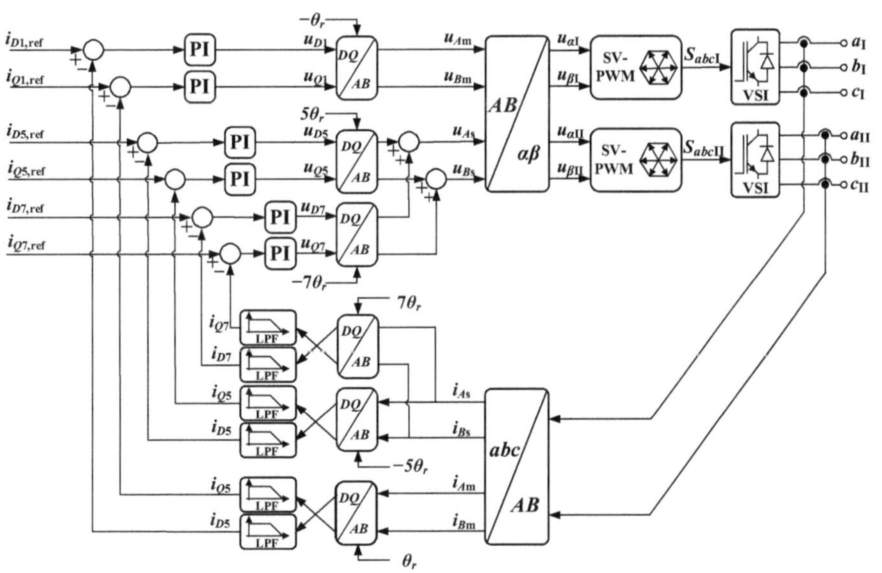

Figure 8.12 Overall control diagram based on SRF-PI control for DTP machines.

After the transformation, the 5th and 7th current harmonics in the z_1z_2 subspace can be expressed as:

$$\begin{bmatrix} i_{5-dz1} \\ i_{5-dz2} \end{bmatrix} = \begin{bmatrix} I_5 + I_7\cos 12\theta \\ I_5 - I_7\sin 12\theta \end{bmatrix} \tag{8.9}$$

$$\begin{bmatrix} i_{7-dz1} \\ i_{7-dz2} \end{bmatrix} = \begin{bmatrix} I_5\cos 12\theta + I_7 \\ -I_5\sin 12\theta + I_7 \end{bmatrix} \tag{8.10}$$

where I_5 and I_7 are the amplitudes of the 5th and 7th stator currents.

It can be seen from (8.9) and (8.10) that the 5th current harmonic is the DC component in the 5th order synchronous reference frame, while the 7th current harmonic is decoupled with the 5th current harmonic, and is still the AC component in the 5th order synchronous reference frame. Similarly, the 7th current harmonic is the DC component in the 7th order synchronous reference frame, while the 5th current harmonic is decoupled with the 7th current harmonic and becomes the AC component in the 7th order synchronous reference frame.

To extract the DC component easily, a low-pass filter (LPF) is required here. Both the first-order $H_{1\text{st-LPF}}$ and the second-order $H_{2\text{nd-LPF}}$ LPFs are widely used, of which the transfer functions are expressed as:

$$H_{1\text{st}-\text{LPF}}(s) = \frac{\omega_c}{s + \omega_c} \tag{8.11}$$

$$H_{2\text{nd}-\text{LPF}}(s) = \frac{{\omega_n}^2}{s^2 + 2\zeta\omega_n s + {\omega_n}^2} \tag{8.12}$$

where $\omega_c = 2\pi f_c$, f_c is the cut-off frequency of the first-order LPF; ω_n is the undamped natural frequency of the second-order LPF, and ζ is the damping ratio of the second-order LPF.

The Bode diagrams for the first-order and second-order LPFs are presented in Figure 8.13.

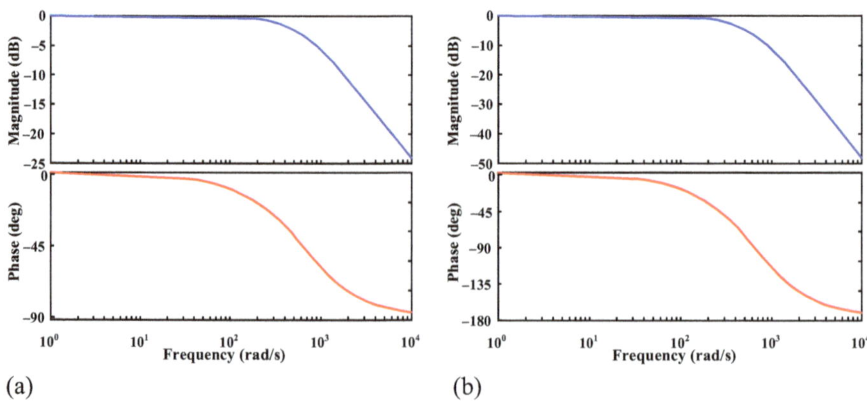

Figure 8.13 Bode diagrams of first-order and second-order LPFs. (a) First-order LPF. (b) Second-order LPF.

As can be seen, the second-order LPF can attenuate high-frequency signals more effectively, resulting in a sharper cut-off frequency than the first-order LPF. Additionally, the second-order filter can provide a smoother transition from passband to stopband compared to a first-order filter.

Overall, the key difference between the first-order and second-order low-pass filters is their roll-off characteristics and simplicity. The first-order LPF is easy to implement and simple to design but has a limited roll-off slope, which means that it attenuates high-frequency signals at a slower rate than a higher-order filter. Hence, in order to filter the high-frequency signals, the cut-off frequency of the first-order LPF is often set very low to achieve high-frequency signal elimination. However, it will cause a significant phase/time delay in the real system. In comparison, the second-order LPF is more suitable in SRF-PI control for DTP machines, since it has a steeper slope and exhibits more effective attenuation of high-frequency signals.

For instance, the experimental results without/with SRF-PI control for harmonic compensation are shown in Figure 8.14 [3]. As can be seen, SRF-PI control exhibits excellent performance in the reduction of current harmonics in the z_1z_2 subspace, while its dynamic response is slow due to employing LPFs, as described in [4].

Besides, the schemes of using a dual PI controller in the synchronous frame [5] are introduced in the DTP machine to suppress the 5th and 7th harmonic components (AC components), where the control structures are illustrated in Figure 8.15. This method is used to suppress positive and negative components in the z_1z_2 subspace, which are caused by the imbalance between the two sets of 3-phase windings, or the imbalance in one of the 3-phase windings.

Virtual multiple reference synchronous frame (MRSF) method

As described above, LPFs are used to reduce the AC components and the extracted DC components serve as the feedback signals of the PI regulators. Hence, LPFs are necessary for the synchronous PI controller method. However, the LPFs can cause a significant delay in the current detection, and reduce the dynamic performance of the current regulation, as well as the stability of the control loops.

The ideal method is to decompose the fundamental and major harmonic components into a different orthogonal subspace so that the harmonics can be independently regulated without using an LPF. In the DTP machine system, as shown in Figure 8.16(a), the sum of the 5th and 7th current harmonics is decomposed in z_1z_2 subspace, while the fundamental, 11th and 13th current harmonics are coupled in $\alpha\beta$ subspace. Obviously, the major current harmonics are not completely separated from each other, and hence the ideal decomposition cannot be realized in the DTP case. In a triple 3-phase (TTP) system with 20° electrical degrees shifting between the adjacent sets, see Figure 8.16(b), the VSD technique merely decomposes the fundamental current into $\alpha\beta$ subspace, both the 5th and 13th into z_1z_2 subspace, and both the 7th and 11th into z_3z_4 subspace, respectively. However, the 5th and 13th current harmonics are still mixed with each other in z_1z_2 subspace, while the 7th and 11th are mixed in z_3z_4 subspace.

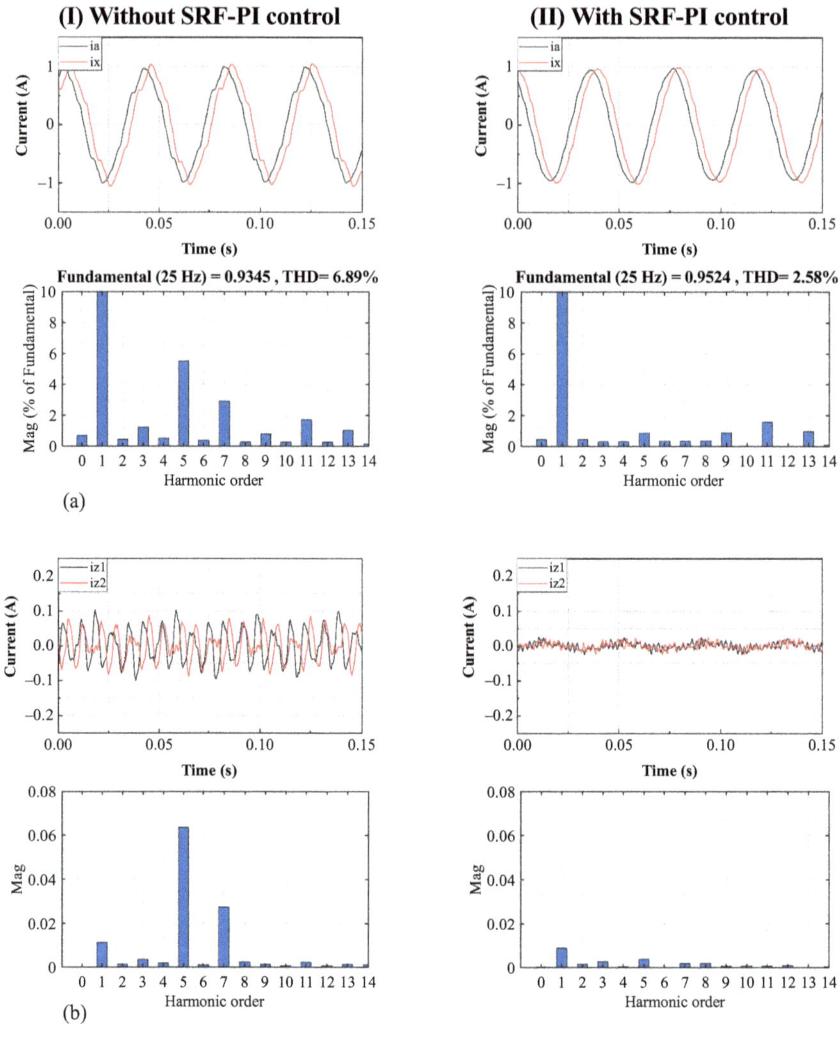

Figure 8.14 Experimental results without and with SRF-PI control for harmonic compensation. (a) Phase currents. (b) Currents and spectra in z_1z_2 subspace.

As the number of three-phase sets increases, the QTP machine, which shifts 15° electrical degrees between the adjacent sets, can improve the decomposition capability due to the complete separation of the fundamental, 5th and 7th current harmonics, as shown in Figure 8.16(c). As a result, it is possible to individually regulate the fundamental, 5th and 7th current harmonics in three independent current loops without using LPFs, but unfortunately impossible for the 11th and 13th current harmonics. Finally, the pentuple 3-phase (PTP) system with 12° electrical

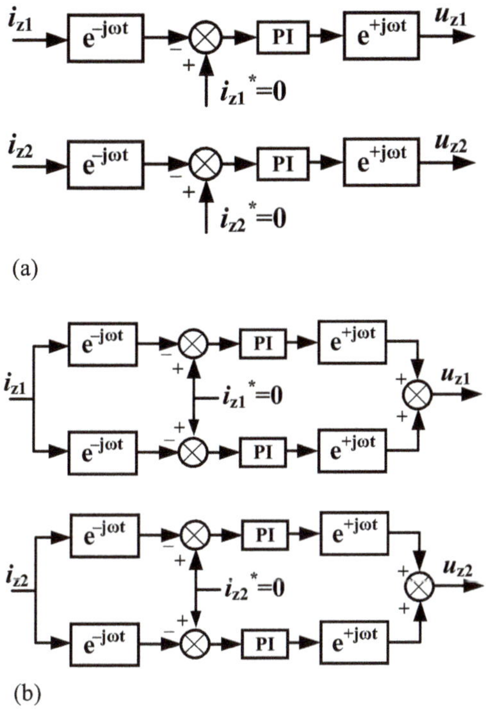

Figure 8.15 Regulators for VSD current control in DTP PMSMs. (a) Single PI in a synchronous frame. (b) Dual PIs in synchronous frame.

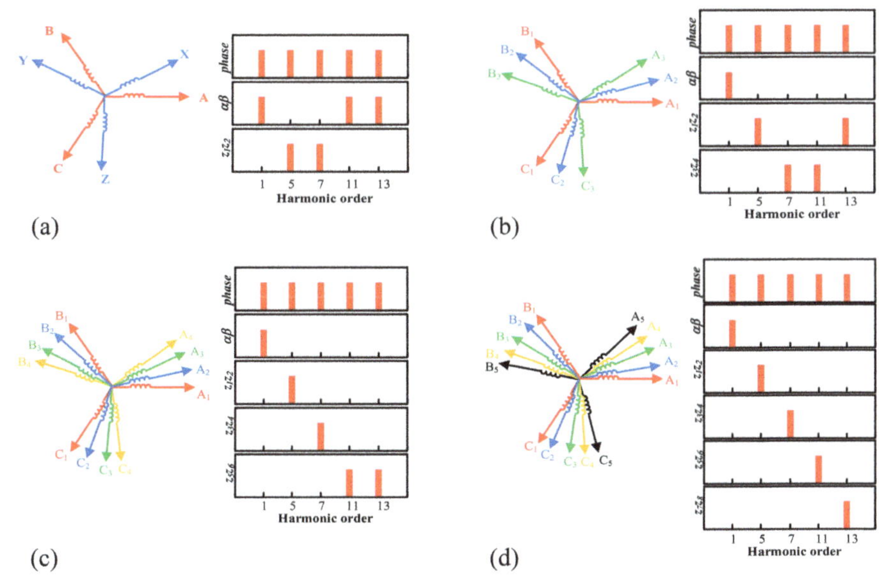

Figure 8.16 Multi three-phase winding configurations and harmonic mapping of VSD techniques. (a) DTP, (b) TTP, (c) QTP, and (d) PTP.

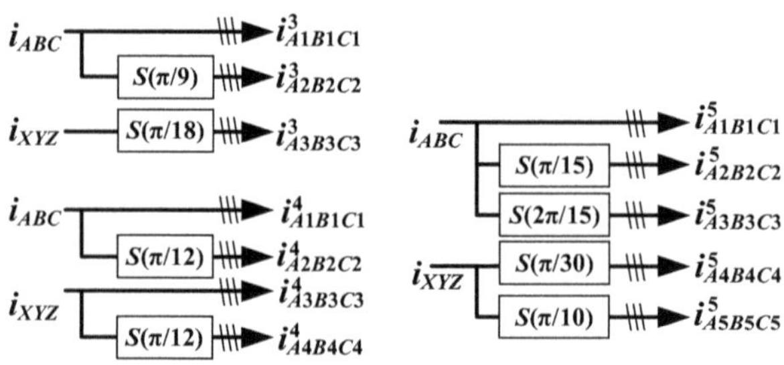

Figure 8.17 Virtual multi-three-phase current reconstruction based on phase shifting operation (S(i) represents the phase shifting operation).

degree shifting has the ideal decomposition capability since the fundamental and major current harmonics (5th, 7th, 11th, and 13th) are, respectively, mapped into five subspaces according to Figure 8.16(d), which means their current loops can also be isolated in each subspace.

Based on the above analysis, a virtual multi-3-phase system is reconstructed by the time fundamental phase shifting between the currents of two adjacent sets as $\pi/6$, $\pi/9$, $\pi/12$, and $\pi/15$. The appropriate phase shifting of the original physical currents in DTP machines can be utilized to reconstruct virtual multi-three-phase (more than two three phases) currents, as shown in Figure 8.17, where "***S***( )" represents the phase shifting operation.

For example, the physical phase currents of the DTP machine, namely $\boldsymbol{i_{ABC}}$ and $\boldsymbol{i_{XYZ}}$, can be regarded as the virtual currents of the first and third sets in the QTP case, i.e., $\boldsymbol{i^4_{A1B1C1}}$ and $\boldsymbol{i^4_{A3B3C3}}$, where the superscript "4" represents the QTP case and is used to distinguish the currents from those in the other multi 3-phase cases. Then, the $\pi/12$ shifting of currents $\boldsymbol{i_{ABC}}$ and $\boldsymbol{i_{XYZ}}$ can be, respectively, regarded as the virtual phase currents of the second and fourth sets, i.e., $\boldsymbol{i^4_{A2B2C2}}$ and $\boldsymbol{i^4_{A4B4C4}}$. Similarly, the $\pi/9$ shifting of $\boldsymbol{i_{ABC}}$ and the $\pi/18$ shifting of $\boldsymbol{i_{XYZ}}$ can be regarded as the virtual phase currents of the second and third sets in the TTP case. To reconstruct the virtual PTP system, the $\pi/15$ and $2\pi/15$ shifting of $\boldsymbol{i_{ABC}}$ are the virtual currents of the second and third sets, and then the $\pi/30$ and $\pi/10$ shifting of $\boldsymbol{i_{XYZ}}$ can be regarded as the virtual currents of the fourth and fifth sets.

By utilizing virtual multiple synchronous reference frame (MSRF) techniques, the current harmonics in DTP machines can be decomposed separately, i.e., the 5th, 7th, 11th, and 13th harmonics can be mapped into four different orthogonal subspaces. After transforming the corresponding coordinates, the AC harmonics can be transferred as DC components. Then, the PI controller can be employed to suppress all order current harmonics. The overall control diagram is presented in Figure 8.18. The experimental results [4] are also shown in Figure 8.19. It can be clearly seen from the waveform variations of the phase A current in different periods, after one suppression works, the selected current harmonic is wiped out in the phase current without interference with the other current harmonics. The individual regulations of the major current harmonics are achieved in the virtual MSRF method.

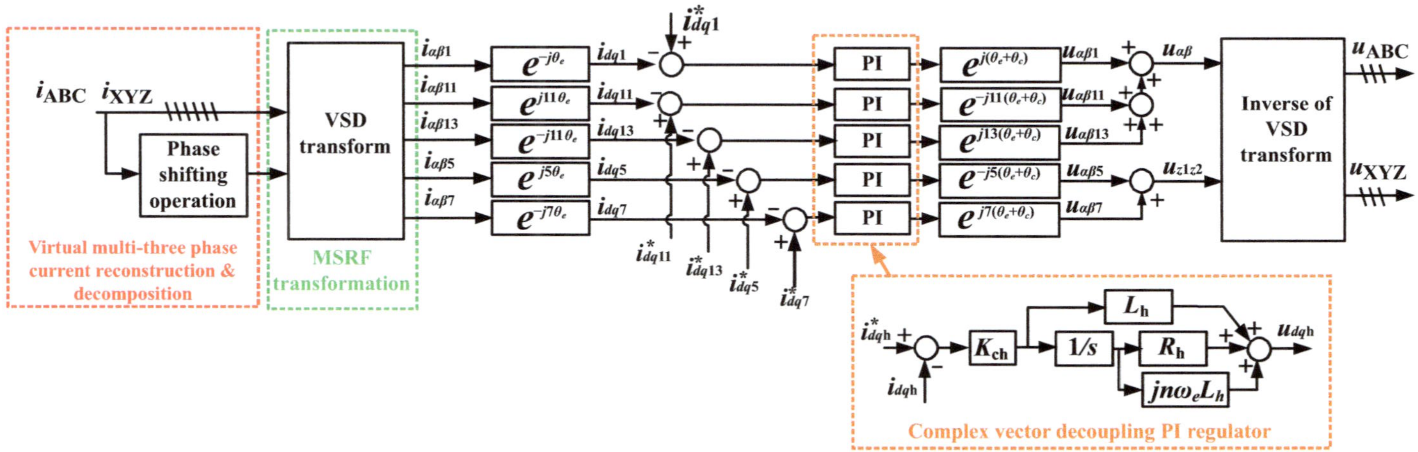

Figure 8.18 Control diagrams of current harmonic regulation based on virtual MSRF.

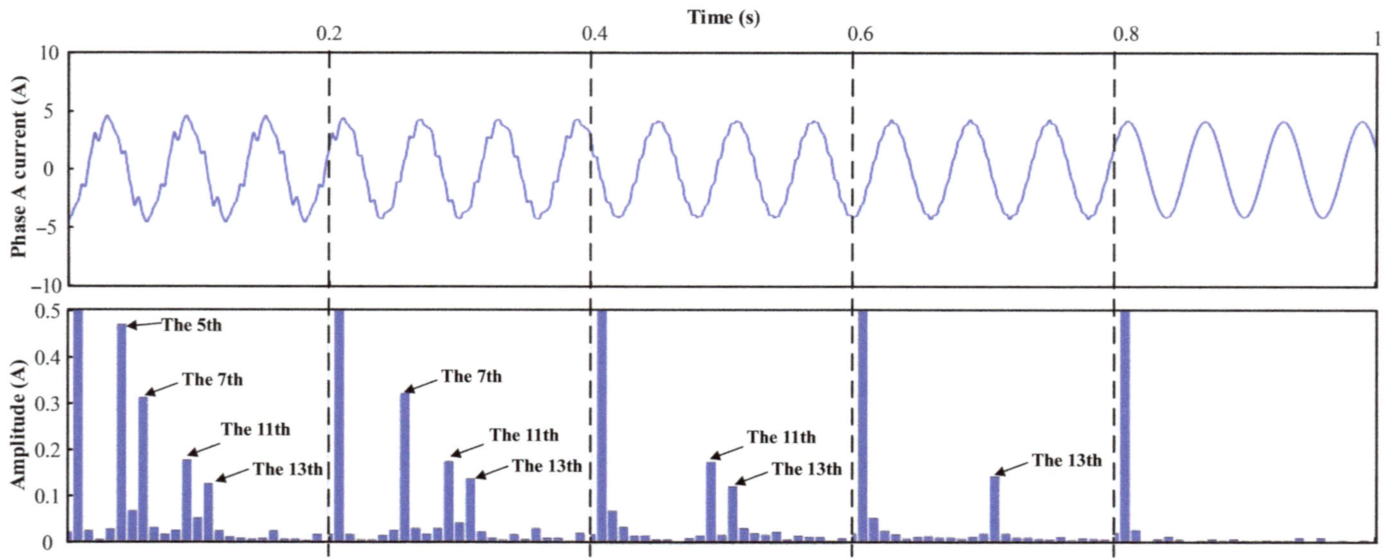

Figure 8.19 *Experimental results of successive individual harmonic suppression. (From left to right: without current harmonic suppression; only with the 5th harmonic suppression; with both the 5th and 7th harmonic suppression; with the 5th, 7th, and 11th harmonic suppression; with all major harmonic suppression.)*

Virtual impedance method

As mentioned above, the active machine impedance can suppress the current harmonics effectively by increasing the equivalent machine impedance, whereas it will introduce extra loss and decrease the DC-link voltage utilization. Hence, the virtual resistance and inductance terms are utilized to replace the real resistors and inductors and further improve the control performance of the current regulation. Figure 8.20 shows the current regulator with virtual impedance in [6]. Similar to the real impedance, the virtual impedance also acts as a virtual voltage drop, which is calculated online by using the virtual resistance, virtual inductance, and currents. After introducing the virtual impedance, the resistance and inductance of the machine are equivalently changed. The key point of this method is that since the regulator design is based on the machine model, the virtual inductance and

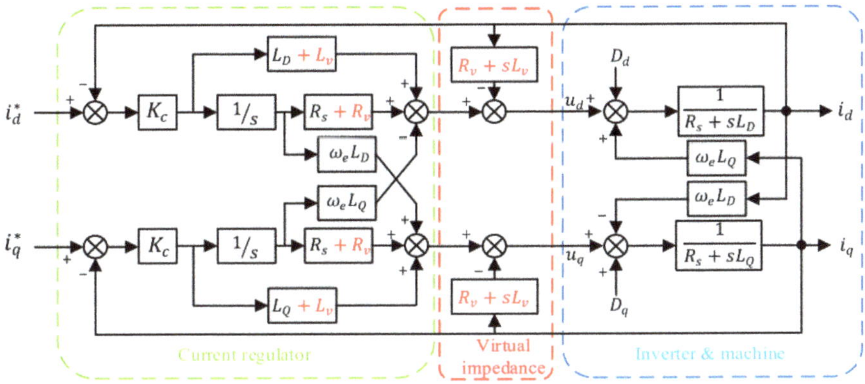

(a)

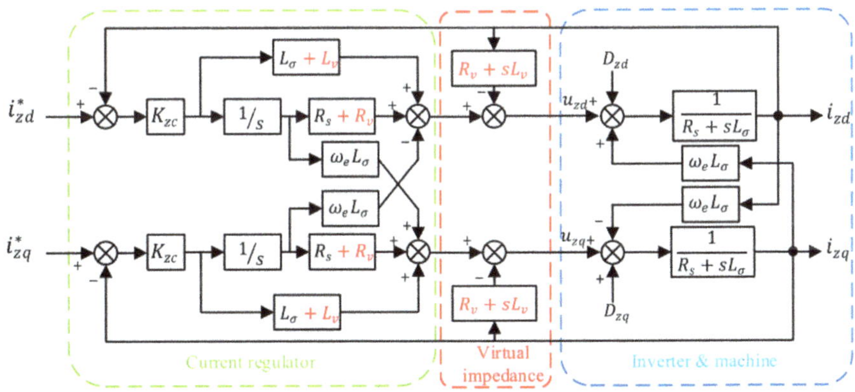

(b)

Figure 8.20 Current loops with virtual impedance. (a) $\alpha\beta$ subspace. (b) z_1z_2 subspace.

resistance are respectively added to the proportional and integral coefficients to cater to that change and maintain the zero-pole cancellation principle.

The transfer functions from reference to current feedback and disturbance to current feedback can be conducted as

$$H_{VI}(s) = \frac{s}{s + K_p} \cdot \frac{1}{(L_s + L_v)s + (R_s + R_v) + j\omega_e L_s} \tag{8.13}$$

where K_p is the proportional gain coefficient; the subscript "s" indicates the system with real stator impedance; the subscript "v" indicates the system with virtual impedance.

The experimental results in [6] are presented in Figure 8.21. It shows that the virtual impendence method can be used to enhance the disturbance rejection capability and suppress the current harmonics with a wide range of frequencies. Meanwhile, the virtual impedance will also strengthen the current regulation by eliminating the overshoot, cross-coupling effect, and pulsating currents due to machine parameter mismatch.

8.2.1.2 Torque enhancement by current harmonics injection

Another interesting research activity is related to torque enhancement. In STP PMSMs, the torque enhancement method is generally the PM rotor shaping with the 3rd harmonics, as shown in Figure 8.22. A PM shaping with the third harmonics can achieve higher fundamental EMFs than the conventional sinusoidal PMs when the minimum air gap length and maximum magnet thickness remain the same.

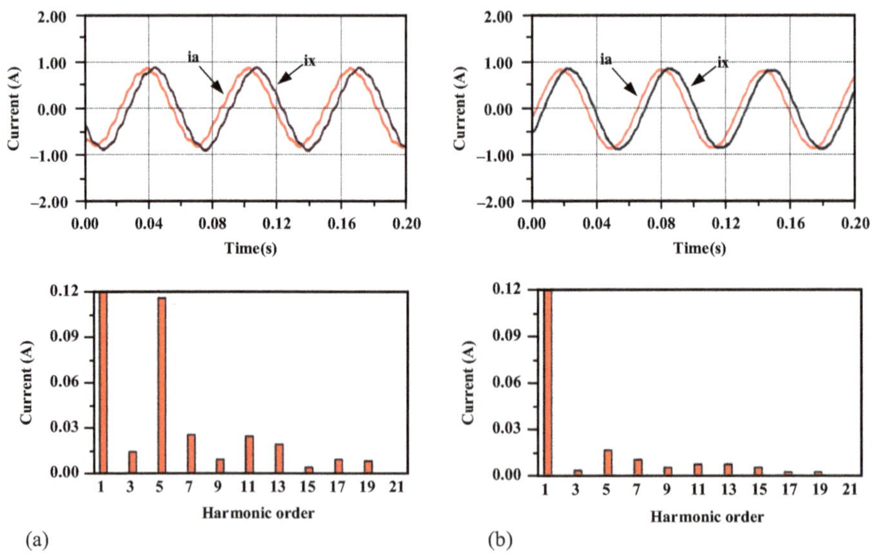

Figure 8.21 Measured currents and spectra without and with virtual impendence. (a) Without virtual impendence. (b) With virtual impendence.

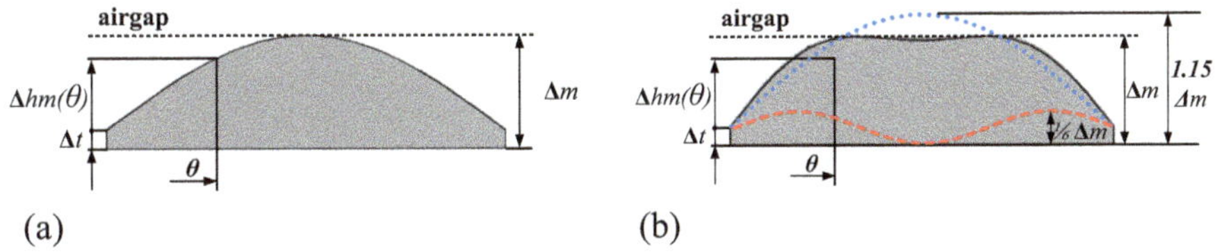

Figure 8.22 PM shaping with harmonics. (a) Sinusoidal PMs. (b) Sinusoidal + 3rd PMs.

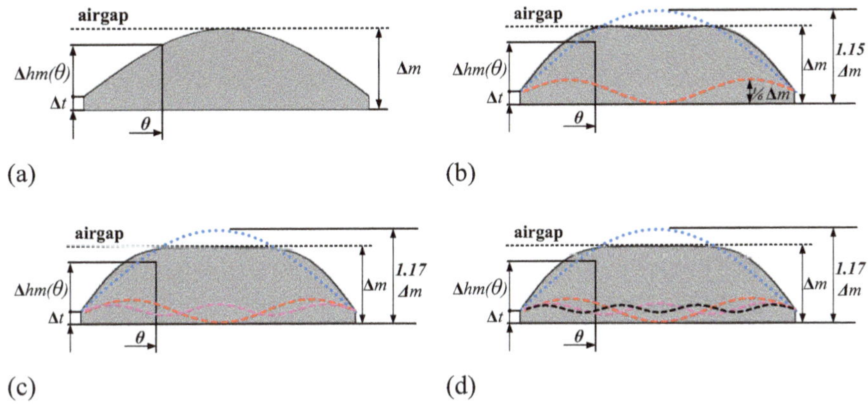

Figure 8.23 PM shapings with harmonics in DTP machines. (a) Sinusoidal PMs. (b) Sinusoidal +3rd PMs. (c) Sinusoidal +3rd+5th PMs. (d) Sinusoidal +3rd+5th+7th PMs.

Therefore, the average torque increases with PM shaping. It is demonstrated in [7] that the optimal third harmonic is 1/6 of the fundamental one. The average torque of the machine of a Sine shaping with an optimal third harmonic injected can be improved by >9%, while the torque ripple is almost similar to that of the one with Sine shaping.

In DTP PMSMs, the torque enhancement methods can be classified as: (1) PM shaping with harmonics; (2) Harmonic currents injection [8]. Different from the STP machines, the PM shaping with the third harmonic, as well as the 5th and 7th harmonics, are all effective in torque enhancement, as shown in Figure 8.23.

Compared to the Sine shape rotor machine, the torque of the sine+3rd, sine +3rd+5th, and sine+3rd+5th+7th PMs machines increases by 12.5% and 15.7%, i.e., an extra 3.2% torque improvement is contributed by the 5th harmonic, but there is no contribution from the 7th harmonic, as shown in Figure 8.24.

Besides, injecting harmonic currents is also an effective way to enhance the average torque. The third-harmonic current, which interacts with third-harmonic EMFs, can be injected into a winding to improve the output torque capability. In STP machines, the winding structure is generally balanced using Y-connected

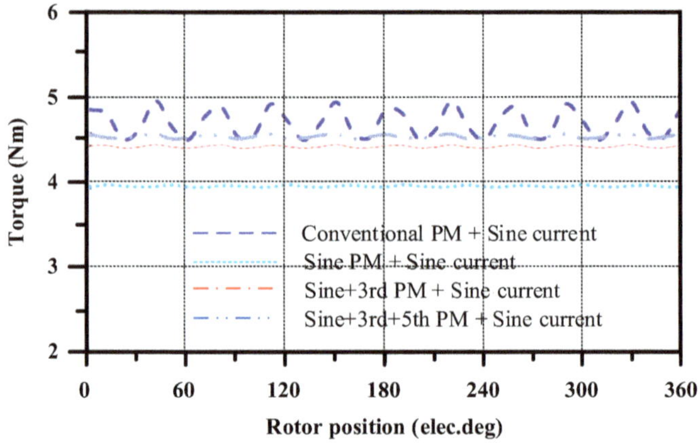

Figure 8.24 Variation of torque with rotor position with different combinations of "magnet shaping" under sinusoidal currents.

windings, and hence, the third-harmonic current cannot be injected. However, the 3rd harmonic current can also be injected in a DTP machine topology with the neutral point of the two sets of windings connected. Figure 8.25 shows the finite element analysis (FEA) and experimental results with or without the third current harmonic injection in a DTP machine. It is obvious that the torque improves significantly in both 12-slot/8-pole and 12-slot/10-pole machines, and the experimental results validate the torque enhancement as well.

Furthermore, the 5th and 7th current harmonics injection is effective in both topologies with/without connected neutral points. The current waveforms with the 5th and 7th harmonics are shown in Figure 8.26. The 5th and 7th harmonic currents interact with the 5th and 7th harmonic EMFs and can be injected into a winding to improve the output torque capability. It can be seen as an alternative way to enhance the torque in the isolated neutral point topology. In [9], it is concluded that the improvement of the overload torque for the 5th and 7th harmonic injection is 7%, whereas the torque enhancement for the third harmonic current injection can be up to 15%.

When the PM rotor shaping and harmonic current injection are utilized simultaneously, the torque can be improved further. It is reported in [10] that the third harmonic current injection with PM shaping can improve the torque by 15.7%, as shown in Figure 8.27.

8.2.2 Field-oriented control for other multiphase machines

FOC is not only commonly used in STP and DTP machines but also popular in multiphase machines with the same objectives, i.e., reducing current harmonics and enhancing output torque.

In a 5-phase machine, different from DTP machines, in which the 5th and 7th harmonics coupling are mapped in the same subspace, only the 3rd current harmonics are mapped in the z_1z_2 subspace, while the fundamental components are

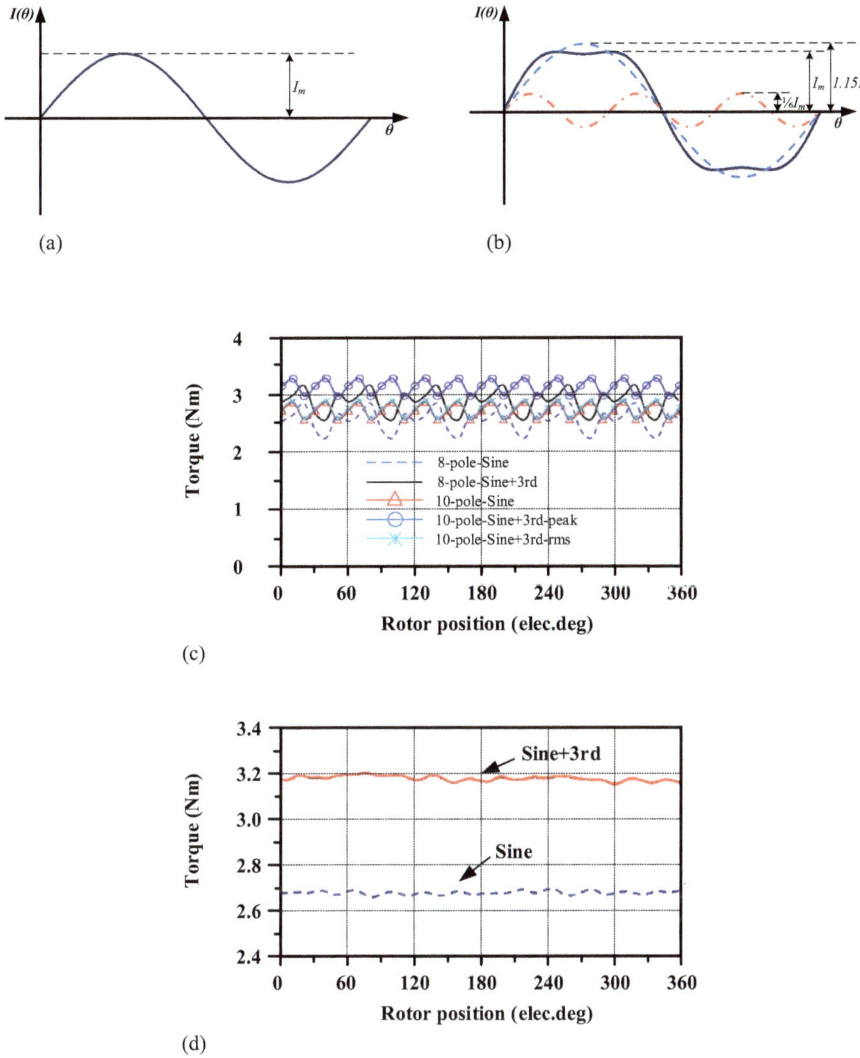

Figure 8.25 Current waveforms, and FEA and experimental torque results for third harmonic current injection. (a) No harmonic injection current waveforms (Sine). (b) Third harmonics injection current waveforms (Sine+3rd). (c) FEA results of torque for 12-slot/8-pole and 12-slot/10-pole DTP SPMSM. (d) Measured torque of a 12-slot/10-pole DTP SPMSM.

mapped into the $\alpha\beta$ subspace. It makes reducing current harmonics much easier in the 5-phase machines.

The synchronous reference frame PI controller, which transfers the 3rd current harmonics to DC components and controls them with a PI controller, can be simply

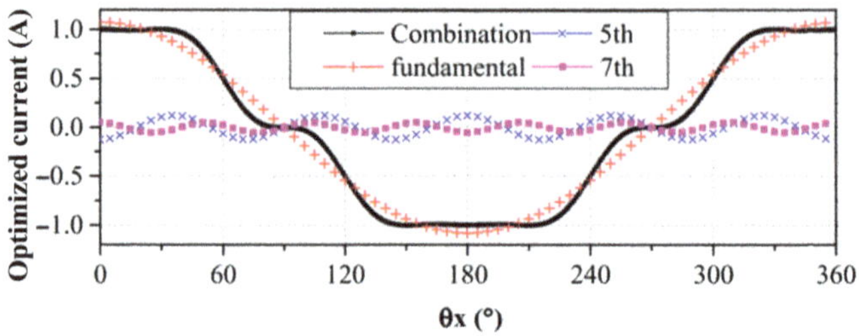

Figure 8.26 Current waveforms with the 5th and 7th harmonic currents.

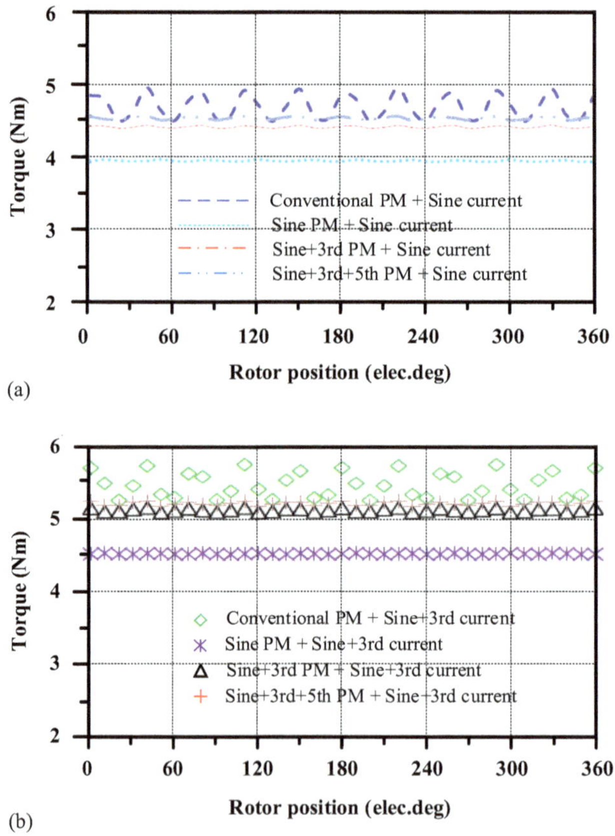

Figure 8.27 Variation of torque with rotor position with different combinations of "magnet shaping" + "injected current". (a) Sinusoidal current. (b) Sinusoidal + 3rd current injection.

utilized in 5-phase machine drives. A resonant controller can also be employed to avoid the coordinate transformation. The SRF-PI controller is employed in 5-phase machines [11] to reduce the current harmonics, of which the control structure is illustrated in Figure 8.28. The experimental results are shown in Figure 8.29. It can

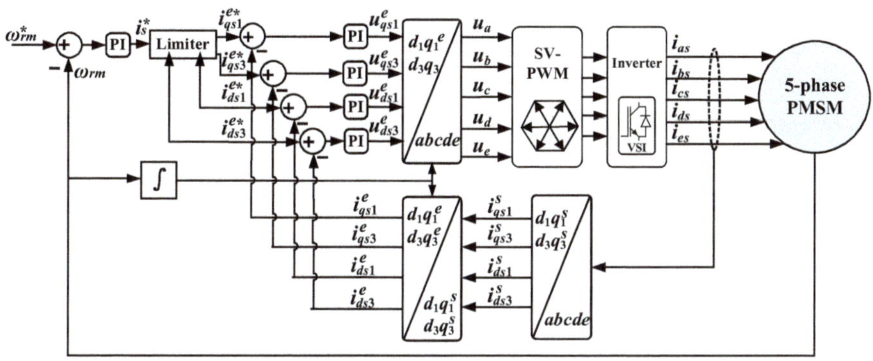

Figure 8.28 Control block for phase current harmonics reduction.

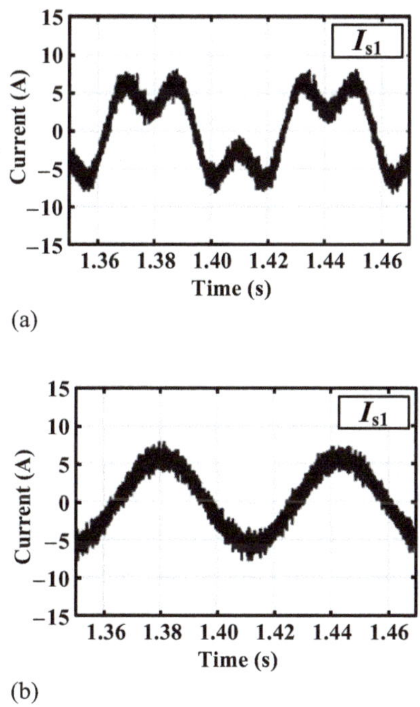

Figure 8.29 Phase currents of the 5-phase machine. (a) Without harmonic control. (b) With harmonic control.

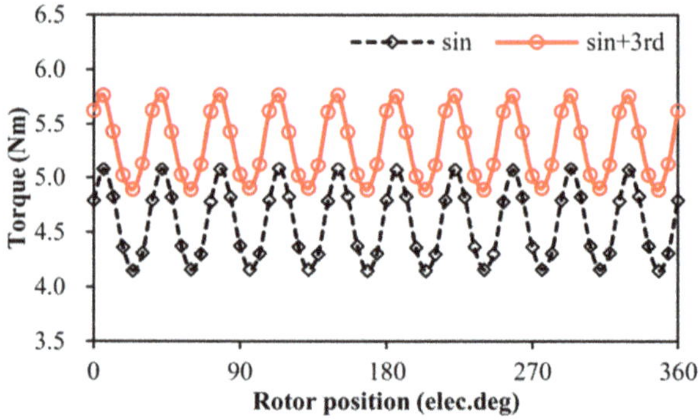

Figure 8.30 Output torque with and without 3rd harmonic current injection under the same amplitude constraint.

be seen that the SRF-PI controller is useful to reduce the 3rd current harmonics in 5-phase machines.

Similarly, for other multiphase topologies, such as 7-, 9-, 12-, 15-, etc., phase machines, the methods described in Section 8.2.1 can all be utilized to reduce the current harmonics. In principle, an n-phase drive with a single neutral point requires $(n-1)$ current controllers in the independent subspaces. It should be noted that the current control is based on the VSD method for multiphase machines, as described in Chapter 5.

For enhancing the output torque, it can also be summarized as two aspects in multiphase machines: (1) PM rotor shaping; (2) Current harmonics injecting. In [12], the 3rd current harmonics are injected in 5-phase machines. When the 3rd harmonic current is injected under the same amplitude constraint, the output torque is improved remarkably by 15.7%, as shown in Figure 8.30. For the other multiphase machines, injecting current harmonics is also an effective way to improve the output torque. Although there is no literature related to other multiphase machine's (such as 7-phase and 9-phase machines) torque enhancement, it is reasonable to infer that injecting harmonics is an effective way to increase torque. Besides, in addition to the third harmonic, other harmonics may also increase the torque.

8.3 Direct torque control for multiphase machines

DTC was first introduced to three-phase machines in 1986 by Takahashi [13] and Depenbrock [14]. DTC has the merits of excellent transient performances, fast dynamic response, and good robustness with respect to the variations of the parameters of the machine. Compared to the FOC and MPC, it is also less sensitive to parametric variations of the machine. DTC was first extended to multiphase machines in 2000 by Toliyat [15], where a 5-phase machine was adopted, and a DTC for DTP machines was also proposed in 2005 by Bojoi [16]. In this section, the DTC for DTP machines is discussed, and the main challenges for DTC in DTP

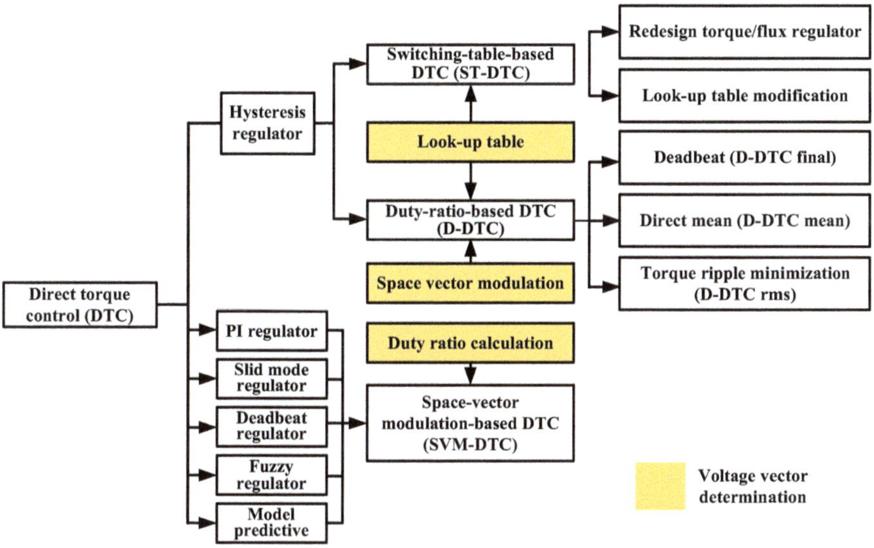

Figure 8.31 Illustration of DTC methods.

machines can be summarized as three aspects: (1) Reduction of current harmonics; (2) Reduction of torque ripples; (3) The issue of a variable switching frequency (which will cause challenges for harmonic filtering and lead to switch loss increase and significant high-order harmonics). In addition, the applications of DTC in other multiphase machines are also described.

As discussed in Sections 7.3 and 7.4 in Chapter 7, DTC aims to directly control the stator flux and the torque by selecting the appropriate inverter state. As illustrated in Figure 8.31, DTC can be classified into three categories according to its basic structure:

- The switching-table-based DTC (ST-DTC)
- The duty-ratio-based ST-DTC (D-DTC)
- The space vector modulation DTC (SVM-DTC)

8.3.1 Switching-table-based DTC for DTP machines

The switching-table-based DTC for DTP machines is discussed in detail in Section 7.4. The concepts of DTC are only described briefly in this section. The switching-table-based DTC, i.e., the classical DTC, employs only flux and torque hysteresis regulators and a look-up table to control machine flux and torque, and its basic structure for DTP machines is illustrated in Figure 8.32. As can be seen, it consists of torque and flux estimators, torque/flux hysteresis regulators, and a look-up table. The switching table directly selects the most appropriate switching states by exploiting the outputs of torque/flux regulators and stator flux vector position. A decoupled control of torque and stator flux is established to permit fast instantaneous control.

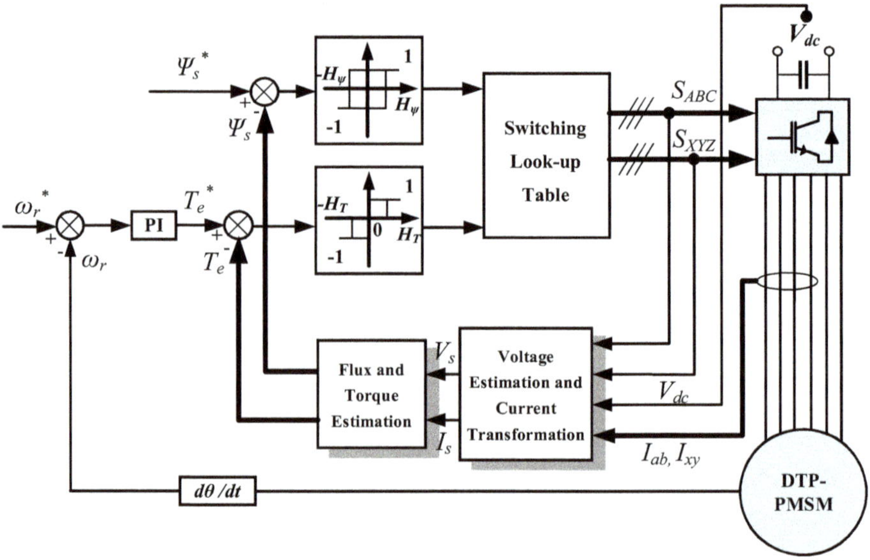

Figure 8.32 Basic structure for DTC in DTP machines.

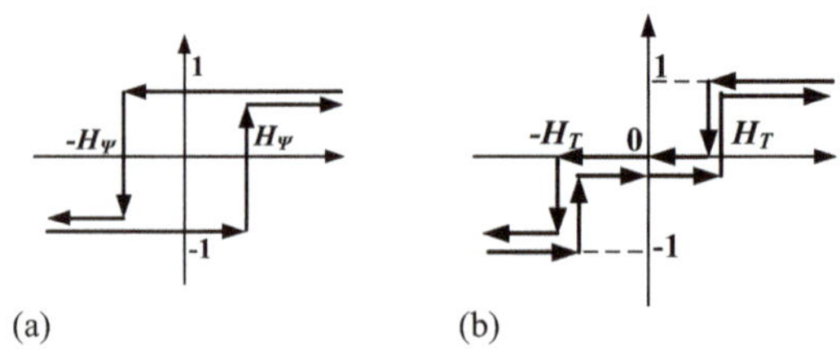

Figure 8.33 Hysteresis regulators in DTC for DTP machines. (a) 2-level flux regulator. (b) 3-level torque regulator.

In general, the stator flux is controlled using a 2-level hysteresis regulator, while the torque is controlled using a 3-level hysteresis regulator, as shown in Figure 8.33.

The outputs of the torque/flux hysteresis regulator, along with sector flux information are used to index the look-up table, to select the appropriate voltage vectors (VVs) to simultaneously control the stator flux and the torque, which are essential for DTC implementation.

For DTP machines, the torque and stator flux estimations under the VSD scheme are calculated in stationary reference frame as follows:

$$\begin{cases} \psi_{s\alpha} = \int (V_{s\alpha} - R_s i_{s\alpha}) dt \\ \psi_{s\beta} = \int (V_{s\beta} - R_s i_{s\beta}) dt \end{cases} \tag{8.14}$$

$$\begin{cases} \psi_s = \sqrt{\psi_{s\alpha}^2 + \psi_{s\beta}^2} \\ \gamma_s = \arctan\left(\dfrac{\psi_{s\beta}}{\psi_{s\alpha}}\right) \end{cases} \tag{8.15}$$

$$T_e = 3p\left(\psi_{s\alpha} i_{s\beta} - \psi_{s\beta} i_{s\alpha}\right) \tag{8.16}$$

According to Section 7.4, there are various methods focused on the modulation strategy, e.g., the VV/Virtual VV (VVV) selection strategy to improve the steady-state performance, particularly the reductions of current harmonics and torque ripple. However, the pre-calculated switching tables used in table-based DTC are based on assumptions and simplifications that may not accurately reflect the behavior of the machine in all operating conditions and cannot be easily modified or adapted to changes in the machine or operating conditions. Consequently, it may result in reduced accuracy and control performance, such as large torque and flux ripples, and limit the flexibility of the control strategy. Besides, the switching frequency of ST-DTC is variable, which would lead to an increase in switching losses and electromagnetic interference (EMI) issues in the machine drive.

8.3.2 Duty-ratio-based DTC for DTP machines

The duty-ratio-based DTC (D-DTC) divides one sampling period into two intervals and applies active vectors and zero vectors together to reduce flux and torque ripples. D-DTC calculates the duty ratio of the active vector and zero vector based on the desired torque and flux. By adjusting the action time durations of both the active vectors and the zero vector, specific control objectives, such as minimum torque ripple or flux ripple, can be achieved. The main advantage of duty ratio-based DTC is its ability to provide precise flux and torque control for the machine, as the duty ratio can be adjusted in real time based on the feedback of torque and currents from the machine. The structure of D-DTC is illustrated in Figure 8.34.

As can be seen, D-DTC still uses hysteresis regulators to compare the actual and desired torques and fluxes. Its difference from ST-DTC can be summarized as: (1) D-DTC uses active VV and one zero VV in each control period, while ST-DTC utilizes one or two active VVs during one sampling interval; (2) The duty ratio of the active vector and the zero vector in D-DTC is adjustable, while in the ST-DTC it is fixed and unchangeable. Since the D-DTC still utilizes the hysteresis regulator, the merits of ST-DTC, such as fast response and excellent dynamic performances, can still persist in this method. However, the issue of variable switching frequency still exists in D-DTC due to the utilization of the switching table and hysteresis regulators.

In DTP machines, the torque ripple is usually considered the most important criterion of DTC performance. The torque variation rate with the active VV and zero VV in DTP machines is expressed as:

$$\frac{dT_e}{dt} = -\frac{R_s}{L_s} T_{e0} + \frac{3}{2}\frac{p}{L_s}\left(-\Psi_{r\beta} V_{s\alpha} + \Psi_{r\alpha} V_{s\beta} - \omega_1 \left|\Psi_{r\alpha\beta}\right| \left|\Psi_{s\alpha\beta}\right| \cos\delta\right) \tag{8.17}$$

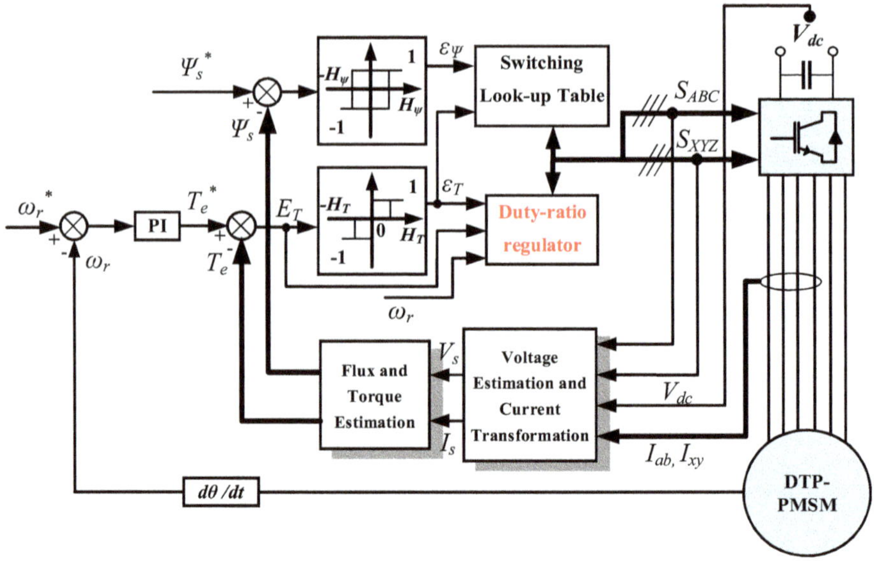

Figure 8.34 Structure diagram of D-DTC in DTP machines.

$$\frac{dT_e}{dt} = -\frac{R_s}{L_s}T_{e0} + \frac{3}{2}\frac{p}{L_s}\left(-\omega_1\left|\Psi_{r\alpha\beta}\right|\left|\Psi_{s\alpha\beta}\right|\cos\delta\right) \tag{8.18}$$

The value of dT_e/dt in (8.17) is generally positive, which implies that the active VV increases the torque. While the value of dT_e/dt in (8.18) is negative, which implies that the zero voltage decreases the torque. It is expected that by introducing the zero-VV, the torque performance should be improved since the zero voltage contributes to a torque decrease. The key point of this strategy is the duty ratio determination, which leads to different control objectives. There are three typical categories of D-DTC, i.e., D-DTC-final, D-DTC-mean, and D-DTC-RMS. One of the most commonly used D-DTC is to force the actual torque to be equal to the reference value at the end of the cycle (D-DTC-final), as shown in Figure 8.35, which satisfies:

$$T_e(k+1) = T_e^*(k) \tag{8.19}$$

In [17], the control target is to force the mean torque equal to its reference value in each control period (D-DTC-mean), expressed as:

$$\frac{1}{T_s}\int_{kT_s}^{(k+1)T_s}\left(T_e^*(k) - T_e\right)dt = 0 \tag{8.20}$$

In [18], it aims to make the square of the RMS torque ripple T_{e_ripple} during the sampling period T_s to be minimal (D-DTC-RMS), expressed as

$$T_{e\_ripple}^2 = \frac{1}{T_s}\int_{kT_s}^{(k+1)T_s}\left(T_e^*(k) - T_e\right)^2 dt \rightarrow \min \tag{8.21}$$

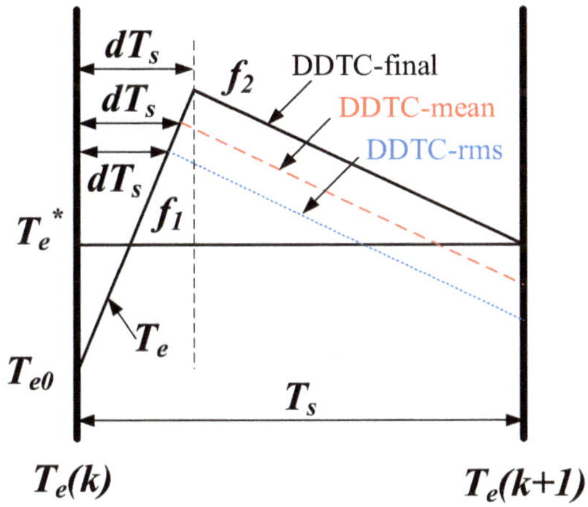

Figure 8.35 Illustration for different duty ratio methods.

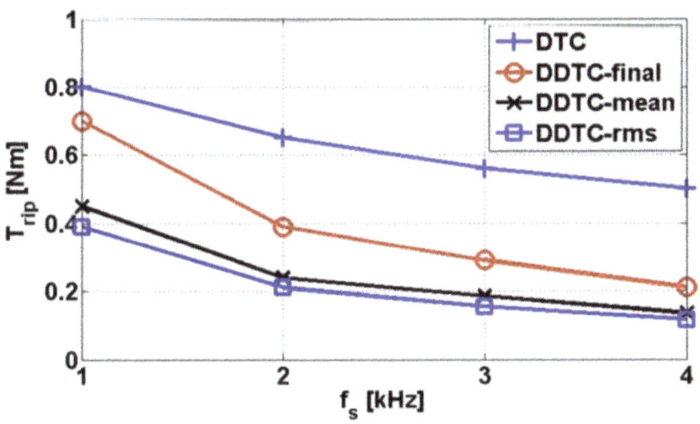

Figure 8.36 Torque ripple for different duty ratio determination strategies.

Finally, as reported in [19], the duty ratio methods lead to different torque ripples. In DTP machine systems, the torque ripple of different duty ratio determination strategies under the same switching frequency is illustrated in Figure 8.36.

8.3.3 Space-vector-based DTC for DTP machines

DTC with space vector modulation, i.e., space-vector-modulation-based DTC (SVM-DTC), is proposed to overcome the demerits of switching frequency variation and large torque ripple. It can synthesize a continuous rotating VV to regulate the stator flux and torque more accurately. Hence, it can achieve torque/flux ripple reduction. The key point of this strategy is the command VV determination. One of the most widely used SVM-DTC schemes is to employ the PI torque, flux

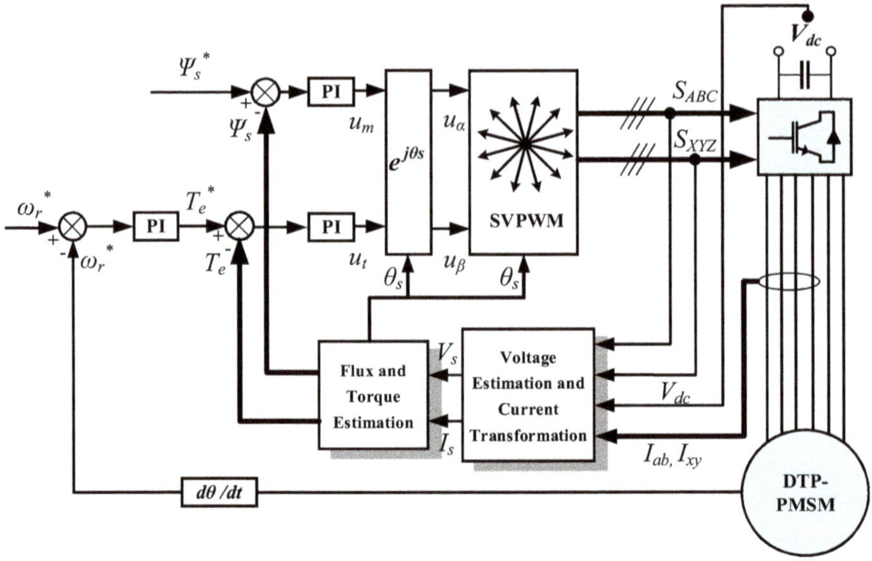

Figure 8.37 SVM-DTC basic structure for DTP machines.

controllers, and space vector modulation to replace the hysteresis regulators and the look-up table in ST-DTC to obtain the command VV, as shown in Figure 8.37.

8.3.3.1 Deadbeat controller

The deadbeat controller can be employed to replace the PI controller in SVM-DTC to generate the reference voltage as the input of an SVM modulator, as proposed in [20]. The deadbeat controller is composed of two steps. First, the reference stator flux vector is obtained based on the principle of deadbeat torque control, and then the reference stator VV to achieve deadbeat control of the stator flux which is obtained based on the stator voltage equation. The key advantage of deadbeat control is its ability to achieve very fast and accurate torque and flux control. Since the VV is applied for a single control cycle, the controller can predict the machine's response in advance and adjust the VV accordingly, resulting in a rapid and accurate response. Since the deadbeat control requires the machine model equation, it is sensitive to the machine parameters and thus, the robustness of the control method may deteriorate.

The basic principle of the deadbeat controller is described as follows.

First, the electromagnetic torque produced by the dual three-phase machine can be described as:

$$T_e = 3p\mathrm{Im}(\psi_s \times I_s) \tag{8.22}$$

where Im() stands for the imaginary part, × is the cross product.

Therefore, the instantaneous torque variation can be expressed as:

$$\frac{dT_e}{dt} = 3p\ \mathrm{Im}\left(\frac{d\psi_s}{dt} \times I_s + \psi_s \times \frac{dI_s}{dt}\right) \tag{8.23}$$

The discretized expression of the torque variation during two adjacent control periods is as follows:

$$\frac{T_e(k+1)-T_e(k)}{T_s}$$
$$= 3p\mathrm{Im}\left(\frac{\psi_s(k+1)-\psi_s(k)}{T_s} \times I_s(k) + \psi_s \times \frac{I_s(k+1)-I_s(k)}{T_s}\right) \tag{8.24}$$

where $T_e(k)$ and $T_e(k+1)$ are the electromagnetic torque of the current and next control period, respectively. $\psi_s(k)$ and $\psi_s(k+1)$ are the stator flux of the current and next control period, respectively. $I_s(k)$ and $I_s(k+1)$ are the stator current of the current and next control period, respectively.

Therefore, (8.23) can be rewritten as:

$$\frac{\Delta T_e}{T_s} = 3p\mathrm{Im}\left(\frac{\Delta\psi_s}{T_s} \times I_s(k) + \psi_s \times \frac{\Delta I_s}{T_s}\right) \tag{8.25}$$

The deadbeat algorithm aims at obtaining the reference values of torque and stator flux in one sampling time, and therefore,

$$\begin{cases} T_e^* = T_e(k+1) \\ \psi_s^* = \psi_s(k+1) \\ I_s^* = I_s(k+1) \end{cases} \tag{8.26}$$

In order to keep the simplicity of the DTC strategy, the deadbeat algorithm is generally implemented in the synchronous reference frame (MT) aligned with the stator flux vector, as expressed in (8.27) and shown in Figure 8.38.

$$\begin{cases} \psi_{sM} = \psi_s \\ \psi_{sT} = 0 \end{cases} \tag{8.27}$$

Therefore, by ignoring the effect of stator resistance on the variation of the stator flux, the *M*-axis reference value of voltage can be expressed as

$$V_{sM}^* = \frac{\Delta\psi_{sM}}{T_s} = \frac{\psi_s^* - \psi_s}{T_s} = \frac{\Delta\psi_s}{T_s} \tag{8.28}$$

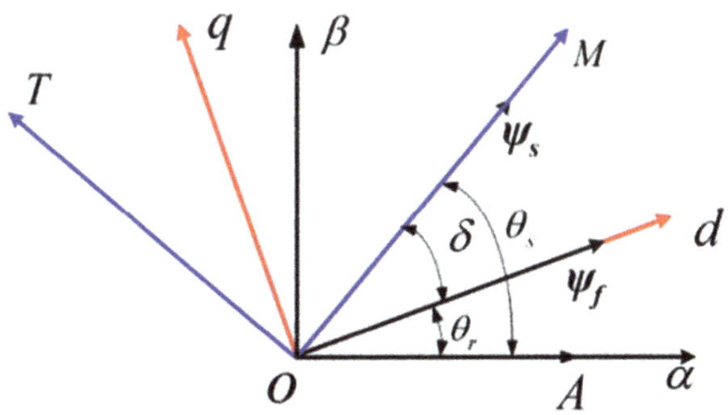

Figure 8.38 Reference frame of DTC for DTP machines.

Afterwards, the T-axis reference voltage for a surface-mounted DTP machine can be expressed as

$$V_{sT}^{*} = \frac{\frac{L_s \Delta T_e}{3pT_s} - L_s V_{sM}^{*} I_{sT} + \psi_s e_{sT}}{\psi_s - L_s I_{sM}} \tag{8.29}$$

where e_{sT} is the back-EMF component in the synchronous reference frame.

$$V_{s\alpha\beta}^{*} = V_{sMT}^{*} e^{j\theta_s} + R_s I_{s\alpha\beta} \tag{8.30}$$

8.3.3.2 Fuzzy logic controller

A fuzzy logic controller (FLC), or fuzzy logic regulator, can also be applied to replace the PI controller in the typical SVM-DTC. FLC is based on the concept of fuzzy set theory, which allows for the use of linguistic variables to represent the uncertainty and vagueness in the input signals. The FLC takes the error between the desired torque and the actual torque, and the error between the desired flux and the actual flux, which are used as input variables to calculate the appropriate switching states for the inverter.

The FLC in DTC typically consists of the following components:

1. **Fuzzy inference system**: This component consists of a set of fuzzy rules that map the input variables (errors in torque and flux) to the output variable (switching state of the inverter). The rules are defined using linguistic variables that represent the degree of membership of the input variables in different fuzzy sets.
2. **Membership functions**: These functions define the degree of membership of the input variables in the fuzzy sets. The membership functions are typically triangular or trapezoidal in shape.
3. **Fuzzification**: This process involves converting the crisp input values (errors in torque and flux) into fuzzy sets using the membership functions.
4. **Rule evaluation**: This process involves applying the fuzzy rules to the fuzzy sets obtained from fuzzification to generate the output fuzzy set.
5. **Defuzzification**: This process involves converting the output fuzzy set into a crisp output value (switching state of the inverter) using a defuzzification method such as the centroid or max-min.

A torque controller that is composed of a fuzzy logic regulator and an MPC is developed in [21] for induction machines. It should be noted that the fuzzy logic controller contributed to the reduction of the torque ripple with a significant decrement of the parameter dependency and limitation of the inrush current. The structure of FLC in [22] is shown in Figure 8.39. The FLC system is designed by choosing a minimum number of linear membership functions (MFs) for each input and using the Takagi-Sugeno (T-S) fuzzy inference system (FIS) technique for calculating the control decision. The FLC in DTC is designed to provide robust and fast control of the torque and flux of the machine, even in the presence of disturbances and uncertainties. But it still has a lot of drawbacks, such as complex

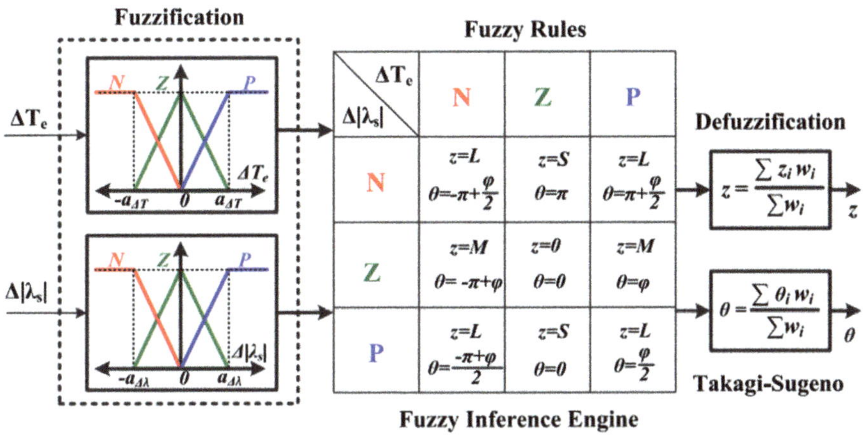

Figure 8.39 Block diagram of FLC for DTC.

design, tuning difficulties, lack of mathematical foundation, high computation requirements, etc.

8.3.3.3 Sliding mode controller

Sliding mode controller (SMC) is a nonlinear and robust control method that can handle uncertainties and disturbances in the system. In DTC, SMC is used to regulate the stator flux and the electromagnetic torque of the machine by controlling the switching states of the inverter. The basic idea of SMC is to design a sliding surface that the state variables of the system follow. The sliding surface is a function of the state variables and their derivatives and is designed to drive the system toward a desired state. Once the system reaches the sliding surface, the controller drives the system along the sliding surface to the desired state. The sliding mode controller in DTC consists of the following components: (1) Sliding surface: The sliding surface is a function of the state variables and their derivatives that drive the system towards the desired state. (2) Control law: The control law is designed to stabilize the sliding surface and drive the system towards the desired state. The control law is typically discontinuous and switches between different control actions based on the position of the state variables with respect to the sliding surface. (3) Discretization: The continuous control law is discretized to obtain the switching states of the inverter.

In [22], a sliding mode controller is utilized in DTC, and the control structure is illustrated in Figure 8.40. Since the conventional SMC suffers from large chattering, a super-twisting sliding mode (STSM) algorithm, which is a second-order scheme, is developed for the system. The sliding mode function $S = [S_1 \ S_2]^{\mathrm{T}}$ is defined as:

$$\begin{cases} S_1 = \psi_s^* - \psi_s \\ S_2 = T_e^* - T_e \end{cases} \tag{8.31}$$

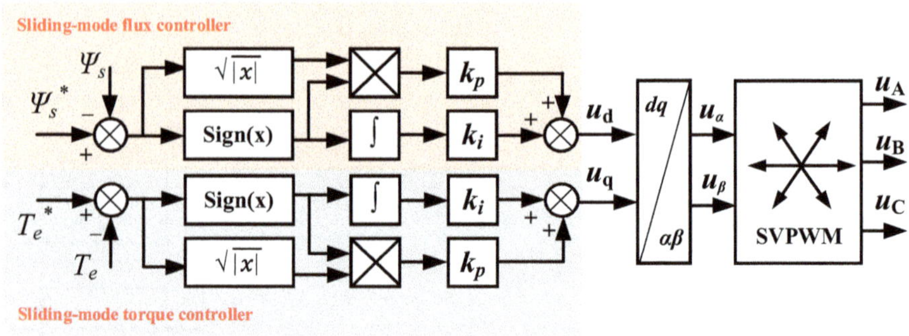

Figure 8.40 Sliding mode torque/flux controller in a sliding-mode-based DTC.

A finite-time convergence STSM torque controller can be obtained with a control constant $r = 0.5$, as expressed below

$$\begin{cases} u_q = k_p\sqrt{|S_2|}\text{sign}(S_2) + u_{q1} \\ \dot{u_{q1}} = k_i\text{sign}(S_2) \end{cases} \tag{8.32}$$

where k_p and k_i are the parameters of the SMC controller.

The advantages of sliding mode control in DTC can be summarized as:

Robustness: SMC is a robust control method that can handle uncertainties and disturbances in the system.

Fast response: SMC can provide fast control response and high tracking accuracy, especially in systems with high dynamics.

Insensitivity to parameter variations: SMC is insensitive to parameter variations and can maintain control performance even in the presence of uncertainties and variations in the system parameters.

Simple implementation: SMC is relatively easy to implement compared to other advanced control methods.

However, SMC suffers from the chattering issue, i.e., a phenomenon that can cause high switching frequency of the control signal, which can lead to increased machine noise and reduced efficiency. Therefore, careful design and tuning of the controller are required to achieve good performance and reduce chattering.

8.3.3.4 Summary

The steady-state and dynamic performance of different DTCs for DTP machines under the same operating condition is presented in Figures 8.41 and 8.42, respectively.

In terms of steady-state performance, the ST-DTC suffers from the largest torque ripple. In comparison, D-DTC reduces the torque ripple, and SVM-DTC has the best performance. It should be noted that better steady-state results can be obtained by optimizing the VV selection strategy, but the vector selection strategy in [23] is still used here for a fair comparison with D-DTC, and the vector selection strategy in [18] for D-DTC is the same as in [23]. Although the phase currents are

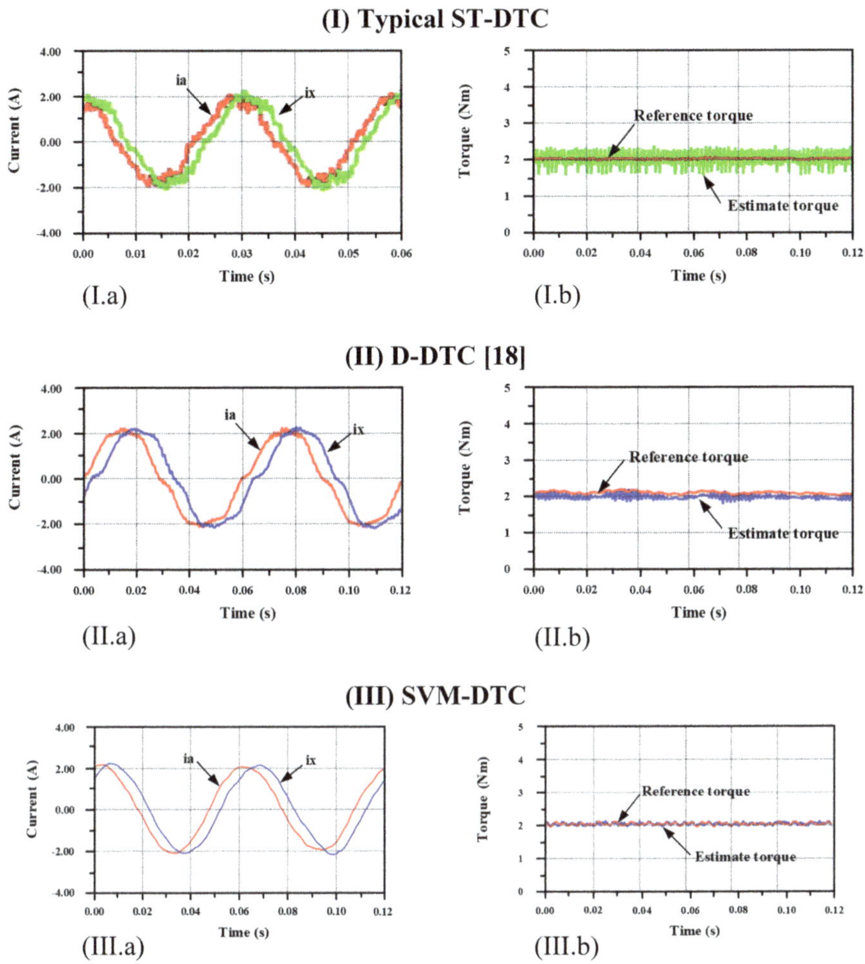

Figure 8.41 Steady-state performance of three DTC methods in DTP machines. (I) Typical ST-DTC. (I.a) Current waveforms. (I.b) Torque waveforms. (II) D-DTC [17]. (II.a) Current waveforms. (II.b) Torque waveforms. (III) SVM-DTC. (III.a) Current waveforms. (III.b) Torque waveforms.

distorted due to an unoptimized VV selection strategy (the same as [23]), the torque ripples are reduced significantly compared to the results of ST-DTC. Overall, it can be concluded that the duty ratio technique is effective in reducing the torque ripple.

In terms of dynamic performance, the ST-DTC and D-DTC share the best and fastest response time due to the hysteresis regulator. In comparison, the SVM-DTC has the worst dynamic performance but the best steady-state performance.

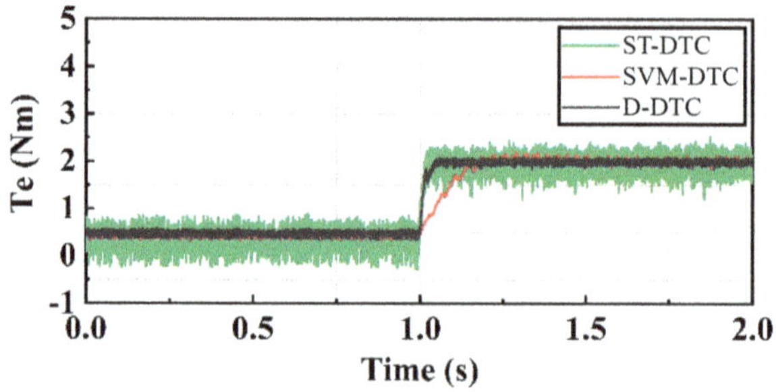

Figure 8.42 Dynamic performance of three DTC methods in DTP machines.

Table 8.1 Comparison of three DTCs in DTP machines

	ST-DTC	D-DTC	SVM-DTC
Steady-state performance	Low	Normal	High
Dynamic performance	Fast	Normal	Slow
Switching frequency	Variable	Variable	Fixed
Implementation complexity	Simple	Normal	Complex
Parameter sensitivity	Low	Low	High
PWM modulator	No	No	Yes
Computation burden	Low	Normal	High
Robustness	High	High	Normal

In summary, the above three DTC strategies have different characteristics, which are compared and summarized in Table 8.1.

8.3.4 Space-vector-based DTC for multiphase machines

The DTC was first extended to a 5-phase induction machine in [24], of which the basic structure of the DTC is the same as the DTC for DTP machines, as shown in Figure 8.43. The difference between the DTC of 5-phase machines from the DTP machine is the VV selection, which is shown in Figure 8.44.

The ST-DTC is employed in [24], where the VVs in 5-phase machines are utilized as much as possible, and even high-level torque and flux hysteresis regulators are applied to fully utilize all possible VVs. Therefore, all drawbacks of ST-DTC still exist. It can be expected that the VVV method in ST-DTC, described in Section 7.4 in Chapter 7, can be adopted to improve the performance.

Besides, the SVM-DTC is also extended into 5-phase machines in [25], where the PI controller replaces the hysteresis regulator, and the space vector modulation strategy replaces the vector look-up table, as presented in Figure 8.45, which is almost the same as the SVM-DTC in DTP machines.

In other multiphase machines, DTC is also an effective way to obtain fast response and good dynamic performance. For example, in the 9-phase, a VVV DT-DTC method

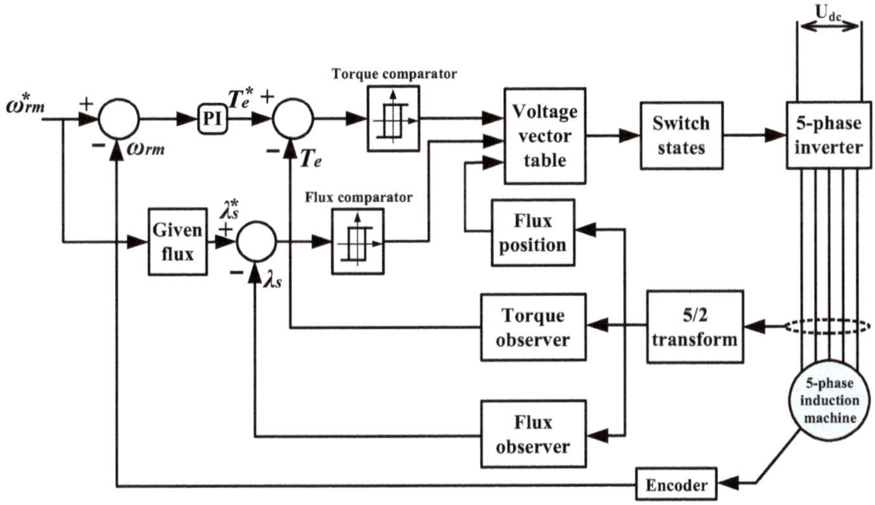

Figure 8.43 Basic structure of DTC for 5-phase machines.

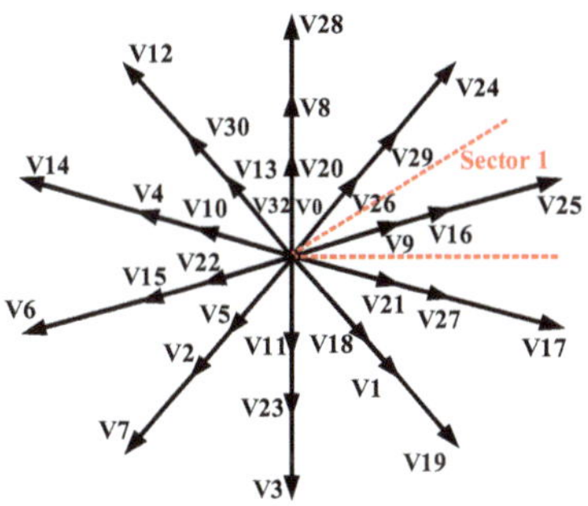

	V_6	V_{14}	V_{12}	V_{28}	V_{24}	V_{25}	V_{17}	V_{19}	V_3	V_7	$V_{0(32)}$
$\lambda_{s(1)}$	↓↓↓	↓↓	↓	↑	↑↑	↑↑↑	↑↑	↑	↓	↓↓	↑↓
$T_{e(1)}$	↓	↑↑	↑↑↑	↑↑↑	↑↑	↓	↓↓	↓↓↓	↓↓↓	↓↓	↓

	V_{15}	V_4	V_{30}	V_8	V_{29}	V_{16}	V_{27}	V_1	V_{23}	V_2	$V_{0(32)}$
$\lambda_{s(2)}$	↓↓↓	↓↓	↓	↑	↑↑	↑↑↑	↑↑	↑	↓	↓↓	↑↓
$T_{e(2)}$	↓	↑↑	↑↑↑	↑↑↑	↑↑	↓	↓↓	↓↓↓	↓↓↓	↓↓	↓

	V_{22}	V_{10}	V_{13}	V_{20}	V_{26}	V_9	V_{21}	V_{18}	V_{11}	V_5	$V_{0(32)}$
$\lambda_{s(3)}$	↓↓↓	↓↓	↓	↑	↑↑	↑↑↑	↑↑	↑	↓	↓↓	↑↓
$T_{e(3)}$	↓	↑↑	↑↑↑	↑↑↑	↑↑	↓	↓↓	↓↓↓	↓↓↓	↓↓	↓

Figure 8.44 VV selection of DTC of 5-phase machines.

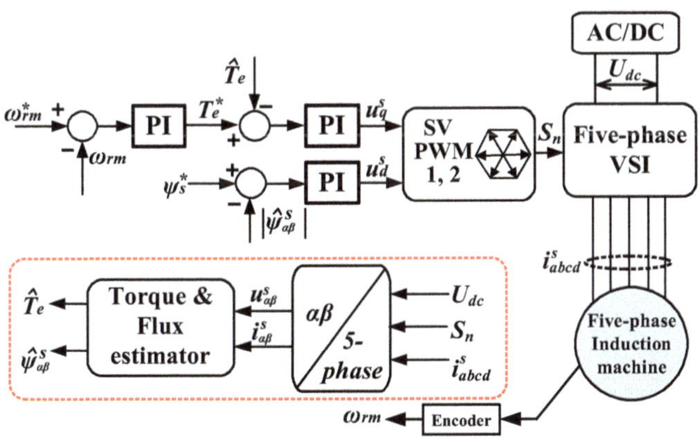

Figure 8.45 Structure of SVM-DTC for 5-phase machines.

is adopted in [26], which exhibits the same characteristics compared with DTC in DTP machines. Similarly, the D-DTC and SVM-DTC can also be extended to other multiphase systems in the same way as DTP and 5-phase systems. It should be noted that as the number of phases increases, the available VVs also increase, so higher order torque and flux regulators are required to efficiently use more vectors. This is a point that needs special attention in DTC for multiphase machines.

8.4 Model predictive control for multiphase machines

8.4.1 Model predictive control for 3-phase machines

MPC was proposed in the late 1970s. The publication, along with earlier papers reporting similar ideas with L. Richalet in 1978 [27], can be considered the earliest MPC publication. After several decades of development, it is regarded by many as one of the most important techniques in machine control. MPC is based on a model of the real system, also called a "predictive model," used to predict its future evolution. The prediction is carried out for possible switching states to determine which one minimizes a defined cost function. The basic idea of MPC is to predict the future behavior of the controlled system over a finite time horizon and compute an optimal control output while ensuring the satisfaction of given system constraints, which is defined as a cost function. In terms of machine control, the control output—specifically, the PWM signal—is calculated by solving the machine model at each sampling instant over a finite horizon. The first part of the resulting optimal output is then applied to the system until the next sampling instant when the horizon shifts and the whole procedure is repeated.

The basic structure of the MPC in a PMSM is shown in Figure 8.46. It consists of a predictive model, cost function, output optimization, and update.

8.4.1.1 Predictive model of 3-phase machines

The predictive modeling process can be divided into two steps: i.e., 3-phase PMSM modeling and state estimation. First, the 3-phase PMSM model can be written in the *dq* reference frame as:

$$\begin{cases} u_d = Ri_d + L_d\dfrac{di_d}{dt} - \omega_e L_q i_q \\ u_q = Ri_q + L_q\dfrac{di_q}{dt} + \omega_e L_d i_d + \omega_e \psi_f \end{cases} \tag{8.33}$$

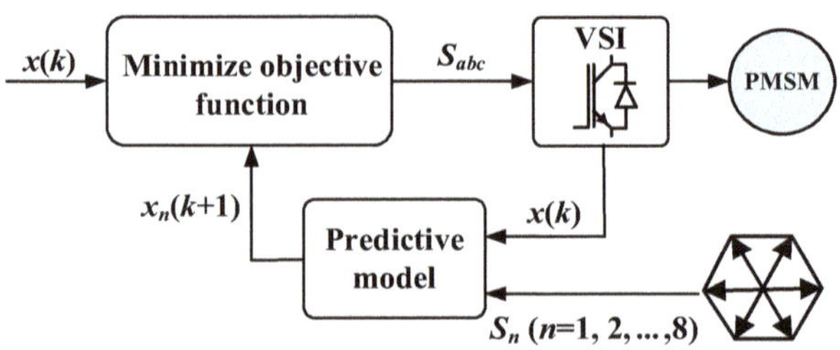

Figure 8.46 Typical MPC diagram for 3-phase PMSMs.

$$T_e = \frac{3}{2} p i_q \left[i_d (L_d - L_q) + \psi_f \right] \tag{8.34}$$

$$J \frac{d\omega_m}{dt} = T_e - T_L - B_m \omega_m \tag{8.35}$$

$$\omega_e = p\omega_m \tag{8.36}$$

where u_d, u_q, i_d, i_q, L_d, and L_q are the d- and q-axis components of the stator voltage, the current, and the inductance in the rotor reference frame, respectively. R_s, ψ_f, and ω_e represent the stator resistance, the flux linkage, and the electrical rotor speed, respectively. p is the number of pole pairs of the machine. J is the system inertia and B_m is the system friction coefficient. T_L and ω_m represent the load torque and the mechanical speed, respectively.

For current predictive control, the machine equation can be rewritten in state space as:

$$\begin{cases} \dfrac{dx(t)}{dt} = Ax(t) + Bu(t) + C \\ \&y = Dx(t) \end{cases} \tag{8.37}$$

where the controllable state variable $x = \begin{bmatrix} i_d(k) \\ i_q(k) \end{bmatrix}$, the control input variable $u = \begin{bmatrix} u_d(k) \\ u_q(k) \end{bmatrix}$, y is the controlled output variable, and $A = \begin{bmatrix} -\dfrac{R_s}{L_d} & \dfrac{L_q}{L_d}\omega_e \\ -\dfrac{L_d}{L_q}\omega_e & -\dfrac{R_s}{L_q} \end{bmatrix}$,

$B = \begin{bmatrix} \dfrac{1}{L_d} & 0 \\ 0 & \dfrac{1}{L_q} \end{bmatrix}$, $C = \begin{bmatrix} 0 \\ -\omega_e \dfrac{\psi_f}{L_q} \end{bmatrix}$, $D = \begin{bmatrix} 1 & 0 \\ 0 & 1 \end{bmatrix}$.

According to the forward Euler equation, (8.37) can be written in discrete form as follows:

$$\begin{cases} x(k+1) = A_0 x(k) + B_0 u(k) + C_0 \\ \&y(k+1) = Dx(k+1) \end{cases} \tag{8.38}$$

where k and $k+1$ represent the values of the current moment and the next moment, respectively, and

$$A_0 = \begin{bmatrix} 1 - \dfrac{T_s R_s}{L_d} & T_s \omega_e \dfrac{L_q}{L_d} \\ -T_s \omega_e \dfrac{L_d}{L_q} & 1 - \dfrac{R_s T_s}{L_q} \end{bmatrix}, B_0 = \begin{bmatrix} \dfrac{T_s}{L_d} & 0 \\ 0 & \dfrac{T_s}{L_q} \end{bmatrix}, C_0 = \begin{bmatrix} 0 \\ -\dfrac{T_s \omega_e \psi_f}{L_q} \end{bmatrix}. \tag{8.39}$$

where T_s is the sampling time.

8.4.1.2 Cost function of optimization

Because the sampling time is very short, it can be considered that the reference value of the current time $x^*(k)$ is equal to the reference value of the next time $x^*(k+1)$. The objective function can be defined as follows:

$$g = \lambda_1 \|x^*(k) - x(k+1)\|_2 + \lambda_2 \|y^*(k) - y(k+1)\|_2 + \lambda_3 \Delta g + \cdots \tag{8.40}$$

where λ_1, λ_2, λ_3 are the weighting coefficients of control objectives, and Δg is the other control constraint. "$\| \, \|_2$" is the square root of the sum of squares.

In PMSM drive systems, common multi-constraint problems include consideration of phase current constraints, torque and stator flux constraints, speed, switching frequency, and stability control. The classical cost function in MPC is model predictive current control (MPCC), model predictive torque control (MPTC), and model predictive speed control (MPSC). The cost functions for MPCC, MPTC, and MPSC are expressed as below:

$$g_{MPCC} = \left| i_\alpha^*(k) - i_\alpha(k+1) \right| + \left| i_\beta^*(k) - i_\beta(k+1) \right| \tag{8.41}$$

$$g_{MPTC} = \left| T_e^*(k) - T_e(k+1) \right| + \lambda \left| \Psi_s^*(k) - \Psi_s(k+1) \right| \tag{8.42}$$

$$g_{MPSC} = \left| \omega_r^*(k) - \omega_r(k+1) \right| + \lambda \left| \Psi_s^*(k) - \Psi_s(k+1) \right| \tag{8.43}$$

The absolute operation may be replaced by a square operation which is applicable to all cost functions which will be described later.

8.4.1.3 Optimization of output voltage vector

The conventional MPC makes n times predictions in one control cycle, which depends on the number of active VVs. Taking a 2-level voltage source inverter as an example, there are eight switching states S_n (n = 1,2,...,8) and these switching states calculate the predicted value $x_n(k+1)$. Finally, the optimal switching pulse sequence S_{abc} is output to the inverter by minimizing the objective function. Similarly, in the next control loop, the above steps are repeated, and the optimization process of each loop does not correlate.

It should be noted that the ergodic calculation and prediction bring a great computational burden and hence high requirements on the microprocessor. Meanwhile, the delay caused by the computational burden will also deteriorate the control performance of the system. In a multiphase system or multilevel inverter, the computation burden will be heavier with the increase of ergodic calculation and prediction. This is a challenge of MPC for multiphase machine systems.

8.4.2 Model predictive control for DTP machines

The modeling for DTP machines is introduced in Chapter 5 in detail. After VSD transformation, the fundamental components, and the harmonic components are mapped into different subspaces. The basic structure of MPCC for DTP machines is shown in Figure 8.47.

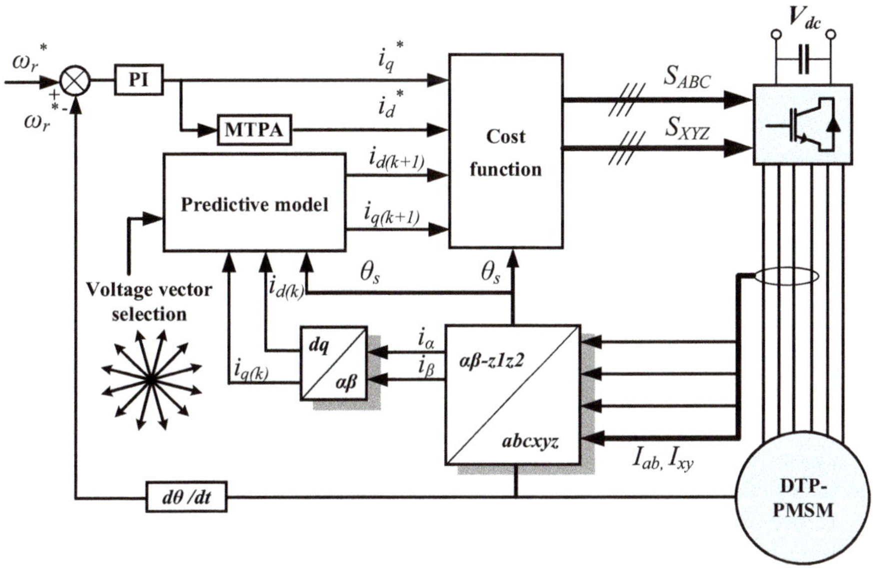

Figure 8.47 Structure diagram of MPCC in DTP machines.

The MPCC was first extended to DTP drives in [28], where only the 12 outer vectors of 64 VVs (the largest ones) are employed to reduce the computation burden. It should be noted that the 5th and 7th current harmonics are inherently significant in DTP machines, and thus, the current harmonic components should be considered in the cost function and expressed as:

$$g = \left|i_\alpha^*(k) - i_\alpha(k+1)\right| + \left|i_\beta^*(k) - i_\beta(k+1)\right| + \lambda\left(\left|i_{z1}^*(k) - i_{z1}(k+1)\right| + \left|i_{z2}^*(k) - i_{z2}(k+1)\right|\right) \tag{8.44}$$

In [29], a simple cost function is utilized in DTP machines, as expressed in (8.44). In this case, only the current harmonics mapped in the z_1z_2 subspace are focused and optimized to reduce current harmonics in DTP machines.

$$g = \sqrt{\left[i_{z1}^*(k) - i_{z1}(k+1)\right]^2 + \left[i_{z2}^*(k) - i_{z2}(k+1)\right]^2} \tag{8.45}$$

Besides, it reduces the times of the ergodic calculation for the VVs by introducing the stator flux angle, which is proposed in [30] and called "reference VV (RVV)-based". The position of the RVV is calculated as:

$$\begin{cases} v_\alpha^{ref} = R_s i_\alpha(k+1) + \dfrac{L_\alpha}{T_s}\left(i_\alpha^{ref} - i_\alpha(k+1)\right) + e_\alpha \\ v_\beta^{ref} = R_s i_\beta(k+1) + \dfrac{L_\beta}{T_s}\left(i_\beta^{ref} - i_\beta(k+1)\right) + e_\beta \end{cases} \tag{8.46}$$

$$\theta_{ref} = \arctan\left(v_\alpha^{ref}/v_\beta^{ref}\right) \tag{8.47}$$

Also, the concept of VVV, described in Section 7.5 in Chapter 7, can be employed in MPC for DTP machines. To optimally select the proper VVV, a cost function related to the VV selection is proposed in [31], as

$$g = \left|\left|v^{ref}\right| - |vv_i|\right| \tag{8.48}$$

where vv_i represents the selected VVV.

The history of VV selection in MPCC can be summarized as follows. First, the selected VV combined with a zero VV is introduced in [32], i.e., one-step modulation predictive current control (OSPC) or PWM-MPCC, where the current harmonics can be suppressed, and thus, the steady-state performance can be improved correspondingly. Similar to [32], other papers, including [28,31,33,34] utilize the largest VVs as the active vectors, and consequently, they all suffer from the non-negligible current harmonics caused by uncontrolled harmonic-related components.

Hence, a VVVs strategy, synthesized by the largest and the second largest actual vectors (synthesized by two groups of VVs), is introduced in MPCC for DTP machines, as described in [31,35]. Similar to the VVV method in the DTC method, the virtual vectors ensure null z_1z_2 voltages on average during the sampling period, and thus, reduce the current harmonics. For instance, two groups of virtual vectors (synthesized by three groups of VVs) aiming at harmonic current reduction are introduced in [35], and the deadbeat current control method is also employed to reduce the computing time considering the increased number of VVs. This technique, i.e., two synthesized groups of virtual vectors, is further refined in [36,37], where a two-step procedure and a dual VV are involved to compensate for the harmonics caused by machine asymmetries, dead time effects, and back-EMF harmonics. Besides, some MPCC methods aiming at other optimal objectives are also developed to reduce the current harmonics. By way of example, its application in torque ripple reduction and DC-link voltage utilization is reported in [38]. Furthermore, an autoregressive with an exogenous variable model and a disturbance observer are employed to compensate for the effects of parameter mismatch errors and un-modeled dynamics, as detailed in [39,40], respectively.

The basic structure of the MPTC for the DTP machine is presented in Figure 8.48. It generally aims at minimizing the torque ripple and flux ripple, and the cost function can be expressed as:

$$g = \left|T_e^* - T_e^{k+1}\right| + \lambda\left|\psi_s^* - \psi_s^{k+1}\right| \tag{8.49}$$

In general, the global optimization for MPTC focuses on the optimization in three processes, i.e., predictive model, cost function, and output VV selection.

8.4.2.1 Optimization of predictive model

By way of example, a simplified deadbeat solution is developed in [41] to simplify the predictive model of MPTC for DTP machines. In comparison to classical

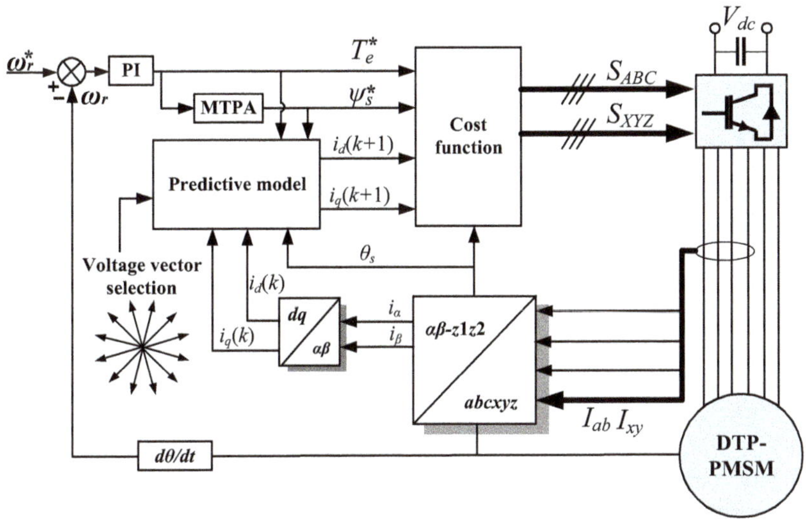

Figure 8.48 Structure diagram of MPTC in DTP machines.

deadbeat modeling, which needs to solve an equation set containing a linear equation and a quadratic equation, the simple modeling in [41] is expressed as:

$$\begin{cases} Au_d(k+1) + Bu_q(k+1) = \Delta T_e + AP - BQ \\ Cu_d(k+1) + Du_q(k+1) = \Delta \psi_s^2 - CP + DQ \end{cases} \tag{8.50}$$

where k represents the kth sampling step, i.e., the current period, $\Delta T_e = \Delta T_{e-\text{ref}} - \Delta T_e(k+1)$, $\Delta\psi_s^2 = \Delta\psi_{s-\text{ref}}^2 - \Delta\psi_s^2(k+1)$. $\Delta T_{e-\text{ref}}$ and $\Delta\psi_{s-\text{ref}}$ are the reference values of the torque and the flux amplitude in $(k+2)$th step, respectively. The coefficients A, B, C, D, P, and Q are expressed as:

$$\begin{cases} A = \left[3pT_s(L_d - L_q)/L_d\right] i_q(k+1) \\ B = 3pT_s\psi_{PM}/L_q + \left[3pT_s(L_d - L_q)/L_d\right] i_d(k+1) \\ C = 2T_s\psi_f + 2T_sL_di_d(k+1) \\ D = 2T_sL_qi_q(k+1) \\ P = \omega_e(k+1)L_qi_q(k+1) - R_si_d(k+1) \\ Q = \omega_e(k+1)L_di_d(k+1) + \omega_e(k+1)\psi_{PM} + R_si_q(k+1) \end{cases} \tag{8.51}$$

where $i_d(k+1)$ and $i_q(k+1)$ are the predicted stator current in $(k+1)$th step, which can be calculated by the sampling value of current and voltage in the (k)th step, expressed as:

$$\begin{cases} i_d(k+1) = (1 - R_sT_s)i_d(k) + \omega_eL_qT_si_q(k) \\ i_q(k+1) = -\omega_eL_dT_si_d(k) + (1 - R_sT_s)i_q(k) \end{cases} \tag{8.52}$$

8.4.2.2 Optimization of cost function

For instance, to eliminate the current components in the z_1z_2 subspace, the cost function for MPTC in [42,43] is written as:

$$g = \left|T_e^* - T_e^{k+1}\right| + \lambda_1\left|\psi_s^* - \psi_s^{k+1}\right| + \lambda\left(\left|\psi_{z1}^* - \psi_{z1}^{k+1}\right| + \left|\psi_{z2}^* - \psi_{z2}^{k+1}\right|\right) \quad (8.53)$$

where both the current harmonics and the torque ripple could be reduced simultaneously.

8.4.2.3 Fixed switching frequency

The VVV strategy in MPTC is the same as the VVV method in DTC and MPC (see Chapter 7), and it is employed in [44] to reduce current harmonics for DTP machines. Similarly, the selected VV combined with a zero vector is also developed in [45] to reduce the torque and flux ripples, with a fixed switching frequency in MPTC, where the zero vector is inserted in the VV output and a duty ratio optimization process is introduced to achieve a fixed switching frequency, as illustrated in Figure 8.49. The cost function is the same as the classical MPTC for DTP machines, as written in (8.46).

There has been limited research on MPSC in DTP machine systems. One reason for this is that the speed fluctuation in such systems is not significant, and the speed performance requirements are usually not stringent. However, it is foreseeable that the MPSC will need to be studied in specific situations, particularly in scenarios involving low or zero speed.

8.4.2.4 Examples of voltage vector selection

Various VV combinations can be applied to MPCs for DTP PMSM drives. In this case, six typical combinations are evaluated, i.e., 2*L* VVs, 4*L* VVs, 2 VVVs [46], 4 VVVs [47], *L*+*ML* VVs, and *L*+*M* VVs, where *L*, *ML*, *M*, and *S* refer to the groups of voltage vectors from the largest to the smallest magnitude in the $\alpha\beta$ subspace as shown in Figure 7.18 and Table 7.3 in Chapter 7. In this section, their capabilities for suppressing z_1z_2 harmonic currents are evaluated experimentally. In all the

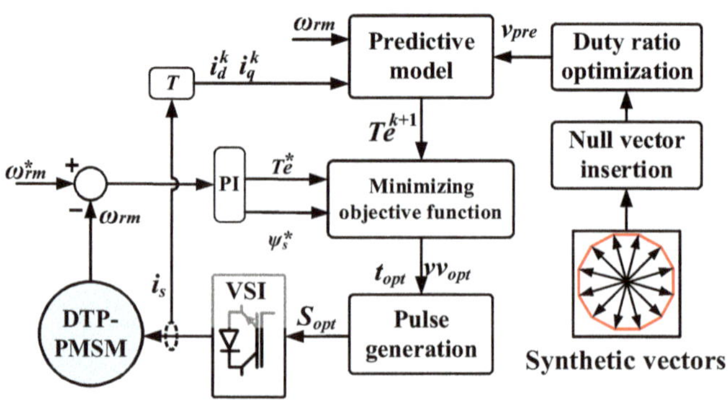

Figure 8.49 Structure diagram of MPTC with duty ratio optimization.

experiments, the MPCC is employed based on the prototype DTP IPMSM (Machine 3 in Appendix A), where the reference currents in the *d*- and *q*-axes are set to 0A and 2.5A, respectively, and the rotor speed is fixed to 300 r/min.

To illustrate the effectiveness of MPCs under different z_1z_2 harmonic currents, the prototype machine is operated without (original) and with an injection of an additional 5th EMF harmonic (2.5V) in the supply voltage, as shown in Figure 8.50. The control diagram for the MPCC is shown in Figure 8.47. The cost function used in all six MPCs for suppressing the current harmonics is given in (8.44), where the weighting factor, λ, is chosen as 0.5. The measured results for six MPCCs are shown in Figures 8.51–8.53 for the cases when the 5th back-EMF harmonic is high and low (with and without injecting the extra 5th EMF harmonic).

Several phenomena can be observed.

1. The MPCC based on 4VVVs, *L*+*ML* VVs, and *L*+*M* VVs can all obtain good control performance under both low and high z_1z_2 EMF harmonics, as shown in Figures 8.52 and 8.53.
2. For the 4*L* VVs, the current distortion is obvious under high z_1z_2 EMF harmonics, as shown in Figure 8.51, due to the small z_1z_2 amplitude of the *L* VVs.

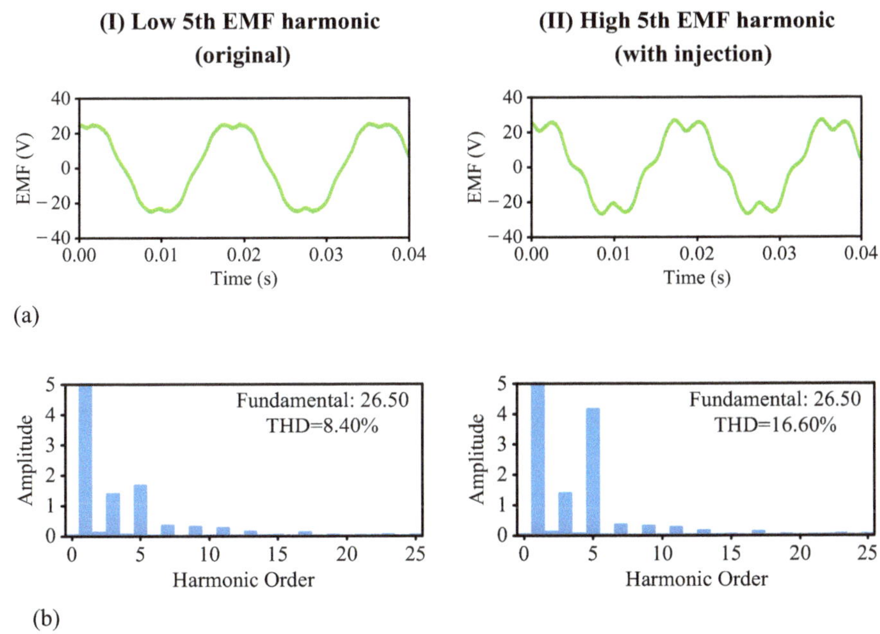

Figure 8.50 Back EMF waveforms and spectra at 300 r/min. (Low 5th EMF harmonic: the original prototype machine. High 5th EMF harmonic: mimicked and realized by injecting 2.5 V in supply voltage.) (a) Waveforms. (b) Spectra.

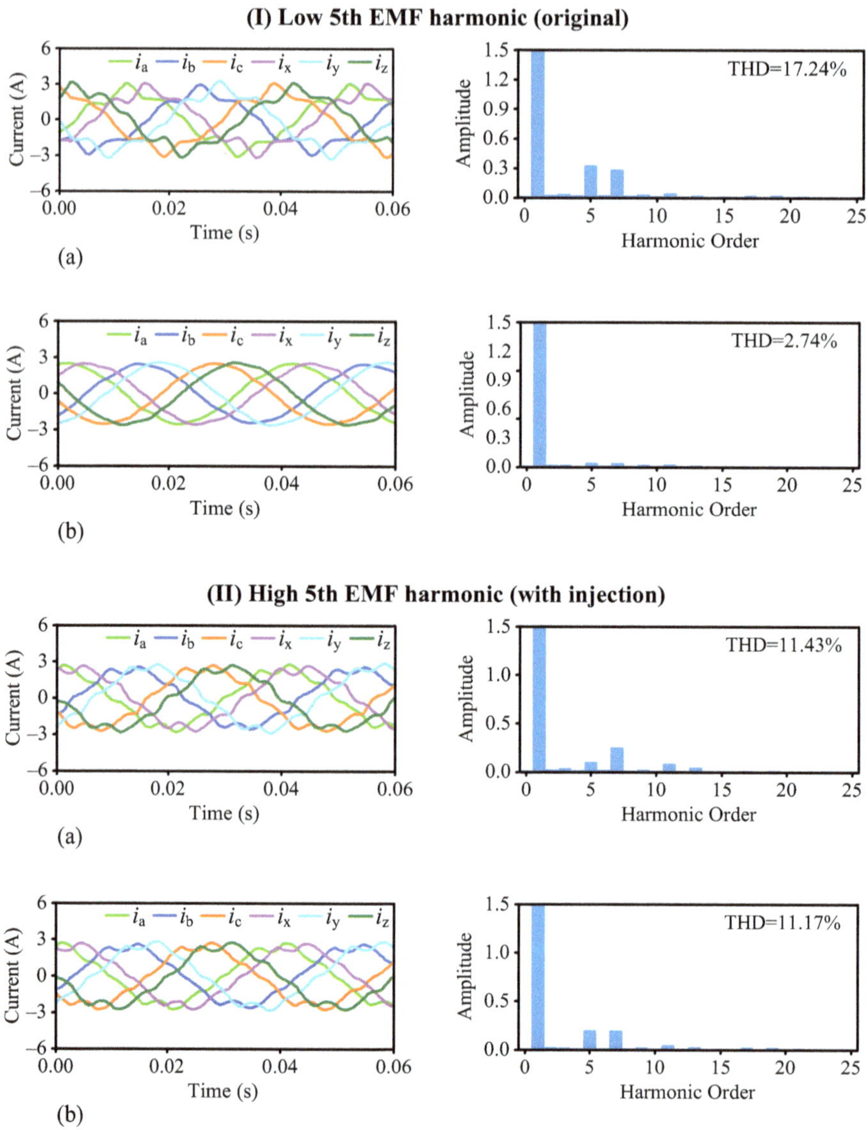

Figure 8.51 Measured phase currents and spectra with MPCC based on 2L and 4L VVs. (a) 2L VVs. (b) 4L VVs.

Nevertheless, when the z_1z_2 EMF harmonics are low, good current harmonic suppression control performance (the same as 4 VVVs, L+ML VVs, and L+M VVs) can also be achieved by 4L VVs, as can be seen in Figure 8.51. It can be explained due to the small impedance for z_1z_2 currents in this case, and thus, even L VVs can effectively suppress the z_1z_2 current harmonics.

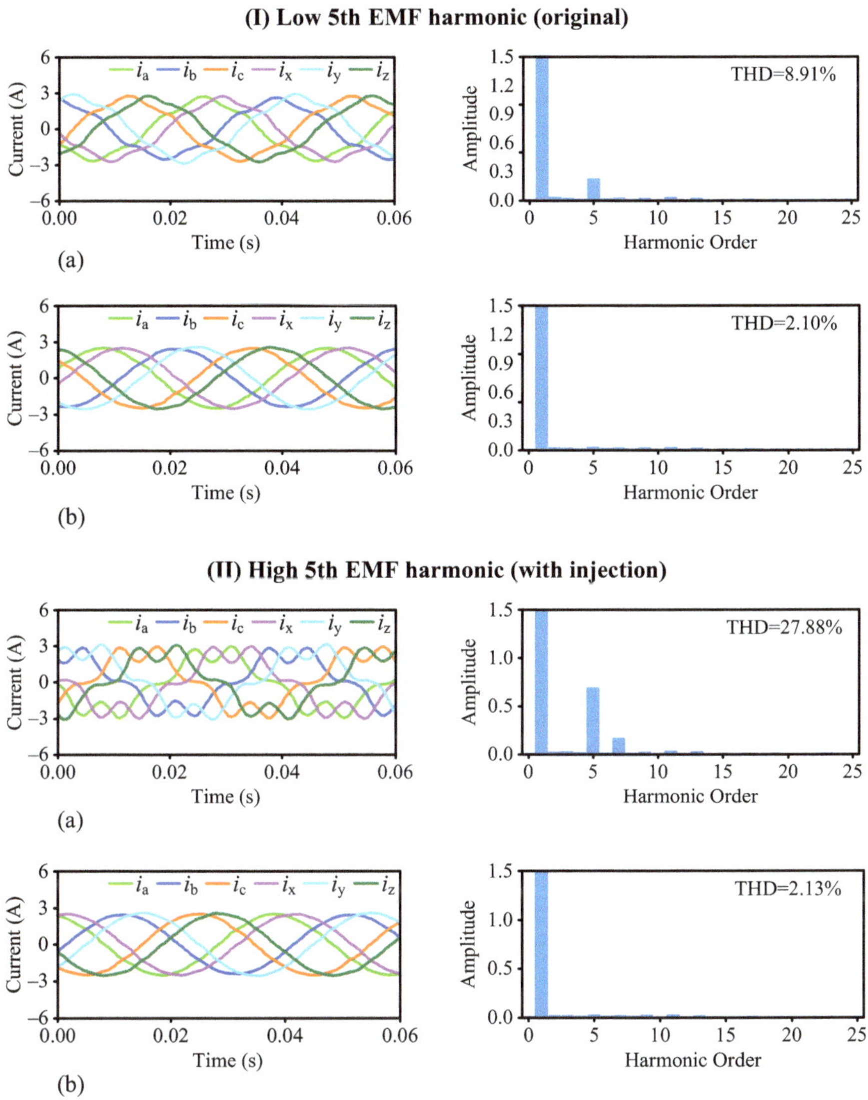

Figure 8.52 Measured phase currents and spectra with MPCC based on 2 VVVs and 4 VVVs. (I) Low 5th EMF harmonic (original) (a) 2L VVVs. (b) 4L VVVs.

3. The 4 VVVs and 4L VVs exhibit better current harmonic suppression control performance than the 2 VVVs and 2L VVs, respectively. The current THD of 2L VVs under high z_1z_2 EMF harmonics is lower than that under low z_1z_2 EMF harmonics, as shown in Figure 8.51. It is attributable that the phasors of the z_1z_2

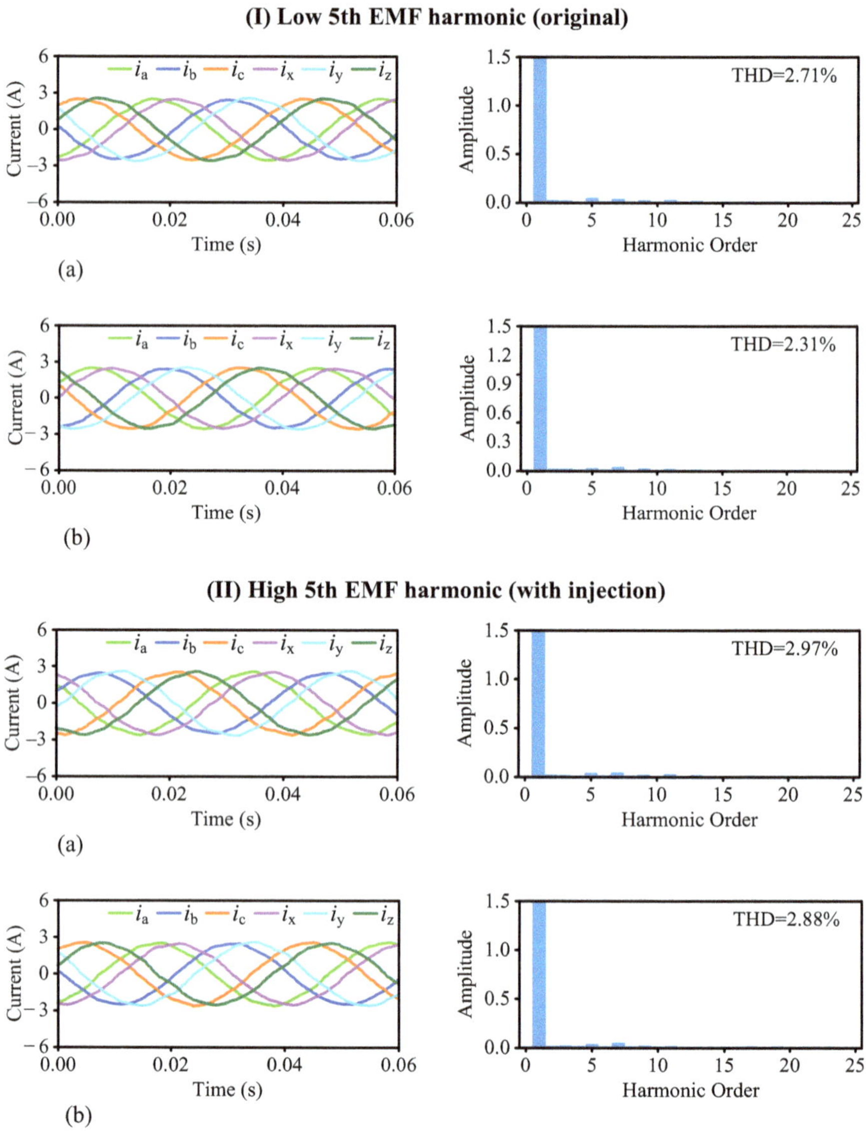

Figure 8.53 Measured phase currents and spectra under MPCC based on L+ML VVs and L+M VVs. (a) L+ML VVs. (b) L+M VVs.

current harmonics introduced by $2L$ VVs and high z_1z_2 EMF harmonics are canceled out with each other.

4. For the 2 VVVs, they do not introduce any z_1z_2 current harmonics due to zero z_1z_2 projections of VVVs. Consequently, the current THD of the 2 VVVs under high z_1z_2 EMF harmonics is larger than that under low z_1z_2 EMF harmonics, as shown in Figure 8.52.

8.4.3 Model predictive control for other multiphase machines

The MPC of other multiphase machines is similar to that of DTP machines. It should be noted that the multiphase machine system provides more VVs to select, and consequently, the output VV optimization is a challenging topic. For instance, the VVV technique, introduced in Section 7.4, is extended into the 5-phase machines in [48]. The structure of the MPC for the 5-phase machine is similar to the MPC for DTP machines, as shown in Figure 8.54.

Accordingly, the cost function is expressed in (8.54) and (8.55), since the third harmonics are reduced automatically by adopting the VVV technique. Similarly, this method is also effective in MPC for 5-phase machines, and the 3rd current harmonics are reduced significantly by applying the VVV technique.

$$g = \left|i_{d1}^{*}(k) - i_{d1}(k+1)\right| + \left|i_{q1}^{*}(k) - i_{q1}(k+1)\right| + \left|i_{d3}^{*}(k) - i_{d3}(k+1)\right| + \left|i_{q3}^{*}(k) - i_{q3}(k+1)\right| \tag{8.54}$$

$$g = \left|i_{d1}^{*}(k) - i_{d1}(k+1)\right| + \left|i_{q1}^{*}(k) - i_{q1}(k+1)\right| \tag{8.55}$$

Similarly, the basic structure of the MPC of the 7-phase motor is like that of the DTP and the 5-phase machine, as shown in Figure 8.55. The MPC for a 7-phase machine system was first investigated in [49], and the optimization target is to reduce the current harmonics. The classical VVV technique is applied, and an accordingly optimized VV selection strategy is employed in the MPC for 7-phase machines. Theoretically, the higher the number of phases, the more space vectors can be selected, so it is more difficult to select the optimal space VV.

Besides, in 9-phase machine systems, $2^9 = 512$ active VVs are available. Although the VVV technique is useful to reduce the complexity of vector selection, the number of virtual vectors is still very large. In [50], the VVV technique is applied in a 9-phase machine system and the current harmonics, as well as the common-mode voltage, are set to the optimized objective. To reduce the complexity of all virtual vectors, Garcia-Entrambasaguas *et al.* [50] only discuss the virtual vectors generated by several groups of vectors with the largest amplitudes, i.e., V_2 and V_3, as shown in Figure 8.56, where

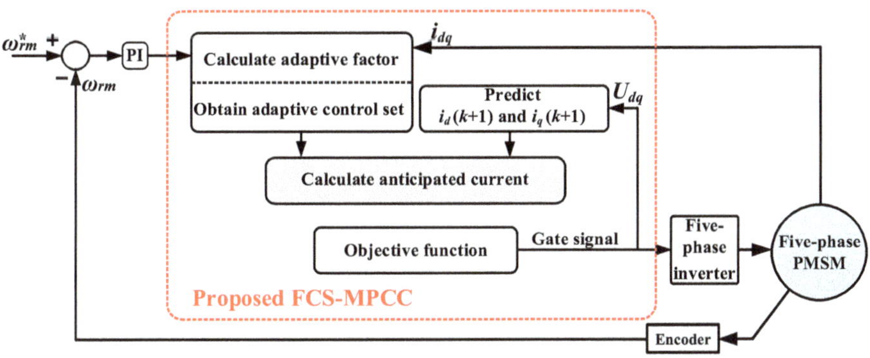

Figure 8.54 Basic structure for MPC of the 5-phase machine system.

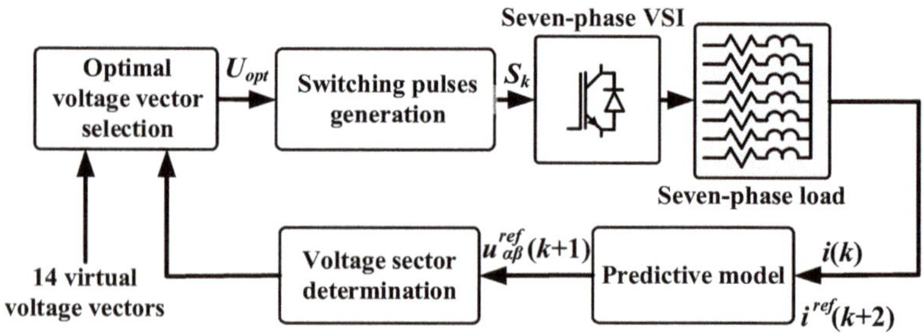

Figure 8.55 Basic structure for MPC of the 7-phase machine system.

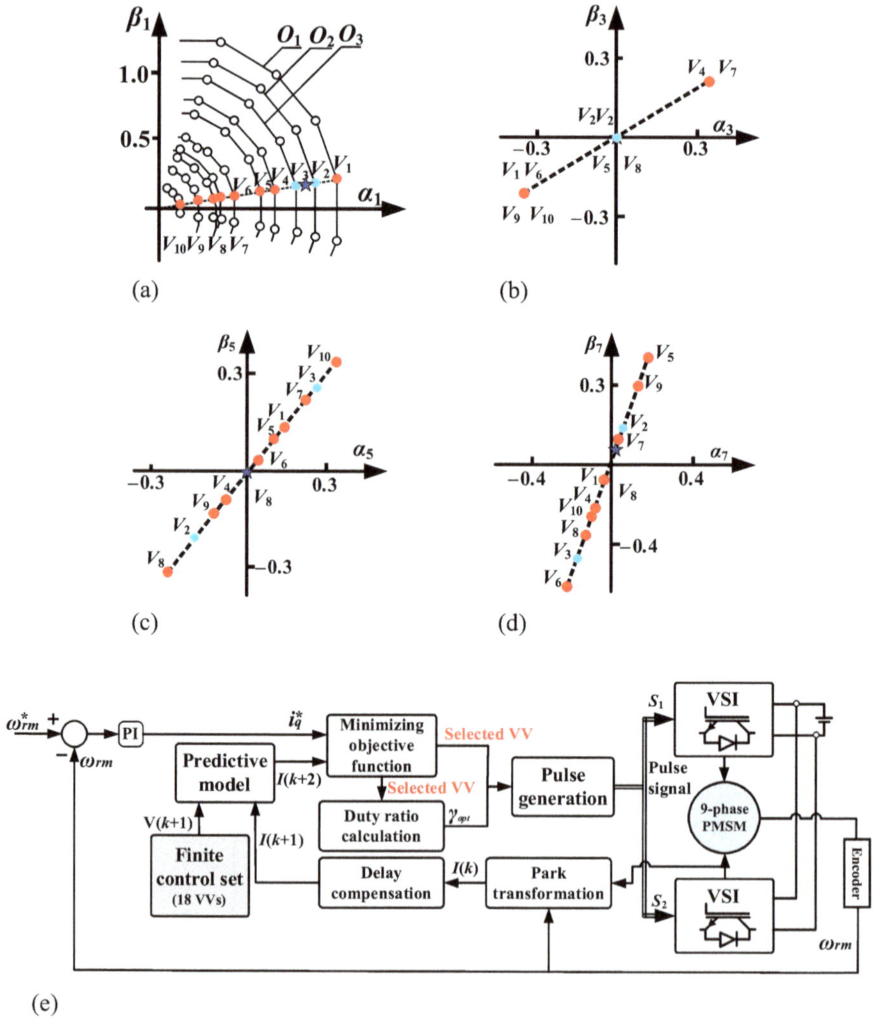

Figure 8.56 VVV techniques and basic structure for MPC in 9-phase machines. (a) α_1–β_1. (b) α_3–β_3. (c) α_5–β_5. (d) α_7–β_7. (e) MPC in 9-phase machines.

good performance in terms of current harmonics and common-mode voltage can be achieved.

Although there are only a few papers conducted for multiphase machines with higher phases, it can be extrapolated that the research on MPC for multiphase machines can be compared and generalized. For instance, the VVV technique can be employed in the MPC vector selection optimization of any multiphase machine. Additionally, optimizing the model is crucial, given that the multiphase machine model is intricate, necessitating a more precise mathematical model to forecast the machine's state parameters in the future. Simplifying the MPC algorithm for multiphase machines is also essential, as demonstrated by [51]. For complex multiphase systems, if the MPC algorithm is not simplified, the computational load would be significantly increased, making it unacceptable in the current real-time control field.

8.5 Summary

This chapter describes various control strategies for multiphase machines, including FOC, DTC, and MPC, which are analyzed for both DTP machines and other multiphase machines.

First, FOC, as a widely used control strategy for multiphase machines, regulates the current in the *dq* axes and thus allows independent control of both torque and flux. The advantages of FOC include fast and accurate transient response, high efficiency, and good steady-state performance. However, FOC is sensitive to parameter variations, and thus, accurate machine parameters are required.

Second, DTC is another popular control strategy for multiphase machines, which directly controls the torque and flux of the machine, without requiring the machine's parameters. The advantages of DTC include fast dynamic response, robustness to parameter variations, and simple implementation. However, DTC has a high torque ripple and may exhibit significant steady-state error.

Third, MPC is a control strategy that predicts the future behavior of the machine and optimizes the control inputs accordingly. MPC provides good tracking performance, high accuracy, and robustness to parameter variations. It also allows for the inclusion of constraints on the control inputs and outputs. However, MPC has a high computational burden and may be sensitive to modeling errors.

Finally, these control strategies are compared and summarized in Table 8.2. In conclusion, the choice of control strategy for multiphase machines depends on the specific application requirements, the steady- and dynamic-state performances, and the trade-off between performance and complexity. FOC and DTC are the most widely used control strategies for multiphase machines, while MPC is also a promising strategy for multiphase machines.

Table 8.2 Characteristics for different control schemes in multiphase machines

	FOC	DTC	MPC
Internal controller	Two PI regulators (current loop)	Two hysteresis regulators (torque and flux loops)	One cost function (depending on optimized objective)
External controller (speed loop)	PI	PI	PI
PWM modulator	Yes	No	No
Sensorless	No	Yes	No
System constraint inclusion	Difficult	Difficult	Easy
Steady-state performance	High	Low	Normal
Dynamic performance	Slow	Fast	Fast
Switching frequency	Fixed	Variable	Variable
Implementation complexity	Normal	Simple	Complex
Parameter sensitivity	Normal	Normal	High
Computation burden	Low	Low	High
Robustness	High	High	Low

References

[1] Y. Hu, Z. Q. Zhu, and K. Liu, "Current control for dual three-phase permanent magnet synchronous motors accounting for current unbalance and harmonics," *IEEE J. Emerg. Sel. Topics Power Electron.*, vol. 2, no. 2, pp. 272–284, 2014.

[2] Z. Ruan, W. Song, and Y. Yan, "Current harmonic suppression for dual three-phase permanent magnet synchronous motor drives," *IEEE Access*, vol. 7, pp. 143888–143898, 2019.

[3] J. Karttunen, S. Kallio, P. Peltoniemi, P. Silventoinen, and O. Pyrhönen, "Decoupled vector control scheme for dual three-phase permanent magnet synchronous machines," *IEEE Trans. Ind. Electron.*, vol. 61, no. 5, pp. 2185–2196, 2014.

[4] L. Yan, Z. Q. Zhu, J. Qi, *et al.*, "Suppression of major current harmonics for dual three-phase PMSMs by virtual multi three-phase systems," *IEEE Trans. Ind. Electron.*, vol. 69, no. 6, pp. 5478–5490, 2022.

[5] H. Che, E. Levi, M. Jones, W. Hew, and N. Rahim, "Current control methods for an asymmetrical six-phase induction motor drive," *IEEE Trans. Power Electron.*, vol. 29, no. 1, pp. 407–417, 2014.

[6] L. Yan, Z. Q. Zhu, J. Qi, *et al.*, "Enhancement of disturbance rejection capability in dual three-phase PMSM system by using virtual impedance," *IEEE Trans. Ind. Appl.*, vol. 57, no. 5, pp. 4901–4912, 2021.

[7] K. Wang, Z. Q. Zhu, Y. Ren, and G. Ombach, "Torque improvement of dual three-phase permanent-magnet machine with third-harmonic current injection," *IEEE Trans. Ind. Electron.*, vol. 62, no. 11, pp. 6833–6844, 2015.

[8] K. Wang, J. Y. Zhang, Z. Y. Gu, H. Y. Sun, and Z. Q. Zhu, "Torque improvement of dual three-phase permanent magnet machine using zero sequence components," *IEEE Trans. Magn.*, vol. 53, no. 11, pp. 1–4, 2017.
[9] Y. Hu, Z. Q. Zhu, and M. Odavic, "Torque capability enhancement of dual three-phase PMSM drive with fifth and seventh current harmonics injection," *IEEE Trans. Ind. Appl.*, vol. 53, no. 5, pp. 4526–4535, 2017.
[10] K. Wang, Z. Y. Gu, Z. Q. Zhu, and Z. Z. Wu, "Optimum injected harmonics into magnet shape in multiphase surface-mounted PM machine for maximum output torque," *IEEE Trans. Ind. Electron.*, vol. 64, no. 6, pp. 4434–4443, 2017.
[11] L. Parsa and H. A. Toliyat, "Five-phase permanent-magnet motor drives," *IEEE Trans. Ind. Appl*, vol. 41, no. 1, pp. 30–37, 2005.
[12] K. Wang, Z. Y. Gu, C. Liu, and Z. Q. Zhu, "Design and analysis of a five-phase SPM machine considering third harmonic current injection," *IEEE Trans. Energy Convers.*, vol. 33, no. 3, pp. 1108–1117, 2018.
[13] I. Takahashi and T. Noguchi, "A new quick-response and high efficiency control strategy of an induction machine," *IEEE Trans. Ind. Appl.*, vol. 22, pp. 820–827, 1986.
[14] M. Depenbrock, "Direct self-control (DSC) of inverter-fed induction machine," *IEEE Trans. Power Electron.*, vol. 3, pp. 420–429, 1988.
[15] H. A. Toliyat and H. Xu, "A novel direct torque control (DTC) method for five-phase induction machines," in *Proc. IEEE APEC*, New Orleans, LA, 2000, pp. 162–168.
[16] R. Bojoi, F. Farina, G. Griva, F. Profumo, and A. Tenconi, "Direct torque control for dual three-phase induction motor drives," *IEEE Trans. Ind. Appl.*, vol. 41, no. 6, pp. 1627–1636, 2005.
[17] M. Pacas and J. Weber, "Predictive direct torque control for the PM synchronous machine," *IEEE Trans. Ind. Electron.*, vol. 52, no. 5, pp. 1350–1356, 2005.
[18] Y. Ren, Z. Q. Zhu, J. E. Green, Y. Li, S. Zhu, and Z. Li, "Improved duty-ratio-based direct torque control for dual three-phase permanent magnet synchronous machine drives," *IEEE Trans. Ind. Appl.*, vol. 55, no. 6, pp. 5843–5853, 2019.
[19] F. Niu, K. Li, and Y. Wang, "Direct torque control for permanent-magnet synchronous machines based on duty ratio modulation," *IEEE Trans. Ind. Electron.*, vol. 62, no. 10, pp. 6160–6170, 2015.
[20] B. H. Kenny and R. D. Lorenz, "Stator- and rotor-flux-based deadbeat direct torque control of induction machines," *IEEE Trans. Ind Appl.*, vol. 39, no. 4, pp. 1093–1101, 2003.
[21] A. Berzoy, J. Rengifo, and O. Mohammed, "Fuzzy predictive DTC of induction machines with reduced torque ripple and high-performance operation," *IEEE Trans. Power Electron.*, vol. 33, no. 3, pp. 2580–2587, 2018.
[22] W. Wang, H. Lin, H. Yang, W. Liu, and S. Lyu, "Second-order sliding mode-based direct torque control of variable-flux memory machine," *IEEE Access*, vol. 8, pp. 34981–34992, 2020.
[23] Y. Ren and Z. Q. Zhu, "Enhancement of steady-state performance in direct torque controlled dual-three phase permanent magnet synchronous machine drives with modified switching table," *IEEE Trans. Ind. Electron.*, vol. 62, no. 6, pp. 3338–3350, 2015.

[24] H. A. Toliyat and H. Xu, "A novel direct torque control (DTC) method for five-phase induction machines," *15th Annual IEEE Applied Power Electronics Conf. Expos.*, vol. 1, New Orleans, LA, USA, 2000, pp. 162–168.

[25] U. R. Muduli, B. Chikondra, and R. K. Behera, "Space vector PWM based DTC scheme with reduced common mode voltage for five-phase induction motor drive," *IEEE Trans. Power Electron.*, vol. 37, no. 1, pp. 114–124, 2022.

[26] P. Garcia-Entrambasaguas, I. Zoric, I. González-Prieto, M. J. Duran, and E. Levi, "Direct torque and predictive control strategies in nine-phase electric drives using virtual voltage vectors," *IEEE Trans. Power Electron.*, vol. 34, no. 12, pp. 12106–12119, 2019.

[27] J. Richalet, A. Rault, J. L. Testud, and J. Japon, "Model predictive heuristic control: application to industrial processes," *Automatica*, vol. 14, no. 5, pp. 413–428, 1978.

[28] F. Barrero, M. R. Arahal, R. Gregor, S. Toral, and M. J. Duran, "A proof of concept study of predictive current control for VSI-driven asymmetrical dual three-phase AC machines," *IEEE Trans. Ind. Electron.*, vol. 56, no. 6, pp. 1937–1954, 2009.

[29] Y. Luo and C. Liu, "Model predictive control for a six-phase PMSM motor with a reduced-dimension cost function," *IEEE Trans. Ind. Electron.*, vol. 67, no. 2, pp. 969–979, 2020.

[30] Y. Luo and C. Liu, "Elimination of harmonic currents using a reference voltage vector based-model predictive control for a six-phase PMSM motor," *IEEE Trans. Power Electron.*, vol. 34, no. 7, pp. 6960–6972, 2019.

[31] I. Gonzalez-Prieto, M. J. Duran, J. J. Aciego, C. Martin, and F. Barrero, "Model predictive control of six-phase induction motor drives using virtual voltage vectors," *IEEE Trans. Ind. Electron.*, vol. 65, no. 1, pp. 27–37, 2018.

[32] F. Barrero, M. Arahal, R. Gregor, S. Toral, and M. Duran, "One-step modulation predictive current control method for the asymmetrical dual three-phase induction machine," *IEEE Trans. Ind. Electron.*, vol. 56, no. 6, pp. 1974–1983, 2009.

[33] M. Duran, J. Prieto, F. Barrero, and S. Toral, "Predictive current control of dual three-phase drives using restrained search techniques," *IEEE Trans. Ind. Electron.*, vol. 58, no. 8, pp. 3253–3263, 2011.

[34] F. Barrero, J. Prieto, E. Levi, *et al.*, "An enhanced predictive current control method for asymmetrical six-phase motor drives," *IEEE Trans. Ind. Electron.*, vol. 58, no. 8, pp. 3242–3252, 2011.

[35] J. J. Aciego, I. González Prieto, and M. J. Duran, "Model predictive control of six-phase induction motor drives using two virtual voltage vectors," *IEEE J. Emerg. Sel. Top. Power Electron.*, vol. 7, no. 1, pp. 321–330, 2019.

[36] P. F. Goncalves, S. M. Cruz, and A. M. Mendes, "Bi-subspace predictive current control of six-phase PMSM drives based on virtual vectors with optimal amplitude," *IET Elect. Power Appl.*, vol. 13, no. 11, pp. 1672–1683, 2019.

[37] P. F. C. Gonçalves, S. M. A. Cruz, and A. Mendes, "Multistage predictive current control based on virtual vectors for the reduction of current harmonics in six-phase PMSMs," *IEEE Trans. Energy Convers.*, vol. 36, no. 2, pp. 1368–1377, 2021.

[38] J. J. Aciego, I. Gonzalez-Prieto, M. J. Duran, M. Bermudez, and P. Salas-Biedma, "Model predictive control based on dynamic VVs for six-phase

induction machines," *IEEE J. Emerg. Sel. Topics Power Electron.*, vol. 9, no. 3, pp. 2710–2722, 2021.

[39] M. Bermudez, M. R. Arahal, M. J. Duran, and I. Gonzalez-Prieto, "Model predictive control of six-phase electric drives including ARX disturbance estimator," *IEEE Trans. Ind. Electron.*, vol. 68, no. 1, pp. 81–91, 2021.

[40] P. F. C. Gonçalves, S. M. A. Cruz, and A. M. S. Mendes, "Disturbance observer based predictive current control of six-phase PMSMs for the mitigation of steady-state errors and current harmonics," *IEEE Trans. Ind. Electron.*, vol. 69, no. 1, pp. 130–140, 2022.

[41] W. Wang, C. Liu, S. Liu, and H. Zhao, "Model predictive torque control for dual three-phase PMSMs with simplified deadbeat solution and discrete space-vector modulation," *IEEE Trans. Energy Convers.*, vol. 36, no. 2, pp. 1491–1499, 2021.

[42] Y. Luo and C. Liu, "A flux constrained predictive control for a six-phase PMSM motor with lower complexity," *IEEE Trans. Ind. Electron.*, vol. 66, no. 7, pp. 5081–5093, 2019.

[43] Y. Luo and C. Liu, "A simplified model predictive control for a dual three-phase PMSM with reduced harmonic currents," *IEEE Trans. Ind. Electron.*, vol. 65, no. 11, pp. 9079–9089, 2018.

[44] Y. Luo and C. Liu, "Model predictive control for a six-phase PMSM with high robustness against weighting factor variation" *IEEE Trans. Ind. Appl.*, vol. 55, no. 3, pp. 2781–2791, 2019.

[45] Y. Luo and C. Liu, "Multi-vector-based model predictive torque control for a six-phase PMSM motor with fixed switching frequency," *IEEE Trans. Energy Convers.*, vol. 34, no. 3, pp. 1369–1379, 2019.

[46] J. J. Aciego, I. Gonzalez Prieto, and M. J. Duran, "Model predictive control of six-phase induction motor drives using two virtual voltage vectors," *IEEE J. Emerging Sel. Top. Power Electron.*, vol. 7, no. 1, pp. 321–330, 2019.

[47] P. F. C. Goncalves, S. M. A. Cruz, and A. M. S. Mendes, "Multistage predictive current control based on virtual vectors for the reduction of current harmonics in six-phase PMSMs," *IEEE Trans. Energy Convers.*, vol. 36, no. 2, pp. 1368–1377, 2021.

[48] W. Wang, Y. Fan, S. Chen, and Q. Zhang, "Finite control set model predictive current control of a five-phase PMSM with virtual voltage vectors and adaptive control set," *CES Trans. Elec. Mach. Syst.*, vol. 2, no. 1, pp. 136–141, 2018.

[49] H. C. Vu and H. H. Lee, "Model-predictive current control scheme for seven-phase voltage-source inverter with reduced common-mode voltage and current harmonics," *IEEE J. Emerg. Sel. Topics Power Electron.*, vol. 9, no. 3, pp. 3610–3621, 2021.

[50] P. Garcia-Entrambasaguas, I. Zoric, I. González-Prieto, M. J. Duran, and E. Levi, "Direct torque and predictive control strategies in nine-phase electric drives using virtual voltage vectors," *IEEE Trans. Power Electron.*, vol. 34, no. 12, pp. 12106–12119, 2019.

[51] H. Wang, X. Wu, X. Zheng, and X. Yuan, "Virtual voltage vector based model predictive control for a nine-phase open-end winding PMSM with a common DC bus," *IEEE Trans. Ind. Electron.*, vol. 69, no. 6, pp. 5386–5397, 2022.

Chapter 9
Sensorless control of multiphase PMSMs

9.1 Introduction

Rotor position information is required for controlling conventional single three-phase (STP) and multiphase PMSMs. It is desirable to use sensorless control techniques to replace physical rotor position sensors, such as a resolver or encoder, to improve the drive system's reliability and reduce the size and cost. High-frequency (HF) carrier signal injection techniques are usually employed for zero-speed and low-speed operation. Once the operation speed is above a certain level, the electromotive force (EMF) is detectable. Therefore, fundamental model-based sensorless control techniques, which usually fail at zero and low speeds due to low signal-to-noise ratio, can be employed at medium and high speeds [1]. In this chapter, both fundamental model-based and saliency tracking-based techniques, as well as the transition between them, will be described for dual three-phase (DTP) permanent magnet synchronous motors (PMSMs), as shown in Figure 9.1. For simplicity, those for STP PMSMs will be described first.

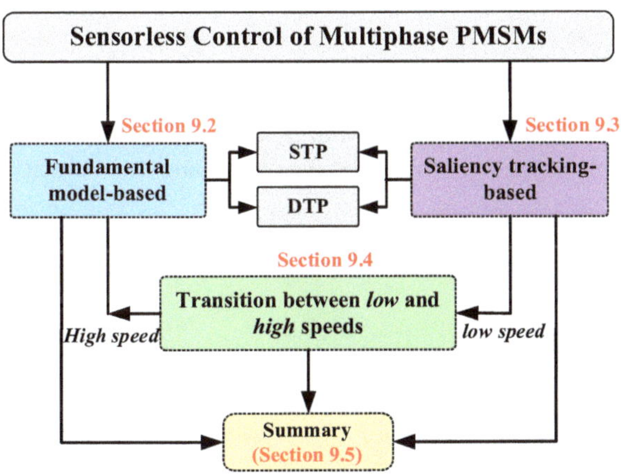

Figure 9.1 Structure of Chapter 9.

9.2 Fundamental model-based sensorless control methods

Fundamental model-based sensorless control methods use electrical signals (voltages and currents) to estimate EMF or flux-linkage which contains the rotor position information based on the fundamental mathematical model of PMSMs, as shown in Figure 9.2 [2–4].

In Figure 9.2, u_s^* is the stator voltage. i_s is the stator current. $e_{\alpha\beta}$ and e_{dq}^e are the EMFs in $\alpha\beta$-axes and estimated dq-axes. $\psi_{\alpha\beta}$ and ψ_{dq}^e are the flux linkage in $\alpha\beta$-axes and estimated dq-axes. θ_r^e and ω_r^e are the estimated rotor position and speed. "Par." denotes the parameters. In Part A, a variety of fundamental model-based sensorless methods are used to estimate EMF or PM flux-linkage containing the rotor position information. Then, Part B estimates rotor position and speed based on the estimated EMF or PM flux-linkage, which may include direct calculation, phase-locked loop (PLL) based methods, and observer-based methods. Both EMF and flux-linkage-based sensorless control methods are well-developed for PMSMs with and without saliency. Since the technical characteristics of the EMF and flux linkage-based sensorless control methods and their transformations from PMSMs with no saliency to PMSMs with saliency are similar, and therefore, only the EMF-based sensorless control method will be discussed in detail in this section. More detailed techniques for flux-linkage-based sensorless control methods can be found in [1,5].

9.2.1 Single three-phase PMSMs

9.2.1.1 EMF-based method

The voltage equations of STP PMSMs without saliency in synchronous rotating reference frame can be written as:

$$\begin{bmatrix} u_d \\ u_q \end{bmatrix} = \begin{bmatrix} R_s + pL_s & -\omega_r L_s \\ \omega_r L_s & R_s + pL_s \end{bmatrix} \begin{bmatrix} i_d \\ i_q \end{bmatrix} + \begin{bmatrix} 0 \\ \omega_r \psi_f \end{bmatrix} \tag{9.1}$$

where p is the differential operator. u_d, u_q and i_d, i_q represent the dq-axis stator voltages and currents, respectively. L_s and R_s are the stator winding inductance and resistance, respectively. ω_r is the rotor electrical angular velocity. ψ_f is the PM flux-linkage.

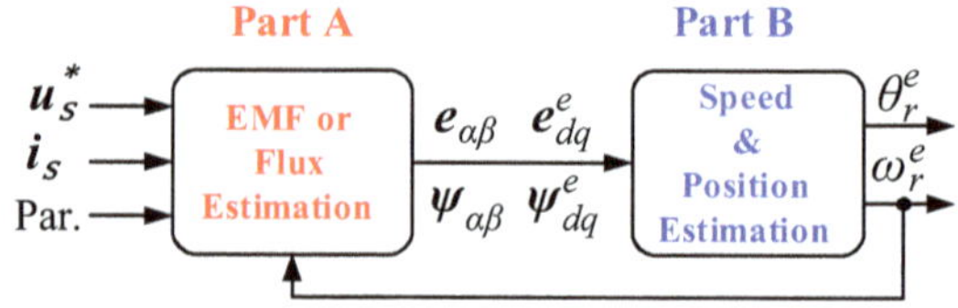

Figure 9.2 Typical framework of fundamental model-based sensorless methods.

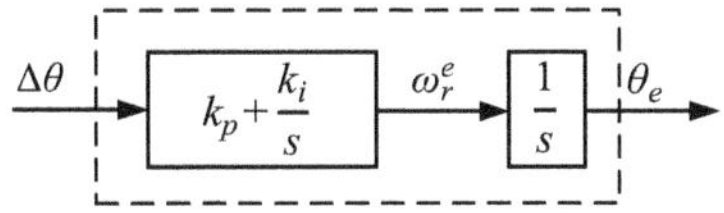

Figure 9.3 Block diagram of PI type observer [7].

For PMSMs with saliency, it can be written as [4]:

$$\begin{bmatrix} u_d \\ u_q \end{bmatrix} = \begin{bmatrix} R_s + pL_d & -\omega_r L_q \\ \omega_r L_q & R_s + pL_d \end{bmatrix} \begin{bmatrix} i_d \\ i_q \end{bmatrix} + \begin{bmatrix} 0 \\ E_{ex} \end{bmatrix} \tag{9.2}$$

where E_{ex} is the extended EMF (E-EMF) and is given by

$$E_{ex} = (L_d - L_q)(\omega_r i_d - p i_q) + \omega_r \psi_f \tag{9.3}$$

and L_d and L_q are the d- and q-axis inductances, respectively. The E-EMF model has been widely used for PMSMs with saliency [6]. For PMSM without saliency, $L_d = L_q = L_s$ and $E_{ex} = \omega_r \psi_f$.

Assuming the estimated speed is sufficiently close to the actual one, (9.2) can be transformed into the estimated synchronous rotating reference frame [4]:

$$\begin{bmatrix} u_d^e \\ u_q^e \end{bmatrix} = \begin{bmatrix} R_s + pL_d & -\omega_r L_q \\ \omega_r L_q & R_s + pL_d \end{bmatrix} \begin{bmatrix} i_d^e \\ i_q^e \end{bmatrix} + \begin{bmatrix} E_{ex,d} \\ E_{ex,q} \end{bmatrix}$$

$$\begin{bmatrix} E_{ex,d} \\ E_{ex,q} \end{bmatrix} = E_{ex} \begin{bmatrix} -\sin \Delta\theta \\ \cos \Delta\theta \end{bmatrix} \tag{9.4}$$

The superscript "e" denotes the value in the estimated dq-axes. $\Delta\theta$ is the estimated position error, which can be derived as:

$$\Delta\theta = \tan^{-1}\left(-\frac{E_{ex,d}}{E_{ex,q}}\right) \tag{9.5}$$

Then, the estimated position error $\Delta\theta$ is applied to the PI-type observer as shown in Figure 9.3 for further speed and position estimation.

Equation (9.2) can also be transformed into the stationary reference frame ($\alpha\beta$-axis) [3]

$$\begin{bmatrix} u_\alpha \\ u_\beta \end{bmatrix} = \begin{bmatrix} R_s + pL_d & \omega_r (L_d - L_q) \\ -\omega_r (L_d - L_q) & R_s + pL_d \end{bmatrix} \begin{bmatrix} i_\alpha \\ i_\beta \end{bmatrix} + \begin{bmatrix} E_{ex,\alpha} \\ E_{ex,\beta} \end{bmatrix}, \quad \begin{bmatrix} E_{ex,\alpha} \\ E_{ex,\beta} \end{bmatrix}$$

$$= \quad E_{ex} \begin{bmatrix} -\sin(\theta_r) \\ \cos(\theta_r) \end{bmatrix} \tag{9.6}$$

Afterwards, the rotor position θ_r can be obtained by:

$$\theta_r = \tan^{-1}\left(-\frac{E_{ex,\alpha}}{E_{ex,\beta}}\right) \tag{9.7}$$

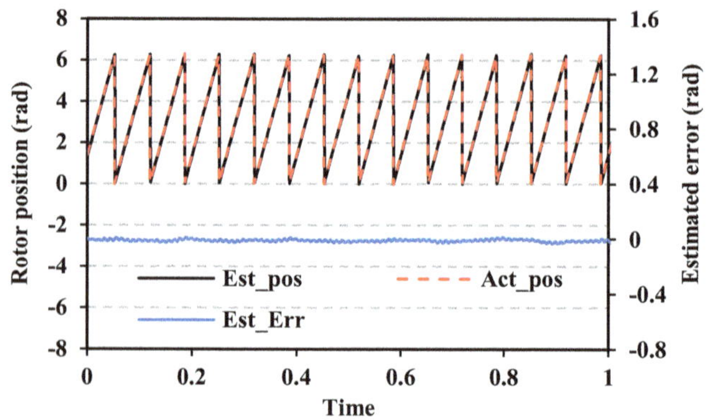

Figure 9.4 Measured estimated position error of E-EMF-based sensorless control of an STP IPMSM.

Compared with forcing $\Delta\theta$ to zero in the estimated synchronous rotating reference frame, the "arctan" operation in (9.7) may introduce noises in the estimated rotor position [1]. Therefore, the sensorless control method in the estimated synchronous rotating reference frame, (9.2)–(9.5), is more widely used.

By way of example, Figure 9.4 shows the experimental result of position error with E-EMF-based sensorless control of an STP IPMSM (interior PMSM) in the estimated synchronous rotating reference frame. The parameters of the prototype machine are shown in Appendix A (Machine 4).

9.2.1.2 Other fundamental model-based sensorless methods

In addition to the EMF and flux-linkage based sensorless control methods [8–10], many modern control theory-based methods can also be used to estimate the rotor position or speed with the fundamental mathematical model of PMSM [1], e.g., sliding mode observer, model reference adaptive system, extended Kalman filter based full-order or reduced-order observer.

9.2.2 Dual three-phase PMSMs

For the DTP PMSM, the same analysis for STP PMSM can be applied. As discussed above, for the E-EMF-based method in STP PMSMs, the "arctan" operation introduces more estimated position noise when the voltage equation model is transformed into the stator stationary reference frame. Therefore, for DTP PMSMs (see Figure 9.5), only the E-EMF-based method in the estimated synchronous rotating reference frame is taken for example.

The double *dq* mathematical model of DTP PMSMs in the synchronously rotating reference frame can be expressed as [11]:

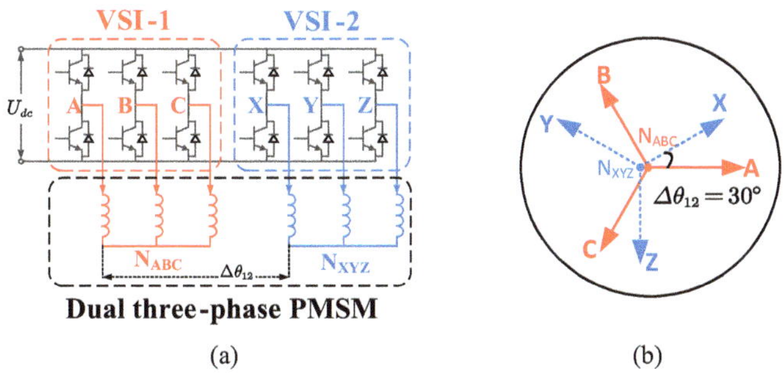

Figure 9.5 DTP PMSMs system [2]. (a) Inverter drive configuration of DTP PMSMs. (b) Machine phasor diagram of a typical DTP PMSMs.

$$\begin{bmatrix} u_{d1} \\ u_{q1} \end{bmatrix} = \begin{bmatrix} R_{s1} + pL_{d1} & -\omega_r L_{q1} \\ \omega_r L_{d1} & R_{s1} + pL_{q1} \end{bmatrix} \begin{bmatrix} i_{d1} \\ i_{q1} \end{bmatrix} + \begin{bmatrix} 0 \\ \omega_r \psi_f \end{bmatrix} + \underbrace{\begin{bmatrix} pM_{d21} & -\omega_r M_{q21} \\ \omega_r M_{d21} & pM_{q21} \end{bmatrix} \begin{bmatrix} i_{d2} \\ i_{q2} \end{bmatrix}}_{\text{Coupling voltage caused by Set 2}} \tag{9.8}$$

$$\begin{bmatrix} u_{d2} \\ u_{q2} \end{bmatrix} = \begin{bmatrix} R_{s2} + pL_{d2} & -\omega_r L_{q2} \\ \omega_r L_{d2} & R_{s2} + pL_{q2} \end{bmatrix} \begin{bmatrix} i_{d2} \\ i_{q2} \end{bmatrix} + \begin{bmatrix} 0 \\ \omega_r \psi_f \end{bmatrix} + \underbrace{\begin{bmatrix} pM_{d12} & -\omega_r M_{q12} \\ \omega_r M_{d12} & pM_{q12} \end{bmatrix} \begin{bmatrix} i_{d1} \\ i_{q1} \end{bmatrix}}_{\text{Coupling voltage caused by Set 1}} \tag{9.9}$$

where the number in subscript represents the different set of the machine, u_{d1}, u_{q1}, u_{d2}, u_{q2} and i_{d1}, i_{q1}, i_{d2}, i_{q2} represent the dq-axis stator voltages and currents, respectively. R_{s1}, R_{s2} are the stator resistances, while L_{d1}, L_{d2} and L_{q1}, L_{q2} denote the d- and q-axis self-inductances for Set-1 and Set-2, respectively. M_{d12}, M_{d21} and M_{q12}, M_{q21} are the d- and q-axis mutual inductances between the two sets, respectively. ψ_f is the PM flux-linkage and ω_r is the rotor's electrical angular speed.

For simplicity, only the model of Set-1 is presented here as the model for Set-2 is the same. The model (9.8) can be rewritten in the E-EMF formation [4]:

$$\begin{bmatrix} u_{d1} \\ u_{q1} \end{bmatrix} = \begin{bmatrix} R_s + pL_D & -\omega_r L_Q \\ \omega_r L_Q & R_s + pL_D \end{bmatrix} \begin{bmatrix} i_d \\ i_q \end{bmatrix} + \begin{bmatrix} 0 \\ E_{ex1} \end{bmatrix} \tag{9.10}$$

The E-EMF E_{ex1} can be expressed as:

$$E_{ex1} = \left(L_D - L_Q\right)\left(\omega_r i_d - p i_q\right) + \omega_r \psi_f \tag{9.11}$$

where $L_D = L_{d1} + M_{d21} = L_{d2} + M_{d12}, L_Q = L_{q1} + M_{q21} = L_{q2} + M_{q12}$ are defined as the equivalent inductances of both sets by considering the two three-phase windings to be identical, as well as the phase resistance $R_s = R_{s1} = R_{s2}$. Moreover,

$i_d = i_{d1} = i_{d2}$ and $i_q = i_{q1} = i_{q2}$, due to the two sets being generally operated at the same condition for the double dq control strategy [12].

Then, assuming that the error between the estimated speed and actual speed is negligible [13], (9.10) can be transformed into the estimated synchronous rotating reference frame:

$$\begin{bmatrix} u_{d1}^e \\ u_{q1}^e \end{bmatrix} = \begin{bmatrix} R_s + pL_D & -\omega_r L_Q \\ \omega_r L_Q & R_s + pL_D \end{bmatrix} \begin{bmatrix} i_d^e \\ i_q^e \end{bmatrix} + \begin{bmatrix} e_{d1}^e \\ e_{q1}^e \end{bmatrix}, \begin{bmatrix} e_{d1}^e \\ e_{q1}^e \end{bmatrix} = E_{ex1} \begin{bmatrix} -\sin \Delta\theta_1 \\ \cos \Delta\theta_1 \end{bmatrix} \tag{9.12}$$

The estimated rotor position error $\Delta\theta_1$ of the Set-1 can be calculated from the estimated E-EMF by:

$$\Delta\theta_1 = \tan^{-1}\left(-\frac{e_{d1}^e}{e_{q1}^e}\right) \tag{9.13}$$

The same as with STP PMSMs, $\Delta\theta_1$ can also be controlled to zero by the PI type observer as shown in Figure 9.3. In this case, the estimated synchronous rotating reference frame can be aligned with the actual synchronous rotating reference frame and the position information can be obtained.

A DTP IPMSM and its parameters along with the drive system are shown in Appendix A (DTP-IPMSMs platform). The control system contains two VSIs, which control two sets of the DTP IPMSM independently. The switching frequency and the sampling frequency are both 10 kHz. The DC-link voltage is set to 40 V. The actual rotor position is measured by an encoder, which is not used for control purposes but only for comparison with the estimated position. The overall configuration of the E-EMF-based sensorless control system is shown in Figure 9.6. Figure 9.7 shows the corresponding measured rotor position errors.

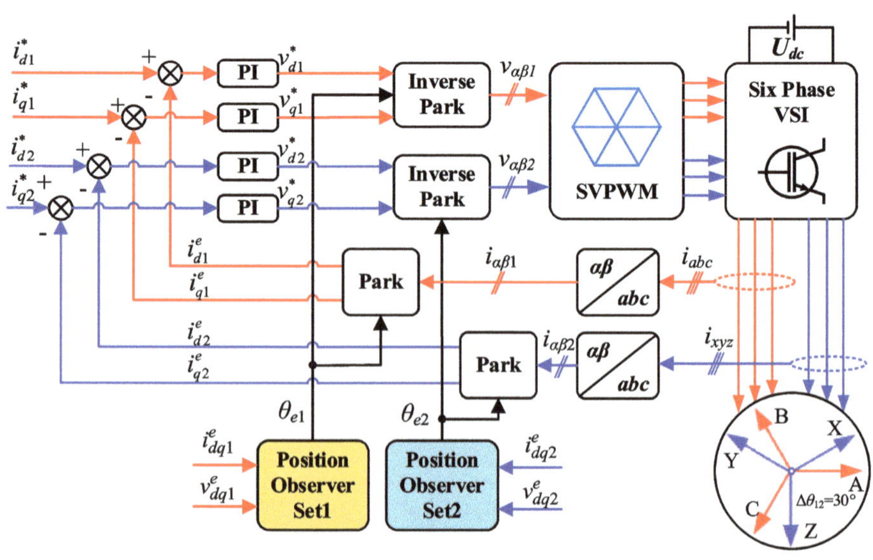

Figure 9.6 Overall block diagram of DTP PMSMs control system with E-EMF based sensorless control method.

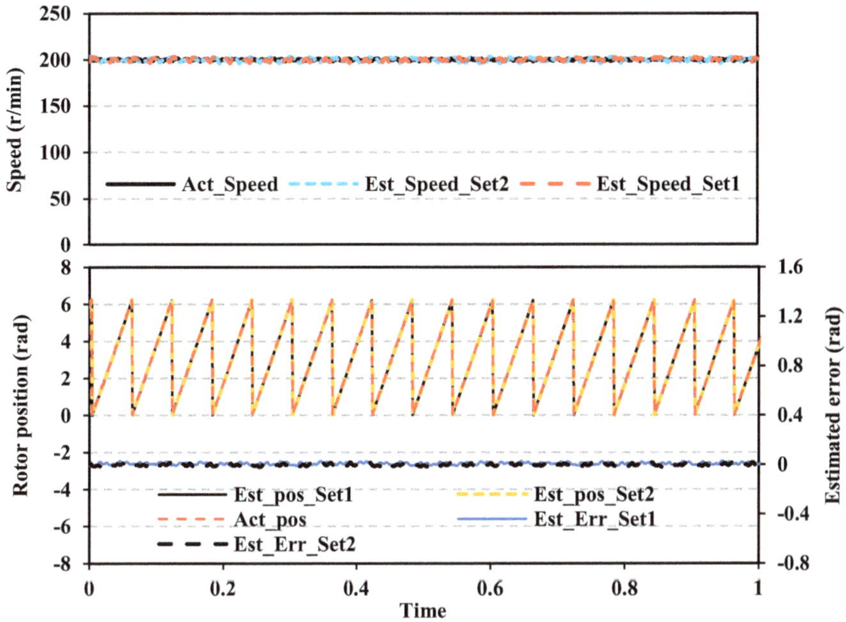

Figure 9.7 Measured rotor position estimation error with E-EMF-based sensorless control method.

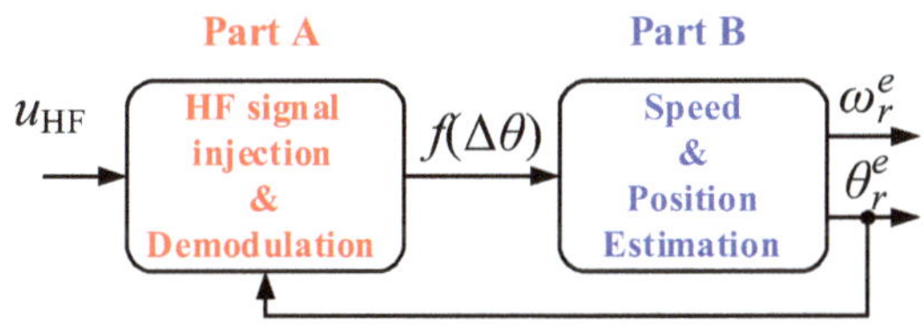

Figure 9.8 Typical frame of saliency tracking-based sensorless control methods.

9.3 Saliency tracking-based sensorless control methods

Saliency tracking-based sensorless control methods are developed to achieve a high rotor estimation accuracy both at a standstill and at low speed. These sensorless control methods rely on machine saliency. The typical frame of saliency tracking-based sensorless control methods is shown in Figure 9.8.

In Figure 9.8, u_{HF} is the HF injected voltage. $f(\Delta\theta)$ is the position error signal function. In Part A, an HF carrier voltage is normally superimposed in a continuous manner onto the fundamental excitation. A rotor carrier current, which is position-dependent, is produced and demodulated thereafter from the interaction of the machine's spatial saliency with the injected carrier voltage. The rotor position could then be extracted from the position-dependent carrier current response through the use of a tracking observer [14] as shown in Part B. The two major methods of continuous carrier signal injection are pulsating signal injection [15–18] and rotating signal injection [19,20].

In this section, the classical sensorless control strategies based on HF carrier signal injections, i.e., pulsating signal injection and rotating signal injection, are applied to STP and DTP PMSMs to investigate the sensorless control performance under steady-state conditions.

9.3.1 Single three-phase PMSMs

9.3.1.1 Machine model of PMSM under high-frequency signal injection

If the cross-coupling magnetic saturation effect is considered, the mathematical model under HF signal injection can be given as [18]:

$$\begin{bmatrix} u_{dh} \\ u_{qh} \end{bmatrix} = \begin{bmatrix} L_{dh} & L_{dqh} \\ L_{qdh} & L_{qh} \end{bmatrix} p \begin{bmatrix} i_{dh} \\ i_{qh} \end{bmatrix} \tag{9.14}$$

where L_{dh} and L_{qh} are the HF incremental self-inductances of d- and q-axes, respectively. u_{dh}, u_{qh}, and i_{dh}, i_{qh} are the HF voltages and currents in dq-axes. L_{dqh} and L_{qdh} are the HF incremental mutual inductances between d- and q- axes due to the cross-coupling saturation effect.

Conventionally, the saliency tracking-based method could be implemented in the stationary reference frame ($\alpha\beta$) or estimated synchronous rotating reference frame ($d^e q^e$). The relationship between these two reference frames and the actual synchronous rotating reference frame is shown in Figure 9.9.

Estimated synchronous rotating reference frame

For sensorless control purposes, since the actual rotor position is not available, the estimated synchronous rotating reference frame is established for further investigation. Therefore, the voltage model (9.14) in the actual synchronous rotating reference frame, i.e., dq-axes, can be transformed into the estimated synchronous rotating reference frame as [15,21,22]:

$$\begin{bmatrix} u^e_{dh} \\ u^e_{qh} \end{bmatrix} = \begin{bmatrix} L_{\mathrm{ave}} - L_{\mathrm{dif}}\cos(2\Delta\theta) - L_{dqh}\sin(2\Delta\theta) & -L_{\mathrm{dif}}\sin(2\Delta\theta) + L_{dqh}\cos(2\Delta\theta) \\ -L_{\mathrm{dif}}\sin(2\Delta\theta) - L_{dqh}\cos(2\Delta\theta) & L_{\mathrm{ave}} + L_{\mathrm{dif}}\cos(2\Delta\theta) + L_{dqh}\sin(2\Delta\theta) \end{bmatrix} \cdot \frac{d}{dt} \begin{bmatrix} i^e_{dh} \\ i^e_{dh} \end{bmatrix} \tag{9.15}$$

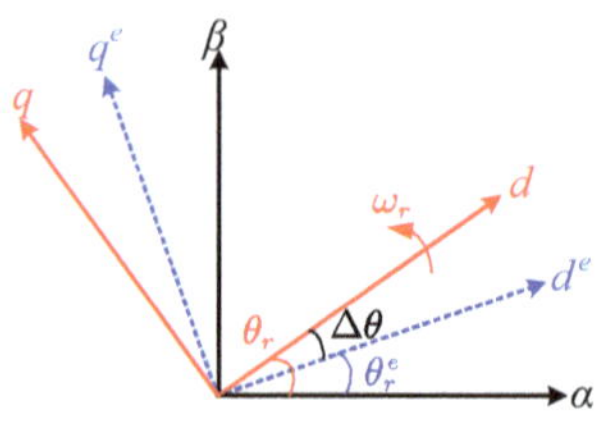

Figure 9.9 Relationship between three reference frames.

where $\Delta\theta$ is the position error between real and estimated reference frames. L_{ave} and L_{dif} are the average and difference of dq-axis incremental inductances, which are defined as:

$$\begin{cases} L_{\text{ave}} = \dfrac{L_{qh} + L_{dh}}{2} \\ L_{\text{dif}} = \dfrac{L_{qh} - L_{dh}}{2} \end{cases} \tag{9.16}$$

Then, by rewriting (9.15), the differentiation of HF currents in the estimated synchronous rotating reference frame can be expressed as:

$$\frac{d}{dt}\begin{bmatrix} i_{dh}^{e} \\ i_{qh}^{e} \end{bmatrix} = \begin{bmatrix} \dfrac{1}{L_p} + \dfrac{1}{L_n}\cos(2\Delta\theta + \theta_m) & \dfrac{1}{L_n}\sin(2\Delta\theta + \theta_m) \\ \dfrac{1}{L_n}\sin(2\Delta\theta + \theta_m) & \dfrac{1}{L_p} - \dfrac{1}{L_n}\cos(2\Delta\theta + \theta_m) \end{bmatrix} \cdot \begin{bmatrix} u_{dh}^{e} \\ u_{qh}^{e} \end{bmatrix} \tag{9.17}$$

where L_p and L_n can be given as:

$$\begin{cases} L_p = \dfrac{L_{dh}L_{qh} - L_{dqh}^2}{L_{\text{ave}}} \\ L_n = \dfrac{L_{dh}L_{qh} - L_{dqh}^2}{\sqrt{L_{\text{dif}}^2 + L_{dqh}^2}} \end{cases} \tag{9.18}$$

and θ_m is the position error caused by the cross-coupling saturation effect, as:

$$\theta_m = \tan^{-1}\left(-\frac{L_{dqh}}{L_{\text{dif}}}\right) \tag{9.19}$$

From (9.17), it can be seen that the position information $\Delta\theta$, which can be utilized for rotor position estimation, is contained in the carrier current responses.

Stationary reference frame

Similarly, the mathematical model in (9.14) can also be transformed into the stationary reference frame as [14,20]:

$$\begin{bmatrix} u_{\alpha h} \\ u_{\beta h} \end{bmatrix} = \begin{bmatrix} L_{\text{ave}} - L_{\text{dif}}\cos(2\theta_r) - L_{dqh}\sin(2\theta_r) & -L_{\text{dif}}\sin(2\theta_r) + L_{dqh}\cos(2\theta_r) \\ -L_{\text{dif}}\sin(2\theta_r) - L_{dqh}\cos(2\theta_r) & L_{\text{ave}} + L_{\text{dif}}\cos(2\theta_r) + L_{dqh}\sin(2\theta_r) \end{bmatrix} \cdot \frac{d}{dt}\begin{bmatrix} i_{\alpha h} \\ i_{\beta h} \end{bmatrix} \tag{9.20}$$

Then, from (9.20), the differentiation of HF currents in the stationary reference frame can be derived as:

$$\frac{d}{dt}\begin{bmatrix} i_{\alpha h} \\ i_{\beta h} \end{bmatrix} = \begin{bmatrix} \dfrac{1}{L_p} + \dfrac{1}{L_n}\cos(2\theta_r + \theta_m) & \dfrac{1}{L_n}\sin(2\theta_r + \theta_m) \\ \dfrac{1}{L_n}\sin(2\theta_r + \theta_m) & \dfrac{1}{L_p} - \dfrac{1}{L_n}\cos(2\theta_r + \theta_m) \end{bmatrix} \cdot \begin{bmatrix} u_{\alpha h} \\ u_{\beta h} \end{bmatrix} \tag{9.21}$$

From (9.21), the rotor position information θ_r can also be found in the current responses of the stationary reference frame.

9.3.1.2 HF signal injection methods

Pulsating signal injection—sinusoidal wave

The pulsating voltage signal can be injected into the *d*- or *q*-axis of the estimated synchronous rotating reference frame [15,23,24]. When an HF sinusoidal voltage signal is injected into the *d*-axis of the estimated synchronous rotating reference frame, the voltage signals are expressed as:

$$\begin{bmatrix} u_{dh}^e \\ u_{qh}^e \end{bmatrix} = V_{\mathrm{HF}} \begin{bmatrix} \cos(\omega_{\mathrm{HF}} t) \\ 0 \end{bmatrix} \tag{9.22}$$

The estimated *d*-axis pulsating injection can be briefly depicted in Figure 9.10.

After the injection, the HF carrier current responses in the estimated synchronous rotating reference frame can be obtained by substituting (9.22) into (9.17), as:

$$\begin{aligned} \begin{bmatrix} i_{dh}^e \\ i_{qh}^e \end{bmatrix} &= \begin{bmatrix} \dfrac{V_{\mathrm{HF}}}{\omega_{\mathrm{HF}} L_p} + \dfrac{V_{\mathrm{HF}}}{\omega_{\mathrm{HF}} L_n} \cos(2\Delta\theta + \theta_m) \\ \dfrac{V_{\mathrm{HF}}}{\omega_{\mathrm{HF}} L_n} \sin(2\Delta\theta + \theta_m) \end{bmatrix} \cdot \sin(\omega_{\mathrm{HF}} t) \\ &= \begin{bmatrix} I_p + I_n \cos(2\Delta\theta + \theta_m) \\ I_n \sin(2\Delta\theta + \theta_m) \end{bmatrix} \cdot \sin(\omega_{\mathrm{HF}} t) \end{aligned} \tag{9.23}$$

From (9.23), it can be clearly seen that the HF carrier current responses contain the rotor position information. To demodulate the amplitude that is relevant to the rotor position information in i_{qh}^e, both sides of (9.23) can be multiplied by $2\sin(\omega_{\mathrm{HF}} t)$, and then apply the product to a low pass filter (LPF), which can be expressed by [25]:

$$\begin{bmatrix} |i_{dh}^e| \\ |i_{qh}^e| \end{bmatrix} = \mathrm{LPF}\left\{ \begin{bmatrix} i_{dh}^e \\ i_{qh}^e \end{bmatrix} 2\sin(\omega_{\mathrm{HF}} t) \right\} = \begin{bmatrix} I_p + I_n \cos(2\Delta\theta + \theta_m) \\ I_n \sin(2\Delta\theta + \theta_m) \end{bmatrix} \tag{9.24}$$

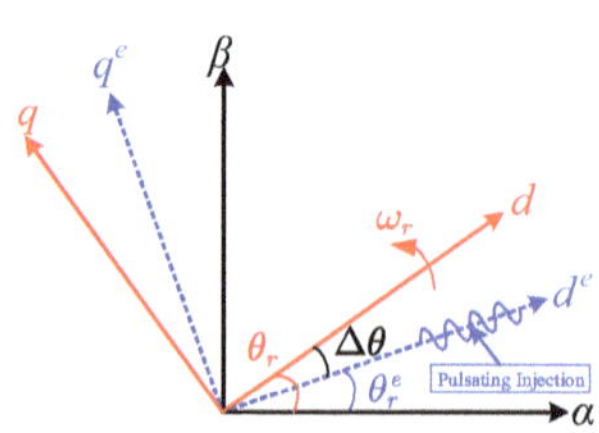

Figure 9.10 Pulsating signal is injected into the estimated d-axis.

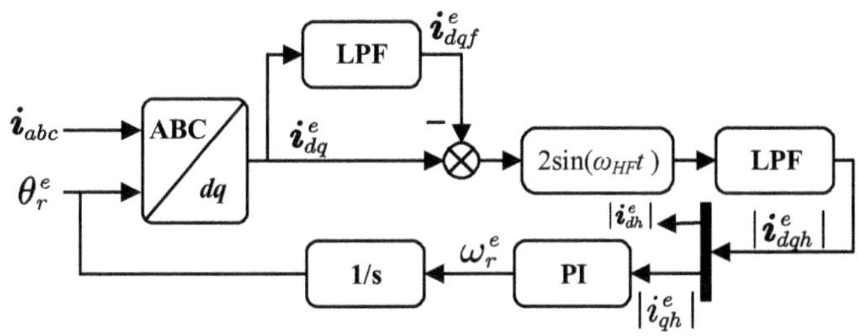

Figure 9.11 Block diagram of carrier signal demodulation and position estimation for pulsating signal injection method.

Then, the amplitude of q-axis carrier current, which contains the rotor position information, is used as the input of the rotor position estimation method.

$$f(\Delta\theta) = \left|i_{qh}^e\right| = I_n \sin(2\Delta\theta + \theta_m) \tag{9.25}$$

A typical pulsating signal injection-based rotor position estimation procedure is shown in Figure 9.11 [2].

With the position observer, $f(\Delta\theta)$ is forced to be zero in the steady-state condition. The rotor position can hence be obtained with the load-dependent position estimation error ($-\theta_m/2$), which occurs due to the cross-coupling saturation effect. The cross-coupling saturation effect can be suppressed by online or offline compensation methods [1]. An offline measured coupling factor λ based compensation method is proposed in [18], where a new error signal accounts for the cross-coupling saturation effect is established as:

$$\varepsilon = I_p + \lambda I_n \approx 2I_n \cos\theta_m \Delta\theta \text{ and } \lambda = \frac{L_{dqh}}{L_{qh}} \tag{9.26}$$

where I_p and I_n are the amplitudes of the HF current responses in the HF signal injection-based sensorless control method, which can be defined as:

$$I_p = \frac{V_{\text{HF}}}{\omega_{\text{HF}} L_p} \text{ and } I_n = \frac{V_{\text{HF}}}{\omega_{\text{HF}} L_n} \tag{9.27}$$

V_{HF} is the amplitude of the injected HF voltage and ω_{HF} is the frequency of the injected HF voltage. When the new error signal ε is controlled to zero, the estimated position error caused by the cross-coupling effect can be compensated for. By injecting HF voltage into the actual d-axis with the help of an encoder, the coupling factor λ can be obtained offline from the measured values of the HF current responses:

$$\lambda = -\frac{i_{qh}}{i_{dh}} = \frac{L_{dqh}}{L_{qh}} \tag{9.28}$$

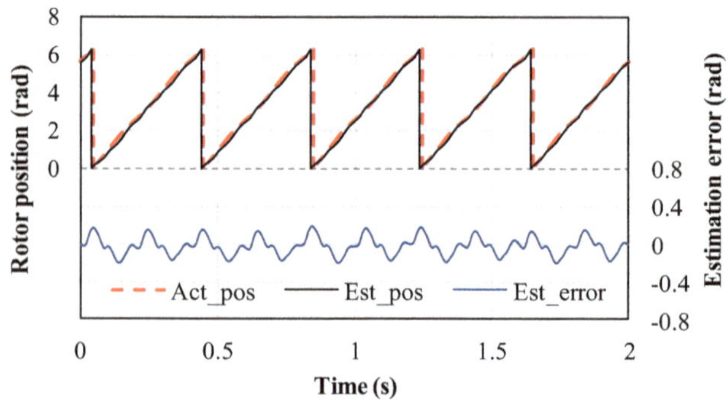

Figure 9.12 Steady-state performance of pulsating sinusoidal signal injection-based sensorless control of STP IPMSM (with cross-saturation compensation, i_q = 2 A, 30 r/min).

In order to validate the effectiveness of sensorless control performance, the pulsating sinusoidal carrier signal injection method is applied to the STP IPMSM (Set 1 of Machine 4 in Appendix A). An HF voltage with an amplitude of 8 V and frequency of 550 Hz is injected into the *d*-axis in the estimated *dq* reference frame, and the performance of the rotor position estimation using the *q*-axis carrier current response under steady-state conditions is shown in Figure 9.12, together with the estimation error [26]. It can be seen that the estimated rotor position exhibits good position-tracking performance with the cross-saturation compensation.

Pulsating signal injection—square wave

Besides the conventional sinusoidal signal, the square-wave-type signal can also be utilized for HF signal injections. The square wave signal can be injected into the stationary reference frame [27] or the estimated synchronous rotating reference frame [28,29]. By applying the HF square wave injection, the frequency that is injected could be much higher than the sinusoidal signal injection, which has advantages in HF signal demodulation [1]. To be more specific, the LPFs, which are widely used in the demodulation procedure of conventional sinusoidal HF signal injection-based methods, are not required in HF square wave injection-based methods. Due to the inherent time delay of LPFs, which degrades the sensorless control performance, the square wave injection can be used instead of sinusoidal signal injection to increase the bandwidths of current, speed, and position controllers [28]. Take the method with a pulsating square wave signal injected into the estimated *d*-axis as an example, which is shown in Figure 9.13 [2], the basic principle of the method is introduced below.

The injected square wave voltage can be described as:

$$u_{dh}^e = \begin{cases} V_{\mathrm{HF}} & \text{half duty} \\ -V_{\mathrm{HF}} & \text{otherwise} \end{cases} \quad (V_{\mathrm{HF}} > 0)$$
$$u_{qh}^e = 0 \tag{9.29}$$

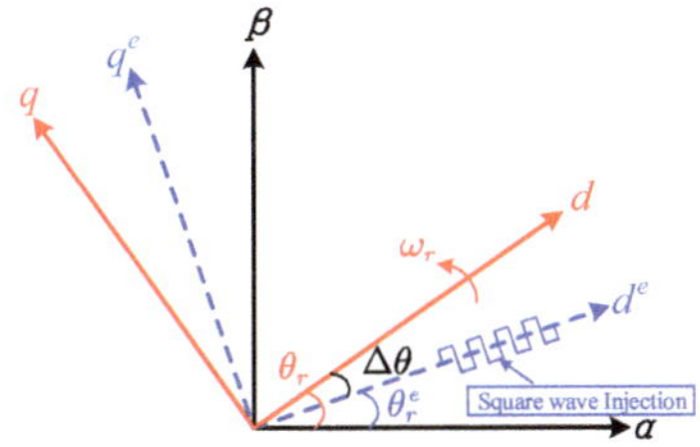

Figure 9.13 Square wave type signal injected into estimated d-axis.

Then, due to the square wave injection, the difference between the present value and the previous value of the carrier current responses can be expressed as:

$$\begin{bmatrix} \Delta i_{dh}^e \\ \Delta i_{qh}^e \end{bmatrix} = \begin{cases} V_{\mathrm{HF}}\Delta T \begin{bmatrix} \dfrac{1}{L_p} + \dfrac{1}{L_n}\cos(2\Delta\theta + \theta_m) \\ \dfrac{1}{L_n}\sin(2\Delta\theta + \theta_m) \end{bmatrix}, u_{dh}^e > 0 \\ -V_{\mathrm{HF}}\Delta T \begin{bmatrix} \dfrac{1}{L_p} + \dfrac{1}{L_n}\cos(2\Delta\theta + \theta_m) \\ \dfrac{1}{L_n}\sin(2\Delta\theta + \theta_m) \end{bmatrix}, u_{dh}^e < 0 \end{cases} \tag{9.30}$$

where ΔT is the half period of the injected signal. By considering the polarity of the injected voltage, (9.30) can be rewritten as:

$$\Delta i_{dh}^{e\,\prime} = \begin{cases} \Delta i_{dh}^e, \text{ if } u_{dh}^e > 0 \\ -\Delta i_{dh}^e, \text{otherwise} \end{cases}$$
$$\Delta i_{qh}^{e\,\prime} = \begin{cases} \Delta i_{qh}^e, \text{if } u_{dh}^e > 0 \\ -\Delta i_{qh}^e, \text{otherwise} \end{cases} \tag{9.31}$$

where $\Delta i_{dh}^{e\,\prime}$ and $\Delta i_{qh}^{e\,\prime}$ are the current responses after polarity normalization, which can be further expressed as (9.22).

$$\begin{bmatrix} \Delta i_{dh}^{e\,\prime} \\ \Delta i_{qh}^{e\,\prime} \end{bmatrix} = V_{\mathrm{HF}}\Delta T \begin{bmatrix} \dfrac{1}{L_p} + \dfrac{1}{L_n}\cos(2\Delta\theta + \theta_m) \\ \dfrac{1}{L_n}\sin(2\Delta\theta + \theta_m) \end{bmatrix} \tag{9.32}$$

From (9.32), it can be seen that the rotor position information is contained in $\Delta i_{qh}^{e\,\prime}$. Similar to the conventional sinusoidal HF signal injection method, the information of the rotor position can thus be obtained by forcing $\Delta i_{qh}^{e\,\prime}$ to zero through a PI regulator [1].

Then, the pulsating square wave carrier signal injection method is applied to an STP IPMSM (Machine 4 in Appendix A). An HF voltage, with an amplitude of 8 V and frequency of 1,000 Hz, is injected into the *d*-axis in the estimated *dq* reference frame, and the performance of the rotor position estimation using the q-axis carrier current response under steady-state conditions is shown in Figure 9.14, together with the estimation error. It can be seen that the estimated rotor position exhibits good position-tracking performance with the cross-saturation compensation.

Rotating signal injection

Generally, the rotating signal injection-based methods inject balanced HF sinusoidal voltage signals into the stationary reference frame [20,24,30], which is illustrated in Figure 9.15.

The HF rotating voltage signal can be defined as:

$$\begin{bmatrix} u_{\alpha h} \\ u_{\beta h} \end{bmatrix} = V_{\mathrm{HF}} \begin{bmatrix} \cos(\omega_{\mathrm{HF}} t) \\ \sin(\omega_{\mathrm{HF}} t) \end{bmatrix} \tag{9.33}$$

Then, by substituting (9.33) into (9.21), the carrier current responses in the stationary reference frame can be derived as:

$$\begin{bmatrix} i_{\alpha h} \\ i_{\beta h} \end{bmatrix} = \begin{bmatrix} I_p \cos\left(\omega_{\mathrm{HF}} t - \dfrac{\pi}{2}\right) + I_n \cos\left(-\omega_{\mathrm{HF}} t + 2\theta_r + \theta_m + \dfrac{\pi}{2}\right) \\ I_p \sin\left(\omega_{\mathrm{HF}} t - \dfrac{\pi}{2}\right) + I_n \sin\left(\omega_{\mathrm{HF}} t + 2\theta_r + \theta_m + \dfrac{\pi}{2}\right) \end{bmatrix} \tag{9.34}$$

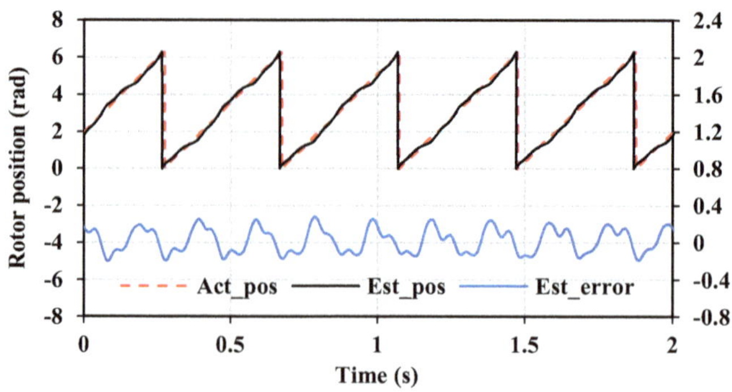

Figure 9.14 Steady-state performance of pulsating square wave signal injection-based sensorless control of a STP IPMSM (with cross-saturation compensation, $i_q = 2$ A, 30 r/min).

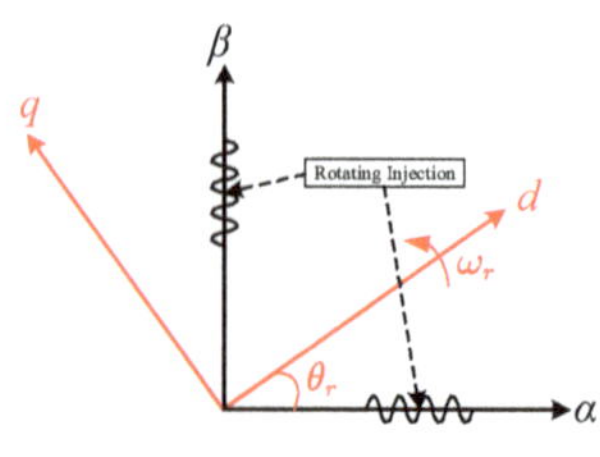

Figure 9.15 Demonstration of rotating signal injection in the stationary reference frame.

By using the form of a complex vector, (9.34) can be rewritten as:

$$\boldsymbol{i}_{\alpha\beta h} = I_p e^{j\left(\omega_{\mathrm{HF}}t-\frac{\pi}{2}\right)} + I_n e^{j\left(-\omega_{\mathrm{HF}}t+2\theta_r+\theta_m+\frac{\pi}{2}\right)} \tag{9.35}$$

From (9.35), it has been noticed that the response of carrier current consists of a positive sequence component (PSC), and a negative sequence component (NSC). The PSC has the same rotating frequency as the voltage signal of injection, while the NSC contains the information on the rotor position θ_r. Consequently, the NSC in carrier current response is generally used for the rotor position estimation.

The rotor position could be acquired from the NSC of the carrier current response, and the most typical solution is the synchronous reference frame filter (SRFF) method [24,30,31], which can be expressed as (9.36) from (9.35):

$$\boldsymbol{i}_n = I_n e^{j\left(-\omega_{\mathrm{HF}}t+2\theta_r+\theta_m+\frac{\pi}{2}\right)} \tag{9.36}$$

$\boldsymbol{i}_n$ can be transformed into the reference frame synchronized with the estimated rotor position as:

$$\boldsymbol{i}_{ndq}^e = I_n e^{j\left(-\omega_{\mathrm{HF}}t+2\theta_r+\theta_m+\frac{\pi}{2}\right)} e^{j\left(\omega_{\mathrm{HF}}t-2\theta_r^e-\frac{\pi}{2}\right)} = I_n e^{j(2\Delta\theta+\theta_m)} \tag{9.37}$$

which can be further rewritten as:

$$\begin{bmatrix} i_{nd}^e \\ i_{nq}^e \end{bmatrix} = \begin{bmatrix} I_n\cos(2\Delta\theta+\theta_m) \\ I_n\sin(2\Delta\theta+\theta_m) \end{bmatrix} \tag{9.38}$$

Therefore, it is easy to find out that i_{nq}^e could be applied for the acquisition of the rotor position, and when i_{nq}^e is forced to zero by the position observer, the estimated position aligns with the actual position. The block diagram of a typical rotating HF signal injection method is shown in Figure 9.16.

The HF rotating signal injection method is applied to the STP prototype PMSM (Machine 4 in Appendix A) as illustrated in Figure 9.17. The position-dependent negative carrier current response is employed to detect the information

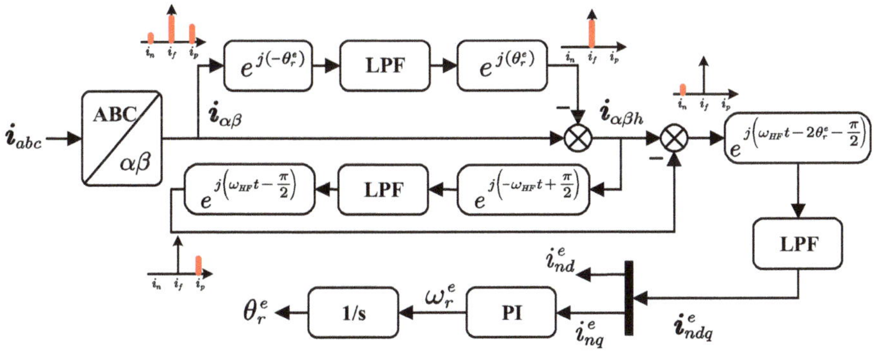

Figure 9.16 Block diagram of carrier signal demodulation and position estimation for the rotating signal injection method [14].

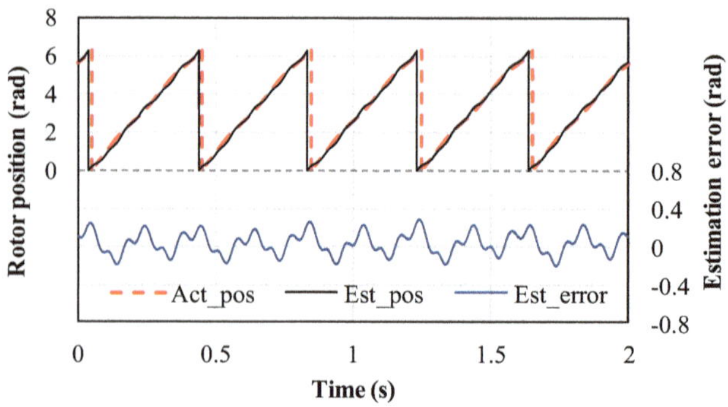

Figure 9.17 Steady-state performance of rotating signal injection-based sensorless control of STP IPMSM (with cross-saturation compensation, i_q = 2 A, 30 r/min).

Table 9.1 Comparison of typical HF signal injection-based methods [32,33]

Reference frame	**Stationary**		**Estimated synchronous**	
Carrier voltage injection	Rotating	Pulsating	Pulsating sinusoidal	Pulsating square wave
Injected axis	$\alpha\beta$-axis	α-axis	d-axis	d-axis
Carrier current response	Phase-modulated	Amplitude-modulated	Amplitude-modulated	Amplitude-modulated
Demodulation	Complex	Medium	Medium	Simple
Torque ripple	Large	Medium	Medium	Small
Bandwidth	Low	Medium	Medium	High
Audible noise	Large	Large	Medium	Medium
Position error of cross-saturation	$-\theta_m/2$	$-\theta_m/2$	$-\theta_m/2$	$-\theta_m/2$

of the rotor position. A HF voltage, with an amplitude of 8 V and frequency of 550 Hz, is injected into the stationary reference frame, and the sensorless control performance is investigated under steady-state conditions, together with the estimation error. A good position tracking performance of the estimated rotor position is observed when considering the compensation of the cross-saturation effect.

The characteristics of some typical HF signal injection methods are compared in Table 9.1.

9.3.2 Dual three-phase PMSMs

It should be highlighted that the abovementioned analyses and expressions are related to STP PMSM. For DTP PMSMs, the HF carrier voltage signals are injected into the two stator winding sets independently, and the same analysis for STP

PMSM is applied. Generally, the models of the DTP PMSMs can be classified into two different models either the double *dq* model, in which the two stator winding sets are controlled separately [34], or the vector space decomposition model (VSD) where the two stator winding sets are mutually decoupled [35]. The two models are well explained in Chapter 5. In this section, the pulsating carrier signal and the rotating carrier signal injection methods are extended to be applied to a DTP machine, and the sensorless control performance under steady-state conditions are examined for both models (i.e., double *dq* model and VSD model).

9.3.2.1 Pulsating signal injection

Double dq model

For this model, the *d*-axis pulsating carrier voltage signal, with an amplitude of 8 V and frequency of 550 Hz, is injected on both stator winding sets independently. By applying the *d*-axis pulsating carrier injection technique on both winding sets independently, the comparison of estimated and actual rotor position under steady-state conditions with compensation of cross-saturation effect is illustrated in Figure 9.18. It can be found that the rotor position estimation performance for the steady-state condition shows good position tracking with the cross-saturation effect compensation [26].

Vector space decomposition model

The model based on VSD is decomposed into three orthogonal subspaces: $\alpha\beta$-subspace, z_1z_2-subspace, and o_1o_2-subspace. Since only the $\alpha\beta$-subspace is related to electromechanical energy conversion [35] as explained in Chapter 5, the *d*-axis pulsating carrier voltage signal, with an amplitude of 8 V and frequency of 550 Hz, is injected into this subspace after it is transformed to the estimated *dq* reference frame. Then, the actual and estimated rotor positions together with position estimation error and compensation of the cross-saturation effect are obtained as illustrated in Figure 9.19. It should be noted that the position estimation

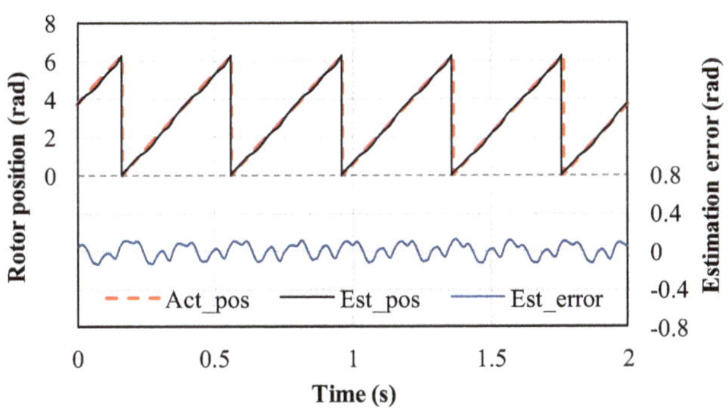

Figure 9.18 Steady-state performance of a pulsating sinusoidal signal injection-based sensorless control of DTP IPMSM (double dq model, with cross-saturation compensation, i_q = 2 A, 30 r/min).

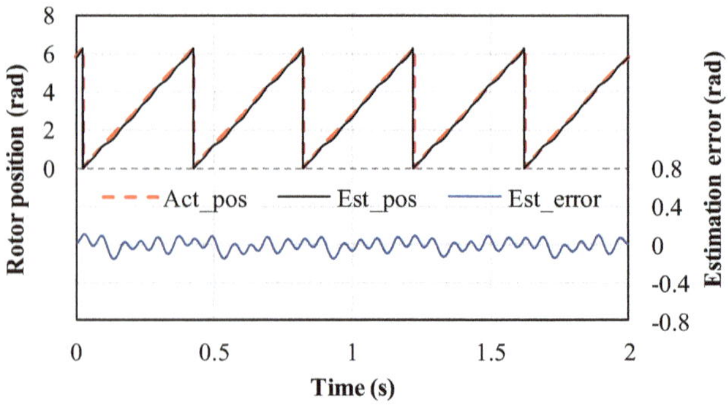

Figure 9.19 Steady-state performance of a pulsating sinusoidal signal injection-based sensorless control of DTP IPMSM (VSD model, with cross-saturation compensation, $i_q = 2$ A, 30 r/min).

error, as well as the estimation error oscillation, are decreased when injecting an HF signal with the aid of the VSD modeling method, this is due to the mutual coupling between the two winding sets being eliminated since the decoupling between the two winding sets is considered in the transformation of this method, while it is not considered in the double *dq* modeling method [35].

9.3.2.2 Rotating signal injection

The rotating carrier signal injection method is extended to a DTP machine, and the sensorless control performance under steady-state and dynamic conditions is carried out for the double *dq* model and the VSD model.

Double dq model

With the aid of the double *dq* model, two independent HF rotating voltage signals, with an amplitude of 8 V and frequency of 550 Hz, are injected into both stator winding sets. Then, the performances of the rotor position estimation with cross-saturation effect compensation under steady-state conditions are demonstrated in Figure 9.20 [26]. For the steady-state condition, good position tracking can be obtained when considering the cross-saturation effect compensation.

Vector space decomposition model

For the VSD model, the rotating carrier voltage signal, with an amplitude of 8 V and frequency of 550 Hz, is injected into $\alpha\beta$-subspace. At the steady-state condition, the comparison of the actual and estimated rotor positions with compensation of cross-saturation is illustrated in Figure 9.21 together with the position estimation error [26]. It is worth noting that the estimation error oscillation decreases when the vector space decomposition modeling method is employed due to the mutual decoupling in this method.

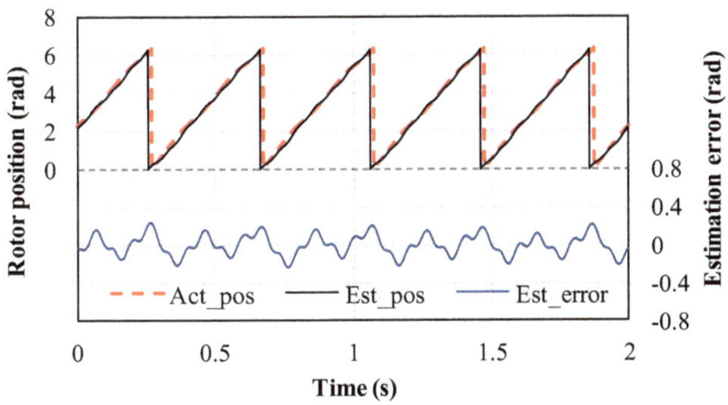

Figure 9.20 *Steady-state performance of rotating signal injection-based sensorless control of DTP IPMSM (double dq model, with cross-saturation compensation, i_q = 2 A, 30 r/min).*

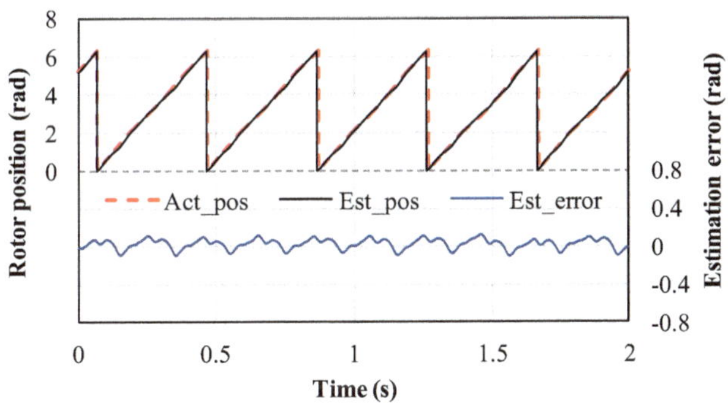

Figure 9.21 *Steady-state performance of rotating signal injection-based sensorless control of DTP IPMSM (VSD model, with cross-saturation compensation, i_q = 2 A, 30 r/min).*

9.3.3 Comparison of sensorless control techniques based on carrier signal injection

This section discusses the commonly used conventional sensorless control strategies based on HF carrier signal injections, i.e., pulsating HF carrier voltage signal injection into the estimated synchronous rotating reference frame, and rotating HF carrier voltage signal injection into the stationary reference frame. These techniques are applied to single and DTP PMSMs to investigate the sensorless control performance under steady-state conditions. Table 9.2 illustrates the characteristics of sensorless control methods.

In terms of the accuracy of the rotor position estimation for both STP and DTP PMSMs, it has been concluded that the oscillation on the rotor position estimation

Table 9.2 Comparison of sensorless control methods based on carrier signal injections for single and DTP PMSMs [26]

Machine	**STP PMSM**		**DTP PMSM**			
Methods	**Pulsating**	**Rotating**	**Double *dq* Model**		**VSD**	
			Pulsating	**Rotating**	**Pulsating**	**Rotating**
Reference frame	Estimated synchronous	Stationary	Two estimated synchronous	Two stationary	Estimated synchronous	Stationary
Injected signal	Pulsating carrier voltage	Rotating carrier voltage	Two independent pulsating carrier voltages	Two independent rotating carrier voltages	Pulsating carrier voltage	Rotating carrier voltage
Carrier current	Amplitude-modulated	Phase-modulated	Amplitude-modulated	Phase-modulated	Amplitude-modulated	Phase-modulated
Cross-saturation angle	$-\theta_m/2$	$-\theta_m/2$	$-\theta_m/2$	$-\theta_m/2$	$-\theta_m/2$	$-\theta_m/2$
Oscillation error	Medium	Medium	Medium	Medium	Low	Low
Load effect	Sensitive	Sensitive	Sensitive	Sensitive	Sensitive	Sensitive
Steady-state performance	Good	Medium	Good	Medium	Very Good	Very Good
Dynamic-state performance	Good	Medium	Good	Medium	Good	Medium

error for DTP PMSM employing the VSD modeling method is significantly reduced for both pulsation and rotating injections compared to that of either of the STP PMSM and DTP PMSM employing double *dq* modeling method. This is due to the mutual coupling between the two winding sets being eliminated in the VSD method.

9.4 Transition between low speed and high speed

Due to the drawbacks of HF carrier signal injection strategies in high-speed range operation such as extra losses, transient disturbances, torque ripple, additional current harmonics, and noise produced from the injection carrier voltage, the HF carrier signal injection techniques are usually considered to be suitable at zero speed and low-speed operation. On the other hand, the fundamental model-based sensorless control methods would fail to estimate the rotor position at low speed, but such methods show a decent performance at high-speed regions [1]. Therefore, it is desirable to combine the merits of the HF signal injection method and the fundamental model-based method to realize full-speed range operation. This combination should consider the smooth transition between various techniques without reducing the dynamic performance. In this section, the investigation is carried out with the HF pulsating signal injection method combined with the flux-linkage-based sensorless control method.

To achieve a smooth transition between the combined methods, a function similar to hysteresis is applied for a certain speed range with optimum weighted gains for both estimated positions as expressed in (9.39) [5], where the HF injection method can work for zero- to low-speed range, and for high-speed range, the flux-linkage based sensorless control method will take over.

$$\widehat{\theta}_r = k\,\widehat{\theta}_{r\mathrm{HF}} + (1-k)\,\widehat{\theta}_{r\mathrm{FO}} \tag{9.39}$$

where $\widehat{\theta}_r$ is referred to as the final estimated rotor position, $\widehat{\theta}_{rHF}$ denotes the estimated rotor position of the HF carrier signal injection method, while the estimated rotor position of the flux-linkage-based sensorless control method is represented by $\widehat{\theta}_{r\mathrm{FO}}$, and k is the weighting gain which can be calculated as

$$k = \begin{cases} 1 & \text{if } \omega_{\mathrm{prev.}} < \omega_{\mathrm{low}} \\ \dfrac{1}{\omega_{\mathrm{low}} - \omega_{\mathrm{high}}}\left(\omega_{\mathrm{prev.}} - \omega_{\mathrm{high}}\right) & \text{if } \omega_{\mathrm{low}} < \omega_{\mathrm{prev.}} < \omega_{\mathrm{high}} \\ 0 & \text{if } \omega_{\mathrm{prev.}} > \omega_{\mathrm{high}} \end{cases} \tag{9.40}$$

where $\omega_{\mathrm{prev.}}$ is the speed of the previous step, ω_{low} is referred to as the low-speed border of the area, and the high-speed border of the area is denoted by ω_{high}.

It is worth mentioning that the injection of the HF signal should be gradually reduced until it reaches zero to reduce the adverse effects of the injected signal on the system when the speed is higher than the high-speed border.

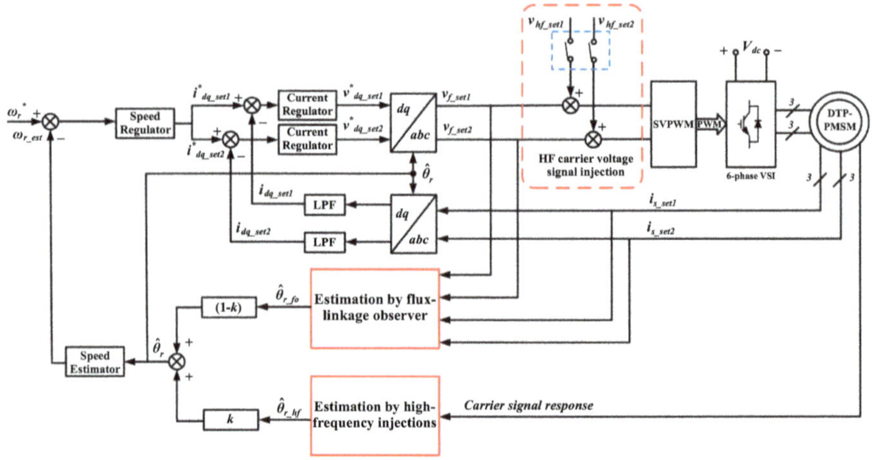

Figure 9.22 Overall sensorless control system.

In order to verify the combined sensorless control method, experimental tests are carried out on a prototype DTP PMSM, of which the specifications are listed in Appendix A (Machine 3), with the aid of the dSPACE platform. The sampling frequency is set to 10 kHz, while the DC-link voltage is 40 V. In the experimental tests, the low-speed border ω_{low} is chosen as 60 r/min, and the high-speed border is chosen as 150 r/min. The overall sensorless control system is illustrated in Figure 9.22 [26].

In the low-speed range, an HF voltage signal, with an amplitude of 8 V and frequency of 500 Hz, is injected into the system. It will disappear when the machine runs in the high-speed range, and the estimator of the flux-linkage-based sensorless control method starts to work. The performance of the rotor position estimation with different combinations for DTP PMSM is investigated and shown as follows.

In this case, the flux-linkage-based sensorless control method is combined with the conventional pulsating injection using a carrier current response [26]. When the rotor's initial speed is zero and the step changed to 80 r/min and then to 250 r/min, the rotor's speed response is shown in Figure 9.23(a), and the comparison of the actual and estimated rotor positions together with the estimation error is illustrated in Figure 9.23(b).

Furthermore, the performance of the combined method [26] is shown in Figure 9.24 when the speed changes from −50 r/min to 200 r/min. Both results show good position tracking with a smooth transition between the two sensorless control methods. From the experimental results, it can be found that a smooth transition from the HF carrier signal injection at low speeds to the flux-linkage-based sensorless control method strategy at higher speeds has been achieved, and the system's stability is ensured.

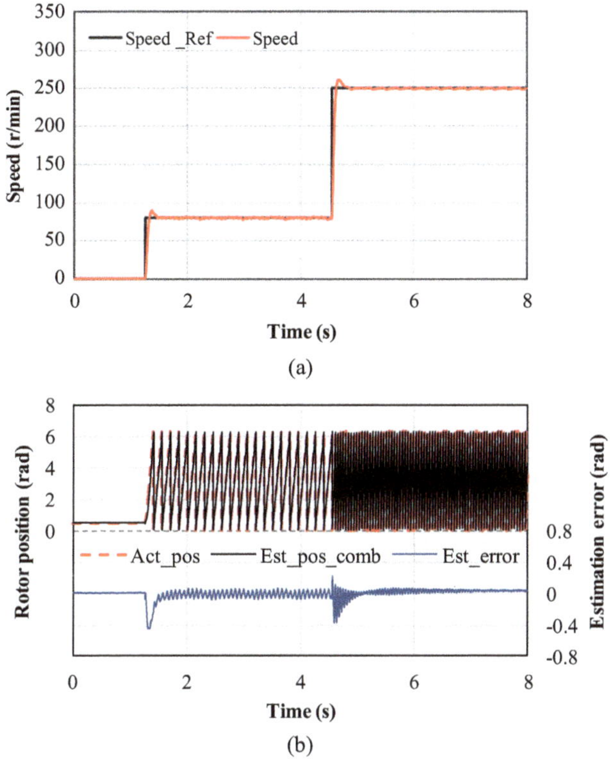

Figure 9.23 Sensorless control performance from low- to high-speed range, 0-80-250 r/min. (a) Speed response. (b) Rotor position estimation.

9.5 Summary

This chapter provides a comprehensive description of sensorless control techniques for both STP and DTP PMSMs. The fundamental model-based sensorless control method relies on the mathematical model of PMSMs. This method is well-suited for high-speed operation. On the other hand, the saliency tracking-based sensorless control technique involves injecting a signal to track the saliency of PMSMs, enabling rotor position estimation at low and zero speeds. In the transition between low-speed and high-speed operation, where a seamless handover between the saliency tracking-based and fundamental model-based sensorless control methods are necessary, a function similar to hysteresis is used to ensure stable and reliable operation across the entire speed range. The key issues and solutions in sensorless control for STP and DTP PMSMs are summarized in Table 9.3.

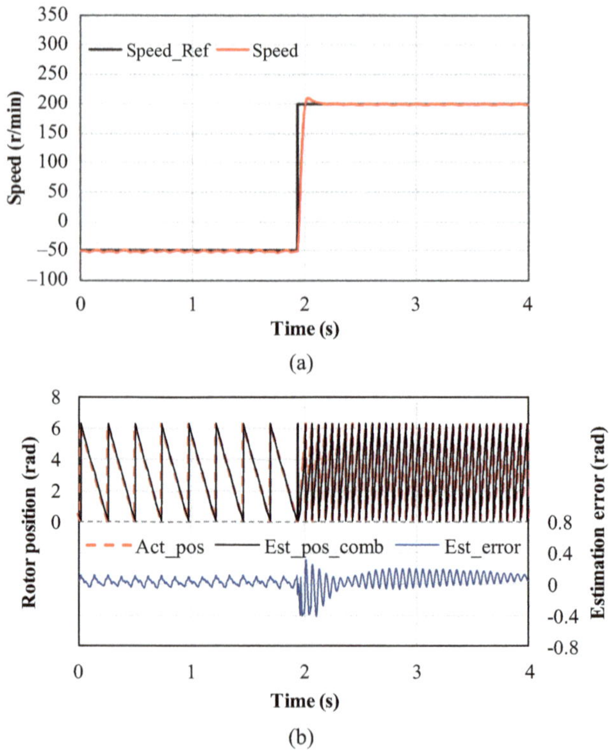

Figure 9.24 Sensorless control performance from low- to high-speed range, −50 to +200 r/min. (a) Position estimation. (b) Speed response.

Table 9.3 Key issues and solutions in sensorless control for STP and DTP PMSMs

Issues	**Solutions**
Cross-coupling effects for both STP and DTP PMSMs	• Offline look-up tables (LUT) [36] • Offline measurement of coupling factor [22]. • Online measurement of coupling inductances based on three HF signal injection stages [37]
Parameter mismatch in fundamental model-based sensorless control for DTP PMSMs	• Parameter deviation correction technique [38]
Coupling effect in saliency tracking based sensorless control for DTP PMSMs	• Saliency tracking-based sensorless control methods with VSD control method [26]
Smooth transition between low-speed and high-speed sensorless control methods	• Implement hysteresis bands for transition [26]

References

[1] Z. Q. Zhu and X. M. Wu, *Sensorless Control of Permanent Magnet Synchronous Machine Drives*. New York: Wiley & IEEE Press, 2023.

[2] T. Y. Liu, "Sensorless control of PM wind power generators with parameter variation and asymmetries," *PhD thesis*, University of Sheffield, 2021.

[3] Z. Chen, M. Tomita, S. Doki, and S. Okuma, "An extended electromotive force model for sensorless control of interior permanent-magnet synchronous motors," *IEEE Trans. Ind. Electron.*, vol. 50, no. 2, pp. 288–295, 2003.

[4] S. Morimoto, K. Kawamoto, M. Sanada, and Y. Takeda, "Sensorless control strategy for salient-pole PMSM based on extended EMF in rotating reference frame," *IEEE Trans. Ind. Appl.*, vol. 38, no. 4, pp. 1054–1061, 2002.

[5] C. Silva, G. M. Asher, and M. Sumner, "Hybrid rotor position observer for wide speed-range sensorless PM motor drives including zero speed," *IEEE Trans. Ind. Electron.*, vol. 53, no. 2, pp. 373–378, 2006.

[6] Z.Q. Zhu, Y. Li, C. Bingham, D. Stone, and D. Howe, "Influence of machine topology and cross-coupling on rotor position estimation accuracy in extended back-EMF based sensorless PM brushless AC drives," *Proc. IEEE Ind. Appl. Soc. Annu. Meet.*, pp. 2378–2385, 2007.

[7] N. Matsui, "Sensorless PM brushless DC motor drives," *IEEE Trans. Ind. Appl.*, vol. 43, pp. 300–308, 1996.

[8] I. Boldea, M. C. Paicu, and G.-D. Andreescu, "Active flux concept for motion-sensorless unified AC drives," *IEEE Trans. Power Electron.*, vol. 23, no. 5, pp. 2612–2618, 2008.

[9] I. Boldea, M. C. Paicu, G.-D. Andreescu, and F. Blaabjerg, ""Active flux" DTFC-SVM sensorless control of IPMSM," *IEEE Trans. Energy Convers.*, vol. 24, no. 2, pp. 314–322, 2009.

[10] Y. Park and S. Sul, "Sensorless control method for PMSM based on frequency-adaptive disturbance observer," *IEEE J. Emerg. Sel. Top. Power Electron.*, vol. 2, no. 2, pp. 143–151, 2014.

[11] Y. Hu, Z. Q. Zhu, and M. Odavic, "Comparison of two-individual current control and vector Space decomposition control for dual three-phase PMSM," *IEEE Trans. Ind. Appl.*, vol. 53, no. 5, pp. 4483–4492, 2017.

[12] S. Kallio, M. Andriollo, A. Tortella, and J. Karttunen, "Decoupled dq model of double-star interior PM synchronous machines," *IEEE Trans. Ind. Electron.*, vol. 60, no. 6, pp. 2486–2494, 2013.

[13] Y. Lee, Y.-C. Kwon, and S.-K. Sul, "Comparison of rotor position estimation performance in fundamental-model-based sensorless control of PMSM," *Proc. IEEE Energy Convers. Congr. Expo.*, pp. 5624–5633, 2015.

[14] L. M. Gong, "Carrier signal injection based sensorless control of permanent magnet brushless AC machines," *PhD thesis*, University of Sheffield, UK, 2011.

[15] M. J. Corley and R. D. Lorenz, “Rotor position and velocity estimation for a salient-pole permanent magnet synchronous machine at standstill and high speed,” *IEEE Trans. Ind. Appl.*, vol. 34, no. 4, pp. 784–789, 1998.

[16] T. Aihara, A. Toba, T. Yanase, A. Mashimo, and K. Endo, “Sensorless torque control of salient-pole synchronous motor at zero-speed operation,” *IEEE Trans. Power Electron.*, vol. 14, no. 1, pp. 202–208, 1999.

[17] J. H. Jang, S. K. Sul, J. I. Ha, K. Ide, and M. Sawamura, “Sensorless drive of surface-mounted permanent-magnet motor by high-frequency signal injection based on magnetic saliency,” *IEEE Trans. Ind. Appl.*, vol. 39, no. 4, pp. 1031–1039, 2003.

[18] Y. Li, Z. Q. Zhu, D. Howe, and C. M. Bingham, “Improved rotor position estimation by signal injection in brushless AC motors, accounting for cross-coupling magnetic saturation” *IEEE Trans. Ind. Appl.*, vol. 45, no. 5, pp. 1843–1850, 2009.

[19] L.M. Gong and Z.Q. Zhu, “Improved rotating carrier signal injection method for sensorless control of PM brushless AC motors, accounting for cross-saturation effect,” *8th Int. Conf. Power Electron. - ECCE Asia.*, pp. 1132–1139, 2011.

[20] P. L. Jansen and R. D. Lorenz, “Transducerless position and velocity estimation in induction and salient AC machines,” *IEEE Trans. Ind. Appl.*, vol. 31, no. 2, pp. 240–247, 1995.

[21] S.C. Yang, T. Suzuki, R.D. Lorenz, and T.M. Jahns, “Surface permanent magnet synchronous machine design for saliency tracking self-sensing position estimation at zero and low speeds,” *IEEE Trans. Ind. Appl.*, vol. 47, no. 5, pp. 2103–2116, 2011.

[22] Y. Li, “Sensorless control of permanent magnet brushless AC motors accounting for cross-coupling magnetic saturation,” *PhD thesis*, University of Sheffield, UK, 2009.

[23] M. Linke, R. Kennel, and J. Holtz, “Sensorless position control of permanent magnet synchronous machines without limitation at zero speed,” *2002 IEEE 28th Annu. Conf. Ind. Electron. Soc. (IECON 02)*, vol. 1, pp. 674–679, 2002.

[24] D. Raca, P. Garcia, D.D. Reigosa, F. Briz, and R.D. Lorenz, “Carrier-signal selection for sensorless control of PM synchronous machines at zero and very low speeds,” *IEEE Trans. Ind. Appl.*, vol. 46, no. 1, pp. 167–178, 2010.

[25] Z. Q. Zhu, “Sensorless operation of permanent magnet brushless DC and AC drives-basic and novel machine design and control aspects,” *8th Int. Conf. Power Electron. – ECCE Asia*, The Shilla Jeju, Jeju, Korea, 2011.

[26] A. H. Almarhoon, “Sensorless control of dual three-phase permanent magnet synchronous machine drives,” *PhD thesis*, University of Sheffield, 2016.

[27] G. Wang, D. Xiao, G. Zhang, C. Li, X. Zhang, and D. Xu, “Sensorless control scheme of IPMSMs using HF orthogonal square-wave voltage injection into a stationary reference frame,” *IEEE Trans. Power Electron.*, vol. 34, no. 3, pp. 2573–2584, 2019.

[28] Y.-D. Yoon, S.-K. Sul, S. Morimoto, and K. Ide, “High-bandwidth sensorless algorithm for AC machines based on square-wave-type voltage injection,” *IEEE Trans. Ind. Appl.*, vol. 47, no. 3, pp. 1361–1370, 2011.
[29] D. Kim, Y.-C. Kwon, S.-K. Sul, J.-H. Kim, and R.-S. Yu, “Suppression of injection voltage disturbance for high-frequency square-wave injection sensorless drive with regulation of induced high-frequency current ripple,” *IEEE Trans. Ind. Appl.*, vol. 52, no. 1, pp. 302–312, 2016.
[30] M. W. Degner and R. D. Lorenz, “Using multiple saliencies for the estimation of flux, position, and velocity in ac machines,” *IEEE Trans. Ind. Appl.*, vol. 34, no. 5, pp. 1097–1104, 1998.
[31] P. Garcia, F. Briz, M. W. Degner, and D. D. Reigosa, “Accuracy, bandwidth, and stability limits of carrier-signal-injection-based sensorless control methods,” *IEEE Trans. Ind. Appl.*, vol. 43, no. 4, pp. 990–1000, 2007.
[32] X. M. Wu, “Novel sensorless control of surface mounted permanent magnet synchronous machines,” *PhD thesis*, University of Sheffield, 2012.
[33] J. M. Liu and Z. Q. Zhu, “Novel sensorless control strategy with injection of high-frequency pulsating carrier signal into stationary reference frame,” *IEEE Trans. Ind. Appl.*, vol. 50, no. 4, pp. 2574–2583, 2014.
[34] R. Bojoi, M. Lazzari, F. Profumo, and A. Tenconi, “Digital field-oriented control for dual three-phase induction motor drives,” *IEEE Trans. Ind. Appl.*, vol. 39, no. 3, pp. 752–760, 2003.
[35] Y. Zhao and T. A. Lipo, “Space vector PWM control of dual three-phase induction machine using vector space decomposition,” *IEEE Trans. Ind. Appl.*, vol. 31, no. 5, pp. 1100–1109, 1995.
[36] Z.Q. Zhu, Y. Li, D. Howe, and C.M. Bingham, “Compensation for rotor position estimation error due to cross-coupling magnetic saturation in signal injection based sensorless control of PM brushless AC motors,” 2007 *IEEE Int. Electr. Mach. & Drives Conf.*, pp. 208–213, 2007.
[37] B. Shuang and Z. Q. Zhu, “A novel method for estimating the high frequency incremental dq-axis and cross-coupling inductances in interior permanent magnet synchronous machines,” *IEEE Trans. Ind. Appl.*, vol. 57, no. 5, pp. 4913–4923, 2021.
[38] T. Liu, Z. Q. Zhu, Z.-Y. Wu, D. Stone, and M. Foster, “A simple sensorless position error correction method for dual three-phase permanent magnet synchronous machines,” *IEEE Trans. Energy Convers.*, vol. 36, no. 2, pp. 895–906, 2021.

Chapter 10
Fault-tolerant control for multiphase PMSMs

10.1 Introduction

Fault-tolerant control has become one of the major control techniques for modern electrical machine drive systems, particularly in applications where reliability and uninterrupted operation are critical. One such area of interest is multiphase electrical machine drives, which are widely employed in aerospace, industrial, and transportation systems.

This chapter describes various aspects of fault-tolerant control strategies for multiphase electrical machines, focusing on fault classification, fault diagnosis, and fault-tolerant control strategies. First, the identification and categorization of different types of faults that may occur in multiphase electrical machines and associated components are discussed. Then, different fault diagnosis algorithms in multiphase electrical machine systems are described, which utilize various signals and measurements from the machine and its environment to analyze the system behavior and detect the presence of faults. Subsequently, this chapter introduces various fault-tolerant control strategies for multiphase electrical machines, including FOC- and DTC-based fault-tolerant control. These control strategies aim to mitigate the effect of those machine faults with the most frequent occurrence and ensure sustained operation during fault operation, while efficiency, torque ripple, or system complexity need to be compromised as well. In summary, the structure of the chapter is illustrated in Figure 10.1.

10.2 Fault classification in multiphase electrical machine drives

A machine drive system typically consists of several components working together to control the speed, rotating direction, and torque of the machine.

These components include:

a) **Electrical machine:** converts electrical energy into mechanical energy, or vice versa.
b) **Electric drive inverter:** facilitates the switching of high-current signals required to drive the machine.
c) **Sensors:** measure a machine's electrical/mechanical quantities, such as speed, current, and temperature and provide feedback to the machine controller.

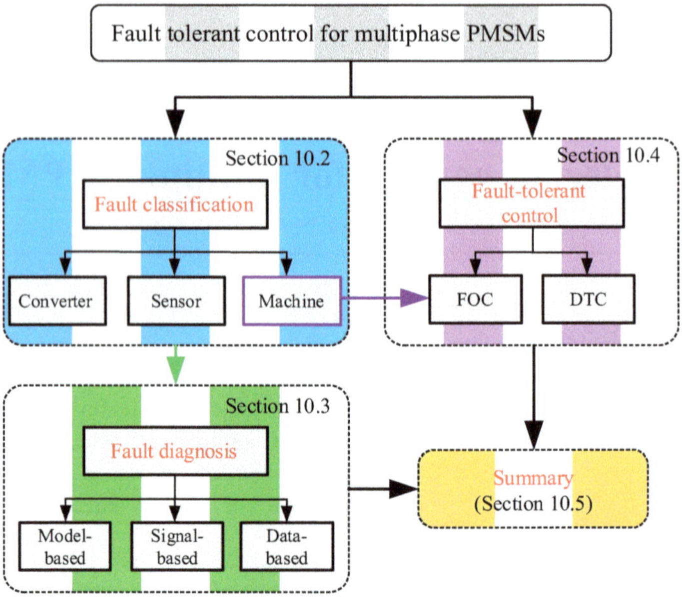

Figure 10.1 Structure of Chapter 10.

d) **Controller circuit:** generates the PWM signals to be sent to the electric drive inverter according to machine parameters monitored from sensors.

In electrical machine control systems, the most common faults occur in the inverter, machine, and sensors, i.e., inverter faults, machine faults, and sensor faults.

10.2.1 Inverter faults in multiphase electrical machine drives

The most general topology of an inverter is the two-level three-phase voltage source inverter (VSI), as shown in Figure 10.2. As can be seen, the faults in the inverter are mainly electric faults [1], which can be classified as switching open-circuit (OC) faults, short-circuit (SC) faults, DC-bus SC faults, and single line-to-ground (SLG) faults.

10.2.1.1 Open-circuit (OC) switching fault

Solid-state switching devices are susceptible to open-circuit (OC) faults, which can arise from various deteriorations such as bond wire lift-off or soldering cracking, which is often caused by thermal-mechanical stress and/or significant changes in junction temperature. Additionally, OC faults may also result from gate driver malfunction, electrical disconnections, and even the lack of a discharge circuit for a DC-link capacitor. In general, OC faults can lead to significant distortion in voltages, currents, and torque, in both single three-phase and multiphase electrical machine drive systems.

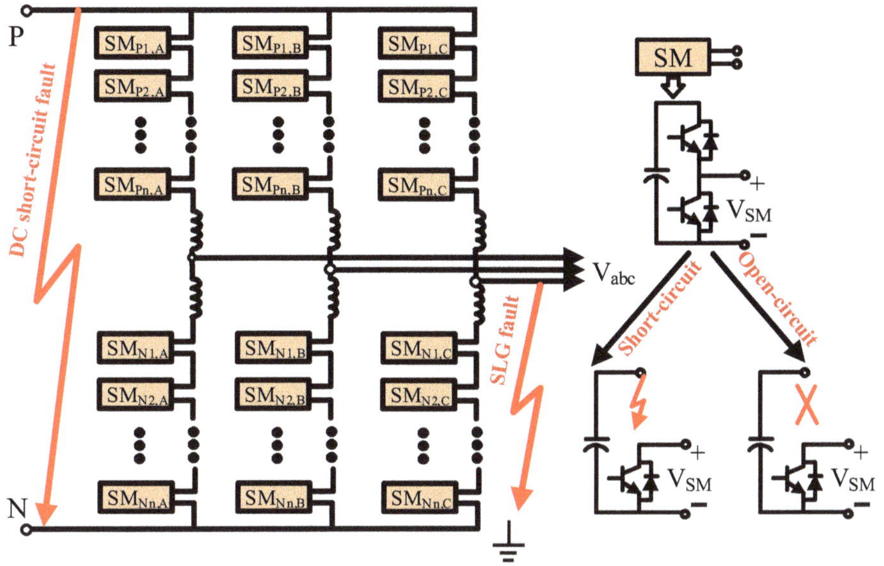

Figure 10.2 Classification of two-level three-phase VSI faults in machine drives.

10.2.1.2 Short-circuit (SC) switching fault

Short-circuit (SC) faults may occur due to parasitic turn-on (e.g., cross-talk), gate driver malfunction, overcurrent, overvoltage, overheating, and other reasons.

10.2.1.3 Short-circuit (SC) fault on the DC side

Like all other VSIs, half-bridge-based modular multilevel inverters (HB-MMCs) are not immune to DC SC fault (i.e., pole-to-pole fault). Once such a fault occurs, the capacitors in the sub-modules (SMs) will discharge, and the fault current in the inverter arms and DC-link will mainly derive from the discharging of SM capacitors.

10.2.1.4 Single line-to-ground (SLG) fault on the AC side

Single line-to-ground fault could occur between any of the AC lines on the ground, as shown in Figure 10.2. When a single line-to-ground (SLG) fault happens, AC voltage fluctuations and a sudden drop in power are both possible in the inverter. Additionally, the DC-link voltage might be subjected to a pulsation of twice the fundamental frequency, which would increase the switch losses and stress of power devices of the inverter, and even cause malfunction of the protection system, such as overvoltage protection circuit faults.

10.2.2 Sensor faults in multiphase electrical machine drives

In a typical PM machine drive system, one position sensor (encoder or Hall sensor), one DC-link voltage, and at least two alternating-current sensors are required to ensure high control performance. Any sensor failure would cause deteriorated machine performance. The sensor faults can be classified as position sensor faults (or speed sensor faults), DC-link voltage sensor faults, and current sensor faults.

10.2.2.1 Position sensor faults

The position/speed sensor is typically used to obtain the rotor position and speed in electrical machine drives. One of the most commonly used speed transducers is the photoelectric incremental encoder, which generates electrical pulses when the machine rotates. In the worst-case speed-sensor fault, all pulse signals disappear. Consequently, the measured speed will be decreased sharply and maintained at zero. Subsequently, the speed proportional integral (PI) controller will become saturated consistently, and the actual speed will increase to the maximum value limited by the DC-link voltage. It should be noted that the uncontrollable high speed will cause continuous overload and dangerous damage.

10.2.2.2 Current sensor faults

The faults of current sensors are mainly divided into the following three types.

a) **Gain fault** occurs when the gain of the amplifier circuitry, which processes the current signal, deviates from its expected value, resulting in an inaccurate measurement of the machine's current. This can happen due to various reasons, such as changes in temperature or the aging of components.
b) **Zero-offset fault** occurs when the output of the sensor does not correspond to zero when there is no current flowing through the machine. Possible reasons include residual magnetization in the magnetic core of the sensor or offset voltage in the amplifier circuitry.
c) **Signal-loss fault** occurs when the current signal from the machine is lost or significantly reduced before reaching the sensor, resulting in little or no output from the sensor. Signal loss may be caused by cable breakage, loose connections, or a damaged sensor or transducer. If the signal loss ratio is too high, the machine current cannot be accurately measured, leading to instability and potential damage to the machine and drive system.

10.2.2.3 DC-link voltage sensor faults

An accurate and reliable DC-link voltage measurement is essential for the operation of machine drives, especially for position estimation in sensorless machine drives. The accurate value of the DC-link voltage under real-time conditions is used to avoid performance deterioration or even worse consequences caused by the fluctuation in the DC-link voltage. The fluctuations in the DC-link voltage may occur due to several reasons:

a) **Downsizing of DC-link capacitors**: DC-link capacitors are often large in size. However, downsizing of DC-link capacitors may be desirable for the compactness of the machine drive system. However, in the systems equipped with a diode bridge passive front end, downsizing can lead to a reduction in stiffness and oscillation in the DC-link voltage at high loads.

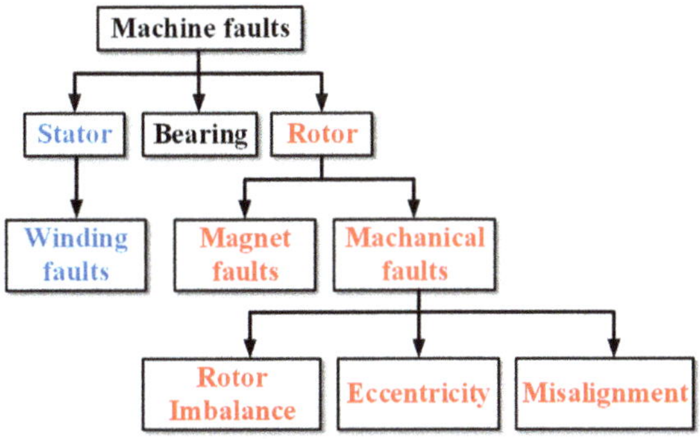

Figure 10.3 Classification of machine faults for multiphase electrical machines.

b) **Weak grid conditions**: Weak grids are susceptible to voltage dips and surges, especially if faults occur in the vicinity.
c) **Battery-powered applications**: DC-link voltage variations can occur in battery-powered servo-drive and propulsion applications as a function of the battery state of charge.

Similar to faults in current sensors, faults of DC-link voltage sensors are mainly divided into gain faults, zero-offset faults, and signal-loss faults. In the absence of a reliable bus voltage measurement, the discrepancy between the reference voltage and output inverter voltage manifests as torque error and ripples in the machine, adversely impacting the load.

10.2.3 Machine faults in multiphase electrical machine drives

The faults in multiphase electrical machine systems can be classified into three typical categories: mechanical faults, magnetic faults, and electrical faults. According to the locations of failure, they can also be divided into stator faults (related to winding and stator core), rotor faults (demagnetization, eccentricity, and imbalance), bearing faults, and shaft faults [2]. The classification of different machine faults can be summarized in Figure 10.3.

According to statistics from IEEE and EPRI [3,4], stator winding failures are one of the most common reasons for AC machine failures, representing between 36% and 66% of all failures, while bearing faults represent between 13% and 41%, respectively, depending on the type and size of the machine, as shown in Figure 10.4.

10.2.3.1 Mechanical faults

The multiphase electrical machine may commonly have mechanical issues, including imbalance, eccentricity, misalignment, and bearing failures. Bearing failures, both rolling and sliding, are the most prevalent mechanical issue, accounting for 41% of all low-voltage machine defects. These failures are often

(a) (b)

Figure 10.4 Faults percentages by various components. (a) Low-voltage electric machines. (b) High-voltage electric machines.

caused by incorrect assembly, inadequate lubrication, machine overloading, and aging.

Eccentricity, specifically rotor eccentricity, occurs when the centers of the rotor and the stator do not coincide. Stresses or defects, such as misalignment and bearing wear, might amplify the eccentricity. For instance, a lateral bearing ball injury might result in shaft inclination and rotor eccentricity. Prolonged misalignment bends the shaft and causes rotor eccentricity as well. The increased rotor eccentricity results in an imbalanced magnetic pull, greater cogging torque, excessive vibration, and higher rotor temperature. The rotor is gradually pulled to a more eccentric position because the direction of the unbalanced magnetic pull is generally consistent with the direction of eccentricity. It eventually leads to rotor-stator friction, lamination damage, and sparks. Furthermore, a significant eccentricity in an inverter-driven PMSM might produce substantial torque oscillations and inverter failures.

An imbalance of the rotor and misalignment leads to distortion in the machine's magnetic field, degrading its dynamic properties, generating additional components in mechanical vibration spectra, increasing noise, and causing torque pulsation.

10.2.3.2 Magnetic faults

Magnetic damage occurs only in PM machines and is concerned with the PMs. It typically arises from mechanical damage and/or demagnetization. The thermal process, corrosion, and aging process are common contributors to the demagnetization of PMs. Meanwhile, PM materials are sensitive to operating temperature. This is due to the PM materials being made from powder technology, which has the problem of oxidation of the elements with increasing temperature. At high temperatures, the magnetic properties of the material may change, and the magnet may become weaker or even lose its magnetism entirely.

For the most used neodymium iron boron magnets, their magnetic strength decreases significantly at temperatures above 80 °C. Above 150 °C, they may even start to demagnetize. The rise in PM temperature can be attributed to various factors, including load conditions, particularly a sudden increase in phase current when an SC fault occurs.

PM demagnetization can significantly reduce the air-gap magnetic density, thereby reducing torque production. To maintain the same torque output, a higher phase current above the rated value is required. However, this increase in phase current also leads to higher copper losses and higher PM temperature. The increase in PM temperature further exacerbates demagnetization, necessitating even higher stator currents and further reducing efficiency. Furthermore, demagnetization may also lead to uneven rotor flux, which, together with overload, results in UMP (unbalanced magnetic pull). UMP gives rise to undesirable noise and damaging vibrations, which accelerates bearing wear and causes rotor damage.

10.2.3.3 Electric faults

Electric faults in stator windings may lead to SC or OC faults in multiphase electrical machines. The faults generally occur because of connection damage or inter-turn winding faults. In terms of SC faults, the inter-turn SC (ITSC) faults are the most common failure in both single three-phase machines and multiphase electrical machines. ITSC faults are usually caused by insulation damage between two adjacent turns of the same coil. Mechanical damage, manufacturing defects, mechanical stress, overheating of the windings, overvoltage, and overcurrent may lead to insulation breakdown, thereby causing ITSC faults in machines. This fault can lead to a variety of problems, including reduced machine performance, overheating, and even machine failure.

OC fault occurs when there is a break or discontinuity in the electrical path in the machine circuit. Mechanical stress, vibration, corrosion, loose connection, or overheating may lead to OC faults. Different from the current surge caused by SC faults, OC faults may generally lead to the machine stopping functioning or operating at reduced capacity.

10.3 Fault diagnosis for multiphase electrical machines

Fault diagnosis plays a critical role in ensuring reliable operation and maintenance of multiphase electrical machines. It involves the detection, localization, and identification of faults within the machine, inverter, and sensor components. Effective fault diagnosis enables timely corrective action by the control system, thereby preventing catastrophic failures. However, since the multiphase electrical machines have a higher number of variables and parameters due to the increased number of phases, this makes fault diagnosis more complex and necessitates the use of advanced techniques for data analysis and modeling. Typically, approaches for fault diagnosis in multiphase electrical machines include model-based fault diagnosis, signal-based fault diagnosis, and data-based fault diagnosis, which will be introduced in this section.

10.3.1 Model-based fault diagnosis methods

Model-based fault diagnosis was first proposed by Beard in 1971 to replace hardware redundancy [5]. In model-based methods, fault diagnosis methods such as observer and parameter estimation are developed to monitor the residuals between the measured and the model-predicted signals.

In terms of observer approaches, extended Kalman filters (EKF) are the most popular parameter estimator-based approaches for detecting PMSM machine problems in the non-linear machine control systems, which are utilized to diagnose ITSC faults [6,7], open phase failure [8], and demagnetization failure [9].

The extended Kalman filter is widely used in fault diagnosis for nonlinear systems, which is extended from the Kalman filter. The general expression of a nonlinear system is shown as:

$$\begin{cases} \dot{X}(t) = f(X(t), \lambda(t), U(t)) + W(t) \\ Y(t) = h(X(t), \lambda(t), U(t)) + V(t) \end{cases} \tag{10.1}$$

where f and h correspond to the state and output equations of the nonlinear model, respectively. $W(t)$ and $V(t)$ are the state and measurement noise vectors with covariance matrices Q and R, respectively.

To estimate one or several parameters (λ) in the model, additional state variables are added to (10.1), resulting in the following extended nonlinear system:

$$\begin{cases} \dot{X}_e(t) = \begin{bmatrix} \dot{X}(t) \\ \dot{\lambda}(t) \end{bmatrix} = \begin{bmatrix} f(X_e(t), U(t)) \\ 0 \end{bmatrix} + \begin{bmatrix} W_X(t) \\ W_\lambda(t) \end{bmatrix} \\ Y(t) = h(X_e(t), U(t)) + V(t) \end{cases} \tag{10.2}$$

where X_e is the extended state vector. The additional equation $\dot{\lambda}(t) = W_\lambda(t)$ reflects that the dynamic response of the estimated parameters is assumed unknown.

For different fault cases in PMSMs, such as OC fault, SC fault, and ITSC fault, the state equations are different. However, the estimated parameter is generally the phase current, which makes the model-based fault diagnosis approaches commonly feasible in machine control systems. For example, it is reported in [6] that an EKF algorithm is used to detect the ITSC fault of a PMSG (Permanent Magnet Synchronous Generator) and estimate the number of shorted turns in the stator winding.

The circuit model for an ITSC fault is described in Figure 10.5. With the occurrence of the ITSC fault, a short circuit path is created in the faulty phase, e.g., phase C. This results in a direct current path without the normal impedance of the faulty phase. As a result, a large current flows through the short circuit path, as $I_{s/c}$.

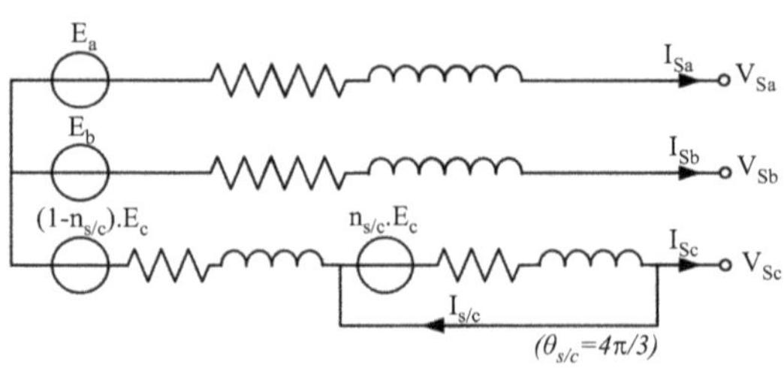

Figure 10.5 Basic model for diagnosis with ITSC fault [6].

Using the expression of the stator voltages and the short circuit loop, the PMSG model, i.e., the state equations with ITSC fault, can be expressed as:

$$\begin{cases} [V_s]_{dq} = -R_s \cdot \left[I_s'\right]_{dq} - \omega \cdot L_s \cdot \begin{bmatrix} 0 & -1 \\ 1 & 0 \end{bmatrix} \cdot \left[I_s'\right]_{dq} - L_s \cdot \left[\dot{I}_s'\right]_{dq} + [E]_{dq} \\ [I_s]_{dq} = \left[I_s'\right]_{dq} - \left[\tilde{I}_{s/c}\right]_{dq} = \left[I_s'\right]_{dq} - \left[Z_{s/c}\right]_{dq}^{-1} \cdot [V_s]_{dq} \end{cases} \tag{10.3}$$

where $\left[\tilde{I}_{s/c}\right]_{dq}$ is the SC current vector in the dq reference frame. $\left[Z_{s/c}\right]_{dq}^{-1}$ is the fault impedance. $P(\theta)$ is the dq transformation matrix. $Q(\theta_{s/c})$ is the fault localization matrix. They are expressed as:

$$\left[\tilde{I}_{s/c}\right]_{dq} = \sqrt{2/3} \cdot n_{s/c} \cdot I_{s/c} \cdot P(\theta) \cdot \begin{bmatrix} \cos(\theta_{s/c}) \\ \sin(\theta_{s/c}) \end{bmatrix} \tag{10.4}$$

$$\left[Z_{s/c}\right]_{dq}^{-1} = \left(2 \cdot n_{s/c} / (3 - 2 \cdot n_{s/c}) \cdot R_s\right) \cdot P(\theta) \cdot Q(\theta_{s/c}) \cdot P(\theta)^T \tag{10.5}$$

$$P(\theta) = \begin{bmatrix} \cos\theta & -\sin\theta \\ \sin\theta & \cos\theta \end{bmatrix} \tag{10.6}$$

$$Q(\theta_{s/c}) = \begin{bmatrix} \cos^2(\theta_{s/c}) & \cos(\theta_{s/c})\sin(\theta_{s/c}) \\ \cos(\theta_{s/c})\sin(\theta_{s/c}) & \sin^2(\theta_{s/c}) \end{bmatrix} \tag{10.7}$$

Extending this model to each phase, three SC impedances ($Z_{s/cA}$, $Z_{s/cB}$, $Z_{s/cC}$ for $\theta_{s/c} = 0$, $2\pi/3$, and $4\pi/3$, respectively) are added to the PMSG model. These fault impedances depend on the three short-circuited turns ratios n_{s/c_A}, n_{s/c_B}, and n_{s/c_C}, which are the estimated parameters in the extended system. Thus, the extended state vector X_e in [6] used is expressed as

$$X_e = \begin{bmatrix} X \\ \lambda \end{bmatrix} = \left[I_{Sd}' | I_{Sq}' | n_{s/c\_A} | n_{s/cB} | n_{s/cC}\right]^T \tag{10.8}$$

The state noise covariance matrix Q and the measurement noise covariance matrix R are tuned based on the dynamic response requirement. According to [7], Q can be separated into two matrices: q_x (for measured states with q_x as a diagonal component) and q_λ (for the estimated parameters with q_λ as a diagonal component). Thus, Q and R can be expressed as

$$\begin{cases} Q = q_x \cdot \begin{bmatrix} I_2 & 0_{2\times3} \\ 0_{3\times2} & \dfrac{q_\lambda}{q_x} \cdot I_3 \end{bmatrix} \\ R = r \cdot I_2 \end{cases} \tag{10.9}$$

On the one hand, q_x and r have been experimentally determined by measuring the variance of noises on input (σ_u^2) and output signals (σ_y^2), i.e.,

$$\begin{cases} q_x = \left(\frac{\partial f_k}{\partial u_k}\right)^2 \cdot \sigma_u^2 = \left(\frac{T_s}{L_s}\right) \cdot \sigma_u^2 \\ r = \left(\frac{\partial h_k}{\partial X_k}\right)^2 \cdot \sigma_y^2 = \sigma_y^2 \end{cases} \tag{10.10}$$

where f_k and h_k are the state and output equations of the discrete model, respectively; and T_s is the sampling period.

On the other hand, the ratio q_λ/q_x is set to satisfy the fast dynamic response requirement for the parameter estimation, i.e.,

$$\sqrt{\frac{q_\lambda}{q_x}} \approx \frac{1}{\frac{\tau}{3R_s} \cdot \sqrt{\left(\frac{R_s}{L_s}\right)^2 + \omega^2} \cdot \sqrt{V_{sd}^2 + V_{sq}^2}} \tag{10.11}$$

In general, τ is set to one electrical period to ensure a response dynamic close to one electrical period, e.g., $\tau = 20$ ms when the electrical period is at 50 Hz. The ratio q_λ/q_x has been defined by the target machine parameters (resistance R_s and inductance L_s) and operating point (speed and voltage conditions).

Additionally, several improved Kalman filter approaches for fault diagnosis have been developed, such as unscented Kalman filters (UKFs), adaptive Kalman filters, and augmented state Kalman filters. The UKF can more accurately capture the real mean and covariance by relying on a more precise stochastic approximation, such as the unscented transform, which improves diagnostic performance [10,11]. In order to obtain an accurate defect diagnosis, adaptive Kalman filters can be used to modify the measurement noise covariance matrix or the process noise covariance matrix [12,13]. System states and fault signals can be concurrently estimated using augmented state Kalman filters [14].

10.3.2 Signal-based fault diagnosis methods

Signal-based fault diagnosis methods in multiphase electrical machines analyze electrical signals to detect and diagnose faults. These methods rely on measuring and analyzing various electrical signals, such as currents and voltages, to identify deviations from normal operating conditions and infer the presence of faults. There are several advantages to using signal-based fault diagnosis methods for multiphase electrical machines. (a) Early fault detection: they enable the early detection of faults, allowing for proactive maintenance, and reducing the risk of machine failures. (b) Non-intrusive: they are non-intrusive as they rely on external measurements without the need for physical access to the machine. (c) Easy integration and automation: they can be automated and integrated into a monitoring system, allowing continuous and real-time fault diagnosis.

According to the extracted signal types, the signal-based fault diagnosis methods can be classified into two categories: time and frequency domain-based methods.

10.3.2.1 Time-domain signal-based method

Extraction of time-domain signals for fault diagnosis is natural when monitoring a continuous dynamical system. The voltage and current signals are the most widely used signals in machine fault diagnosis when their actual values vary with the operating point. For example, the reference voltage vector components are used as a diagnostic signal in [15]. The instantaneous value of the angle between the reference voltage components has been used directly as an efficient fault indicator to distinguish an ITSC fault and a partial irreversible demagnetization fault. δ is the angle between the dq-axis voltage vectors, i.e., voltage angle. When a partial irreversible demagnetization fault occurs, the voltage angle will increase, and the faulty angle δ_f will be larger than the health angle δ_h, i.e., $\delta_f > \delta_h$. However, when an ITSC fault occurs, the voltage angle will decrease, and the faulty angle δ_f will be smaller than the health angle δ_h, i.e., $\delta_f < \delta_h$.

In [16], current signals are used to identify the switch faults in PMSMs, as outlined in Figure 10.6, where three reference current errors e_n are calculated from the measured i_n and reference i_n^* machine phase currents (n can be either a, b, or c, representing quantities of phases A, B, and C, respectively). As the inverter is operating normally, the three-phase or multiphase currents can track their reference signals i_n^*, i.e., $e_n = 0$, resulting in zero diagnostic variable (d_n). The diagnostic variables d_n are determined by $d_n = e_n/|i_n|$.

In the presence of a fault, the machine currents deviate from their reference signals. For instance, a fault in the phase a upper switch significantly affects the current in this phase, leading to zero current during its positive half-cycle. Therefore, the diagnostic variables d_n become non-zero. Thus, if a defined threshold value k_f is set, when d_n is larger than k_f, the output will be 1 [16]. When d_n is smaller than $-k_f$, the output will be -1. Otherwise, the output will be 0. In this way, the output of fault detection can be illustrated as in Table 10.1.

10.3.2.2 Frequency-domain signal-based method

The signal-based method in the frequency domain aims to detect changes or faults by using spectrum analysis tools, such as discrete Fourier transformation. The spectrum analysis tools involve transforming the time-domain signals into the frequency domain using techniques such as Fourier analysis or wavelet transform.

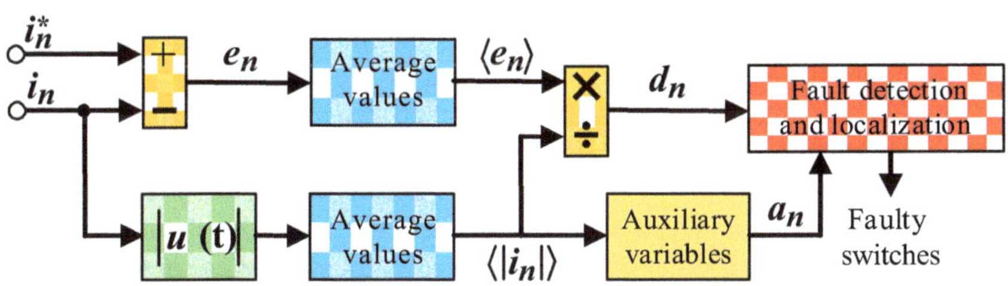

Figure 10.6 Diagram scheme of time-domain signal-based method for detecting switch faults.

Table 10.1 Switch fault detection and localization [16]

d_a	d_b	d_c	**Faulty switch**
1	0	0	High side switch-Phase A
−1	0	0	Low side switch-Phase A
0	1	0	High side switch-Phase B
0	−1	0	Low side switch-Phase B
0	0	1	High side switch-Phase C
0	0	−1	Low side switch-Phase C

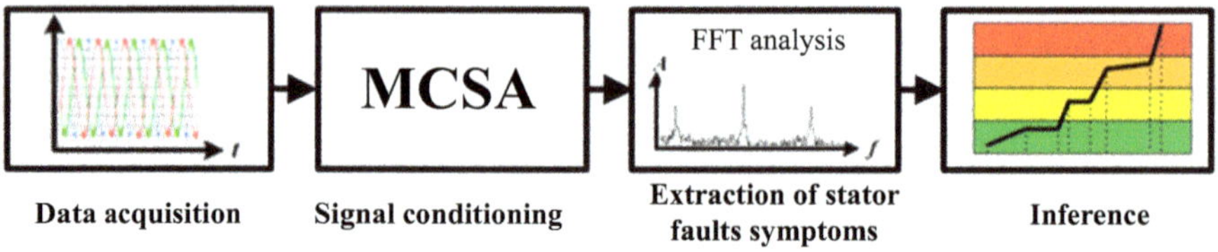

Figure 10.7 Diagnostic information conditioning for spectral analysis of stator currents.

By analyzing the frequency content, specific frequency components associated with faults in the machine can be identified.

In general, the most widely used frequency-domain method is machine current signature analysis (MCSA). MCSA methods usually employ the fast Fourier transform (FFT) of the stator current for the extraction of the fault characteristic frequencies. The simple FFT allows describing any signal by its frequency spectrum expressed by

$$x(f) = \int_{+\infty}^{-\infty} x(t)e^{-j2\pi f \cdot t}dt \tag{10.12}$$

The analysis of a signal is conducted over a finite period, and therefore, it is necessary to choose a weighting window for the analysis (rectangular window, Hann, Hamming) to correct the effects of finite temporal windowing. In addition, the size of the window influences the maximum achievable resolution. Indeed, frequency accuracy depends on the sampling frequency f_s and the number N of samples.

The diagnostic signal processing diagram is presented in Figure 10.7 [17]. First, the chosen signals are measured and then properly conditioned. A representation of the frequency domain, either graphical or numerical, is calculated by FFT with a selected window, e.g., flat-top. To isolate the characteristic harmonics specific to the analyzed stator winding failures, sufficient resolution in spectral analysis is required. Subsequently, the damage symptoms are extracted, involving the amplitudes of these fault-related harmonics.

For different faults in multiphase electrical machines, the stator current harmonics with specific frequency components are summarized in [2]. For example, the misalignment faults will lead to $f_s \pm kf_r$ order current harmonics. The rotor unbalance will lead to $f_s \pm f_r$ order current harmonics. The demagnetization faults will lead to $f_s(1 \pm k/p)$ ($k = 1, 3, 5, \ldots$) order current harmonics. Then, the ITSC faults will lead to large odd harmonics, especially $3f_s$ current harmonics in the drive system.

Besides, mechanical vibration signals are also used in detecting some specific fault categories in multiphase electrical machines. For example, bearing faults will lead to $kf_u \pm l.f_r$ ($k = 1, 2, 3\ldots$, $l = 0, 1$) order harmonics in the mechanical vibration signals. The eccentricity faults will lead to $2pk \pm 1$ ($k = 1,2,3\ldots$) order harmonics in the mechanical vibration signals. The rotor unbalance faults will lead to f_r-order harmonics in the mechanical vibration signals.

Some other signals, such as noise, airgap search coil voltage, stator voltage, speed, load torque, d-axis stator inductance, back EMF, and unbalanced magnetic pull, can also be used to detect specific faults. However, since the stator current contains the most information and can be easily classified according to different frequencies when different faults occur, MCSA is the most used frequency-domain signal-based method in multiphase electrical machine fault diagnosis.

In some cases, both time- and frequency-domain signal-based methods are used simultaneously, known as the time-frequency signal-based method. Time-frequency analysis can identify the signal frequency components and reveal their time-variant features effectively. Several time-frequency signal-based methods, such as short-time Fourier transform (STFT), wavelet transforms (WT), Hilbert–Huang transform (HHT), and Wigner–Ville distribution (WD), analyze the characteristics of time and frequency together to detect the faults in multiphase electrical machines.

The time-frequency signal-based method can also overcome the drawbacks of sole time-domain and sole frequency-domain signal-based methods. For example, sole frequency-domain signal-based methods require the signals to be stationary, which is challenging to meet in modern drive systems. Time-domain signal-based methods require a relatively long measurement time for the diagnostic signals, and their processing is associated with the loss of information about the moment of occurrence of a given fault component. Hence, the time-frequency signal-based method becomes the most promising control method among all signal-based fault diagnosis methods in multiphase electrical machine drive systems.

By way of example, the experimental results of an ITSC fault for a 24 slot/8-pole SPMSM (Machine 5 in Appendix A) are shown in Figures 10.8 and 10.9. As can be seen, with an 11.25% SC ratio (defined as the ratio of SC turn number to the turn number of the entire phase), the SC current reaches ~8 A at 600 r/min. The major frequency components in the SC current include the fundamental and 3rd harmonic components.

As can be observed, although the SC current already reaches about four times that of the phase current amplitudes, there is no significant change in the waveforms of the phase currents. Considering that phase currents are usually measurable in a drive system instead of an SC current, it is critical to extract fault-related information from the phase current waveforms. Besides, comparing the spectra of phase currents at faulty conditions with those at healthy conditions, it is obvious

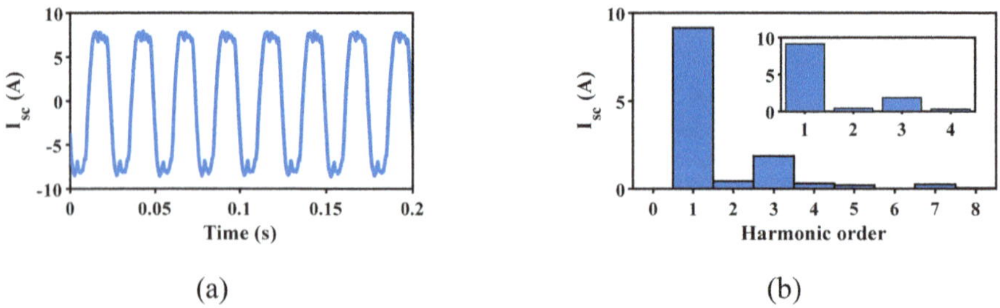

Figure 10.8 *(a) Waveform and (b) spectrum of SC current at ITSC condition (600 r/min, SC ratio 11.25% at Phase A).*

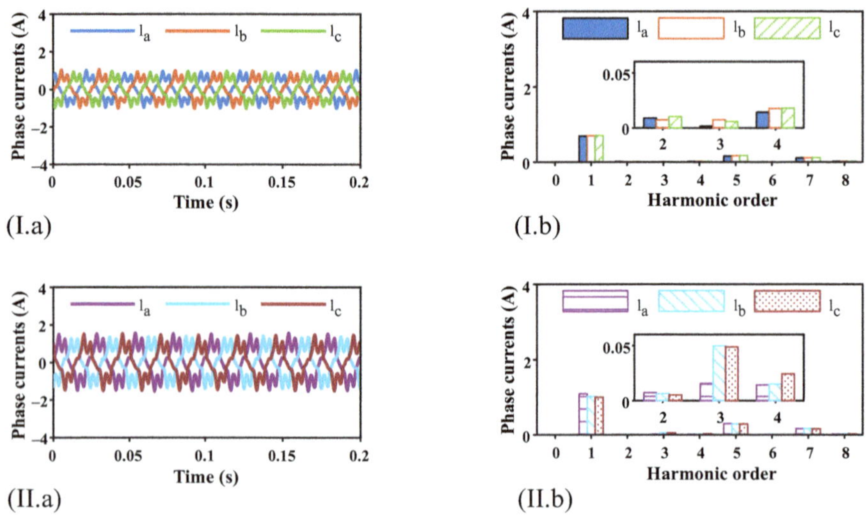

Figure 10.9 *Waveforms and spectra of phase current at (I) healthy and (II) ITSC conditions (600 r/min, SC ratio 11.25% at Phase A). (I.a) Waveforms at healthy condition. (I.b) Spectra at healthy condition. (II.a) Waveforms at faulty condition. (II.b) Spectra at faulty condition.*

that the 3rd harmonics in all three phase currents increase significantly at faulty conditions. This phenomenon has been widely used as the fault indicator for an ITSC fault, as introduced previously.

10.3.3 Data-based fault diagnosis methods

Data-based fault diagnosis methods, also known as knowledge-based fault diagnosis methods, differ from model-based methods and signal-based approaches. As introduced previously, the model-based methods rely on known models of the system or predefined signal patterns to detect faults, while signal-based approaches analyze the characteristics of signals to detect faults. In contrast, data-based fault diagnosis starts with a large volume of historical data without requiring an *a priori* known model or signal patterns.

To analyze and process large volumes of historical data, various artificial intelligence (AI) techniques, such as machine learning (ML) or deep learning (DL), are applied to fault diagnosis processing. These methods enable observation of patterns and trends from historical data and facilitate fault diagnosis decisions with the aid of classifiers. Besides, it should be noted that while all three approaches (model-based, signal-based, and data-based) utilize real-time data for online fault diagnosis, data-based diagnosis methods specifically rely on historical data. Therefore, it is often referred to as data-driven fault diagnosis, emphasizing its reliance on historical data for data extraction and decision-making.

AI plays a crucial role in data-driven technologies, and its application has been extensively studied in various industrial applications. One typical AI technique is artificial neural networks (ANN), which utilize activation functions to model interactions between artificial neurons, mimicking biological neural systems. ANN can effectively extract nonlinear characteristics and perform classification and regression tasks in machine drive systems.

For instance, ANN is employed in [18] to track and diagnose external faults in three-phase induction machines. The network was trained using data from stator voltage, current, and machine speed. Similarly, supervised and unsupervised neural networks are used in [19] for the identification and diagnosis of induction machine faults. The application of neural networks in classifying partial discharges in machine insulation is investigated in [20], where the fault of the induction machine is determined by the imbalanced inductor current. Moreover, the state estimation of induction machines is focused on [21] by using classical ML algorithms with nonlinear state estimators based on stator current and rotor angular velocity measurements.

Furthermore, DL techniques have proven effective in complex fault diagnosis. Various DL techniques, including Autoencoder (AE), Recurrent Neural Network (RNN), Generative Adversarial Network (GAN), Convolutional Neural Network (CNN), and Deep Belief Network (DBN), have been successfully utilized for machine fault diagnosis. Traditional ML can adapt to a wide range of data quantities, particularly in circumstances with small data volumes.

In comparison, the DL is a sub-domain of machine learning. It usually has much better representative capability than traditional ML algorithms, or significant advantages in dealing with certain tasks such as time-sequence features extraction RNN, image-like data processing CNN, and unsupervised generative tasks GAN. Therefore, if the amount of data increases rapidly, then the effect of DL will be more prominent. The flow charts of traditional ML and DL are summarized in Figure 10.10.

Take the ANN as an example, which is a typical ML method in fault diagnosis. As the word "neural" suggests, these systems are brain-inspired and aim to replicate human learning. There are input and output layers in neural networks, as well as one or several hidden layers of units that transform the inputs into something that the output layer can use. Figure 10.11 shows various layers used in ANN, which consist of input layers, several hidden layers, and an output layer.

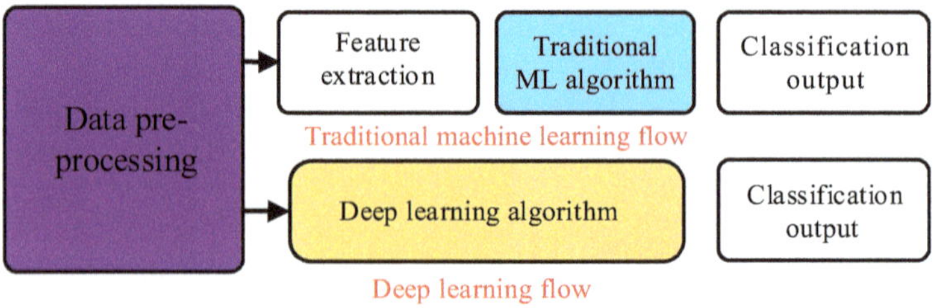

Figure 10.10 Flowchart of traditional ML and Deep learning [22].

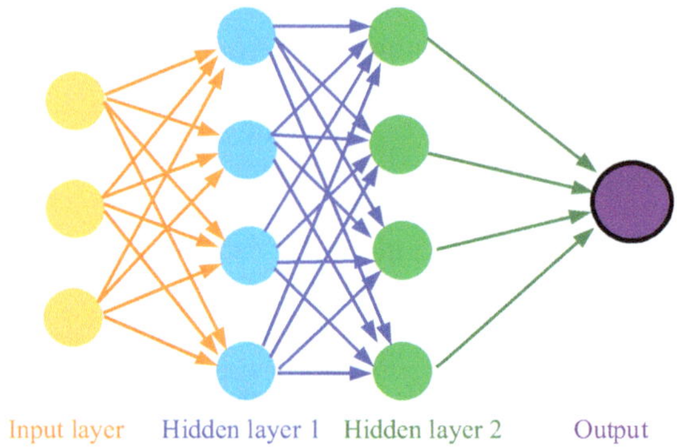

Figure 10.11 Structure of ANNs.

More specifically, the input layer receives the initial input data and serves as the entry point of information into the network. Each neuron in the input layer represents a feature or input variable. The number of neurons in the input layer is determined by the dimensionality of the input data. Hidden layers are intermediate layers between the input and output layers. They play a crucial role in extracting and transforming the input data into a format that is useful for making predictions or classifications. ANNs can have multiple hidden layers, and the number of neurons in each hidden layer can vary based on the complexity of the problem. The neurons in hidden layers apply weights and activation functions to the inputs, allowing the network to learn complex patterns and representations. The output layer provides the final output of the network. The number of neurons in the output layer depends on the nature of the problem being addressed. For example, in a binary classification problem, there may be one neuron representing each class, while in a regression problem, the output layer might consist of a single neuron representing the predicted continuous value.

For ANN in fault diagnosis, the process begins with collecting relevant data from the machine's control system, including sensor measurements and fault labels. This data undergoes pre-processing to ensure its quality and compatibility with the ANN architecture. Next, the architecture of the ANN is designed, specifying the number of layers, neurons, and activation functions. The network is then trained using optimization algorithms like backpropagation to minimize the difference between its predictions and actual fault labels. The performance of the trained ANN is validated using a separate testing dataset to assess its generalization capability. Afterward, ANN can be used for real-time fault diagnosis, taking input from the machine's control system and producing fault diagnosis outputs. In addition, periodic performance evaluation helps monitor the accuracy and effectiveness of the ANN in diagnosing faults. Through this process, ANNs adaptively learn to recognize patterns and relationships in the data, enabling them to accurately diagnose faults in machine control systems.

10.4 Fault tolerant control for multiphase electrical machines

10.4.1 Fault tolerant control for DTP machines

In dual three-phase (DTP) systems, fault tolerance is focused on electric machine and switch faults, typically classified as SC and OC faults. Fault-tolerant control for SC faults is only feasible in DTP machines with modular structures and winding arrangements [23,24]. Under SC faults, fault-tolerant control isolates the faulty sets of three-phase windings and allows the operation of the remaining sets of windings. More recently, the most widely investigated cases are post-fault control strategies under OC fault due to more freedom and potential with the redundant available phases. Particularly, one phase-open fault tolerant has attracted a lot of interest, and the performance of post-fault control strategies against OC fault can be found in [25–31].

The basic idea of fault tolerant control is to generate a rotating field using the residual healthy phases. The effectiveness of the control strategies depends on the degrees of freedom of the machine. According to the connection of neutral points, the DTP machines can be divided into two types, i.e., the isolated and connected neutral point topologies. Table 10.2 outlines the key characteristics of these two different topologies under healthy and faulty operations.

The isolated topology of the system offers advantages, such as higher DC-link voltage utilization and high efficiency under healthy operation. On the other hand, the connected topology demonstrates better performance in managing faulty conditions and offers greater flexibility in faulty operation. However, the isolated neutral point topology can address certain faulty conditions due to the individual nature of each set of three-phase windings. Fault scenarios for both isolated and connected neutral point topologies are shown in Tables 10.3 and 10.4, where the affected phases are highlighted in red.

It should be pointed out that the faulty cases, such as A-B open fault, are feasible in the connected neutral point topology, as the machine can operate as a four-phase machine. However, it is unavailable in isolated neutral point topology since no closed circuit in phase windings ABC. Similarly, the three-phase-open faults like A-B-Z open are also unavailable in isolated neutral point topology, while

Table 10.2 Characteristics for different neutral types under healthy and faulty operations

Topology	Health operation	Fault operation
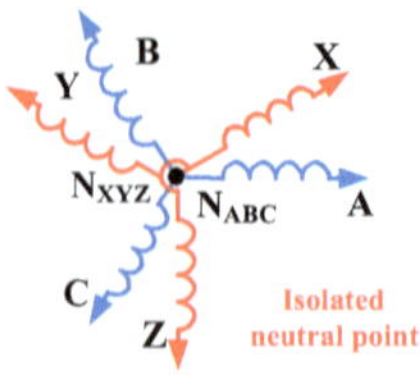	• No zero-sequence path • High dc-link utilization • High efficiency • Simple current control	• Only applicable for some faulty cases • Low torque range • Reconstructed model required
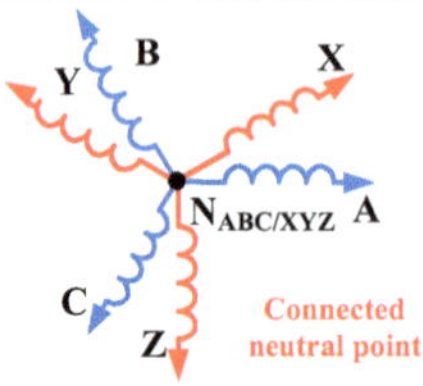	• Large zero-sequence currents • Low dc-link utilization • Low efficiency • Complex current control	• Applicable for all kinds of faulty cases • High torque range • Simple modelling

Table 10.3 Different faulty cases for isolated neutral point topology

	Isolated neutral point		
One phase-open		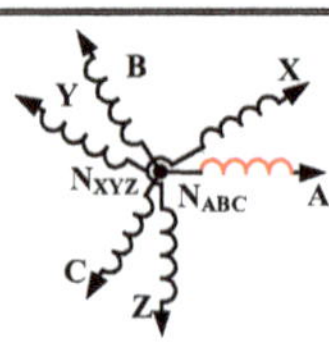 **(A open)**	
Two phase-open	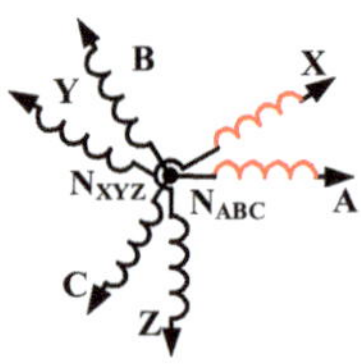 **(A-X open)**	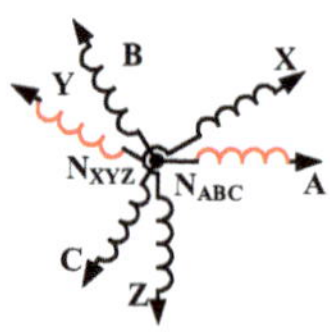 **(A-Y open)**	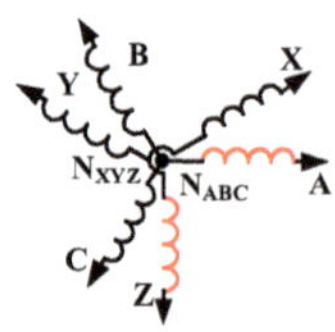 **(A-Z open)**

they can operate in connected neutral point topology with the remaining three-phase windings. Besides, for one-phase open cases, phase A is considered as the faulty phase to simplify the explanation. Faults in other phases can readjust the order of the machine phases to establish the faulty power leg as phase "A".

Table 10.4 Different faulty cases for connected neutral point topology

	Isolated neutral point	
One phase-open	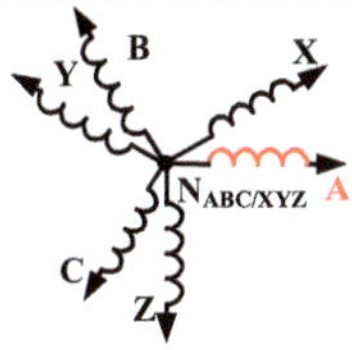 (A open)	
Two phase-open	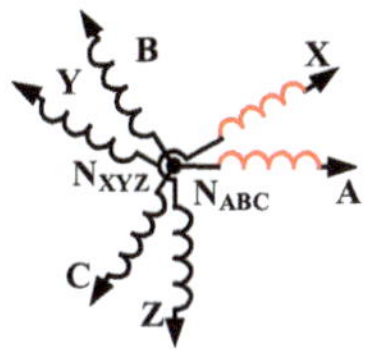 (A-X open)	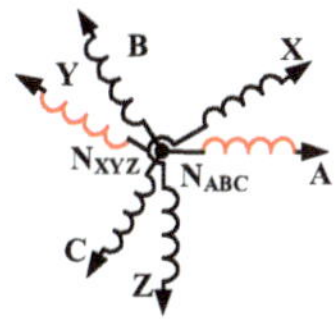 (A-Y open)
	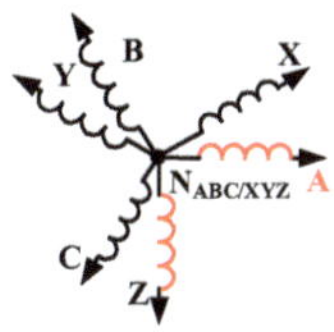 (A-Z open)	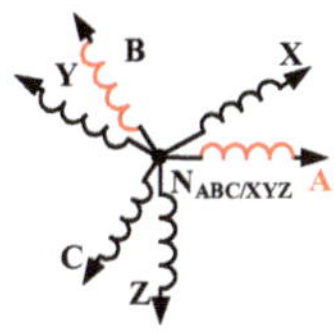 (A-B open)
Three-phase-open	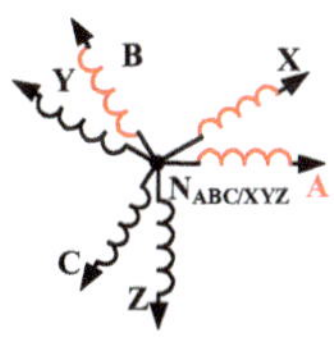 (A-B-X open)	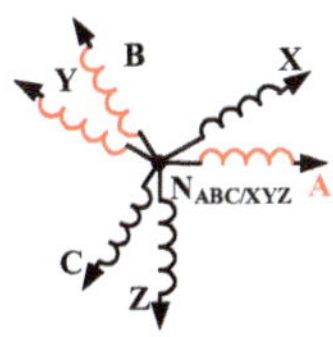 (A-B-Y open)
	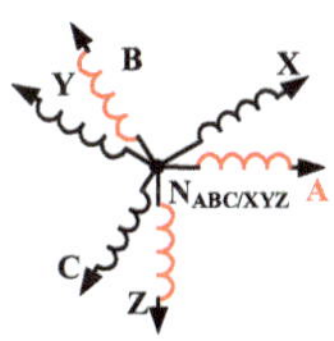 (A-B-Z open)	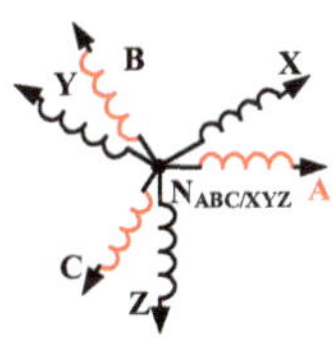 (A-B-C open)

Similar to the normal operation of DTP-PMSMs, the static decoupling transformation matrix under open-phase faults can be expressed as [32,33]:

$$\begin{bmatrix} v_\alpha & v_\beta & v_z & v_{o1} & v_{o2} \end{bmatrix}^T = T_{5s} \cdot \begin{bmatrix} v_a & v_b & v_c & v_x & v_y \end{bmatrix}^T \tag{10.13}$$

$$T_{5s} = \frac{1}{3}\begin{bmatrix} 1 & -\frac{1}{2} & -\frac{1}{2} & \frac{\sqrt{3}}{2} & -\frac{\sqrt{3}}{2} \\ 0 & \frac{\sqrt{3}}{2} & -\frac{\sqrt{3}}{2} & 0 & 0 \\ 1 & -\frac{1}{2} & -\frac{1}{2} & -\frac{\sqrt{3}}{2} & \frac{\sqrt{3}}{2} \\ 1 & 1 & 1 & 0 & 0 \\ 0 & 0 & 0 & 1 & 1 \end{bmatrix} \tag{10.14}$$

where the fundamental components of the machine variables are mapped into subspace, which are related to torque production. The harmonic components are located in the z subspace, which contributes to loss generation.

10.4.1.1 FOC-based fault tolerant control

In terms of the isolated type, the most simple and common way is to cut off the three-phase set. This method does not need a specifically designed postfault algorithm if each set is individually regulated, which is quite simple to implement and practical in industrial applications, such as railway traction [34] and ship propulsion [35]. Meanwhile, its disadvantage is obvious, in that the second set is simply abandoned, and thus, the currents of the first set should be doubled to compensate for the lost torque. Otherwise, the available torque and power are reduced to half of the rating in this case. The healthy phases in the fault set, together with the corresponding switching devices, are not utilized, and the system utilization is not maximized.

To improve the system utilization of the machine with isolated neutral points, the current should be redistributed to the remaining healthy phases. However, the isolated neutral point can only accommodate certain faulty scenarios. When two sets of three-phase windings have two isolated neutral points, the absence of a zero-sequence current path in the o_1o_2 subspace restricts the current control for DTP PMSMs with one phase open fault to only the $\alpha\beta$ and z subspaces.

Typically, fault-tolerant control for DTP PMSMs has three different modes according to different coefficients, as concluded in [30], i.e., minimum stator loss mode, maximum torque mode, and single three-phase mode. In the minimum stator loss mode, i_z^* is set to zero to minimize the copper loss in the stator windings, which is defined as

$$P_{cu} = R_s\left(i_\alpha^2 + i_\beta^2 + i_z^2 + i_{o1}^2 + i_{o2}^2\right) \tag{10.15}$$

where P_{cu} is the copper loss.

In the maximum torque mode (maximum $i_{\alpha\beta}$), i_z^* is set to - i_α^* to produce the largest $|I_{\alpha\beta}|$ with the maximum phase current amplitude maintained at the rated value. The simulation results in Figures 10.12 and 10.13 illustrate the behavior of the system under various conditions, including healthy conditions and Z phase OC fault, both with and without fault-tolerant control. As can be seen, when one phase-open fault occurs, the current waveforms exhibit obvious distortions, and the speed

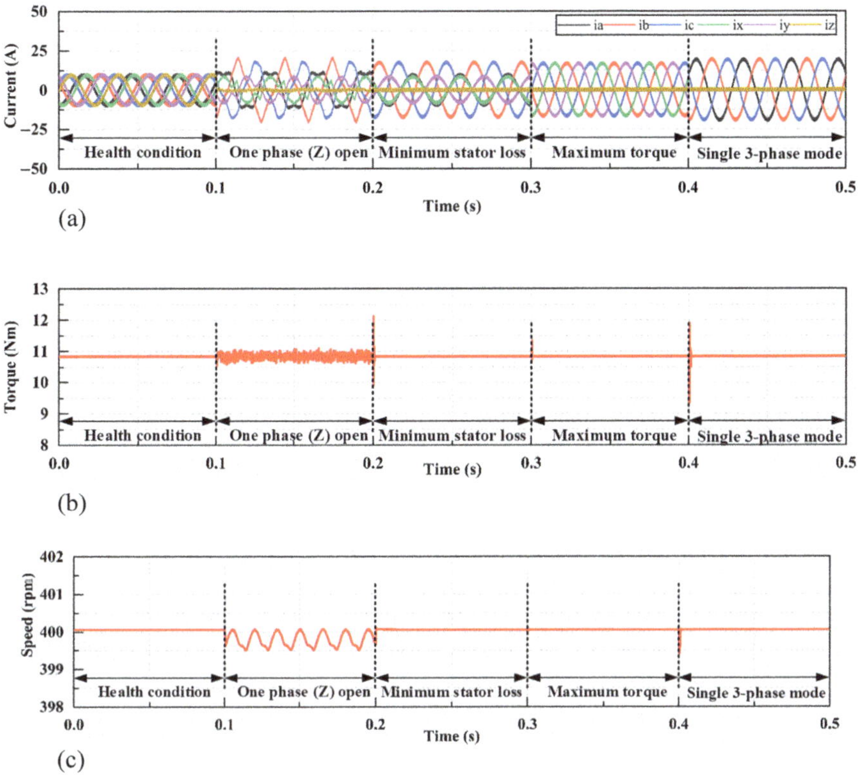

Figure 10.12 Simulation results for different post-fault control modes. (a) Phase current. (b) Torque. (c) Speed.

and torque have obvious pulsations. For minimum stator loss control with one open phase, the secondary pulsation of the speed is suppressed significantly, it can be seen from the current waveform that the amplitude and phase of each phase current have changed. For maximum torque control with one open phase, the amplitudes of the remaining phases are basically equal, but only four phases are accessible in this control mode. For single three-phase control mode, it is the same as three-phase machines, and the amplitudes of phase currents are doubled compared with the healthy conditions. For more details, the results are shown in Figure 10.12 and Table 10.5.

The pulsating flux generated by the fault set can be decomposed as positive-sequence and negative-sequence rotating flux. The positive-sequence flux can provide average torque while the negative-sequence flux will generate the second-order torque ripple. The basic idea of this kind of fault tolerant control is to inject a negative-sequence current in the healthy set to counterbalance the negative-sequence flux, and as a result, the average torque remains while the torque ripple is eliminated. For implementation, a genetic algorithm is provided in [26] to optimize

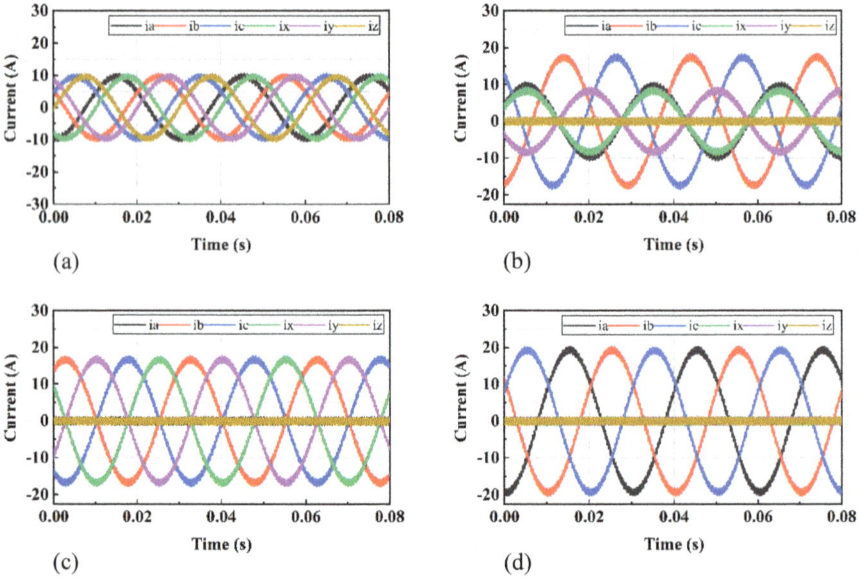

Figure 10.13 Current waveforms under different fault tolerant control modes. (a) Health condition. (b) Minimum stator loss. (c) Maximum torque. (d) Single three-phase.

Table 10.5 Summary of different fault tolerant control modes

Control mode	Copper loss	Peak phase current
Healthy	$6I_m^2R_s$	I_m
Minimum stator loss	$9I_m^2R_s$	$1.803I_m$
Maximum torque	$12I_m^2R_s$	$1.732I_m$
Single three-phase	$12I_m^2R_s$	$2I_m$

the stator currents, thereby maximizing the average torque or minimizing the torque ripple. Also, the copper loss of faulty DTP PMSMs with isolated neutral points is further reduced by minimizing the d-axis current of the healthy three-phase winding [29,36,37].

It is also possible to use modeling techniques for DTP PMSMs under OC fault to achieve different control purposes. Particularly, the VSD technique can decouple machine variables in several orthogonal subspaces, thereby controlling them independently [38]. Because the VSD technique can decompose the negative-sequence current in the z subspace, the control structure in [31] can be simplified, as shown in Figure 10.14. To eliminate the negative-sequence current, a proportional integral resonant (PIR) regulator is employed, and the fundamental electrical

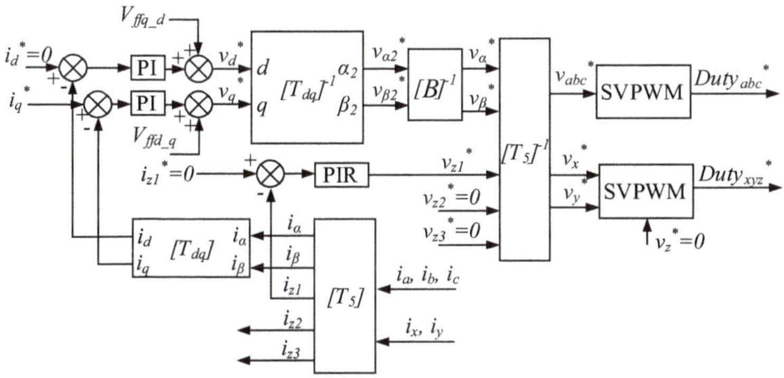

Figure 10.14 FOC-based fault tolerant control of DTP PMSM with one-phase open using VSD approach [31].

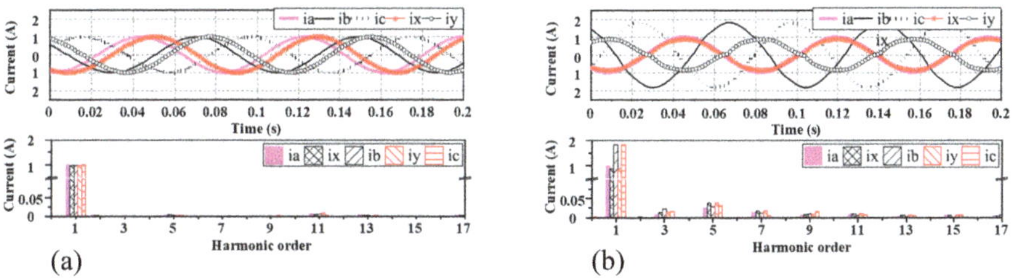

Figure 10.15 Measured currents and spectra. (a) Under healthy condition. (b) Under fault condition.

frequency is utilized as the resonant frequency. The experimental results under healthy and fault conditions are shown in Figure 10.15.

10.4.1.2 DTC-based fault tolerant control

The fault-tolerant DTC for DTP PMSM is outlined, focusing on the "One phase-open VSD" method. The voltage space vectors are mapped into three new subspaces, i.e., $\alpha\beta$ subspace, z subspace, and o_1o_2 subspace. Similarly, since two sets of three-phase windings have two isolated neutral points, there is no zero-sequence current path in the o_1o_2 subspace. Consequently, the current control for a DTP machine with one phase open fault requires only currents in the $\alpha\beta$ and z subspace. The control diagram is shown in Figure 10.16, and the voltage space vectors for the health condition and one phase-open condition are shown in Figure 10.17.

It can be seen from Figure 10.17 that the number of voltage vectors for one phase-open condition is 2^5=32, which is half of that for a healthy condition. According to the virtual voltage vector method mentioned in Chapter 5, new virtual voltage vectors can be obtained. The virtual voltage vector sets the amplitudes in the z subspace to zero to decrease the harmonic current. Then, the new virtual voltage vectors for one phase-open condition are shown in Figure 10.18.

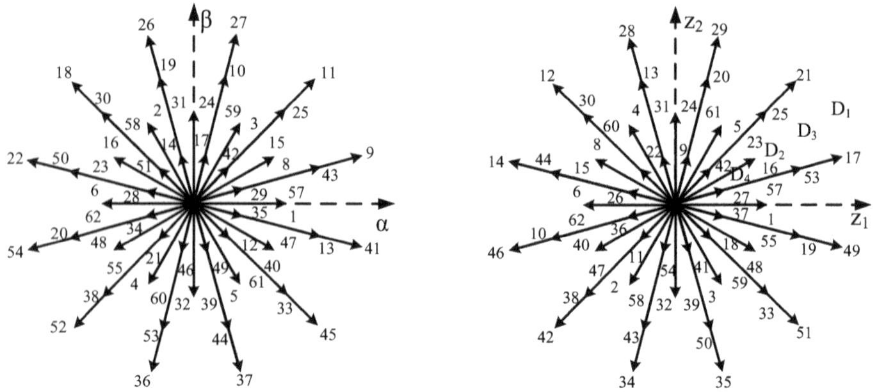

(a) (b)

Figure 10.16 Voltage space vectors for health condition. (a) αβ subspace for health condition. (b) z_1z_2 subspace for health condition.

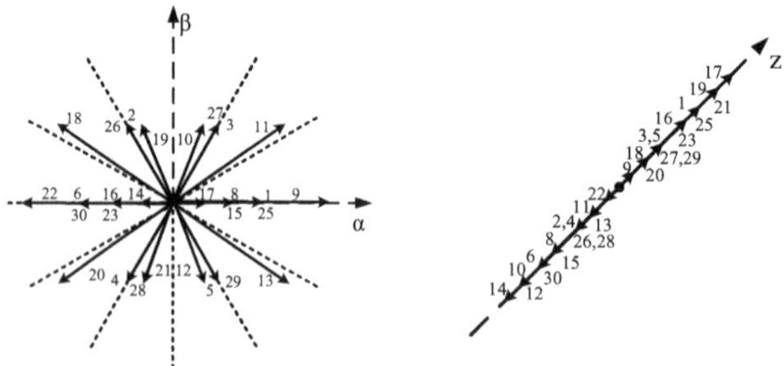

(a) (b)

Figure 10.17 Voltage space vectors for one phase-open (phase Z) condition. (a) αβ subspace for one phase-open (phase Z) condition. (b) z subspace for one phase-open (phase Z) condition.

The experimental results for different control methods are shown in Figures 10.19 and 10.20. A comparison between healthy condition, one phase-open without fault tolerant control, and minimum stator loss control (ML) without and with virtual voltage assessment (VV) is shown in Figure 10.18. It can be seen that the ML with VV has small torque ripples and speed ripples, and lower stator peak current compared to others. Furthermore, the phase current waveforms are shown in Figure 10.19. It can be seen that the ML with VV has a sinusoidal phase current similar to DTC in healthy conditions. However, the control without fault tolerant control and ML without VV suffer from large distortion and current harmonics, which are shown in Figure 10.19(b) and (c).

The type with connected neutral points has more degrees of freedom because all the healthy phases are available. However, the post-fault control algorithm is

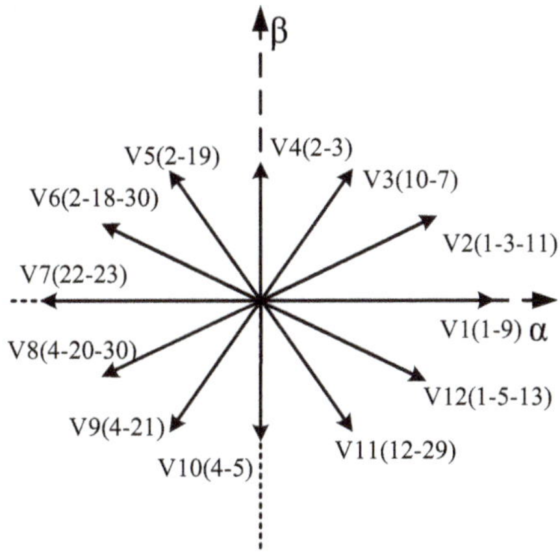

Figure 10.18 Virtual voltage vectors for one phase-open (phase Z) condition.

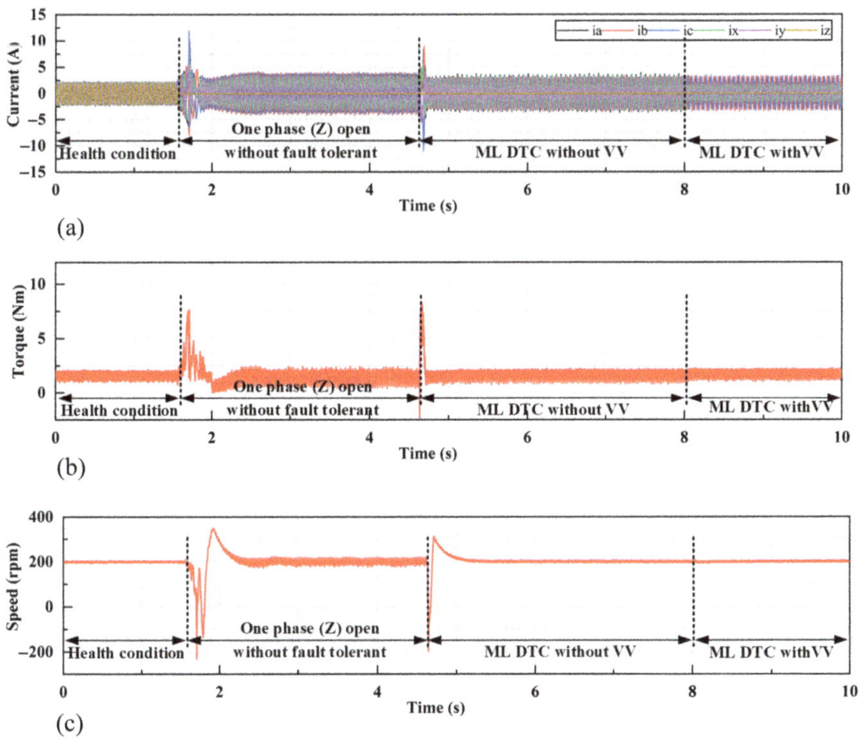

Figure 10.19 Experimental results for different control modes (from left to right: health, one phase-open without fault tolerant control, one phase-open with ML control but no virtual voltage assessment, one phase-open with both ML control and virtual voltage assessment). (a) Phase current. (b) Torque. (c) Speed.

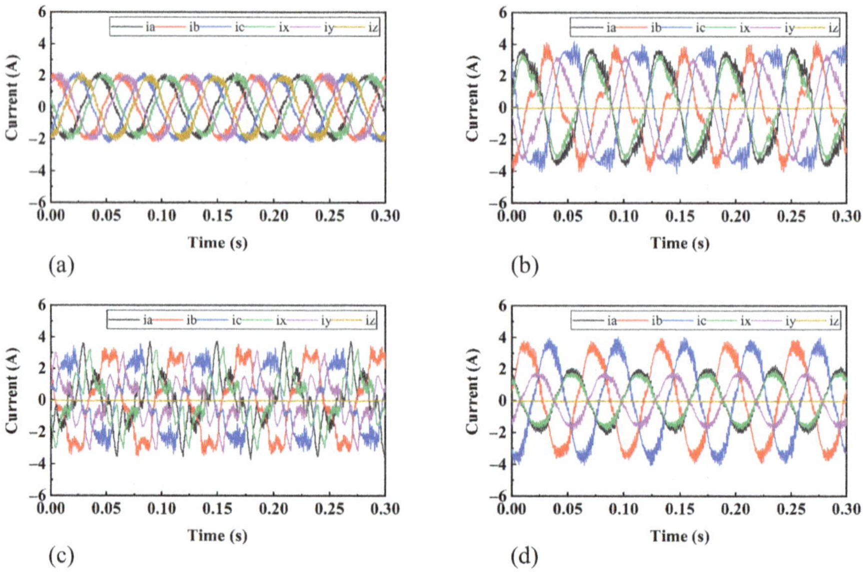

Figure 10.20 Experimental results of current waveforms of different control modes. (a) Health condition. (b) Without fault tolerant. (c) ML fault tolerant without VV. (d) ML fault tolerant with VV.

different from the normal healthy case and should be specifically designed. Thus, the main issue lies in designing the optimal current reference for each healthy phase. The optimization objectives include maximal average torque, minimal torque ripple, minimal copper loss, and uniform distribution of copper loss.

In [37,39], two different optimization criteria, i.e., minimum stator copper loss and maximum torque, are introduced in DTP machines with a voltage source inverter. Additionally, an improved approach to achieve the minimum losses in full torque range with one phase-open fault is also proposed in [40]. The current harmonics may be injected to boost the average torque and reduce the torque ripple of a DTP PMSM under open-fault conditions [27]. It should be noted that these optimal current control approaches emphasize designing the current references, and various current regulators, as studied in FOC, DTC, and MPC, can be utilized to track the designed current references. Further, fault-tolerant controls in DTP machines fed by multi-level inverters are also investigated by considering both OC switch faults and phase faults in [41,42], which provide higher reliability and better fault capability.

10.4.2 Fault tolerant control for other multiphase electrical machines

The fault tolerances for five-phase PMSMs under both OC fault and short circuit fault occurrence are systematically discussed in [43], which is based on the FOC-based algorithm (called current control in this subsection).

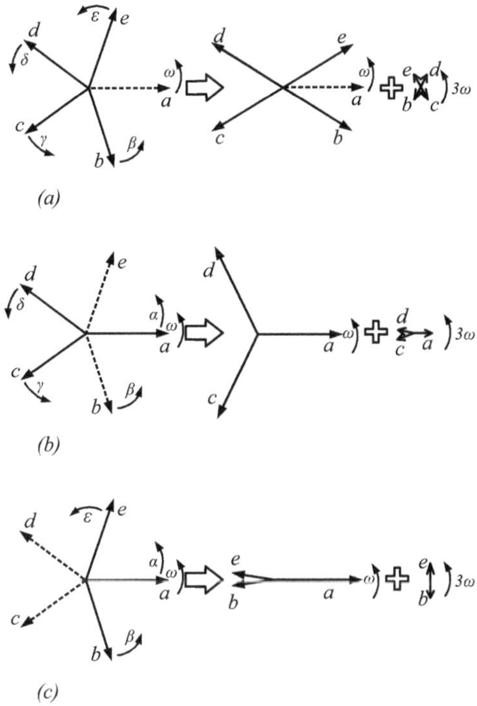

Figure 10.21 Current phasor diagram in the event of OC faults [43]. (a) Only a-phase open. (b) Nonadjacent phase open, i.e., phases b and e. (c) Adjacent phase open, i.e., phases c and d.

For the OC fault of a single phase, as shown in Figure 10.21(a), the proposed current control strategy in [43] maintains current symmetry and avoids zero-sequence currents. The key point for fault tolerant control is to choose the proper current phasor angle when the faults occur. Simultaneously, the torque ripple should also be considered. By injecting a third time-harmonic current, the fourth-order torque harmonic is reduced to zero, resulting in reduced torque ripple. In the case of a two-phase fault, both nonadjacent and adjacent phase faults are considered. The current control strategies are adjusted accordingly to maintain torque performance. However, in the case of an adjacent phase fault, the limitation of having no zero-sequence current can cause a significant decrease in average torque. In addition, the effects of an SC fault are also discussed in [43], where the SC current is limited by the winding inductance and does not compromise the operation of healthy phases.

The DTC-based fault tolerant control is discussed in [44] with a five-phase induction machine. In the case of an open-phase fault, the DTC method faces challenges due to the reduced number of active phases. The proposed technique addresses these challenges by modifying the classic DTC structures using a

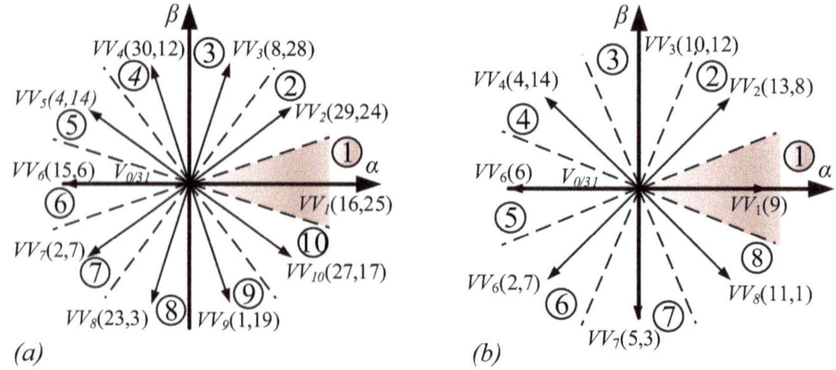

Figure 10.22 Virtual voltage vectors (VV_i) in the α–β subspace. (a) Normal operation. (b) Open-phase post-fault operation.

reduced-order transformation matrix, redefining the utilized virtual voltage vectors based on the available voltage vectors in faulty operation, and elaborating a new look-up table for control purposes [44], as shown in Figure 10.22. Experimental results have shown that the proposed fault-tolerant DTC scheme can maintain the speed, torque, and flux references after the fault occurrence, as long as the applied load torque is within the maximum torque that the post-fault control scheme can manage. The performance of the multiphase drive using the proposed controller has been successfully demonstrated in various experimental tests.

The multiphase electrical machines with an odd number of phases are summarized in [45], including the seven-phase machine. The control scheme utilizes multiple space vector representations of multiphase quantities to develop a fault-tolerant control technique. This approach analytically determines the expression of the stator current in the healthy phases to ensure disturbance-free operation in both transient and steady-state conditions, while minimizing stator copper losses, as demonstrated in the example of the seven-phase machine in [45]. If one phase is open, the control strategy for the minimum stator copper losses leads to the following coefficients as (10.16), which makes the current references in different spaces as (10.17).

$$K_{3\alpha} = -\frac{1}{2} \quad K_{3\beta} = 0 \quad K_{5\alpha} = -\frac{1}{2} \quad K_{5\beta} = 0 \tag{10.16}$$

$$\begin{cases} i_{S3,ref,\alpha} = K_{3\alpha} i_{S1,ref,\alpha} \\ i_{S3,ref,\beta} = K_{3\beta} i_{S1,ref,\beta} \\ i_{S5,ref,\alpha} = K_{5\alpha} i_{S1,ref,\alpha} \\ i_{S5,ref,\beta} = K_{5\beta} i_{S1,ref,\beta} \end{cases} \tag{10.17}$$

The voltage vectors of the seven-phase machine system are analyzed in [46] and a novel SVPWM strategy is developed. It provides a deeper insight into the reconstruction of voltage vectors without using a reference current, ensuring applicability across various fault-tolerant principles. An eight-sector-based sector division method is proposed, and specific nonzero and zero-voltage vectors in each sector are determined as candidates. The symmetrical modulation module can still be utilized without conflict with the switching state, thereby minimizing modifications post-fault occurrence. Although this method is proposed for FOC, it can be adapted to DTC. By analyzing the voltage vector effects during fault occurrence, the performance of both strategies can be significantly improved.

The fault-tolerant control of other multiphase machines is generally similar to the above-mentioned DTP, five-phase, and seven-phase. The basic idea focuses on FOC-based fault-tolerant control, which involves recalculating new current references to achieve different control purposes. For fault-tolerant control based on DTC, it mainly analyses the space voltage vector after the fault, followed by the utilization of different virtual vector strategies to improve the control effectiveness.

10.5 Summary

This chapter has introduced fault-tolerant control strategies for multiphase PMSMs. It began by discussing fault classification in multiphase electrical machine drives, including inverter faults, sensor faults, and machine faults. Afterwards, various fault diagnosis methods for multiphase electrical machines, such as model-based, signal-based, and data-based approaches, have been described. Throughout, the chapter has underscored the importance of striking a balance between performance optimization and fault accommodation in developing optimal fault-tolerant control strategies.

In summary, fault-tolerant control is crucial for ensuring reliability and operational continuity in multiphase electrical machine systems. By accurately diagnosing faults and implementing appropriate control strategies, the performance and efficiency of these machines can be maintained even in the presence of faults.

References

[1] J. He, Q. Yang, and Z. Wang, "On-line fault diagnosis and fault-tolerant operation of modular multilevel converters—A comprehensive review," *CES Trans. Electr. Mach. Syst.*, vol. 4, no. 4, pp. 360–372, 2020.

[2] T. Orlowska-Kowalska, M. Wolkiewicz, P. Pietrzak, M. Skowron, P. Ewert, and G. Tarchala, "Fault diagnosis and fault-tolerant control of PMSM drives—State of the art and future challenges," *IEEE Access*, vol. 10, pp. 59979–60024, 2022.

[3] P. O'Donell, "Report of large motor reliability survey of industrial and commercial installations: Part I–II," *IEEE Trans. Ind. Appl.*, vol. IA, 21, no. 4, pp. 853–872, 1985.

[4] J. He, C. Somogyi, A. Strandt, and N. A. O. Demerdash, "Diagnosis of stator winding short-circuit faults in an interior permanent magnet synchronous machine," *Proc. IEEE Energy Convers. Congr. Expo. (ECCE)*, 2014, pp. 3125–3130.

[5] R. Beard, "Failure accommodation in linear system through self reorganization," *PhD dissertation, MIT*, Cambridge, MA, USA, 1971.

[6] B. Aubert, J. Regnier, S. Caux, and D. Alejo, "Kalman-filter-based indicator for online interturn short circuits detection in permanent-magnet synchronous generators," *IEEE Trans. Ind. Electron.*, vol. 62, no. 3, pp. 1921–1930, 2015.

[7] W. El Sayed, M. A. E. Geliel, and A. Lotfy, "Fault diagnosis of PMSG stator inter-turn fault using extended Kalman filter and unscented Kalman filter," *Energies*, vol. 13, no. 11, pp. 2972, 2020.

[8] B.-G. Park, J.-S. J, T.-S. Kim, and D.-S. Hyun, "EKF-based fault diagnosis for open-phase faults of PMSM drives," *Proc. IEEE 6th Int. Power Electron. Motion Control Conf.*, 2009, pp. 418–422.

[9] X. Xiao, C. Chen, and M. Zhang, "Dynamic permanent magnet flux estimation of permanent magnet synchronous machines," *IEEE Trans. Appl. Supercond.*, vol. 20, no. 3, pp. 1085–1088, 2010.

[10] H. Liu, D. Liu, C. Lu, and X. Wang, "Fault diagnosis of hydraulic servo system using the unscented Kalman filter," *Asian J. Control*, vol. 16, no. 6, pp. 1713–1725, 2014.

[11] M. Laila, N. Naveen, and K. Vishakh, "Comparison of estimation capabilities of extended and unscented Kalman filter in an RLV," *Int. J. Adv. Res. Elect. Electron. Instrum. Eng.*, vol. 2, no. 7, pp. 3480–3488, 2013.

[12] C. Hajiyev and H. Soken, "Robust adaptive Kalman filter for estimation of UAV dynamics in the presence of sensor/actuator faults," *Aerosp. Sci. Technol.*, vol. 28, no. 1, pp. 376–383, 2013.

[13] A. Izadian, "Self-tuning fault diagnosis of MEMS," *Mechatronics*, vol. 23, no. 8, pp. 1094–1099, 2013.

[14] F. Hmida, K. Khemiri, J. Ragot, and M. Goss, "Three-stage Kalman filter for state and fault estimation of linear stochastic systems with unknown inputs," *J. Franklin Inst.*, vol. 349, no. 7, pp. 2369–2388, 2012.

[15] Z. Ullah, S.-T. Lee, and J. Hur, "A novel fault diagnosis technique for IPMSM using voltage angle," in *Proc. IEEE Energy Convers. Congr. Expo. (ECCE)*, 2018, pp. 3236–3243.

[16] J. Estima and A. Cardoso, "A new algorithm for real-time multiple open-circuit fault diagnosis in voltage-fed PWM motor drives by the reference current errors," *IEEE Trans. Ind. Electron.*, vol. 60, no. 8, pp. 3496–3505, 2013.

[17] M. Skowron, M. Wolkiewicz, T. Kowalska, and C. Kowalski, "Application of self-organizing neural networks to electrical fault classification in induction motors," *Applied Science*, vol. 9, no. 4, pp. 1–22, 2019.

[18] E. M. T. Eldin, H. R. Emara, E. M. Aboul-Zahab, and S. S. Refaat, "Monitoring and diagnosis of external faults in three phase induction motors using artificial neural network," *2007 IEEE Power Eng. Soc. Gen. Meet.*, Tampa, FL, USA, 2007, pp. 1–7.

[19] S. Premrudee Preechacharn, T. Utthiyuung, K. Kruepengkul, and P. Puongkaew, "Induction motor fault detection and diagnosis using supervised and unsupervised neural networks," *IEEE Int. Conf. on Ind. Tech. (ICIT)*, Thailand, vol. 1, 2002, pp. 93–96.

[20] R. Natarajan, "Failure identification of induction motors by sensing unbalanced stator currents," *IEEE Trans. Energy Convers.*, vol. 4, no. 4, pp. 585–590, 1989.

[21] S. Aksoy and A. Mühürcü, "Recurrent neural network based nonlinear state estimation for induction motors," *J. Int. Review Electr. Eng.*, vol. 6, no. 2, p. 636, 2011, Papers Part A.

[22] W. Lang, Y. Hu, C. Gong, X. Zhang, H. Xu, and J. Deng, "Artificial intelligence-based technique for fault detection and diagnosis of EV motors: A review," *IEEE Trans. Transp. Electrif.*, vol. 8, no. 1, pp. 384–406, 2022.

[23] M. Barcaro, N. Bianchi, and F. Magnussen, "Faulty operations of a PM fractional-slot machine with a dual three-phase winding," *IEEE Trans. Ind. Electron.*, vol. 58, no. 9, pp. 3825–3832, 2011.

[24] L. Alberti and N. Bianchi, "Experimental tests of dual three-phase induction motor under faulty operating condition," *IEEE Trans. Ind. Electron.*, vol. 59, no. 5, pp. 2041–2048, 2012.

[25] B. Shao, "Direct torque control of dual three phase permanent magnet synchronous machine drives with particular reference to reduction of current harmonics and torque ripple," *PhD dissertation*, University of Sheffield, Sheffield, UK, June 2022.

[26] G. Feng, C. Lai, W. Li, J. Tjong, and N. C. Kar, "Open-phase fault modeling and optimized fault-tolerant control of dual three-phase permanent magnet synchronous machines," *IEEE Trans. Power Electron.*, vol. 34, no. 11, pp. 11116–11127, 2019.

[27] G. Feng, C. Lai, W. Li, Y. Han, and N. C. Kar, "Computation-efficient solution to open-phase fault tolerant control of dual three-phase interior PMSMs with maximized torque and minimized ripple," *IEEE Trans. Power Electron.*, vol. 36, no. 4, pp. 4488–4499, 2021.

[28] Z. Wang, X. Wang, Y. Wang, J. Chen, and M. Cheng, "Fault tolerant control of multiphase multilevel motor drives – Technical review," *Chin. J. Electr. Eng.*, vol. 3, no. 2, pp. 76–86, 2017.

[29] X. Wang, Z. Wang, M. He, Q. Zhou, X. Liu, and X. Meng, "Fault-tolerant control of dual three-phase PMSM drives with minimized copper loss," *IEEE Trans. Power Electron.*, vol. 36, no. 11, pp. 12938–12953, 2021.

[30] H. S. Che, M. J. Duran, E. Levi, M. Jones, W. P. Hew, and N. A. Rahim, "Postfault operation of an asymmetrical six-phase induction machine with single and two isolated neutral points," *IEEE Trans. Power Electron.*, vol. 29, no. 10, pp. 5406–5416, 2014.

[31] Y. Hu, Z. Q. Zhu, and Z. Wu., "Modelling and vector control of dual three-phase PMSM with one-phase open," *IET Electr. Power Appl.*, vol. 15, no. 7, pp. 847–860, 2021.

[32] C. Zhou, W. Tang, X. D. Sun, Z. J. Zhou, G. J. Yang, and J. Y. Su, "Control strategy for dual three-phase PMSM based on reduced order mathematical model under fault condition due to open phases," *J. Eng.*, vol. 2018, no. 13, pp. 489–494, 2018.

[33] Y. Hu, Y. Feng, and X. Li, "Fault-tolerant hybrid current control of dual three-phase PMSM with one phase open," *IEEE J. Emerg. Sel. Topics Power Electron.*, vol. 10, no. 3, pp. 3418–3426, 2022.

[34] S. Mantero, A. Monti, and S. Spreafico, "DC-bus voltage control for double star asynchronous fed drive under fault conditions," *Proc. IEEE Annual Conf. on Power Elect. Specialists, PESC*, Galway, Ireland, 2000, pp. 533–538.
[35] M. Qiao, C. Jiang, Y. Zhu, and G. Li, "Research on design method and electromagnetic vibration of six-phase fractional-slot concentrated winding PM motor suitable for ship propulsion," *IEEE Access*, vol. 4, pp. 8535–8543, 2016.
[36] M. Shamsi-Nejad, B. Nahid-Mobarakeh, S. Pierfederici, and F. Meibody-Tabar, "Fault tolerant and minimum loss control of double-star synchronous machines under open phase conditions," *IEEE Trans. Ind. Electron.*, vol. 55, no. 5, pp. 1956–1965, 2008.
[37] W. Wang, J. Zhang, M. Cheng, and S. Li, "Fault-tolerant control of dual three-phase permanent magnet synchronous machine drives under open-phase faults," *IEEE. Trans. Power Electron.*, vol. 32, no. 3, pp. 2052–2063, 2017.
[38] M. A. Fnaiech, F. Betin, G. A. Capolino, and F. Fnaiech, "Fuzzy logic and sliding-mode controls applied to six-phase induction machine with open phases," *IEEE Trans. Ind. Electron.*, vol. 57, no. 1, pp. 354–364, 2010.
[39] A. G. Yepes, J. Doval-Gandoy, F. Baneira, and H. A. Toliyat, "Control strategy for dual three-phase machines with two open phases providing minimum loss in the full torque operation range," *IEEE Trans. Power Electron.*, vol. 33, no. 12, pp. 10044–10050, 2018.
[40] I. González-Prieto, M. J. Duran, and F. J. Barrero, "Fault-tolerant control of six-phase induction motor drives with variable current injection," *IEEE Trans. Power Electron.*, vol. 32, no. 10, pp. 7894–7903, 2017.
[41] Z. Wang, X. Wang, J. Cao, M. Cheng, and Y. Hu, "Direct torque control of T-NPC inverters-fed double-stator-winding PMSM drives with SVM," *IEEE Trans. Power Electron.*, vol. 33, no. 2, pp. 1541–1553, 2018.
[42] X. Wang, Z. Wang, Z. Xu, J. He, and W. Zhao, "Diagnosis and tolerance of common electrical faults in T-type three-level inverters fed dual three-phase PMSM drives," *IEEE Trans. Ind. Electron.*, vol. 35, no. 2, pp. 1753–1769, 2020.
[43] N. Bianchi, S. Bolognani, and M. Dai Pre, "Strategies for the fault-tolerant current control of a five-phase permanent-magnet motor," *IEEE Trans. Ind. Appl.*, vol. 43, no. 4, pp. 960–970, 2007.
[44] M. Bermudez, I. Gonzalez-Prieto, F. Barrero, H. Guzman, M. J. Duran, and X. Kestelyn, "Open-phase fault-tolerant direct torque control technique for five-phase induction motor drives," *IEEE Trans. Ind. Electron.*, vol. 64, no. 2, pp. 902–911, 2017.
[45] A. Tani, M. Mengoni, L. Zarri, G. Serra, and D. Casadei, "Control of multiphase induction motors with an odd number of phases under open-circuit phase faults," *IEEE Trans. Power Electron.*, vol. 27, no. 2, pp. 565–577, 2012.
[46] G. Yang, S. Li, H. Hussain, J. Zhang, and J. Yang, "A novel SVPWM fault-tolerant strategy for torque ripple reduction of seven-phase induction machines under single-phase open-circuit fault," *IEEE Trans. Power Electron.*, vol. 38, no. 4, pp. 5217–5229, 2023.

Appendix A

Specifications of prototype machines and experimental platforms

A 24-slot/22-pole DTP SPMSM (Machine 1) and a 48-slot/8-pole DTP IPMSM (Machine 2) are employed to illustrate the influences of the different phase shift angles (Chapter 2) and the double-layer short-pitched winding configuration (Chapter 3) on the electromagnetic performance of DTP PMSMs. Their main design parameters are listed in Tables 2.13 and 3.1, respectively. The other DTP IPMSM (Machine 3) and two STP PMSMs (Machines 4 and 5) are used to illustrate various modulation strategies and control methods in Chapters 7–10, and their specifications are listed in Table A.1.

The overall setup of the experimental system is based on the dSPACE DS1005 system, as shown in Figure A.1. The experimental system includes three parts: a control system, a drive system, and a test rig. The control system consists of a host PC, Processor Board DS1005, A/D Sampling Board DS2004, PWM Signals Generation Board DS5101, and Encoder Board DS3001. Processor Board DS1005 features a PowerPC 750GX processor running at 1 GHz, which provides the computing power for the real-time control system and functions as an interface to the I/O boards and the host PC. The analog signals, including phase currents and

Table A.1 Specifications of prototype PMSMs

Parameters	**Machine 3 (DTP-IPMSM)**	**Machine 4 (STP-IPMSM)**	**Machine 5 (STP-SPMSM)**
Applications in Chapters	7, 8, and 9	9	10
Slot number	12-slot	18-slot	24-slot
Pole number	10-pole	6-pole	8-pole
Rated phase current (RMS) [A]	4.0	4	4
Rated power [W]	240	600	150
Rated speed [r/min]	400	1000	400
DC-link voltage [V]	40	158	48
Phase resistance [Ω]	1.096	6.0	0.63
α- and β-axis inductances [mH]	2.142	–	–
z_1- and z_2-axis inductances [mH]	0.875	–	–
d- and q-axis inductances [mH]	–	38.10 and 55.00	0.63
Permanent magnet flux [Wb]	0.734	0.236	0.487

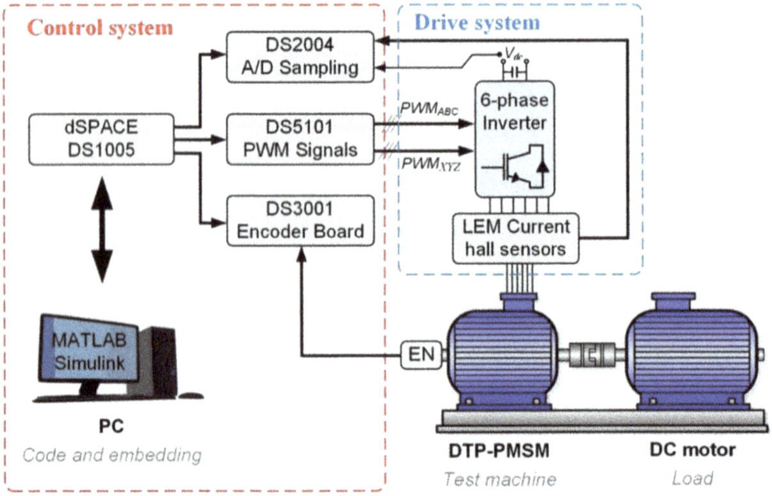

Figure A.1 Overall setup of experimental system.

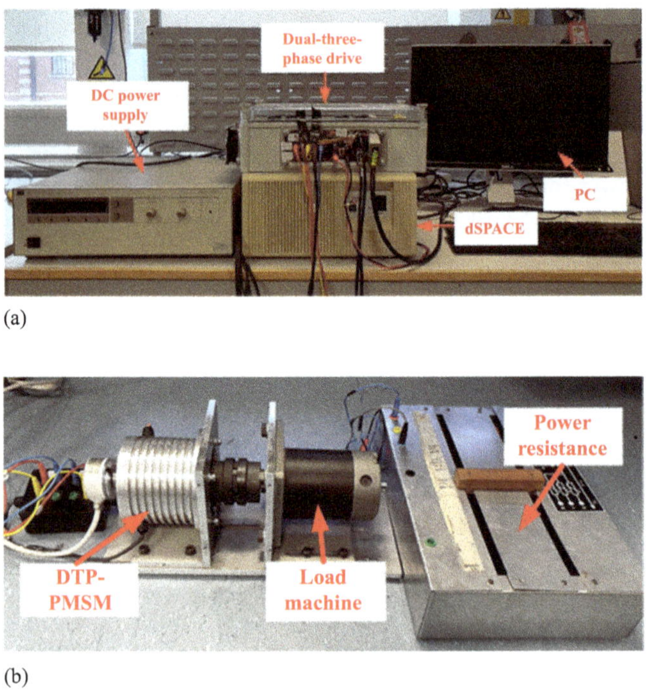

Figure A.2 Experimental system of DTP-IPMSM. (a) Drive control system. (b) Test rig.

DC-link voltage, are converted to digital signals and captured by A/D Sampling Board DS2004. PWM Signals Generation Board DS5101 and Encoder Board DS3001 offer 6-phase PWM signal outputs and incremental encoder signals capture, respectively. The host PC is connected to the dSPACE system by an optical fiber cable, to achieve coding and embedding control program in the dSPACE system, and also capture and monitor data during the motor operation.

The test rig includes the prototype PMSM and a DC machine. The prototype PMSM is connected to the DC machine with an adjustable power resistor as a load. The stator phase currents and the DC-link voltage are measured by several Hall effects current and voltage sensors, and the torque can be estimated accordingly or measured by a torque transducer. The pictures of the experimental setup are shown in Figure A.2. All the experimental results are collected from dSPACE and plotted in MATLAB®.

By way of example, the back EMF (electromotive force) of the prototype DTP-IPMSM (Machine 3) is shown in Figure A.3. It is clear that the back EMF is non-sinusoidal and has large harmonics, in which the 5th and 7th harmonics are 5.23% and 1.77% of the fundamental component, respectively.

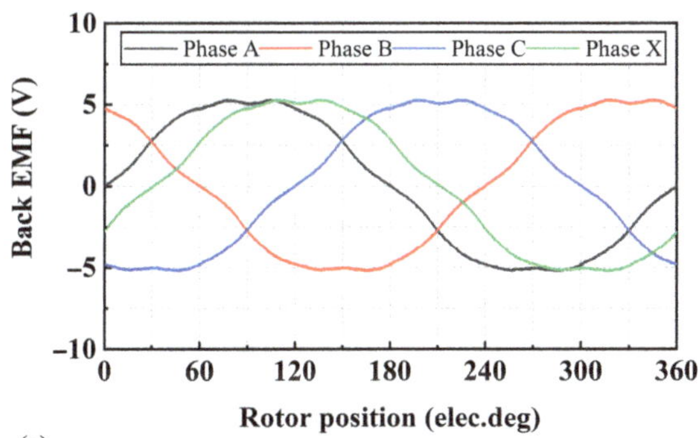

(a)

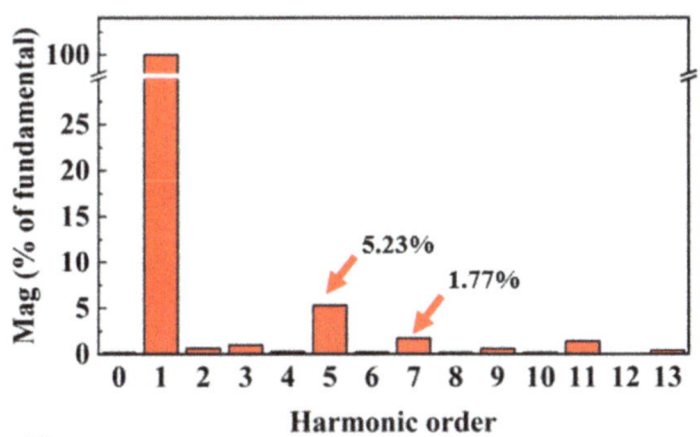

(b)

Figure A.3 Magnitudes, phases, and harmonics for back EMFs of prototype DTP-IPMSM (Machine 3). (a) Waveforms. (b) Spectra.

Index

www.ingramcontent.com/pod-product-compliance
Ingram Content Group UK Ltd.
Pitfield, Milton Keynes, MK11 3LW, UK
UKHW021938240425
457754UK00001B/1